*History of
The American Physiological Society
The First Century, 1887–1987*

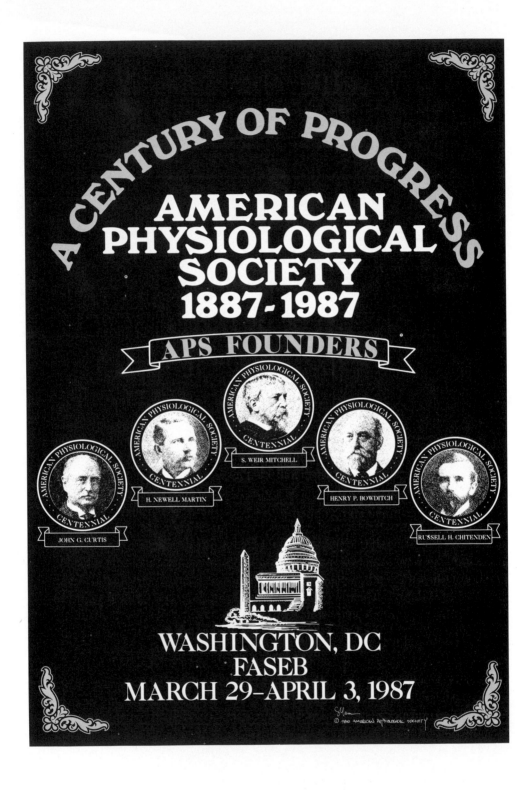

HISTORY OF
THE AMERICAN PHYSIOLOGICAL
SOCIETY

The First Century, 1887–1987

EDITED BY

John R. Brobeck

Orr E. Reynolds

Toby A. Appel

THE AMERICAN PHYSIOLOGICAL SOCIETY, BETHESDA, MARYLAND

© Copyright 1987, American Physiological Society

Library of Congress Catalog Card Number 87-12605

International Standard Book Number 0-683-01067-0

Printed in the United States of America by
    Waverly Press, Inc., Baltimore, Maryland 21202

Distributed for the American Physiological Society by
    The Williams & Wilkins Company, Baltimore, Maryland 21202

Foreword

In its hundred and first year the American Physiological Society finds its members concerned about the identity, the nature, and the future of their science. That these concerns are not new a survey of the scientific backgrounds of the Society's first members demonstrates. Of the thirty "original" members listed by Howell (W. H. Howell, *The American Physiological Society During Its First Twenty-five Years.* Baltimore, MD: Am. Physiol. Soc., 1938, p. 1–89), twenty-one were graduates of medical schools, and of this number nine had been educated in schools that did not have a professor of physiology. Rather their teaching was supervised by a professor of the institutes of medicine. Until he moved to Philadelphia to become professor of medicine at the University of Pennsylvania, William Osler held the chair of the Institutes of Medicine at McGill University (1875–84). His responsibilities included histology and pathology, in addition to physiology. McGill, like Pennsylvania and Jefferson Medical College, had modeled their faculty assignments after Edinburgh, which in turn had adopted them from Leiden. Published in Leiden in 1708, Herman Boerhaave's small textbook, *Institutione Medicae*, had become the eighteenth-century guide for teaching of sciences fundamental to instruction in medicine (G. A. Lindeboom, *Herman Boerhaave. The Man and His Work.* London: Methuen, 1968). From the middle of the nineteenth century, however, the Germanic and French usage, "Physiologie," was assumed by the English and most American medical schools. At least half of the original thirty members of the Society had received a part of their training in Germany, often in Carl Ludwig's laboratory in Leipzig.

The eight original members who earned no medical degree were making their careers in a variety of disciplines—physiological chemistry (Chittenden), embryology/comparative anatomy (Minot), sanitary "engineering" (Sedgwick), experimental neurology (Donaldson), and psychology (Hall and Jastrow), in addition to physiology (Howell and Lee). Many of those holding an M.D. degree were practicing such diverse medical specialties as neurology/psychiatry (Mitchell) and ophthalmology (Ellis), as well as pathology (Welch) and materia medica/therapeutics (Hare, Wood, and possibly Meltzer). We can wonder what a contemporary membership advisory committee would say about the qualifications of such a diversified group of applicants. The only sure conclusion seems to be that if this is physiology, then it is indeed a diffuse kind of science, distributed almost ubiquitously throughout biology and medicine.

In discovering that questions about physiology are longstanding, we do not immediately solve our own problems concerning nature and future. We can find ourselves reassured, however, as we learn that the questions we ask have remained unanswerable for at least a hundred, if not two hundred, years. Perhaps the vitality of the science depends less on categorization and definition than on the lively imagination and dedicated research effort of coming generations of investigators.

Editorial responsibility for this historical volume became at once lighter and more rewarding when Orr Reynolds agreed to serve as coeditor and when Toby Appel joined the staff as archivist and historian. We are indebted to the editorial staff in the Society's office for the careful attention they gave to publication of the book and to Martin Frank and members of his office for much needed assistance from time to time. As editors, we share also a sense of gratitude to the presidents who served while the volume was in preparation and to the Centennial Committee, particularly to Chairman Alfred Fishman. Finally, and perhaps most important, we thank the presidents and past presidents of the most recent twenty-five years and the authors of the several chapters for their willingness to undertake summaries of material that often required days or weeks of painstaking searching and review, as well as literary composition. We believe that the thoroughness and even elegance with which they have accomplished their assignments will add substantially to the celebration of the centennial. Meanwhile their good-natured spirit of cooperation has made our oversight of this history a pleasant task.

John R. Brobeck

Contents

CHAPTER 1

Physiology Flourishes in America

EDWARD F. ADOLPH†

One hundred years ago, twenty-eight Americans agreed to form the American Physiological Society (APS, the Society). Who were they? Eighteen of them taught physiology to students of medicine; twenty had earned M.D. degrees; ten had Ph.D. degrees (three had both degrees); and sixteen had done experimental research on animals. (For detailed information on the early years of APS, see ref. 16, p. 5, and chapts. 2–6 in this volume.)

The stated object of the Society was to promote the advance of physiology and to facilitate personal intercourse among American physiologists. Persons eligible for membership were those who had conducted research or had promoted physiological research and who resided in North America.

In 1887 the population of the United States plus Canada was sixty-five million. The twenty-eight members of APS thus constituted 0.43 per million inhabitants. In 1984 there were 5,800 members in a population of 265 million, or twenty two per million inhabitants.

From a precarious employment in 1887 that sustained 0.43 of one physiologist, firm support now exists for twenty-two. Somehow fifty times as many physiologists today justify their economic support by each million citizens. How did this army of physiologists come to deserve its keep?

By what criteria may the worth of physiologists to the population at large be measured? Physiologists contribute to education, to research, to understanding, and to health. Never before have so many citizens profited from, and indirectly paid for, these contributions, and physiology is only one fraction of the sciences that have equally flourished in this twentieth century.

The physiology taught by the twenty-eight original members of APS was repre-

† Edward F. Adolph died on 15 December 1986.

1

sented in factual textbooks that enumerated the structures of the human body and stated the function of each. But even in 1887 an effort was made to include, among the twenty-eight, a wider representation, namely, a plant physiologist and animal biologists of allied interests. Only a few of the twenty-eight were "full-time" physiologists; medical practice furnished a means of support. In subsequent decades the term physiologist embraced persons of even wider interests (13), some classes of whom will be identified later.

From Europe to America

In classical times, physiology was regarded as a body of knowledge and speculation comparable to anatomy, represented by Aristotle in 340 B.C. (see ref. 2) and by Galen in about A.D. 180 (see ref. 12). The belief that every structure in the animal or human had a use was not rigorously tested until centuries later, when experiments on body organs were gradually devised. Only after 1800 did systematic studies in laboratories reveal specific responses to artificial arousal. Walking, breathing, and digesting were then seen to be controlled processes, each in its particular tissues.

In the early nineteenth century, animal experimentation was practiced in laboratories devoted to it, especially in France (1820–80) by Magendie (see ref. 21) and by Bernard (5), who described his own particular research in which experimental methods were used. The results were so rewarding, in terms of physiological understanding, that practices and techniques of experiment spread.[1] In Germany from 1865 to 1895 Ludwig (23) and others educated a generation of teachers and researchers in laboratories.

Animal experimentation became a refined craft as relevant procedures were discovered. Ether anesthesia was devised in 1847. Aseptic surgery was understood by 1880. However, many studies required neither anesthesia nor surgery. Thus humans and intact animals were subjects for elaborate research on reflex movements, vestibular functions, vision, and tests on blood samples. These methods of study were common in nineteenth-century America, as in Europe. Today these methods are used especially in those countries where domestic animals are not available for students' experiments.

Also in America, research sometimes employed physiological experimentation. Information on the status of physiology among American physicians and medical students in the period up to 1850 resides in the printed theses required of medical students at several schools; the most prominent among these was the University of Pennsylvania. From before 1800 to 1821, 322 theses were printed. Of these, forty-two are now judged to include original experiments (3). Despite this number, it would be difficult to demonstrate that this research appreciably added new understanding of living processes. The chief result, of course, was the educational value of the experiments being performed.

Other modest additions to physiological knowledge came from experiments described by American physicians, as reviewed by Weir Mitchell (20), who was later a principal founder of APS. As Mitchell pointed out, this research, too, did not appreciably benefit physiology as a science, except for the famous study of gastric digestion by William Beaumont (4).

Permanent laboratories for physiological research and instruction were first established in America in the 1870s. These laboratories copied the European practices. Their principal founders were H. P. Bowditch, H. N. Martin, V. C. Vaughan, and R. H. Chittenden (see ref. 16, p. 1 and 46).[2] The urge to study animals in convenient laboratories was in turn promoted by the organization of APS.

The last half of the nineteenth century saw consistent efforts in research by American physicians and others who were aware of the growth of physiology in Europe.

The New Century

In the first quarter century of the Society, those American biologists who were interested in physiology outside of its bearings on medicine exerted no appreciable influence. Two streams of interest by such persons gradually appeared: *1*) the study of animals seen in the wild, and especially at the seashore, and *2*) the appraisal of physical and chemical phenomena in living tissues.

Stream 1 arose most sharply among zoologists of the "naturalist" groups about 1888. The U.S. Bureau of Fisheries and the Marine Biological Laboratory established laboratories at Woods Hole, where a unique course in physiology was offered each summer, beginning in 1892 (17). Varied studies of fish, shellfish, insects, and protozoa were recognized as physiological. Jacques Loeb became a member of APS in 1893; G. H. Parker and H. S. Jennings joined the Society in 1900–01 (see ref. 16, p. 179–180). Their research was represented in the Society's journal, but only rarely in its meeting programs.

Stream 2 arose in studies of the physics and chemistry of living animals. At all stages of APS history, models were proposed; an outstanding one was the model of conduction in iron wire immersed in acid, as investigated by R. S. Lillie (18, 19).

Those physiologists who regarded Woods Hole as their home organized the Society of General Physiologists in 1946. Studies of biophysics were not adequately provided for by APS, and a separate society of biophysicists was formed in 1957. By then the definition of physiologist had come to include these and others as subgroups (13).

How did the early part of the new century (1900–40) become the period of reorganization of physiology in America? American scientists still studied in Europe. American medical schools became parts of universities. The Flexner Report on medical education in 1910 (11) led to drastic changes in the quality of physiological teaching and research. The Ph.D.-type of experience on the part of teachers added new foundations to all the sciences. Successful research in specific fields like nerve conduction (7–9) encouraged independent original work in physiology. Federal and state governments took responsibilities for support and practical applications of sciences (e.g., animal husbandry and health). Philanthropists who supported some types of research efforts appeared. Indeed private philanthropy was directly encouraged to support education and research institutions as a means of avoiding the payment of high income taxes. In short, the economics and the sentiments of the times brought together extended efforts in advanced research and teaching. This recognition of the sciences maintains its momentum in the traditions and actions of the present day.

Changes

What led to the new attitudes toward the sciences? The changes in the material utensils with which scientists worked are easily described and dated. Can the thinking and motivations that elevated physiology from rare experiment to flood of research also be recognized?

Teachers of physiology became professional researchers.[3] From 1910 to 1930 research took on the dignity of an occupation. It attracted approval of colleagues and students. It qualified the scientists to teach. The mental processes, as well as the manipulations, produced personal satisfactions. An emotional drive was established. Friendly rivalries among colleagues, among departments, and among institutions led to public appreciation and to enhanced efforts.

These were some of the factors that led to the fiftyfold multiplication of active physiologists during the span of one century. In turn the number of students acquiring doctoral degrees in physiology increased in American universities, from 17 per year in 1925 to 358 per year in 1970 (1). The practical factors that support researchers became available with the internal drives for doing what scientists do.

Instruments of laboratory teaching and research gave scope to practical physiology. In the nineteenth century, progress had depended greatly on the inductorium for activation of excitable tissues and the kymograph for recording of movements. Those two instruments may be compared with the varied manipulations in use today; electrical stimulations now vary in a dozen parameters. Recorded responses can be amplified even from cellular units. There are flow recorders, fluorometers, telemeters, and nuclear magnetic resonances. Recorders now yield not only amplified information but novel types of information. They supply data that no human sense organ can detect without transducers.

The sudden expansion of physiological research after 1940 was a self-acceleration. Demonstrated accomplishments in research and testing emerged during World War II. Graduate students and young physicians were drawn into the expanding laboratories of that period. Traditions of success arose. Federal support for more individuals and for free choices of areas in education, as well as in research, expanded the number and the self-selection of physiologists.

Mixtures of material factors and motivational factors thus favored the multiplication of research efforts. The availability of radioactive isotopes, chromatography, and other methods promoted new specific types of curiosity. Cultural changes justified the existence of all physiology and its importance in the public eye. Similar opportunities for careers in biological sciences had never before existed.

The New Physiologist

The cultural changes of the twentieth century and the deeds of physiologists in the same period elevated the status of life scientists. Particularly in war times, practical efforts became known.

In World War I, aviation subjected pilots to new stresses of low barometric pressures, low oxygen pressures, high centrifugal forces, and low temperatures, all of which required investigation (14). Poison gases suddenly called for study (14). Circulatory shock became a major problem of physiology (6) and continues to be investigated today.

In World War II, extremes of muscular exertion, inanition, dehydration, heat, and ionizing radiation received intensive study (10). In each area the practical approach led to new concepts and to interrelated research in advance of the public interest in these topics.

The growth of physiology in America could easily be divided into the four quarter centuries that APS has celebrated.

The first quarter century (1887–1912) was characterized by the organized effort to advance teaching and research by mutual cooperation. APS was responsible for three outstanding achievements.

1. Ten members of APS wrote a comprehensive textbook of physiology (15), with W. H. Howell (1897) as chief editor. After two editions, Howell sponsored the entire book through successive editions.

2. The *American Journal of Physiology* was first published (1898), with W. T. Porter as managing editor.

3. Apparatus for use in physiological laboratories was manufactured in America (1901) under the supervision of Porter. All three projects required major efforts.

During the second quarter century (1913–36) physiology was established as a profession. Every medical school and many biology departments regarded a full-time physiologist as an important member of the faculty. Both teaching and research by those physiologists became standard and traditional occupations.

In the third quarter century (1937–62) a remarkable expansion of physiology, and especially of research, was evident. Results of research were now expected of every teacher and graduate student. The topics of the research expanded with the new types of equipment that could be applied and afforded. The concepts of animal function that arose in the minds of experimenters brought a new generation of physiologists into being.

In the fourth quarter century (1963–87) deep specialization of concepts and deeds in research and in teaching developed. Instead of designation as a physiologist, an individual was, for example, a neurophysiologist or a vascular physiologist. This development will be further illustrated in subsequent paragraphs.

Social attitudes toward research have undergone a great revolution in the past one hundred years. In 1887 research in physiology or other disciplines was prompted only occasionally, and by the curiosity of individuals. A few saw it as an indistinct path toward improvement of their technical outlook and a contribution to the teaching of their subject. Others were content with book learning.

Today one's accomplishment in research is an entrance to a career, indispensable to the future university professor, government expert, or industrial analyst. More than that, research is recognized as contributing to the national well-being. Hence the support of students and of laboratories is today an obligation of government. Scientists are a national resource. The career scientist cultivating his profession is even accounted an instrument of international policy. What a change in his position from beggar to commander!

As a result of the support of research, teaching has also changed. A large departmental staff allows for variety in methods of student education, more or fewer contacts by faculty in laboratories, more or less initiative by students. Whereas in 1920 a department of physiology usually consisted of three persons (professor, assistant professor, and instructor), in 1987 a department maintains eight to twelve such groups (each group having associate professor, postdoctoral fellow, predoctoral student, and technician). Each group cultivates a specialty.

What social factors led to this specialization of physiologists after 1945? Research accomplishments of the individual gave him recognition as teacher. Once recognized, the associate professor asked for facilities and privileges for himself and his protegés. Teaching hours were generally reduced. Every physiologist was full time. His desirability for "tenure" was enhanced by his research grants, recognition awards, extracurricular activities, and publications. Status crept in where previously the scientist had been self-educated in semi-isolation.

Within APS what has happened in the last quarter century (1963–87)?

Most communications at meetings became "posters." Those who exhibit posters talk one-to-one to help others assess the significance of their data and concepts.

Symposia are more numerous; organizers of symposia vie with one another to report timely advances.

Members and nonmembers join "sections," which tend to gather those with like topics of scientific interest into specialized sessions.

Program subcommittees compete for general attention to their specialties.

Membership in a section confers on the individual a sense of belonging to a manageable portion of physiology (e.g., cell physiology).

Meetings, even fall meetings, are too large to be accommodated on a university campus.

New categories of membership in APS make it easier for more persons to belong.

Voting for officers is by mail, and thus the business sessions are shortened. *The Physiologist* carries reports of Society activities.

Women members have achieved high status in scientific and Society activities, including its presidency.

The face of physiology has changed. In the last quarter century, new techniques and new concepts have been cultivated. Single neurons are studied in situ, cell components and enzymes can be identified as participants in special processes or actions, whole animals and men go for space walks, and tissues are subjected to zero gravity for long periods.

Not only action potentials are recorded, but also millivolt drifts. There are ion channels, permeability releases, juxtaposed cells, and subcellular compartments. These and many other phenomena are certain to furnish topics of study for decades to come.

It can be said that whereas the past century revealed gross changes in work and concepts of physiologists, even a quarter century is time enough for vast shifts in their outlooks. A single generation of physiologists develops wholly new territories; in turn those fields of endeavor are swallowed as that generation retires. The new generation finds its own areas of interest; it seems intent on superseding in the cherished territory that was the center of effort by predecessors.

New Laboratories

If one of the pioneer physiologists of 1887 were able to return in 1987, he might not recognize the polished rooms as laboratories. What would he notice on coming into a present-day department?

Large number of departmental members
Considerable proportion of women physiologists
Large working space occupied
Rooms set aside for group research
Large number of participants in each research group
Huge, full parking lot
Bright fluorescent lighting
Shining equipment, special to each room
Intercom television
Air conditioning
Bound volumes of departmental research papers
Multiple authorship of each paper
Spacious class laboratory rooms
Cleanliness of animal quarters
Variety of animals living there
Noninvasive techniques
Recording equipment
Computers
Word processors

Along with the revolutions in laboratories are the revolutions in scientists' homes. Ease has been built simultaneously into both laboratories and homes. Fast travel and telephones supplement the new laboratories for electron microscopy and flowmeters, or isotopes and ultracentrifuges, in promoting scholarly work. The efficient use of these conveniences enhances general experience. Still the mind of the researcher, teacher, and student continues to be a limiting factor in his accomplishments.

New Knowledge

The growth in volume of physiological knowledge and understanding is visible to everyone. But the directions and kinds of these "advances" were not predictable. If results of research were measured as information, their kinds appear to arise unforeseen and disconnected. The researcher plans his work as best he can, but the results may appear in an unintended constellation. Workers who adopt a ten-year plan nearly always shift toward brighter prospects as new directions become visible.

Any appraisal of growth in physiology fails if it relies on sheer accumulations of publications or on man-years of employment. In practice there is little agreement on the values of the results reached by research. Eventually some results lead nowhere, whereas others become the bases for brilliant concepts, pathways to future goals.

If future values could be foreseen, more scientists would shift their efforts in preferred directions each year. Though workers search for clear prophecies, most may be better served not to trust them. Perhaps random distributions of efforts produce equal or greater yields for physiology.

The encyclopedic accumulation of physiological knowledge is impressive to the world but frequently is oppressive to students. Students require protection, by learning to pursue their own pathways of search for information.

No one goal in physiological research has come to light as the single aim of this science. Hence each physiologist is entitled to state his private goal in his own way. He is expected to develop the wisdom and the zeal to pursue his aim as he thinks best. He has one lifetime in which to make his contribution.

One minor revolution has occurred in the "planning" of research. A century ago, scientists often tested their ideas in the laboratory before they went to the library to learn what had been done to answer their questions at issue. Today every scientist is compelled, by the need to write a research proposal, to fortify his plans with pertinent knowledge. If the information is in a language other than English, he will wish he could read other languages, as his grandfather scientist could. In any case, his freedom to choose what he does can be defended, since it endows him with maximal power and courage to reach results.

Dependence on libraries adds degrees of freedom to physiologists, often equal to dependence on laboratories. In 1900 the department's subscription to four journals (one volume of each per year) fulfilled most needs. By 1960 the same department would receive twenty or thirty journals. In 1987 even the university's library as a whole cannot subscribe to enough journals (and monographs) to satisfy the needs of all its own physiologists.

Correspondingly, manuscripts for journal publication are usually acceptable only if shorn of details of technical procedures. Even standard methods often are transmitted by interview with workers in a distant laboratory.

Around 1960, interest in the results of scientific work spread to larger populations, "the public." Reporting of experiments and their implications occasionally reached the daily and weekly news. In exceptional research, oral contraceptives were devised as a result of extensive experiments by Gregory Pincus (22) and others; they gave millions of people valued choices in life. In contrast, any one of a hundred other researches that cost equal effort is appreciated only by professional physiologists.

News reporters now visit Woods Hole, Federation meetings, and home laboratories. Scientists appear in television interviews. Prize-winning men and women and their deeds bring to popular attention the sorts of current efforts that were once hidden from laymen's view. Perhaps this attention is part of the feedback through which support for physiologists and for attractive laboratories is now made available.

Summary

The physiologist propelled into 1987 finds himself in a shining world. Instead of gas light and telegraph, he uses fluorescent light and jet travel. Instead of kymograms he preserves magnetic tapes. Instead of synapses he views protein and other chemical receptors. The historical steps by which such methods, results, and concepts were reached, fortunately, are matters of record. By present inference the physiologist can be certain that there is enormously more to be learned by observation and analysis and conceptual thinking. The physiologist of today lives accidentally at one point in a continuum in time that has no visible terminus.

NOTES

[1] Experiment as the critical method of physiology gained acceptance in an unforeseen consequence of a special practice in the Paris Academy of Sciences. As early as 1809, after a physiological paper was read to the academy, a committee of referees might be appointed to report on its soundness. Referees sometimes felt obliged to repeat certain of the procedures used by the original author, which meant that they either borrowed or created laboratory arrangements of their own (see ref. 21, p. 66).

[2] The laboratories developed by Vaughan and by Chittenden were for studies of what is now termed biochemistry. But up to 1906 there was no separate society for that discipline. In fact, for nine successive years, Chittenden was president of APS.

[3] The author (Adolph) entered physiological research as a third-year college student in 1915, early in the second quarter century of APS. Recollection of that time helps him compare how physiologists did and thought, then and now.

REFERENCES

1. *American Doctoral Dissertations.* Ann Arbor, MI: Univ. Microfilms, 1925–70. (Compiled for Assoc. Res. Libraries.)
2. ARISTOTLE. *Parts of Animals,* translated by A. L. Peck. London: Heineman, 1945.
3. ATWATER, E. C. Squeezing Mother Nature: experimental physiology in the United States before 1870. *Bull. Hist. Med.* 52: 313–335, 1978.
4. BEAUMONT, W. *Experiments and Observations on the Gastric Juice and the Physiology of Digestion.* Plattsburgh, NY: Allen, 1933.
5. BERNARD, C. *An Introduction to the Study of Experimental Medicine,* translated by H. C. Greene. New York: Macmillan, 1927, p. 151–154.
6. CANNON, W. B. *Traumatic Shock.* New York: Appleton, 1923.

7. ERLANGER, J., AND H. S. GASSER. A study of the action currents of nerve with the cathode-ray oscillograph. *Am. J. Physiol.* 62: 496–524, 1922.

8. ERLANGER, J., AND H. S. GASSER. *Electrical Signs of Nervous Activity.* Philadelphia, PA: Univ. of Pennsylvania Press, 1937.

9. ERLANGER, J., AND H. S. GASSER. *Nobel Lectures in Physiology or Medicine.* Amsterdam, The Netherlands: Elsevier, 1944, vol. 3, p. 31–74.

10. FENN, W. O. *History of the American Physiological Society: The Third Quarter Century, 1937–1962.* Washington, DC: Am. Physiol. Soc., 1963.

11. FLEXNER, A. *Medical Education in the United States and Canada.* New York: Carnegie Found. Adv. Teach., 1910.

12. GALEN. *Usefulness of the Parts of the Body,* translated by M. T. May. Ithaca, NY: Cornell Univ. Press, 1968, vols. 1 and 2.

13. GERARD, R. W. *Mirror to Physiology.* Washington, DC: Am. Physiol. Soc., 1958.

14. HENDERSON, Y. *Adventures in Respiration.* Baltimore, MD: Am. Physiol. Soc., 1938, chapts. 5 and 8.

15. HOWELL, W. H. (Editor). *An American Textbook of Physiology.* Philadelphia, PA: Saunders, 1897.

16. HOWELL, W. H., AND C. W. GREENE. *History of the American Physiological Society: Semicentennial, 1887–1937.* Baltimore, MD: Am. Physiol. Soc., 1938.

17. LILLIE, F. R. *The Woods Hole Marine Biological Laboratory.* Chicago, IL: Univ. of Chicago Press, 1944.

18. LILLIE, R. S. The passive iron wire model. *Science* 48: 51, 1918.

19. LILLIE, R. S. The passive iron wire model. *Biol. Rev.* 11: 181, 1936.

20. MITCHELL, S. W. Report on the progress of physiology and anatomy. *North Am. Med-Chir. Rev.* 2: 105–131, 1958.

21. OLMSTED, J. M. D. *Francois Magendie.* New York: Schuman, 1944, p. 35 and 75.

22. PINCUS, G. *The Control of Fertility.* New York: Academic, 1965, chapt. 8.

23. SCHRÖER, H. *Carl Ludwig.* Stuttgart, FRG: Wissenschaftliche, 1967, p. 74, 108, and 135.

CHAPTER 2

Founding

TOBY A. APPEL

On Friday, 30 December 1887, at 10:00 A.M., seventeen men met in the physiology laboratory of the College of Physicians and Surgeons, Columbia University, on Fifty-ninth Street in New York City to found the American Physiological Society (APS, the Society). They adopted a constitution, named twenty-eight "original" members, elected a Council, and agreed to meet twice in the following year for the presentation of scientific papers. In founding a new society, these men, perhaps unknown to themselves, were establishing several important precedents. Their society was one of the earliest national disciplinary societies in the sciences, the first society in the biomedical sciences, and the first to require of its prospective members publication of original research.

In America the late nineteenth century through the end of World War I was the great era for the formation of national societies in the various sciences. Almost all the national disciplinary societies corresponding to university departments date from this period. Before this time, scientific societies in America were mostly local organizations that held frequent meetings at fixed locations. In the 1840s two large national organizations that held annual meetings in different cities each year were created—the American Medical Association (AMA), founded in 1847, and the American Association for the Advancement of Science (AAAS), founded in 1848. Both were heterogeneous organizations, which by the 1880s were divided into sections by interest. AMA excluded homeopaths and other sectarians from membership, but otherwise any physician in America could without great difficulty become a member. In AAAS, untrained but enthusiastic amateurs mixed with the best of professional researchers; there was no restriction on membership.

The societies formed at the end of the century were of a different character from societies that had preceded them. Like AMA and AAAS, they were national in scope, but unlike them, they were selective in their membership. Either a fixed number of members or restrictive qualifications for membership were incorporated into their constitutions. The formation of the National Academy of Sciences (NAS) in 1863,

11

with a membership limited to fifty and special ties to the federal government, was a harbinger of this trend. Secondly, the new societies were specialized, concerned with the advancement of a single scientific discipline. It was a time when universities were expanding and reorganizing by departments and when the medical schools in America were just beginning the transition from proprietary schools (i.e., institutions owned by professors and run for a profit) to university-affiliated, research-oriented institutions. APS was one of the earliest of this new type of scientific society to be formed. The physiologists preceded the anatomists, the zoologists, the botanists, the geologists, the psychologists, the mathematicians, the astronomers, and the physicists in forming a national professional society. These other groups, however, were not far behind in forming their own organizations.

The time was therefore in one sense ripe for the formation of societies such as APS, but in another sense, American physiologists were barely ready to create a professional society. Experimental physiology was a relatively new science in America in 1887. The Society was formed when laboratories and training programs in physiology had only recently got underway.

Physiology in the nineteenth century was a broadly used term. When Robley Dunglison defined "biology" in his popular *Medical Lexicon*, he required only a single word, "physiology." He defined physiology as "the science of life" (14). In the popular mind, physiology was often used as nearly a synonym of hygiene. Physiology was intimately linked with hygiene, the science of health, for it was assumed that study of the structure and function of the human body would enlighten the public on how best to live. Thus conceived, physiology taught the virtues of proper exercise, diet, clothing, and daily habits. American textbooks of physiology, beginning with Robley Dunglison's two-volume *Human Physiology* in 1832, varied widely from descriptive works on the human body for medical students to tracts on particular theories of hygiene intended for a popular audience.

Although one can identify many isolated examples of experimental physiology in the first half of the nineteenth century, there was no continuous tradition of experimentation until the 1850s or 1860s (7). William Beaumont (1785–1853), though recognized in America as well as abroad as a pioneer American physiologist, was a unique phenomenon. His experiments on digestion, however ingenious, did not lead immediately to an awakening of experimental physiology in America. Physiology in the early nineteenth century was by no means synonymous with experimental physiology.

It was this broad definition of physiology as hygiene, coupled with a particular program for health reform, that underlay the formation of a society in Boston in 1837 known as the American Physiological Society. Founded by William A. Alcott and Sylvester Graham, this namesake of our present society had as its purpose "to acquire and diffuse a knowledge of the laws of life, and of the means of promoting human health and longevity." Its leaders promoted the virtues of a vegetarian diet with bread as the staple item of consumption. Membership in this first American

Physiological Society numbered at one point over 250, both men and women, but by 1840 the society had failed as the leaders broke with each other over the increasing overemphasis on diet to the exclusion of all other aspects of "physiology." The history of the first American Physiological Society was written in 1937 at the centennial of its founding by two physiologist/historians of the present APS, Hebbel Hoff, a member of APS since 1933, and John F. Fulton (20). Fulton sent copies to Walter J. Meek and to William Henry Howell, an original member of APS who was then engaged in writing the first twenty-five-year history of APS. Both were surprised to learn of the existence of such an earlier precedent. Howell replied to Fulton, "I never heard the Boston men, Bowditch and others, speak about it so I assume that they also were uninformed."[1]

The second American Physiological Society was founded fifty years later on an entirely different understanding of physiology as a professional, experimental science.

Five Founders of APS

By 1887 a handful of physiological laboratories had been established where good research was being carried out and where students might obtain a thorough training. Experimental physiology was a transplant from England, France, and especially Germany, one that had only recently taken root in America. The careers of the five men recognized as founders of APS illustrate the state of physiology in America in 1887. The five founders include the three men who signed the letters of invitation to form APS—S. Weir Mitchell, Henry Pickering Bowditch, and Henry Newell Martin—and two others who consulted in the arrangements for the organizational meeting and played an important role in the early history of the Society—Russell H. Chittenden and John Green Curtis (23, p. 19 and 21).

S. Weir Mitchell (1829–1914), the eldest and most distinguished of the three signers of the letters of invitation to form APS, presided over the organizational meeting in 1887. It was said that he initiated the idea of forming the Society. No doubt, he would have been elected the first president of APS had he not taken himself out of the running. He served instead as second president for two years (1889–90) (22; 23, p. 5–10 and 58).

Born in Philadelphia, Mitchell attended the University of Pennsylvania and Jefferson Medical College, where his father, John Kearsley Mitchell, was a professor. In medical school he worked as an assistant to the physiologist Robley Dunglison. After graduation in 1850 he spent a year studying medicine and physiology in Paris, during which he attended courses given by Claude Bernard and the microscopist Charles Robin. On his return to Philadelphia, he joined his father's medical practice. His hospital experience during the Civil War led him to become an expert in diseases of the nervous system; he acquired an international reputation as the foremost neurologist in America and became a leader in the American medical community. Later in his career he also became widely known for his literary accomplishments; he was the author of some fifteen novels and several volumes of poetry.

Despite the demands of his practice, before the Civil War Mitchell set up a private laboratory for experimental research and became a prolific contributor to the literature of physiology. He published experimental research on a wide variety of topics, including a series of works on the composition and physiological effects of snake venom on the blood and studies on the function of the cerebellum, the knee jerk, and the physiology and pathology of nerves. That Mitchell was twice passed over in attempts to gain a chair of physiology, first at the University of Pennsylvania in 1863 and then at Jefferson in 1868, reflects a time when research accomplishments were not yet the deciding factor in choosing professors (9, 18). Howell wrote that he was "the outstanding physiologist in the United States" in the period before the establishment of physiological laboratories in the 1870s. He described him as

> a man of fine presence, distinctly patrician in his bearing and manners, making the impression of a cultured man of the world, but genial and courteous and deeply interested at all times in new discoveries in science.

His leadership in forming the Society helped give the fledgling organization prestige in the medical world (23, p. 8).

Henry Pickering Bowditch (1840–1911), chosen first president of APS at the organizational meeting in 1887, served in 1888 and again from 1891 to 1895. Bowditch graduated from Harvard College in 1861. After Civil War service, he attended the Lawrence Scientific School at Harvard where he worked in the laboratory of the comparative anatomist Jeffries Wyman. In 1868 he received his M.D. degree from Harvard Medical School. He then spent a year in France and two years in the physiological laboratories of Carl Ludwig in Leipzig. Not long after his arrival he exhibited his talent for mechanical invention by developing an improvement on Ludwig's kymograph for measuring time relations automatically. Based on his research in Leipzig, he wrote a classic paper on the irritability of cardiac muscle in which he demonstrated the phenomenon of the Treppe and the "all-or-none law" of muscular contraction. In another paper he described the influence of variations of arterial blood pressure on the accelerator and inhibitor nerves of the heart.

In 1871 President Charles W. Eliot of Harvard appointed Bowditch assistant professor of physiology at Harvard Medical School. That year, in two small attic rooms, he set up the first university laboratory of physiology in America. Bowditch became full professor in 1876 and George Higginson Professor in 1903. He retired in 1906. From 1883 to 1893, he also served as dean of Harvard Medical School. Several of the original members of APS—Charles Sedgwick Minot, Joseph Weatherhead Warren, Isaac Ott, Granville Stanley Hall, Frederick W. Ellis, and Warren P. Lombard—had worked with Bowditch in his laboratory. Bowditch's research at Harvard touched on a wide variety of subjects, including ciliary motion, the knee-jerk response, the indefatigability of the nerve trunk, and the growth of children (8; 11; 17; 23, p. 10–15).

Howell, who knew him well, wrote of him:

In the Physiological Society his influence in the beginning was predominant. He was facile princeps, not only because of his position as president during the early years, but chiefly, perhaps, because of his personality. He was a man of commanding and distinguished appearance. His beard and heavy mustache and his general carriage were suggestive of his military experiences in the war, and his fine face when in repose had a certain severe dignity which helped to make him an impressive presiding officer.... But business aside and in personal intercourse he was cordial and friendly and, at times, even jovial, with his great hearty laugh and his almost boyish enjoyment of fun.... Needless to say he was loved and respected by all of his fellow members of the Society (23, p. 12–13).

Henry Newell Martin (1848–96), first secretary-treasurer of APS, was born in County Down, Ireland. He was attracted to physiology as a medical student at the University of London by Michael Foster, who was then a physiology instructor there. Martin followed Foster to Cambridge University, and in 1875 he received the first D.Sc. degree in physiology granted by that university. He also served as an assistant to Thomas Henry Huxley and wrote under his supervision the popular textbook, *A Course of Practical Instruction in Elementary Biology* (1875). On Huxley's strong recommendation, Daniel Coit Gilman, president of the newly founded Johns Hopkins University, chose Martin in 1876 to head the Department of Biology. From this strategic position, Martin and his laboratory exerted a unique influence on the development of physiology in America. Martin's department was the first in America to award Ph.D. degrees in physiology. Five of his students—Henry Gustav Beyer, Henry Herbert Donaldson, William Henry Howell, William T. Sedgwick, and Henry Sewall—were among the original members of APS. Two others, Thomas Wesley Mills and Isaac Ott, had also worked in his laboratory. Sewall, Martin's first student, had already gone to the University of Michigan and established the first physiological laboratory there in 1882. As a researcher, Martin is best known for his investigations of cardiac function based on his development of the surgical procedure for isolating and perfusing the mammalian heart.

Martin appears to have been the main author of the APS Constitution adopted on 30 December 1887 (23, p. 57–58). He was especially remembered to have been responsible for the provision in the constitution limiting membership in the Society to those who had conducted and published original research. Martin served as secretary-treasurer of APS through 1892. Because of failing health (a neurological disorder complicated by alcoholism), Martin resigned his position at Johns Hopkins in 1893. His student Howell became the first professor of physiology at the Johns Hopkins Medical School, which opened in 1893. Martin returned to England where he died at the young age of 48 (19; 23, p. 15–18; 28).

Russell Henry Chittenden (1856–1943) was the dean of "physiological chemistry," or biochemistry, in America, a field that he always considered to be a part of physiology. Trained in agricultural chemistry at the Sheffield Scientific School of Yale University, he then went to Germany, where he studied biochemistry and physiology in the laboratory of Willy Kühne at Heidelberg. In 1880 he received his Ph.D. degree from Yale, the first degree in physiological chemistry given by an American university. Two years later he was appointed professor of physiological chemistry at the Sheffield Scientific School, a post he held until his retirement in 1922; he took an active role in the government of the school by becoming its director and treasurer in 1898.

Chittenden's was the first laboratory of physiological chemistry in America. Most of his many students became early members of APS, among them Lafayette B. Mendel, his successor at Yale, and Yandell Henderson, who became professor of physiology at Yale Medical School. His early work dealt with the chemical nature of proteins and the action of enzymes in digestion. From the study of enzymes, he was led to consider general problems of nutrition; his best known contribution to science was his study of the protein requirement of humans. Later work was in the area of toxicology, particularly studies of the effects of arsenic, alcohol, and sodium benzoate.

Chittenden has the distinction of serving as president of APS for the longest period of time, from 1896 through 1904, a total of nine years. He was also a member of the first Editorial Board of the *American Journal of Physiology*. Though he had been among the small number to object to the formation of a separate biochemical society, he was elected the first president of the American Society of Biological Chemists in 1906. He lived long enough to take part in the fiftieth birthday celebration of APS in 1938 (5; 23, p. 18–20; 24; 31; 32).

John Green Curtis (1844–1913), the fifth founder of APS, hosted the Society's organizational meeting. He graduated from Harvard College in 1866, where he also received an M.A. degree in 1869; in 1870 he received his medical degree from the College of Physicians and Surgeons, then only nominally affiliated to Columbia University. After graduation, he began medical practice; he rose to the rank of attending surgeon at Bellevue Hospital but was drawn to the teaching of physiology by John Call Dalton, professor of physiology at Physicians and Surgeons. Dalton, whom Mitchell called the first professional physiologist in America (21), was at age sixty-two the oldest original member of APS. An American student of Claude Bernard, he published experimental research on bile and sugar formation in the liver and was a prolific writer of textbooks of physiology. Curtis began assisting Dalton and, in 1883, when Dalton became dean of the medical school, Curtis succeeded him as full professor of physiology. He retired in 1909.

In 1887, when the medical school moved from its old quarters on Twenty-third Street to a new building on Fifty-ninth Street, Curtis acquired a new physiological laboratory—the site of the organizational meeting of APS—which he had outfitted with the latest and best available equipment, selected after making a tour of European laboratories in 1886. Although not an original investigator himself, Curtis was well versed in the literature of physiology and did much to stimulate physiological research and teaching in America. He was one of the contributors to the two-volume *American Text-book of Physiology*, edited by Howell, in 1896. A master of classical languages and avid enthusiast of the history of physiology, he wrote a scholarly monograph, published posthumously, entitled *Harvey's Views on the Use of the Circulation of the Blood* (1915). Curtis was a member of the first Council of APS, serving through 1893, and was an active participant in the Society's early meetings (12; 23, p. 20–21; 25). Frederic S. Lee, his successor at Columbia, wrote of him:

> He was a man of a lovable personality, of a superior culture, always the gentleman, a lover of high thoughts, a follower of noble ideals, a hater of sham and pretense, generous to others, and considerate of others' shortcomings (25).

The founders of APS based their society on two models, the Physiological Society in Great Britain, founded in 1876, and the American Society of Naturalists, founded in 1883. Martin, who probably drafted the constitution of APS, was a charter member of both organizations. The Physiological Society, the first national physiological society, was established as the direct result of the activities of antivivisectionists and the passing of a national law to regulate animal experimentation. Originally organized as a dining club that met at London restaurants several times a year for social intercourse, the Physiological Society only gradually instituted formal presentation of scientific papers. Despite the differences between the British and American societies, partly because of differences in the size of the countries, many of the provisions of the constitution of the Physiological Society found their way into the first constitution of APS: the statement of purpose, the criteria for honorary membership, the method of election of the Council, and the expulsion clause all reveal their origin in similar provisions in the constitution of the Physiological Society (10, 16, 29).

A second, equally important precedent was the formation of the Society of Naturalists of the Eastern United States in 1883 with 109 original members, known after 1886 as the American Society of Naturalists. The naturalists were said to be the first society to restrict membership to professionals. The constitution of the society stipulated:

> Membership in this society shall be limited to Instructors in Natural History, Officers of Museums and other Scientific Institutions, Physicians and other persons professionally engaged in some branch of Natural History.

New members were to be proposed by existing members, approved by the Executive Council, and then approved by the majority of members at any meeting. Unlike the Physiological Society, the American Society of Naturalists had a chief officer or president. Annual meetings were held during Christmas recess in various eastern cities for the presentation of papers, and dues, as in the early days of APS, were a modest two dollars. Perhaps so as not to compete with AAAS, where naturalists traditionally presented the results of their research, the programs of the society were initially devoted to discussions of methodology, whether of museum exhibits, laboratory techniques, or instruction. Bowditch, Martin, and Chittenden were active members—all three eventually served as president. By 1887 half of the original members of APS were members of the naturalists (1, 27). APS was in a sense the first society to splinter from the naturalists. Other specialized biological societies soon followed. For many years, it will be seen, APS and a number of other societies held their annual meetings in a loose affiliation with the naturalists. To a large extent APS modeled its organization on that of the naturalists, though from the beginning, papers on the substance of research predominated over methodological papers.

The stage was surely set for the formation of a national physiological society, but that alone is not sufficient to account for the founding of APS in December 1887. Although the physiologists would certainly have founded a society eventually, they might well have waited a few more years if it were not for a precipitating event— the founding of a union of medical societies called the Congress of American Physicians and Surgeons, which, beginning in 1888, was to meet in Washington, D.C., every three years.

The Congress of American Physicians and Surgeons was first conceived in 1886 as an alternative to the International Medical Congress to be held in Washington, D.C., in 1887 and under the control of AMA. These two rival medical congresses in Washington—the International Medical Congress of 1887 and the Congress of American Physicians and Surgeons of 1888—reflect a period of great dissension in the medical community between research-minded specialists from the better medical schools, mostly in the East, and the rank and file of practitioners across the country (6).

When the planning for the International Medical Congress was begun in 1884 by AMA, it was placed under the control of a committee of eight well-known medical men, among whom John Shaw Billings, librarian of the Surgeon General's Library and founder of *Index Medicus*, took a leading role. The International Medical Congress had met in various European cities since its founding in 1867. Billings personally presented the invitation to meet in America at the eighth congress in Copenhagen in 1884. The invitation was accepted, and the committee proceeded according to its mandate to enlarge itself and select officers of the congress. Among those given official roles in the congress were four of the five founders of APS— Mitchell, Bowditch, Martin, and Curtis. Dalton was to have been chairman of the Section of Physiology. But in May 1885 at the next AMA meeting, this organization of the congress was torn apart when factions in AMA, feeling that control of the congress had been passed over to an elite group of men not representative of American medicine as a whole, had the mandate of the original committee overturned. All additions to the original committee (among them Mitchell) were removed from their positions, and all appointments were subject to review by a new committee in which the original eight members were swamped by newly elected representatives of each of the states, the Army, and the Navy. As a result of the animosity aroused by this precipitate action, a good many of the leading medical men of the country, including Mitchell, Martin, Bowditch, Curtis, and Dalton, were removed or resigned from their positions and would have nothing further to do with the congress. Although the congress met in Washington only three months before the founding of APS, none of the five founders of APS attended and not one future member of APS was on the program of the Section of Physiology.

The Congress of American Physicians and Surgeons was intended as an alternative to the large and politicized meetings of AMA. It was organized in such a way that it would emphasize medical research and would avoid the squabbles over codes of ethics and regulation of medical practice that characterized AMA meetings. Its delegates would come not from state and county medical societies as in AMA but from the recently created national societies of medical specialists. Those societies represented at the first congress in 1888 and the dates on which they were founded are listed in Table 1 (30). Every three years, beginning in 1888, the participant societies were to hold their annual meetings together in a large national congress in Washington, D.C., to which foreign guests would also be invited. If the physiologists wished to participate in this congress, dedicated to the advance of medical research, they too had to form a national professional society.

Table 1
SOCIETIES REPRESENTED AT THE FIRST CONGRESS OF AMERICAN PHYSICIANS AND SURGEONS

	Year Founded
Am. Ophthalmological Soc.	1864
Am. Otological Soc.	1868
Am. Neurological Assoc.	1875
Am. Gynaecological Soc.	1876
Am. Dermatological Assoc.	1876
Am. Laryngological Assoc.	1878
Am. Surgical Assoc.	1880
Am. Climatological Assoc.	1883
Assoc. Am. Physicians	1885
Am. Assoc. Genito-Urinary Surgeons	1886
Am. Orthopedic Assoc.	1887

On 23 May 1887, Mitchell, Martin, and Bowditch sent a circular letter duplicated by a process of cyclostyling on stationery of the Harvard Physiological Laboratory to an unknown number of physiologists inquiring of their interest in forming a national society.[2]

Dear Sir,

The assembling of the various societies of specialists in Washington in 1888 with a view of forming a national medical association, offers a favourable opportunity for bringing together those who are interested in the pursuit of Physiology, and formation of a Physiological Society which shall, by its annual meetings, afford facilities for professional and social intercourse and supply that mutual aid and encouragement which are so valuable to those who have a common interest in any department of science. In order that the society, if formed, may be ready to join with other societies of specialists in the summer of 1888, it seems desirable that a meeting for organization be held during the coming summer. You are therefore requested to reply to the following questions

1) Are you in favour of the formation of a National Physiological Society?

2) Are you willing to attend during the coming summer a preliminary meeting for the discussion of plans of organization?

3) At what time and place can such a meeting most conveniently be held?

Yours very truly,
S. Weir Mitchell
H. N. Martin
H. P. Bowditch

Please send reply as soon as possible to Dr. H. P. Bowditch, Harvard Medical School, Boston, Mass.

The replies have unfortunately not survived. But in a letter to Mitchell on 1 June 1887,[3] Bowditch explained why the meeting was postponed from summer to December.

There seems to be a good deal of difference of opinion about the best time for a preliminary meeting. It has been suggested that we postpone the meeting till the Christmas recess & then have it either in New Haven or New York just before or just after the meeting of the Soc. of Naturalists. How does this strike you? I think it is very important to have the meeting when both you & Martin can be present.

Mitchell annotated the letter, "Suit me very well."

The naturalists were meeting on 27–29 December 1887 at Yale. Martin was to present an exhibit of "newly devised Physiological Apparatus." The organizational meeting of APS was scheduled for the following day at the College of Physicians and Surgeons in New York. The new medical school building on Fifty-ninth Street had just opened in September 1887, and on 29 December two additional buildings in the complex, the Vanderbilt Clinic and the Sloane Maternity Hospital, had been dedicated. By meeting in New York, the physiologists would have the opportunity to view for the first time Curtis's brand new and well-appointed laboratory, lecture room, and offices that occupied a large portion of the third floor of the medical school building (6).

A second circular letter signed by Mitchell, Martin, and Bowditch, dated 10 November 1887, informed physiologists of the date and place of the organizational meeting (23, p. 4).

Dear Sir:

You are invited to attend a meeting of persons interested in the formation of a National Physiological Society to be held in the Physiological Laboratory of the "College of Physicians and Surgeons" 437 W. 59th St., New York, on Friday, December 30th, 1887, at 10 AM. Please notify Dr. H. P. Bowditch, Harvard Medical School, Boston, whether you will be able to attend this meeting or not.

Yours Truly,
S. Weir Mitchell
H. N. Martin
H. P. Bowditch

In a third letter, signed by Bowditch and dated 2 December 1887, a copy of the proposed constitution was enclosed. Recipients were "invited to suggest any Amendment that may seem to you desirable." Copies of the last two letters were inserted into the first volume of minutes of the Society and were later printed in Howell's history (23, p. 4).

Organizational Meeting

The organizational meeting of APS was held, as Martin wrote in the minutes, "to consider the advisability of instituting a Society of American Physiologists." How many men were invited is not known, but seventeen attended: H. P. Bowditch, R. H. Chittenden, J. G. Curtis, H. H. Donaldson, F. W. Ellis, G. Stanley Hall, W. H. Howell, Joseph Jastrow, W. P. Lombard, H. N. Martin, C. S. Minot, S. W. Mitchell, Isaac Ott, W. T. Sedgwick, H. Sewall, J. W. Warren, and H. C Wood. Mitchell was called to the chair, and according to a published report (4) Bowditch acted as clerk.

The first order of business was the reading and discussion of the proposed constitution clause by clause. The draft constitution included in the circular letter of 2 December has unfortunately not been preserved, so it is not known what changes were made. Inserted into the minutes of the organizational meeting was a printed copy of the constitution as amended and unanimously adopted. It was a document of thirteen clauses. The first three stated the name and purpose of the new Society and the categories of members:

1. This Society shall be named "THE AMERICAN PHYSIOLOGICAL SOCIETY."
2. The Society is instituted to promote the advance of Physiology and to facilitate personal intercourse between American Physiologists.
3. The Society shall consist of ordinary and honorary members.

The most interesting feature of the constitution was the fourth clause defining its membership policy:

4. Any person who has conducted and published an original research in Physiology or Histology (including Pathology and experimental Therapeutics and experimental research in Hygiene), or who has promoted and encouraged Physiological research, and who is a resident of North America, shall be eligible for election as an Ordinary member of the Society.

The membership clause was at the same time both broad and restrictive. It was restrictive because membership in APS was limited to those who had conducted and published original research. APS was probably the first society in America formally to require publications as a condition of election to membership. This was a bold step on the part of the founders of the Society, for in 1887 medical schools did not require research of their professors, and many teachers of physiology at medical schools did not therefore qualify for membership. The membership provision was broad in that the constitution defined physiology to encompass not only physiology proper, but biochemistry, pharmacology, experimental therapeutics, experimental pathology, experimental histology, and experimental hygiene. Among early members were physiological botanists, experimental psychologists, nutritionists, and immunologists. The clause allowed for the possibility of including an important promoter of physiology who was not himself an investigator, but this provision was never used after 1887 and eventually was eliminated.

Bowditch recalled at the 1896 meeting that it was to Martin that the Society was indebted

> for establishing the high standard which the Society has always maintained with regard to the qualifications of the members. It was always Dr. Martin's contention that a candidate for admission to our ranks should be required to demonstrate his power to enlarge the bounds of our chosen science, and not merely to display an interest in the subject and an ability to teach text-book physiology to medical students. To his wise counsel in this matter the present prosperity of the Society is, I think, largely due.

Candidates for membership were then, as now, to be proposed in writing by two ordinary members of the Society and nominated for election by the Council. The names of the nominated candidates, their sponsors, and their qualifications were to be posted on the bulletin board at the first session of the annual meeting and voted on by the members at the last session of the meeting. It was provided that "one blackball in eight shall exclude." In practice the Council's nominees were almost always elected. Honorary membership was reserved by the constitution for "Distinguished men of science who have contributed to the advancement of Physiology" and might include up to three Americans (this provision was eliminated in 1896) as well as foreigners. The only American ever to be elected an honorary member was William T. Porter, founder of the *American Journal of Physiology* and the Porter Fellowship Program, in 1948. Honorary members were to be proposed by Council and elected "only by the unanimous vote of the members present at a meeting." Although the matter of honorary members was brought up several times in the 1890s, choosing them was apparently a difficult and divisive task, and it was not until 1904 that an honorary member was actually elected.

Management of the Society was from the beginning vested in a Council. The Council was to consist of five members elected by ballot for the following year by the ordinary members at each annual meeting. The members of the Council were permitted to appoint from among themselves a president and a secretary-treasurer. Although there was no limit on the number of years an individual might serve on Council, some turnover was provided for by the provision that only four members of Council were eligible for reelection at any annual meeting. Ballots were to be sent by the secretary to the membership before the annual meeting, so that those who were unable to attend could vote by proxy. Otherwise voting would take place at the annual meeting. Voting continued to take place at the business meeting of the Society until the present system was adopted in 1975.

There has never been a nominating committee in APS. It has always been left to the membership at large to select candidates for office. According to the original constitution, any member could nominate a member for Council to appear on the ballot. If the five current members of Council received the most votes, the member with the fewest votes would be dropped and the new candidate receiving the greatest number of votes would be elected. In practice, in the early elections, only one additional nomination appeared on the printed ballots, and he became the new member of Council. In a few years, the attempt at forced rotation was relaxed and the method of electing officers simplified. Members simply voted for five members of Council, and the top five candidates were considered elected. It was not until 1904 that the constitution was amended so that the members directly elected the president, secretary, and treasurer.

The constitution provided for annual meetings to be held between Christmas and New Year's Day and special meetings that could be called at the discretion of the Society. Election of members and alterations of the constitution were reserved for annual meetings.

The thirteenth and final clause of the constitution provided that "if the majority of the Council shall decide that the interests of the Society require the expulsion of a member," the member could be expelled on a two-thirds vote of the members present at an annual meeting. This provision, which remained practically unaltered in the constitution until the entire document was revised in 1953, was used only once in the history of APS. In 1915 Silas P. Beebe of New York, elected in 1904, was expelled for publicly promoting a cancer cure known as autolysin.

With the constitution adopted, the next order of business was to vote for members of Council. On the first ballot, Bowditch, Curtis, Martin, Wood, and Mitchell were elected. Mitchell, who might well have been selected first president if he had been elected to Council, declined to serve, probably because he was planning a trip abroad and would not be able to attend the congress in September 1888 (6). Henry Sewall was elected in his place.

Next came the naming of charter members. It was decided that in addition to those present, all those who were circularized and had indicated interest in the Society would be elected "original members of the Society," namely H. G. Beyer, H. C. Chapman, J. C. Dalton, G. L. Goodale, T. W. Mills, E. T. Reichert, and R. M. Smith. Council was authorized to invite not more than ten others to be ordinary members without the usual formalities. As the final order of business, it was voted to hold a special meeting in Washington, D.C., in September 1888 as part of the Congress of American Physicians and Surgeons.

After the adjournment of the Society meeting, the five members of Council met with Curtis as chairman and balloted for officers. Bowditch was elected first president of APS and Martin first secretary. The commencement of the APS Archives is due to the foresight of Martin who took minutes of the Society and Council meetings in longhand and saved announcements and correspondence to be pasted into the volumes. Two volumes of minutes were begun, one for Society (business) minutes and the other for Council minutes. Under the authority given them by the Society, members of Council elected four more men to be considered original members— H. A. Hare, W. Osler, V. C. Vaughan, and W. H. Welch. None was primarily a physiologist, but all were influential advocates of experimental research. It was decided that the first annual meeting of the Society would be held on the Thursday or Friday after Christmas in Philadelphia. An order of business for future meetings of the Society was decided on, and dues for the following year (due immediately) were set at two dollars.

Original Members

The twenty-eight original members of APS and their positions at the time of the organizational meeting are listed in Table 2. They were, for the most part, young men; over half were under forty. Except for Vaughan and Sewall, all were from the East. Seventeen of these men had only an M.D. degree, for Ph.D. degrees were a recent phenomenon in American higher education. Yet eight original members had only Ph.D. degrees: five from Johns Hopkins, two from Harvard (Minot and Hall), and one from Yale (Chittenden). Three members had both degrees (Vaughan, Beyer, and Martin). The twenty-eight men formed a remarkably coherent group. All but Vaughan had at one time studied or taught in the institutions associated with one or more of the five founders—University of Pennsylvania, Jefferson Medical

Table 2
APS ORIGINAL MEMBERS

	Position
H. G. Beyer (1850–1918)	Past assistant surgeon, U. S. Navy Medical Dept.
H. P. Bowditch (1840–1911)	Professor of physiology, Harvard Medical School
H. C. Chapman (1845–1909)	Professor in institutes of medicine and medical jurisprudence, Jefferson Medical College
R. H. Chittenden (1856–1943)	Professor of physiological chemistry, Sheffield Scientific School, Yale Univ.
J. G. Curtis (1844–1913)	Professor of physiology, College of Physicians and Surgeons, Columbia Univ.
J. C. Dalton (1825–1889)	Dean and emeritus professor of physiology, College of Physicians and Surgeons, Columbia Univ.
H. H. Donaldson (1857–1938)	Associate in psychology, Johns Hopkins Univ.
F. W. Ellis (1857–1948)	Practicing ophthalmologist, Springfield, MA
G. L. Goodale (1839–1923)	Professor of botany, Harvard Univ.
G. S. Hall (1846–1924)	Professor of psychology, Johns Hopkins Univ.
H. A. Hare (1862–1931)	Lecturer in physiology, Biological Dept., Univ. of Pennsylvania
W. H. Howell (1860–1945)	Associate in physiology, Johns Hopkins Univ.
J. Jastrow (1863–1944)	Fellow in psychology, Johns Hopkins Univ.
W. P. Lombard (1855–1939)	Assistant at the College of Physicians and Surgeons, Columbia Univ.
H. N. Martin (1848–1896)	Professor of biology, Johns Hopkins Univ.
T. W. Mills (1847–1915)	Professor of physiology, McGill College
C. S. Minot (1852–1914)	Instructor in histology, Harvard Medical School
S. W. Mitchell (1829–1914)	Practicing physician, Philadelphia, PA
W. Osler (1849–1919)	Professor of clinical medicine, Univ. of Pennsylvania
I. Ott (1847–1916)	Lecturer in physiology, Medico-Chirurgical College, Philadelphia, and practicing physician, Easton, PA
E. T. Reichert (1855–1931)	Professor of physiology, Univ. of Pennsylvania
W. T. Sedgwick (1855–1921)	Professor of biology, Massachusetts Institute of Technology
H. Sewall (1855–1936)	Professor of physiology, Univ. of Michigan
R. M. Smith (1854–1919)	Practicing physician, Philadelphia, PA
V. C. Vaughan (1859–1929)	Professor of physiological chemistry, Univ. of Michigan
J. W. Warren (1849–1916)	Instructor of physiology, Harvard Medical School
W. H. Welch (1887–1934)	Professor of pathology, Johns Hopkins Univ.
H. C Wood (1841–1920)	Professor of therapeutics and clinical professor of diseases of the nervous system, Univ. of Pennsylvania

College, Harvard, Johns Hopkins, Columbia, and Yale. Only somewhat over half were physiologists proper, and some, like Osler, were not experimental scientists at all, but they all had a deep appreciation of experimental physiology and its future promise for biology and medicine.

First Scientific Meetings, 1888

On 23 January 1888 the newly formed APS was accepted for membership in the Congress of American Physicians and Surgeons, the next to the last of twelve societies to join. Bowditch was appointed delegate and Martin alternate to the Executive Committee of the congress. APS was the only biological (as opposed to medical) society to take part. Thus the first meeting of APS for the presentation of scientific research was held in Washington, D.C., on 18–20 September 1888. Some 400 medical researchers and 23 foreign guests attended the meeting, over which John Shaw Billings, president of the congress, and William Pepper, chairman of the Executive Committee, presided. Plenary sessions were held in the Grand Army of the Republic Building on Pennsylvania Avenue. Two physiology sessions, which differed considerably from those announced in the printed program, were held at the newly constructed Army Medical Museum on the Mall. Twelve members of APS attended the congress, though some members of other societies were also present at the physiology sessions (23, p. 61; 30).

By chance, Howell recalled in his history, he had the honor of presenting the very first paper before the Society, "On the Origin and Regeneration of Red Blood Corpuscles." This was followed by "On the Relations of Costal and Diaphragmatic Breathing, with Particular Reference to Phonation," by Henry Sewall; "On Histological Changes Produced in Ganglion Cells," by H. H. Donaldson; and "On the Physiological Action of Uranium Salts," by Russell Chittenden. On the following day the British surgeon and physiologist Victor C. Horsley, one of several foreign scientists to attend the congress, was given a special invitation to present a paper on "Observations on Negative Variation in the Spinal Cord and on the Relation of Clonic Spasms to Cortical Stimulation." Other papers included "A Plethysmographic Study of the Changes of Volume Produced in the Human Finger and Forearm by Various Methods of Electrical Stimulation," by Henry Sewall; "On Fever," by Isaac Ott; "On the Nature of the Knee Jerk," by Warren P. Lombard; "On the Knee Jerk Phenomena," by Joseph W. Warren; and "The Influence of Acetanilide or Antifibrin on Proteid Metabolism," by Chittenden. On 20 September a joint discussion session with the Association of American Physicians was arranged on "The Relation Between Trophic Lesions and Diseases of the Nervous System." Edward C. Seguin of New York acted as "referee" for the Association of American Physicians and Horatio C Wood for the physiologists.

Social events at this first meeting included an evening banquet in honor of the foreign guests at the Willard Hotel, receptions at the U.S. National Museum (now the Arts and Industries Building of the Smithsonian Institution) and the Army Medical Museum, and a special reception at the White House given by President and Mrs. Cleveland. For his presidential address, Billings presented his now classic essay, "On Medical Museums, With Special Reference to the Army Medical Museum at Washington" (2, 3).

The First Annual Meeting of APS, attended by thirteen members, was held in Philadelphia on 29 and 31 December 1888 just after the American Society of Naturalists' meeting in Baltimore. It was the only time the physiologists met on their own, apart from other societies, until the first fall meeting in 1948. Sessions were held on 29 December at Jefferson Medical College and on 31 December at Medical Hall, University of Pennsylvania. At the end of the sessions, those in attendance were invited to visit the physiological laboratories. Six new members were nominated and approved and one candidate deferred. Elected were John Pendleton Campbell, Simon de Jager, George T. Kemp, Frederic S. Lee, Franklin P. Mall, and Samuel J. Meltzer. Of these, Lee and Meltzer were future presidents of APS and Mall was a future president of the anatomists.

Seven papers were presented at the two sessions: "The Excitability of the Different Columns of the Spinal Cord" and "The Rate of Transmission of Nerve Impulses," both by E. T. Reichert; "On Sensory Reinforcements of the Knee Jerk," by J. W. Warren; "On Changes in Ganglion Cells due to Stimulation," by H. H. Donaldson; "The Lethal Temperatures of the Cat's Heart," by H. Newell Martin; "Exhibition of a Specimen Showing the Anatomical Portion of Thermogenic Centers in the Brain," by I. Ott; and "On the Influence of Light upon the Excretion of CO_2 by Frogs Deprived of their Cerebral Hemispheres," also by Martin. After the meeting, everyone was invited to hear "the exposition by Professor Reichert of the construction and mode of working of a new calorimeter."

At this meeting, the First International Physiological Congress to be held at Basel in 1889 was announced, and circulars from the Physiological Society concerning the congress were distributed. The congresses were said to have stemmed from dissatisfaction with the International Medical Congresses, which were too formal and social and did not have an experimental component. A contributory cause was the need for international cooperation in combating the antivivisectionists. The plans to hold a congress were initiated by Michael Foster, Willy Kühne, Hugo Kronecker, Fritiof Holmgren, and Gerald Yeo, secretary of the Physiological Society. The first step was taken in November 1887, a month before the founding of APS, when the Physiological Society instructed Yeo to communicate with Kronecker in Bern to organize an international meeting in Bern during the following summer. At Kronecker's suggestion, the Physiological Society addressed a preliminary proposal to 109 physiologists in March 1888. A meeting was held at Kronecker's home in September 1888 to make final arrangements. As Kronecker's institute, the Hallerium, in Bern was not yet completed, Basel was instead selected as the site for the first congress to be held on 10–12 September 1889. It was agreed that the emphasis of the sessions would be on experimental demonstrations and that greatest informality would prevail. Bowditch was appointed to the Organizing Committee as the American representative, and he and Lombard were among the 123 to 129 physiologists who, according to various accounts, took part in the congress (15).

Also at the first annual meeting of APS, APS's first prize award was announced. To encourage physiological research, Mitchell offered to establish a prize of $200 for the best essay on rates of transmission of nervous impulses in humans and animals. He emphasized that "the prize was in no way to be regarded as a reward for research but rather as a help toward payment of the expenses of persons engaged in such researches." Mitchell's offer was gratefully accepted, and an announcement of the prize was printed and circulated in early February 1889. However, no communications were sent in competition for the prize, and it lapsed. The Society offered a second prize through a circular in February 1890 for the best essay on generation and regeneration of nerves. This time there was a worthy competitor, and the prize was awarded to Howell in 1891.

These two meetings set the pattern for much of the first quarter century. When, in 1932, Velyien Henderson wrote on behalf of the Federation of American Societies for Experimental Biology to all the surviving original members of APS, they all looked back to 1887 with wonder at the great changes in physiology that they had witnessed in their lifetimes. Chittenden wrote:

> I recall very vividly the feelings of doubt and uncertainty that prevailed when a few of us met together in New York to discuss the possibility of a society to encourage research in experimental physiology. At that date there were so few interested in experimental work in the biological sciences, so few laboratories and so little appreciation in the colleges and universities of America of the importance of firsthand knowledge that the skepticism more or less prevalent seemed justified. But today we see a different picture, and we on this side of the water may well feel proud of the part that is being played by the growing multitude of workers in the fields of experimental biology. To have had a part in this development brings a feeling of satisfaction.[4]

To Howell the 1932 meeting at which H. H. Donaldson had given an address on the founding of APS (13) brought back

> memories of the small band who gathered at the early meetings of the Society. Truly it has been a remarkable growth for one generation, but I wonder if those who attend the huge gatherings of today get the same inspiration and companionship that came to our little group thrown together so closely.[5]

NOTES

[1] W. H. Howell, letter to J. F. Fulton, 24 October 1937; W. J. Meek to Fulton, 3 November 1937; Fulton to Meek, 8 November 1937. Obituary files, "American Physiological Society," Yale Univ. Medical School.

[2] S. W. Mitchell, H. N. Martin, and H. P. Bowditch, letter to W. P. Lombard, 23 May 1887. APS Archives, Bethesda, MD.

[3] H. P. Bowditch, letter to S. W. Mitchell, 1 June 1887. S. W. Mitchell correspondence, College of Physicians, Philadelphia, PA.

[4] R. H. Chittenden, letter to V. Henderson, 21 May 1932. APS Archives, Bethesda, MD.

[5] W. H. Howell, letter to V. Henderson, 20 May 1932. APS Archives, Bethesda, MD.

REFERENCES

These chapters on the history of the first seventy-five years of APS are based on the bound Society and Council minutes in the APS Archives, Bethesda, Maryland. Specific citations are given only to material in the archives not found in the minutes.

1. AMERICAN SOCIETY OF NATURALISTS. *Records* 1, 1884–95.
2. ANONYMOUS. Correspondence. The Congress of American Physicians and Surgeons. *Med. News* 53: 341–343, 1888.
3. ANONYMOUS. Correspondence. The recent Medical Congress. *Boston Med. Surg. J.* 119: 319, 1888.
4. ANONYMOUS. Organization of the American Physiological Society. *Am. Naturalist* 22: 372–374, 1888.
5. ANONYMOUS. Russell Henry Chittenden. *Physiologist* 3(1): 3–4, 1960.
6. APPEL, T. A. Biological and medical societies and the founding of the American Physiological Society. In: *Physiology in the American Context, 1850–1940*, edited by G. L. Geison. Bethesda, MD: Am. Physiol. Soc., 1987, chapt. VI, p. 155–176.
7. ATWATER, E. C. "Squeezing mother nature": experimental physiology in the United States before 1870. *Bull. Hist. Med.* 52: 313–335, 1978.
8. BOWDITCH, M. Henry Pickering Bowditch: an intimate memoir. *Physiologist* 1(5): 7–11, 1958.
9. BYNUM, W. F. Silas Weir Mitchell. In: *Dictionary of Scientific Biography.* New York: Scribner, 1974, vol. 9, p. 422–423.
10. BYNUM, W. F. A short history of the Physiological Society, 1926–1976. *J. Physiol. Lond.* 263: 23–72, 1976.
11. CANNON, W. B. Biographical memoir of Henry Pickering Bowditch, 1840–1911. *Mem. Natl. Acad. Sci.* 17: 181–196, 1922.
12. CHITTENDEN, R. H. John Green Curtis. In: *Dictionary of American Biography.* New York: Scribner, 1930, vol. 4, p. 616–617.
13. DONALDSON, H. H. The early days of the American Physiological Society. *Science* 75: 599–601, 1932.
14. DUNGLISON, R. *Medical Lexicon: A Dictionary of Medical Science* (7th ed.). Philadelphia, PA: Lea and Blanchard, 1848, p. 113 and 656.
15. FRANKLIN, K. J. A short history of the International Congresses of Physiologists, 1889–1938. In: *History of the International Congresses of Physiological Sciences, 1889–1968.* Washington, DC: Am. Physiol. Soc., 1968. (Reprinted with original pagination from *Ann. Sci.* 3: 241–335, 1938.)
16. FRENCH, R. D. *Antivivisection and Medical Science in Victorian Society.* Princeton, NJ: Princeton Univ. Press, 1975.
17. FYE, W. B. Why a physiologist?—the case of Henry P. Bowditch. *Bull. Hist. Med.* 56: 19–29, 1982.
18. FYE, W. B. S. Weir Mitchell, Philadelphia's "lost" physiologist. *Bull. Hist. Med.* 57: 188–202, 1983.
19. FYE, W. B. H. Newell Martin—a remarkable career destroyed by neurasthenia and alcoholism. *J. Hist. Med.* 40: 133–136, 1985.
20. HOFF, H. E., AND J. F. FULTON. The centenary of the first American Physiological Society founded at Boston by William A. Alcott and Sylvester Graham. *Bull. Hist. Med.* 5: 687–734, 1937.
21. HOLMES, F. L. John Call Dalton. In: *Dictionary of Scientific Biography.* New York: Scribner, 1978, vol. 15, p. 107–110.
22. [HOWELL, W. H.] Silas Weir Mitchell. *Physiologist* 2(1): 4–7, 1959.
23. HOWELL, W. H., AND C. W. GREENE. *History of the American Physiological Society Semicentennial, 1887–1937.* Baltimore, MD: Am. Physiol. Soc., 1938.
24. IHDE, A. J. Russell Henry Chittenden. In: *Dictionary of American Biography.* New York: Scribner, 1973, suppl. 3, p. 162–164.
25. LEE, F. S. John Green Curtis, M.D. 1870, LL.D. 1904. *Columbia Univ. Q.* 54–57, Dec. 1913.
26. MENDELSOHN, E. Henry Pickering Bowditch. In: *Dictionary of Scientific Biography.* New York: Scribner, 1970, vol. 2, p. 365–368.
27. MINOT, C. S. The relation of the American Society of Naturalists to other scientific societies. *Science* 15: 241–244, 1902.
28. ROSENBERG, C. E. Henry Newell Martin. In: *Dictionary of Scientific Biography.* New York: Scribner, 1974, vol. 9, p. 142–143.

29. SHARPEY-SHAFER, E. *History of the Physiological Society During Its First Fifty Years, 1876–1926.* Cambridge, UK: Cambridge Univ. Press, 1927.

30. *Transactions of the Congress of American Physicians and Surgeons. First Triennial Session Held at Washington, D.C., September 18th, 19th and 20th 1888.* New Haven, CT: Congr. Am. Physicians Surg., 1889.

31. VICKERY, H. B. Russell Henry Chittenden. In: *Dictionary of Scientific Biography.* New York: Scribner, 1970, vol. 2, p. 256–258.

32. VICKERY, H. B. Biographical memoir of Russell Henry Chittenden. *Biogr. Mem. Natl. Acad. Sci.* 24: 95–104, 1945.

CHAPTER 3

First Quarter Century 1887–1912

TOBY A. APPEL

In its first quarter century, the American Physiological Society (APS, the Society) expanded from 28 charter members in 1887 to 188 ordinary members (and 6 honorary members) by 1912. Yet it remained a relatively small, intimate, and informal society. Its chief activity in its first twenty-five years was the scientific meetings at which members could present the results of their research and demonstrate new techniques and equipment. Annual meetings took place each December with the American Society of Naturalists and other recently formed biological societies, and special meetings were held every three years through 1907 in Washington, D.C., with the Congress of American Physicians and Surgeons.

In 1898 the Society added to its responsibilities that of sponsoring a journal, the *American Journal of Physiology*. It was not until the second quarter century, however, that the Society took over full ownership and management. Through the end of the period covered by this chapter, the *American Journal of Physiology* was the property of its founder, William T. Porter, who edited it under contract to the Society.

The first quarter century of the Society's history was a period of transformation in medical education that created ever-greater opportunities for physiologists and physiological research. The expansion of scientific medicine in America also led to the formation of new societies of biochemists and pharmacologists, both offshoots of APS, in 1906 and 1908. These three societies, in their desire to prevent any further separation of their activities, agreed at the APS Twenty-fifth Annual Meeting to unite in a federation. The formation of the Federation of American Societies for Experimental Biology (FASEB, the Federation) on the evening of 31 December 1912 was a fitting culmination to the Society's first twenty-five years and laid the foundation for its next seventy-five years.

Officers and Organizational Changes

For its first twenty-five years, APS had only five presidents. Elected by the Council rather than by the membership, presidents tended to be reelected as long as they were willing to serve. In 1888, when S. Weir Mitchell declined to be a member of the first Council and thus was ineligible to serve as president, Henry P. Bowditch was elected first president. Mitchell was elected president for 1889 and 1890. In 1891 he was to be president of the second Congress of American Physicians and Surgeons and thus asked to step down. Thereupon Bowditch was reelected and served through 1895, when he too asked to be relieved. For nine years, from 1896 through 1904, Russell Chittenden served as president; his was by far the longest incumbency in the history of APS. In 1904, perhaps because it was felt that Chittenden had been in the chair too long, the constitution was amended to allow the membership to elect the president and other officers directly. When given the opportunity to select the president for the first time at the meeting in December 1904, the members chose William Henry Howell, professor of physiology at Johns Hopkins, although Chittenden was to remain on Council for another year. The fifth president, Samuel J. Meltzer, head of the Department of Physiology and Pharmacology of the Rockefeller Institute for Medical Research, was elected in 1910 and served from 1911 through 1913. By the order of seniority adopted in the bylaws of the Federation, he had the honor of becoming the first chairman of the Executive Committee.

Other changes in the organization of the Society—the adoption of a set of bylaws in 1904 and the various other amendments to the constitution and bylaws before 1913—were motivated by problems related to meetings and publications and will be discussed under those headings.

Membership

From its foundation, APS carefully scrutinized the qualifications of prospective members. Most of its members, in addition to appropriate physiological publications, had either a Ph.D. or an M.D. degree or both. In contrast to a somewhat later time, if a physiologist had published at all, he could generally become a member at the next meeting after the degree was awarded.

The membership roll of the Society in its first quarter century included a great many venerable names. Among those elected between 1888 and 1900 were John J. Abel (1891), Albert P. Brubaker (1890), Walter B. Cannon (1900), W. S. Carter (1896), James McKeen Cattell (1895), Arthur Cushny (1893), Percy M. Dawson (1900), George P. Dreyer (1893), William J. Gies (1900), Charles W. Greene (1900), Winfield S. Hall (1898), Yandell Henderson (1900), Theodore Hough (1895), Reid Hunt (1895), Frederic S. Lee (1888), Phoebus A. Levine (1894), Jacques Loeb (1893), Graham Lusk (1892), A. B. Macallum (1896), Samuel J. Meltzer (1888), Albert P. Mathews (1898), Lafayette B. Mendel (1896), William T. Porter (1891), Alfred N. Richards (1900), C. C. Stewart (1898), and George N. Stewart (1893).

Although standards were not as high as they were to become, not everyone who was nominated was accepted. At its first annual meeting in 1888, one candidate, Albert P. Brubaker, was "postponed" and not elected until 1890. In 1895 two candidates were rejected for membership. Beginning in 1904 candidates were rejected or deferred every year. From 1888 to 1912 eight persons were rejected and fourteen others deferred, sometimes for a period of several years.

The first person to be rejected for membership was interestingly a woman, Frances Emily White (1832–1903), professor of physiology since 1876 at Woman's Medical College of Pennsylvania, where she had received her M.D. degree in 1872. She had visited German and British laboratories and was recommended by Mitchell and Horatio C Wood. Her publications, however, although numerous, were almost entirely in popular journals. Her age—sixty-three in 1895—also must have told against her, for she was unlikely to persevere in physiological research (26). As Howell noted in his history, the grounds of her rejection were carefully stated to avoid for the present the potentially divisive issue of women in the Society (18, p. 70–71). Council rejected the nomination "on the ground that the publications submitted by the candidate did not fulfill, from the standpoint of original research, the qualifications of membership demanded by the Society." A man, W. T. Howard, professor of pathology and bacteriology at Western Reserve, was also rejected in 1895 on the grounds of publications (presumably they were not sufficiently physiological). In 1902 he was elected president of the American Association of Pathologists and Bacteriologists.

The issue of women as members was not brought up again until 1902, when Meltzer and Lee proposed Ida Henrietta Hyde (1857–1945) for membership. In this case, her credentials could not be doubted. She had received a Ph.D. degree from the University of Heidelberg, had worked in several German laboratories as well as with Porter at Harvard Medical School, and had several publications to her credit, including an article in the first volume of the *American Journal of Physiology.* Since 1899 she was associate professor of physiology (professor after 1905) at the University of Kansas (14, 28). Although the constitution had not prohibited women, and other professional societies such as the naturalists and American Society of Zoologists had already elected them to membership, the Council felt it necessary to request discussion of the issue. It was voted "that in connection with the nomination of Miss Hyde, the Society be asked to consider specifically the question of admission of women to membership." The minutes simply state, "After a full discussion on motion it was voted that the President be requested to cast the ballot of the Society in favor of the admission of Miss Ida Hyde." Howell in his history prided the Society on its equal treatment of men and women in the matter of election to membership (18, p. 71). However, the most serious barrier to a woman becoming a member of APS was not discrimination at the time of election but rather an inability to obtain a position that would provide resources for the ongoing independent research that the Society demanded of its members. She was the only woman to be elected to membership in the Society's first quarter century (see chapter on women in physiology).

Whereas most of the new members of APS held academic appointments, it is worth noting that APS did not reject out of hand physiologists employed in industry. By contrast, the American Society for Pharmacology and Experimental Therapeutics (ASPET), when it was formed in 1908, prohibited the election of pharmacologists working in industry. Moreover, if any member took an industrial job, he thereby forfeited his membership. APS had no such restriction and thus was able to provide a scientific affiliation for industrial pharmacologists, as well as physiologists in industry, until ASPET's provision was changed in 1941. The first industrial member of APS was Thomas Bell Aldrich, affiliated with Johns Hopkins University when elected in 1895 but employed by Parke Davis and Company in 1898. Elijah Mark Houghton was already employed by Parke Davis when elected to membership in 1901.

In 1907–08, perhaps spurred by the competition of the newly formed American Society of Biological Chemists (ASBC) with its relatively large number of charter members, the problems of meeting with the American Association for the Advancement of Science (AAAS) (see below), and the possibility that a broadly based American Biological Society might be formed, APS considered, but in the end rejected, radical changes in its membership policy. At a business meeting in Chicago in 1907, in conjunction with a proposal by Albert Prescott Mathews to form an American Biological Society, "the question of the future policy of the Society in admitting new members with the possible result of a very large increase in membership, was presented for discussion." A Committee on Policy of ten members was appointed by the president to consider the problem, solicit opinions from the membership, and recommend constitutional changes, if necessary. The various considerations, as outlined for the committee by Howell, are of interest, for they recurred several times in other guises in the later history of the Society.[1]

The argument in favor of liberalizing membership was that the Society had in the past been too restrictive and that "it should attempt to encourage in a broad way the development of an interest in physiology in this country" by enlarging its membership to include teachers of physiology, physicians, and those in allied subjects. By a liberal interpretation of the clause making eligible for membership those who have promoted and encouraged physiological research, the constitution would already provide for such an enlargement. It had been suggested that "by this means the membership might be raised to a thousand or more," that the journal could be supported and enlarged, and that those working in physiology "might exert a wider influence in moulding opinion and in promoting the cause of investigation."

The conservative position and its advantages were

that the Society should adhere strictly to what may be designated as its traditional policy, that is, to admit to ordinary membership only those who have demonstrated their right to be recognized as research workers; to have it understood that the Society is composed only of investigators and to seek to encourage physiological research in this country by thus forming a group of workers who shall meet to discuss their results and to promote personal acquaintance. It was urged in favor of this point of view, that it has been most successful in the past so far as the return to the members themselves is concerned, and it has been stimulating to beginners in research, since it has been understood that membership in the Society constitutes an actual recognition and honor. It was urged also, that for all purposes of propaganda and the encouragement of those not recognized as professional physiologists there exist already two associations, Section K of the Association for the Advancement of Science, and the Section in Physiology and Pathology of the American Medical Association, which are in a position to encourage a large membership without rigid restrictions, and that therefore the American Physiological Society should be maintained as a select body of actual investigators in physiology and the related sciences.

After deliberation, the committee decided to recommend no changes in the criteria for ordinary membership in APS.

The constitution of APS provided for two categories of membership, "ordinary" and "honorary." The selection of people who would merit the title of honorary member of APS was an especially troublesome task, in large part because unanimity of members present at a business meeting was required. The issue was brought up several times, but no decisions were made (18, p. 69–70). In 1888 the "action of examining honorary members" was postponed by Council. At the special meeting in 1891 the advisability of nominating for honorary membership some of the foreign guests expected to attend the Congress of American Physicians and Surgeons was discussed. Again nothing was done. In 1903 Council proposed seven honorary members, but the membership postponed action until the following year. The first honorary members, elected in 1904, were Albert Dastre of Paris, Theodor Wilhelm Englemann of Berlin, Franz Hofmeister of Strasburg, John Newport Langley of Cambridge, Ivan P. Pavlov of Petrograd, and Charles Scott Sherrington of Oxford. Again in 1906 the Society postponed action on a group of six names proposed by Council. In 1907 only three were elected: Olöf Hammarsten of Upsala, Eduard Pflüger of Bonn, and Carl von Voit of Munich. Hugo Kronecker of Bern, Ewald Hering of Leipzig, and Albrecht Kossel of Heidelberg were passed over. In 1912 after the election of Edward Sharpey-Shafer of Edinburgh, no further honorary members were chosen until 1942. Because rules for the selection of honorary members were too stringent, subject to the veto of any single member, the Society lost the opportunity to elect a number of eminent early international leaders of physiology.

Scientific Meetings

Before the founding of FASEB in 1912 at the APS Twenty-fifth Annual Meeting, the Society was loosely affiliated with two other federations of societies. On the medical side, it participated in the Congress of American Physicians and Surgeons held in Washington, D.C., every third year, and on the biological side, it held its annual meeting each year with the American Society of Naturalists and other affiliated societies. Thus, after its first annual meeting in 1888, APS never held meetings entirely on its own, but always in conjunction with other societies. The history of working with other medical and biological societies as part of federations provided the necessary experience for the founding of a more formal and more lasting federation of biomedical societies in 1912.

Special Meetings With the Congress of American Physicians and Surgeons

The Society held seven "special meetings" (1888, 1891, 1894, 1897, 1900, 1903, and 1907) with the Congress of American Physicians and Surgeons, which was organized as a loose federation of societies. According to its bylaws, it was to be composed of "national associations for the promotion of medical and allied sciences" and to be governed by an Executive Committee consisting of one member from each society. This committee elected the president of each congress, who presided over the plenary sessions and delivered a presidential address. Among the early presidents were APS members S. Weir Mitchell (1891), William Henry Welch (1897), and Henry P. Bowditch (1900). Whereas the majority of societies participating in the congress represented clinical specialties, the American Association of Anatomists, formed at the congress in 1888, met with the congress from 1891 through 1903, and the newly formed ASBC held its first meeting for the presentation of scientific papers as part of the congress in 1907 (27).

The meetings of the congress were grand affairs, addressed by high government officials, attended by eminent invited foreigners, and extensively reported on in the newspapers. The physiologists presented their scientific papers at separate sessions which, in the later congresses, were held in the physiology laboratory of Columbian (now George Washington) University. Sometimes they also participated in joint sessions with one of the other societies in the congress. In 1897, for example, APS and the Association of American Physicians took part in a public discussion of internal secretions, and in 1907 the same two societies held a symposium on "acidosis" at the New Willard Hotel. The "special" meetings were in effect regular meetings of the Society, at which business and Council meetings were held and important decisions taken (though new members and officers were elected only at annual meetings).

At the annual meeting in December 1908, Meltzer proposed that the Society should no longer meet with the congress, and after a discussion the motion was adopted. One may guess at some of the reasons for this action. The meetings were expensive affairs for a society like APS with a small membership and annual dues of no more than two dollars. The registration fee for the congress was five dollars (there was no registration fee at APS annual meetings), and the cost of the banquet could be as high as ten dollars. The charge to the participating societies for publishing the *Transactions* (which were of little use to physiologists) placed a considerable burden on the Society's treasury. In 1888, for example, this charge came to $72.90 and produced a $13.63 deficit in Martin's first treasurer's report. The meetings, dominated by clinical interests, were always more poorly attended by APS members than the APS annual meetings. By 1907 the breach between the leaders of American medicine and the American Medical Association (AMA), which had led to the formation of the congress, had healed, and the AMA was reorganized to lead in the reform of medical education (19). Thus the congress lost much of its original justification as an alternative national medical forum. In fact, immediately after the members of APS agreed not to meet further with the congress, it was suggested that a joint meeting with the AMA Section on Pathology and Physiology be considered. In response to a letter of inquiry, AMA issued a special invitation to APS members to attend their meetings, though no joint meeting was ever held.

Annual Meetings With the American Society of Naturalists

For over twenty years APS held its annual meeting each year between Christmas and New Year's Day as part of a second informal federation of societies led by the American Society of Naturalists. In 1888 the naturalists, concerned by the possible loss of membership from the recent formation of societies of physiologists, anatomists, and geologists, decided to invite the new societies to meet with them. The four societies met together in New York in 1889 (13). Through the 1890s and most of the first decade of the new century, APS met "in affiliation" with the naturalists and various shifting combinations of other professional societies, including the American Association of Anatomists, the American Morphological Society (precursor to the American Society of Zoologists), the American Psychological Association, the Society for Plant Morphology and Physiology (precursor to the Botanical Society of America), and the Society of American Bacteriologists (now the American Society for Microbiology). Before 1901, when the combined meeting of biological societies was held at the University of Chicago, all meetings were held on college campuses in eastern cities: New York, Boston, Princeton, New Haven, Baltimore, and Ithaca. APS sessions (there were no simultaneous sessions) were often held right in the physiology laboratories, and demonstrations and tours of the facilities were a valuable part of the meetings. If there was more than one institution in a city, as in Philadelphia and New York, the sessions would be held in different schools on successive days, so that more laboratories might be viewed.

These meetings, at least at first, were quite congenial and provided decided advantages to APS over meeting alone. For the most part the physiologists were satisfied with the naturalists' choices of meeting location, though Princeton in 1892 lacked a physiology laboratory and Ithaca in 1897 failed to attract a quorum to carry on business. Affiliation with the naturalists allowed APS to survive the 1890s as a small society of under a hundred members, fewer than thirty of whom attended the meetings. Much of the work of organizing meetings was done by the naturalists, the combined meetings ensured an attendance, and the officers of APS could concentrate their efforts on the content of the scientific programs. The physiologists could take part in common functions, such as the annual banquet with the presidential address, usually held at a hotel, and invitations to receptions and to social clubs. Financially the arrangement was advantageous, because the combined societies usually provided enough members to obtain reduced fares on the railroads, then a major consideration for the holding of scientific meetings. Meeting together also allowed for the possibility of overlapping membership in the societies (3). Among the presidents of the naturalists in this period were Martin (1890), Chittenden (1893), and Bowditch

(1898), as well as such other charter members of APS as George L. Goodale (1889), Charles S. Minot (1894), and William T. Sedgwick (1901) (2).

In 1893 the naturalists submitted a plan for a more formal affiliation or "amalgamation" of societies. This plan is of interest because it foreshadowed some of the provisions of the Federation. Committees of naturalists, morphologists, and physiologists met on 27 December 1893 and voted to recommend the affiliation of these societies and the anatomists. Each society was to elect its own members, but only "professionals" were to be eligible. Membership in any of the affiliated societies carried with it membership in the American Society of Naturalists. Dues were to be levied by the naturalists and funds held in common, though the affiliated societies could assess special levies for their particular purposes. Meetings were to be held at the same time and place, and a common list of members was to be published (2). At the APS business meeting, Curtis's motion, seconded by Chittenden, that "a closer union with the Societies of Naturalists, Morphologists and Anatomists would be to the advantage of the Society," was "discussed at length." As is common with early APS minutes, the details of the no doubt lively debate were omitted, and only the results were recorded. Meltzer's amendment—"provided the meetings shall always take place at a place where there are laboratory facilities"—was not seconded, and Curtis's original motion went down in defeat, three to seven. Howell instead recommended that the president communicate with the Executive Committee of the naturalists to ask whether the physiologists might participate in selecting the location of meetings. Each year thereafter the physiologists elected a delegate to a Committee on Affiliated Societies to determine the place of the next meeting. In 1894 the question of "organic union" with the naturalists was again considered and was defeated four to nine. Apparently the other societies also viewed the naturalists' proposal with skepticism. The sticking point was probably the preeminence given to the American Society of Naturalists and the required membership in that society. A loose form of affiliation continued as before.

Scientific Programs

Programs for the early meetings of APS were arranged in an informal manner. Those who wished to present a paper simply provided the secretary with a title a few weeks before the meeting. Members could give several papers at a meeting, and there was initially no regulation on the length of the presentation. The programs as recorded in the Society's minutes often differed considerably from the programs as promised. In 1892, for example, at a session attended by only nine members, Meltzer presented a paper and then volunteered papers were requested to fill up the program!

In 1894, however, the first effort was made to regulate and formalize the programs. Council recommended to the secretary, Warren P. Lombard, that short abstracts of the papers read at the meeting be sent to the secretary after the meeting for publication. For three years from 1894, abstracts were published in *Science* and made available to members through reprints. For meetings beginning with 1897, abstracts were published in the *American Journal of Physiology*. In 1894 also came the first attempt to limit the length of presentations. Lombard moved that papers be limited to twenty minutes and readers be warned at the end of fifteen minutes. In 1896 Council recommended that papers be limited to fifteen minutes, a proposition accepted by the Society, but "with a certain degree of latitude."

In 1903 regulations regarding the scientific programs were codified in a set of bylaws that were appended to the constitution. The previous year a committee consisting of Frederic S. Lee, Joseph W. Warren, and Graham Lusk was appointed to consider ways and means of improving the programs of the Society and to formulate bylaws to be presented at the next meeting. These provided that the length of papers be limited to fifteen minutes, that there be no more than two papers per member, and that scientific proceedings be published after the meetings. In 1911 the bylaws were changed so that if a member submitted more than one paper and there was insufficient space in the program, the second paper would be "read by title."

In 1902 the character of the annual meeting was radically changed by the decision of AAAS to switch from meeting in the summer to meeting at the same time as the naturalists, during Christmas vacation. Many of the professional societies that had previously met with AAAS also changed to the new time. Officers of AAAS corresponded with university officials and persuaded them, if they did not already do so, to set aside the week between Christmas and New Year's Day as a time when faculty would have no official duties. The week became known as Convocation Week, the time for meetings of scientific and learned societies. There were great expectations for the impact on the public of such large assemblies of scientists (24).

In 1901, probably in anticipation of the inauguration of Convocation Week, AAAS, long since divided into sections by scientific field, established Section K, Physiology and Medicine (1). At joint meetings of APS and AAAS, a joint session with Section K was generally held, at which the chairman of the section, elected annually and usually, but not always, a member of APS, presided. The existence of Section K, as has been noted, provided an argument for not liberalizing the membership policy of APS, because those who were not eligible for membership in APS could, as members of AAAS (which had no membership restriction), present papers through Section K and attend APS sessions.

For societies like APS and the American Society of Naturalists, AAAS's effort to act as umbrella organization for all scientific societies created difficulties. AAAS was a large national society without restrictions on membership, and it was committed to meeting in the South, Midwest, and West, as well as in the East. Conflicts arose each year over whether to meet with AAAS. APS did so in 1902, 1904, 1906–09, 1911, and 1912, as well as a few times after the Federation was formed, in 1916, 1920, and 1924. The AAAS meetings that APS avoided were in out-of-the-way locations, given the membership of the Society at that time—St. Louis in 1903, New Orleans in 1905, and Minneapolis in 1910.

As research-oriented biologists spread to schools all over the country, whether to meet elsewhere than in the East became a heated topic of discussion. Societies threatened to break into sections along geographic lines. The zoologists and naturalists were for several years divided into eastern and central branches. Even APS responded to the prevailing tensions by providing for sections, not by scientific specialty as in the 1970s, but by region of the country. At the 1905 meeting the president of APS, Howell, discussed before the Society "the possibility of having more than one section of the Society to facilitate meetings in different places, East and West." A committee was established and formulated a new bylaw, adopted in 1906, allowing Council to call special meetings of the Society for the reading of papers at the request of ten members.[2] No such special meetings were in fact called in this period.

Publications: Founding of the *American Journal of Physiology*

When APS was formed, no immediate thought was given to a publication, since the British *Journal of Physiology* was in effect a joint publication of British and American physiologists (9, p. 58). The question of an American publication was first raised in May 1894, at one of the special meetings with the Congress of American Physicians and Surgeons. Lee moved that a committee be appointed to report at the December meeting "as to the feasibility of the establishment of an American Journal of Physiology." The committee, consisting of Lee, Porter, and Henry Herbert Donaldson, reported progress in December and asked to be continued, but by 1895, they had evidently reached an impasse and asked to be discharged. At the special meeting in May 1897, the subject was raised anew by President Chittenden and was discussed by Chittenden, Bowditch, Warren, Meltzer, Lee, Porter, George T. Kemp, and James McKeen Cattell. A new committee, composed of Chittenden, Lee, Porter, Howell, and Arthur Cushny, was instructed to consider the matter and report at the next meeting. By December, Chittenden could report on behalf of the committee "that *The American Journal of Physiology* was an accomplished fact." Publication of the journal was to be undertaken by one of the members of the committee, William T. Porter.

The committee had prepared a formal report, which was printed and sent to the membership in June 1897. It advocated the publication of a physiological journal "at once." The number of investigations by American physiologists published in medical journals, especially the recently founded *Journal of Experimental Medicine* edited by Welch, ensured no lack of material. The committee wrote:

> It is no less evident that the various expedients that have hitherto served us are growing irksome and disadvantageous. The sending of manuscripts across the ocean, often to be printed in foreign languages, has been a necessary and valuable resource. But the time for this is past. . . . There can be no question that the position of our profession and its power for usefulness both at home and abroad would be increased by a publication that should be to us what the great archives of physiology are to Germany, France, England, and Italy. We believe that physiology in this country should occupy a position not less dignified and secure.

Of sixty-two members who received the report and the terms of the contract with Porter, forty-one responded, and all approved of the project.

By the terms of the contract Porter agreed to edit five volumes of the journal, each about 500 pages, at a subscription price of five dollars per volume. Contributors would be guaranteed immediate publication and fifty free reprints. Any profits would accrue to the Society. The Society, in turn, would appoint a Publications Committee that would have control of the editorial management of the journal and would enter into formal contract with Porter in the name of the Society for the publication of the volumes. In the absence of a quorum at the annual meeting in 1897, held in Ithaca, New York, no formal vote could be taken. Formal approval by the Society was given in December 1898, with the text of the agreement amended to read that the journal was "to appear in such a form and at such intervals as are satisfactory to the Committee: provided, that the Society shall be free from all financial responsibility for such publication." The first Editorial Board, also called the Publications Committee, appointed by Chittenden in 1897, consisted of Bowditch, Howell, Lee, Jacques Loeb, Warren P. Lombard, and Porter, with Chittenden an ex officio member. As with many journal boards at the time, the members, other than Porter, were chosen because they represented different university cities, in this case, Boston, Baltimore, New York, Chicago, Ann Arbor, and New Haven. Although Chittenden, when he announced the formation of the journal to the Council, requested that the amount of subscription be included in the annual dues, the proposition was voted down. Instead, in 1899 and several times afterward, circulars urged members to subscribe, but then as now, many members did not require individual subscriptions, and institutions and nonmember subscribers made up a good part of the subscription list.

The first issue of the journal appeared in January 1898. It commenced with a paper from the laboratory of the Sheffield Scientific School by Chittenden and William Gies, "The influence of borax and boric acid upon nutrition with special reference to proteid metabolism." No doubt the most memorable contribution to this inaugural volume was Walter B. Cannon's first full-length paper on the application of X rays, discovered in 1896, to physiology, "The movements of the stomach studied by means of roentgen rays." Among the other authors represented were Howell, Lee, Porter, Charles Wilson Greene, Lafayette B. Mendel, Lusk, Colin C. Stewart, A. N. Richards, and Ida Hyde. The entire first volume consisted of thirty-two papers in four issues and 522 pages and concluded with the publication of the proceedings (abstracts of papers) of the December 1897 APS Meeting. At first issues appeared every two months, but after July 1899, they appeared monthly. As with many journals, the period of time covered by a volume varied depending on how much material was received. Subscription was not by year but by volume at a cost of five dollars per volume, and even in 1898 there was material sufficient for more than one volume a year. There was no formal review process, though Porter was said to have gone over papers thoroughly and returned them to the authors with suggested changes (5). The main function of the Editorial Board was to help attract articles to the journal, not to critique them.

In all, Porter edited thirty-three volumes of the *American Journal of Physiology*, through March 1914. The slight deficit on the first volume was made up when the Society voted to pay fifty dollars to have the proceedings (abstracts of papers) of the 1898 meeting published; they continued this practice in future years. Publication at first generally ran at a loss. The Society contributed $100 to support the journal in 1908 and another $100 toward defraying the cost of an index volume in 1910. The contract with Porter was renewed in 1900 for a period of five years after completion of volume five, again in 1905, after an extended discussion, for another period of five years, and again in 1911. Like most scientific journals in this period, the *American Journal of Physiology* was started and financed by an individual or individuals. There was little alternative, for societies at this time were small, and to retain their membership, dues had to be inexpensive. They simply did not have the resources to put up the capital for a publication venture or to take on the financial responsibility if the journal should fare poorly. The *American Journal of Physiology* differed from the journals associated with other societies, including ASBC and ASPET, in an important respect. Although APS accepted no financial responsibility for the journal and members were not required to subscribe, the journal from the outset had an official connection with the Society. When, in 1914, Porter offered the journal debt free to the Society, APS became one of the first biological societies in America to assume full ownership and management of a journal.

Biographical Sketch of William Townsend Porter

William Townsend Porter, the founder of the *American Journal of Physiology*, has been the Society's greatest benefactor. Born in Plymouth, Ohio, in 1862, Porter received his M.D. degree in 1885 from St. Louis Medical College (later incorporated into Washington University School of Medicine), studied physiological chemistry in Philadelphia, and then went abroad to study physiology in Kiel, Breslau, and Berlin. In 1887 he was appointed to teach physiology at the St. Louis Medical College and established the first physiology laboratory beyond the Mississippi River. Porter was elected to APS at its fourth annual meeting in 1891 and presented a paper before the Society in 1892 on the effects of ligating coronary arteries. His publications on ventricular filling and pressure, control of respiration, coronary circulation, origin of the heartbeat, and the physical and mental development of children brought him to the attention of eminent physiologists, among them Bowditch, who in 1893 invited Porter to join his department and introduce laboratory experimentation as part of the regular physiology course.

Because imported laboratory instruments were far too costly to equip a teaching laboratory, Porter established a machine shop in the department to make simplified and less expensive apparatus. It was found that enough instruments could be produced to equip other physiology laboratories. With the help of Charles W. Eliot, president of Harvard, capital was raised to found the Harvard Apparatus Company in 1901. The company prospered, and in 1920 Porter used the profits to found the Porter Fellowship. On two occasions, Porter offered to give the company to the

Society, but the Society felt unequal to the task of attempting to manage such an enterprise properly and reluctantly declined.

By 1900 Porter had nearly full charge of the teaching of physiology at Harvard Medical School. As his teaching assistant he chose Walter B. Cannon, who received his M.D. degree from Harvard in 1900. Despite Porter's devotion to the reform of the teaching of physiology, he was not popular with the medical students. A strict disciplinarian, his expectations were too high; he failed roughly a third of the students from 1902 to 1904. The students revolted, and when Bowditch retired in 1906, Cannon rather than Porter succeeded him as George Higginson Professor of Physiology. Porter became instead professor of comparative physiology. A breach that lasted many years was created between Cannon and Porter. However, in 1937 Cannon was among those who enthusiastically proposed Porter to be honorary president for the Semicentennial Celebration of APS. Porter was the master of ceremonies for the occasion. In 1948 he was given the distinction of becoming the only American ever to become an honorary member of the Society. Porter retired from Harvard in 1928 but continued working for the Harvard Apparatus Company until his death in 1949. A. C. Barger wrote of him, "Physiology was Porter's religion; he had no other" (4, 5, 7, 21).

Professional Concerns: Education and Defense of Medical Research

Professional concerns such as education and the defense of medical research played a relatively minor role in the activities of APS during its first fifty years. Many members of the Society were actively involved in such concerns, but they generally worked through other institutions. As a small and specialized society, APS chose to concentrate on what it could do best, that is, advancing physiological research through its programs and publications and leaving physiological education and public information to larger, broader-based, and better-funded medical organizations.

However, problems of the profession were not entirely absent from the Society's minutes and proceedings. Several of the demonstrations at the early meetings were of instruments, models, and procedures designed for classroom use. In 1889, for example, John G. Curtis gave a communication on a "Method of Demonstrating to a Large Class of Students the Beating of the Heart of the Calf, in Opened Thorax," and in 1892 C. F. Hodge presented "A Method of Preparing the Eye for Demonstration and Study." G. W. Fitz showed "A Working Model of the Eye" in 1895 and the following year exhibited several pieces of apparatus, including "A Lever System to Illustrate the Action of Muscles in Relation to Joints" and "A Form of Student's Myograph." In addition, the tours of physiology laboratories allowed physiologists to examine a variety of equipment that could be used by students. In the 1890s APS participated in a joint committee with the naturalists and other affiliated societies on problems of biological education (2). In 1893 at Curtis's suggestion, a committee consisting of Howell and Lee was appointed to consider and report on "Should

Physiology be taught in the [secondary] Schools and if so how?" Little is known of the committee's activities. It asked to be discharged in 1896 because "its usefulness was impaired by its inability to carry out its ideas." At the annual meeting of the naturalists in December 1898 a symposium was held on "Advances in Methods of Teaching." Porter's paper advocating laboratory instruction was a classic statement of the importance of learning physiology by hands-on experience (7, 25). At that same meeting, Bowditch, as president of the naturalists, devoted his presidential address to "The Reform of Medical Education" (8), a subject long of deep concern to him.

APS did not feel the need to take a leading role in the defense of animal experimentation, because there were other larger and more effective organizations to take action. In the 1890s antivivisectionist agitation became a serious concern when a bill was introduced in the U.S. Congress to regulate animal experimentation in the District of Columbia by setting up a system of licensing and registration. Both the Congress of American Physicians and Surgeons and the American Society of Naturalists and its affiliated societies became involved. William Henry Welch, a charter member of APS and leader of the campaign in defense of medical research, called on the 1897 Congress of American Physicians and Surgeons, of which he was president, to combat the threat. Under the authority of the congress, he wrote a circular letter inviting members from every state of the union to join a committee. The letters and resolutions solicited by this committee, especially those from family physicians of senators and congressmen, were effective in preventing the bill from ever coming to a vote. Among those who testified before the U.S. Congress was Bowditch (16). The American Society of Naturalists and its affiliated societies likewise set up a committee in the 1890s to respond to the problem. APS did appoint its own ad hoc committee to take action if needed, but none seems to have been required. In the following decade, a vigorous campaign against the antivivisectionists was led by Walter B. Cannon working through the Committee on Defense of Animal Research of AMA (6).

Founding of ASBC and ASPET

The first decade of the twentieth century was one of rapid change in medical education. Many poorer, proprietary schools were shut down, whereas the better or more fortunately placed schools became more firmly integrated into university structures. The standards of admission to medical school were gradually raised to require a bachelor's degree, and the length of the medical program increased from as little as two terms to four years. The celebrated Flexner Report of 1910 only served to accelerate a process of radical transformation that was already well underway. The preclinical sciences, including physiology, anatomy, biochemistry, and pharmacology, were among the beneficiaries of this transformation. These fields now had the advantages of a larger share of the medical school curriculum, better-prepared students, and access to university endowments and other resources. There was a rapid growth in number of full-time positions for well-trained experimental physiologists (20). Physiology, as represented by members of APS, spread from a few centers mainly in the East to schools and related institutions all over the country.

This growth of the biomedical sciences was reflected in the growth of the Society and in the enlargement of its scientific programs and of its journal. In 1906 there were 128 regular members of APS, and the number attending the annual meeting, held in New York, reached 50 for the first time. At the three scientific sessions, one at the physiology laboratory of University and Bellevue Hospital Medical College and two at the physiology laboratory of the College of Physicians and Surgeons, twenty-seven papers were presented and discussed, three papers were read by title, and two demonstrations were given. At a fourth session, held jointly with Section K of AAAS and presided over by Simon Flexner, chairman of Section K, eleven more papers were read and discussed and five more were read by title. That year, twelve issues of the Society's journal, containing eighty-two articles and filling 1,300 pages and nearly three volumes, appeared.

The very success of APS, coupled with the opportunities provided by the reform of medical schools, gave rise to pressures to form new societies to consolidate new disciplines. In 1906 the founding of ASBC represented the first of a succession of offshoots or splinter groups from APS. On each occasion the formation of a new society was greeted by the parent society with regret and fear that physiology would become fragmented. But in each case, physiology has survived and continued to grow and flourish. The founder of ASBC was John J. Abel, professor of pharmacology at Johns Hopkins University School of Medicine and a member of APS since 1891. Although not primarily a biochemist, Abel was responsible for instruction in biochemistry at Johns Hopkins until 1908. With Christian A. Herter and three other men, he founded the *Journal of Biological Chemistry* as a private joint stock company in 1905 and was coeditor of the journal until 1909. [The journal was not officially associated with ASBC until it was given to the society in 1919 (15).]

Biochemistry benefited from the same opportunities for growth as did physiology. Russell Chittenden recalled in his history of the first twenty-five years of ASBC (12, p. 1) that when APS was founded, only two of the charter members, himself and Victor Vaughan, were biochemists. The number increased rapidly in the 1890s as Chittenden's students and other young biochemists were elected to membership. By 1906, at Chittenden's count, the number of APS members active in biochemistry had reached at least thirty-seven. The expansion in the number of biochemists, the increasing number of chemical papers on the crowded APS program, the success of the *Journal of Biological Chemistry*, the efforts of the American Chemical Society to set up a biological section, the prospect of chairs and departments, all contributed to a sentiment, discussed over several years, that a separate society was needed (12, p. 1–3).[3]

On 16 October 1906 Abel sent the twenty-four men whose names appeared on the cover of the *Journal of Biological Chemistry* a circular and proposal to form a new society. Of these men, twenty-one were APS members. On receiving a favorable response, Abel wrote to a longer list of biochemists on 13 December and called for an organizational meeting on 26 December in New York, the day before the commencement of the APS annual meeting. Abel fortunately saved the replies to his letter of inquiry, and they are now among the Abel Papers at Johns Hopkins. Most respondents were entirely enthusiastic. However, a few, including Chittenden, with his student and colleague at Yale, Lafayette B. Mendel, and Graham Lusk at Cornell Medical School, were worried about the effect of the new society on APS. The most significant fear, that of a complete separation of biochemists and physiologists, was allayed when Abel assured correspondents that the two societies would continue to meet at the same time. The attitude of the physiologists, Chittenden recalled, was "that physiology was a study of function in the broadest sense and to divorce physiological chemistry from the parent society would not only weaken the latter, but would produce a wrong impression" (12, p. 2). Although Chittenden regarded biochemistry (which he called physiological chemistry) as a branch of physiology, Abel believed that physiological chemistry formed only one part of biochemistry and that the new society would acquire as members many who were not eligible for or interested in APS. Indeed APS did not welcome as members biochemists who were primarily organic chemists rather than physiological chemists. L. J. Henderson, then an instructor in biochemistry at Harvard, was deferred in 1905 and 1906 and accepted in 1907 only after he had become a charter member of ASBC. The Food and Drug Administration chemist, Harvey Wiley, rejected by APS in 1905, became a member of ASBC in 1907. Gustav Mann and M. X. Sullivan, biochemists rejected by APS, were also accepted into ASBC.

Both Mendel and Chittenden raised with Abel the possibility of creating a biochemical section of APS. Mendel wrote to Abel on 4 December 1906:

> I would like to hear your objections to the division of the Physiological Society into two sections, one of which shall be biochemical. This would satisfy the claims of "die gesamte Physiologie" and at the same time give us due independence. Our programs also have enough titles now to justify such a division. My only reason for suggesting the above again is the fact that no less than eight of our colleagues have expressed a similar view to me, and it therefore seems worthy of some consideration.

He hoped the matter could be discussed at the APS smoker at 8:30 P.M. on 26 December. Abel, however, felt that nothing would be gained from such a discussion and was not amenable.[4,5] The organizational meeting was held at 4:30 P.M. on 26 December, and by the time of the APS smoker, the new society was a fait accompli. Of the twenty-nine men present at the organizational meeting, twenty-two were already members of APS, and two more were elected in 1906. A temporary organization was set up with Chittenden as first president, and articles of agreement were adopted until a constitution could be drafted and approved. Eventually eighty-one persons were given the status of charter member (12, p. 4–8).

When Chittenden officially announced to the members of APS in 1906 the formation of ASBC, he "expressed the hope that its relations with the Physiological Society would be most cordial and that no conflict of interest or purposes should be allowed to arise." Howell, who was president of APS at this time, recalled in his history of APS the feelings of regret at this separation from the parent society but noted:

> As a matter of fact there was no conflict of interests. The two societies arranged to hold their annual meetings together and acted in close affiliation on matters of common concern, until their final amalgamation in 1913 into the Federation (18, p. 76–77).

Abel envisioned the creation of a biochemical society as a first step toward the later creation of a pharmacological society. He took steps to create the latter society as well as a pharmacological journal in 1908. Again the proposal met with concern on the part of physiologists. Meltzer, a good friend of Abel and a charter member of ASPET, replied to Abel's letter of inquiry, "I shall not oppose this organization in the first place because you are the originator of this idea. I am of course sorry to see how the future of the Physiological Society becomes more and more uncertain."[6] ASPET was officially launched on 28 December 1908 when eighteen men, fifteen of them members of APS by 1908, adopted articles of agreement. Abel became first president and served through 1912. By the time of the first annual meeting in 1909, at which a constitution was adopted, fifty-two charter members were named. At the organizational meeting in 1908 Abel announced the formation of a journal, and its first issue was published in June 1909. Although the *Journal of Pharmacology and Experimental Therapeutics* published the abstracts of the society, it had no official connection with the society. It remained the property of Abel, who edited it for twenty-three years, until it was donated to the society in 1933 (11).

The three societies continued to meet each year at a common place and time. Joint sessions of APS and ASBC were held at every meeting from 1907 through 1912. Joint demonstrations with ASPET were held in 1909 and a joint session in 1911. In 1906 and 1907 the physiologists and biochemists collaborated on a report on uniformity in protein nomenclature, a project initiated by committees of the Physiological and Chemical Societies in Great Britain (12, p. 47–48). There was from the beginning and there remained a very large overlap in membership in the three societies. With dues no more than two dollars per society, double or triple membership was not a great burden. Members could attend the meetings of all three societies and shift from one society to another from session to session. An informal federation always existed.

About the time of the formation of the two new societies, APS was experiencing increasing difficulties in arranging its annual meetings, now that AAAS and all its associated societies had become involved. The problem of where and when to meet was frequently discussed in the APS minutes from 1908 to 1911 as well as in the ASBC minutes. Each year the Society had to decide whether to follow the naturalists, AAAS, or neither. Whereas the large meetings with AAAS had advantages such as inexpensive rail fares and the opportunity to meet scientists in widely different fields, they were increasingly unwieldy. In 1908 Council took a vote on a motion to continue meeting with AAAS. The motion lost. It was decided instead that APS should determine each year whether to meet with AAAS but that at least some meetings with AAAS were desirable.

Even before the founding of the Federation, the officers of the three societies worked in concert to determine the place of the annual meeting. The relationship was partially formalized in 1910 when, on the recommendation of APS Council, a joint committee of one member from each of the physiological, biochemical, and pharmacological societies was formed to work out plans for joint sessions of the societies at future annual meetings. The 1909 meeting was held with AAAS and the naturalists in Boston. There was so much dissatisfaction on the part of the naturalists and zoologists with meeting with AAAS that those societies decided to try meeting elsewhere with the hope that other biological societies would join them. They met in Ithaca in 1910 and Princeton in 1911, both unsuitable locations for APS because of the lack of a medical school. In 1910 APS, ASBC, and ASPET met in New Haven, and in 1911 the three societies met partly in Washington, D.C., with AAAS and partly in Baltimore, where Johns Hopkins afforded better laboratory facilities. After consultation among the secretaries, the three societies agreed in 1911 to hold the 1912 meetings with AAAS and the naturalists in Cleveland, where Torald Sollmann and J. J. R. Macleod of Western Reserve could act as hosts.[7]

In 1912, the year the Federation was formed, some twenty-three societies affiliated with AAAS for the Cleveland meeting. They included not only major professional research-oriented societies, such as the Astronomical and Astrophysical Society of America, the American Mathematical Society, the American Physical Society, the American Psychological Association, the American Association of Anatomists, and the zoologists and naturalists, but also such groups as the School Garden Association of America, the Association of Official Seed Analysts, and the American Federation of Teachers of the Mathematical and Natural Sciences. In all, over 1,000 people came to Cleveland. The overall program was cluttered with official addresses, not only of all the sectional officers of AAAS, but also of presidents of many of the societies. It is not to be wondered that small and relatively exclusive societies such as APS, ASBC, and ASPET would desire alternative meeting arrangements.

The idea of forming a federation was being discussed by a number of biological societies at this time. The proposal to form an American Biological Society—a federation of biological societies—first advanced in 1907, was revived at this time and widely publicized (23). One of the major difficulties to be overcome, according to A. J. Carlson (10), was the different standards for membership for the different societies. FASEB was seen by some as the first step in the direction of a more general union. Once the Federation was formed, representatives were to confer with the naturalists and zoologists, but nothing seems to have come of their discussions.

Also promoting affiliation of APS, ASBC, and ASPET was the increase in number of papers submitted for the APS and ASBC programs. The APS program for 1911, for example, consisted of five morning or afternoon sessions (one each joint with ASBC and ASPET) at which sixty papers or demonstrations were presented and six more papers read by title. A sixth session with Section K of AAAS was devoted to a symposium on acapnia and shock at which five papers were presented and discussed. The program for the three-day meeting was filled to the limit of time available. At an "executive session" of the Society, various alternatives were discussed for remedying the congestion. Among them were meeting "in sections" (in more than one session by topic), increasing the length of the meetings, changing the time of meeting to June, making abstracts mandatory, eliminating reading of papers and just discussing them, and limiting the number of papers. Of special interest, in view of subsequent events, was the alternative of "closer affiliation with the Biochemical and Pharmacological Societies." By 1911 ASBC was also having difficulties with its program and adopted a bylaw to regulate length of papers.[7] The Federation was, among other things, a means of relieving congestion of programs by better distribution of papers among the various member societies.

Who exactly originated the move to create a federation is not known. At the 1911 APS meeting, formation of a greater physiological society was discussed informally. The biochemists, concerned that all the societies should be on even footing, appear to have taken some of the initial steps to initiate a federation in the form that was eventually adopted. At the 1911 meeting of ASBC, a motion by Otto Folin providing

> that a committee be appointed by the incoming President to consider the advisability and feasibility of a more formal affiliation or federation with the American Physiological Society and the American Society of Pharmacologists than now exists

was passed. On 11 May 1912 the president of ASBC, A. B. Macallum, appointed Lafayette B. Mendel (chairman), Walter Jones (secretary), Holmes C. Jackson, Torald Sollmann, and H. C. Bradley to the committee and requested a report at the next meeting.[7] Exactly what occurred next is uncertain. Apparently APS Council, then consisting of Meltzer, Carlson, Cannon, Erlanger, and Lee, was consulted. A mimeographed page dated 20 November 1912, sent by Carlson, secretary of APS, to A. N. Richards, secretary of ASBC and a member of the ASBC committee, stated:

> The Council of the American Physiological Society goes on record as being in favor of the formation of a
>
> Federation of American
> Societies for Experimental Biology
>
> The Federation shall consist of the American Physiological, Biochemical and Pharmacological Societies and its object shall be the establishment of a stable connection between the three societies for the purpose of fixing the place and time of the annual meetings, the arranging of joint sessions whenever possible, and in general to establish officially closer scientific and social affiliations between the sister societies while retaining their individual independence.[7]

The ASBC committee then presented an official report in which it was recommended

> that the American Society of Biological Chemists express itself in favor of the formation of a Federation of American Societies for Experimental Biology along the lines recently suggested by the Council of the American Physiological Society, viz.: [above paragraph repeated].

No record of the mid-year deliberations of APS Council leading up to the 20 November document was saved. Probably Meltzer played a prominent role in selecting the name and defining the objects of the Federation, for he was a man with decided convictions on matters of medical research and education and on the interrelationship of scientific societies. He already had to his credit the founding of several societies, among them the Society for Experimental Biology and Medicine (its name reminiscent of the Federation) and the Society for Clinical Investigation. Remembered by Howell as a "kingmaker" (17), Meltzer is known to have initiated, behind the scenes in 1913, the formation of the fourth member of the Federation, the American Society for Experimental Pathology (22, p. 7–8). The first chairman of the Executive Committee of the Federation, he may have been its chief architect, but if so, the historical documentation has not been found.

By the time the annual meetings began in Cleveland on Sunday, 29 December 1912, the Federation was virtually formed. At the APS Council meeting on Sunday, 29 December, the Council recommended the formation of a Federation of American Societies for Experimental Biology, in the words of its memorandum of 20 November. The Council then "voted to recommend appointment of a conference committee of three to meet with conference committees of the other societies with a view to the formation of such a federation." On 30 December Council recommended that the Conference Committee be given "full power to act with the other Committees in matters needing settlement before the next annual meeting." At the business meetings of the various societies, these recommendations were approved and members of the Conference Committee appointed. The physiologists were represented by Meltzer, Lee, and Cannon; the biochemists by Gies, Lusk, and H. Gideon Wells; and the pharmacologists by Sollmann, Loevenhart, and Auer. All but Gies attended the meeting of the committee at the Colonial Hotel on New Year's Eve, 31 December 1912. At this meeting the Federation was officially formed and began operations, although the report of the Conference Committee could not be formally ratified until the 1913 annual meeting. Meltzer was elected temporary chairman of the organizational meeting and Cannon temporary secretary. The following motions embodied in a report signed by Cannon as secretary and dated 2 January 1913 were unanimously adopted:

That a federation of the three societies be hereby established.

That the presidents and secretaries of the constituent societies form the Executive Committee of the Federation.

That the chairmanship of the Executive Committee be held in turn by the presidents of the constituent societies who shall succeed one another annually in the order of seniority of the societies (physiological, biochemical and pharmacological).

That the secretary of the Executive Committee shall be the secretary of the society whose president is chairman.

That the secretaries of the three societies shall consult in preparing the programs of the annual meeting, and that, as far as practicable, and with the author's consent, papers be so distributed as to be read to the society in which they properly belong.

That the programs of the three societies be published by the secretary of the Federation under one cover and that the expense of publication be shared *pro rata* by the societies according to the number of members.

That the official title of the new organization be "The Federation of American Societies for Experimental Biology (Comprising the American Physiological Society, the American Society of Biological Chemists, and the American Society for Pharmacology and Experimental Therapeutics)."

That a common meeting place of the Federation with the anatomists, zoologists and naturalists is desirable but not mandatory.

That in the name of the Federation the International Physiological Congress be invited to meet in the United States in 1916.

That the present Conference Committee delegate all its powers to the Executive Committee of the Federation.

Although the formation of the Federation was ratified by the societies the following year, it was effectively formed as of New Year's Eve 1912, and it soon set to work to organize the 1913 meeting. The agreement to meet as part of a federation made it possible for APS, ASBC, ASPET, and the American Society for Experimental Pathology, which joined the Federation in 1913, to meet on their own apart from the naturalists and AAAS. In 1914 all four societies, as well as the anatomists, went to St. Louis at the invitation of Washington University School of Medicine, while the naturalists and AAAS met in Philadelphia. The formation of the Federation by APS and its offshoots in 1912 was symbolic of the maturation of the biomedical sciences in America.

NOTES

[1] R. Hunt, letter (and enclosure) to J. Erlanger, 21 January 1908. Erlanger Papers, Washington Univ. School of Medicine Archives, St. Louis, MO.

[2] W. H. Howell, letter to W. P. Lombard, 29 January 1906. APS Archives, Bethesda, MD.

[3] Minutes and records of Am. Soc. Biol. Chem., 1906–09. Am. Soc. Biol. Chem., Bethesda, MD.

[4] L. B. Mendel, letter to J. J. Abel, 4 December 1906; Abel to Mendel, 6 December 1906. J. J. Abel Papers, Alan Mason Chesney Medical Archives, Johns Hopkins Univ., Baltimore, MD.

[5] R. H. Chittenden, letter to J. J. Abel, 19 October 1906. J. J. Abel Papers, Alan Mason Chesney Medical Archives, Johns Hopkins Univ., Baltimore, MD.

[6] S. J. Meltzer, letter to J. J. Abel, 17 December 1908. Alan Mason Chesney Medical Archives, Johns Hopkins Univ., Baltimore, MD.

[7] Minutes and records of Am. Soc. Biol. Chem., 1910–15. Am. Soc. Biol. Chem., Bethesda, MD.

REFERENCES

This chapter is based on the minutes of Society and Council meetings and the annual directories of members. Only additional sources are cited here.

1. AMERICAN ASSOCIATION FOR THE ADVANCEMENT OF SCIENCE. *Proceedings* 1901.
2. AMERICAN SOCIETY OF NATURALISTS. *Records* 1, 1884–95; 2, 1896–1913; and 3, 1914–23.
3. APPEL, T. A. Biological and medical societies and the founding of the American Physiological Society. In: *Physiology in the American Context, 1850–1940,* edited by G. L. Geison. Bethesda, MD: Am. Physiol. Soc., 1987, p. 155–176.
4. BARGER, A. C. William Townsend Porter. In: *Dictionary of American Biography.* New York: Scribner, 1974, suppl. 4, p. 675–677. [Reprinted in *Physiologist* 22(5): 20, 1979.]
5. BARGER, A. C. The meteoric rise and fall of William Townsend Porter, one of Carl J. Wiggers' "old guard." *Physiologist* 25: 407–413, 1982.
6. BENISON, S. In defense of medical research. *Harvard Alumni Med. Bull.* 44(1): 16–23, 1970.
7. BORELL, M. Instruments and an independent physiology: the Harvard Physiological Laboratory, 1871–1906. In: *Physiology in the American Context, 1850–1940,* edited by G. L. Geison. Bethesda, MD: Am. Physiol. Soc., 1987, p. 293–321.

8. BOWDITCH, H. P. Reform in medical education. *Boston Med. Surg. J.* 139: 643–646, 1898.

9. BYNUM, W. F. A short history of the Physiological Society, 1926–1976. *J. Physiol. Lond.* 263: 23–72, 1976.

10. CARLSON, A. J. The Federation of American Societies for Experimental Biology. *Science* 39: 217–218, 1914.

11. CHEN, K. K. *The American Society for Pharmacology and Experimental Therapeutics: The First Sixty Years, 1908–1969.* Bethesda, MD: Am. Soc. Pharmacol. Exp. Ther., 1969.

12. CHITTENDEN, R. H. *The First Twenty-five Years of the American Society of Biological Chemists.* New Haven, CT: Am. Soc. Biol. Chem., 1945.

13. COPE, E. D., AND J. S. KINGSLEY. Editor's table. *Am. Naturalist* 23: 32–34, 1889.

14. DEYRUP, I. J. Ida Henrietta Hyde. In: *Notable American Women.* Cambridge, MA: Harvard Univ. Press, 1971, vol. 3, p. 247–249.

15. EDSALL, J. T. The *Journal of Biological Chemistry* after seventy-five years. *J. Biol. Chem.* 255: 8939–8951, 1980.

16. GOSSEL, P. P. William Henry Welch and the antivivisection legislation in the District of Columbia, 1896–1900. *J. Hist. Med.* 40: 397–419, 1985.

17. HOWELL, W. H. Biographical memoir, Samuel James Meltzer, 1851–1920. *Mem. Natl. Acad. Sci.* 21: 15–23, 1926.

18. HOWELL, W. H., AND C. W. GREENE. *History of the American Physiological Society Semicentennial, 1887–1937.* Baltimore, MD: Am. Physiol. Soc., 1938.

19. KING, L. S. *American Medicine Comes of Age, 1840–1920: Essays to Commemorate the Founding of the Journal of the American Medical Association, July 14, 1883.* Chicago, IL: Am. Med. Assoc., 1984.

20. KOHLER, R. E. *From Medical Chemistry to Biochemistry: The Making of a Biomedical Discipline.* Cambridge, UK: Cambridge Univ. Press, 1982.

21. LANDIS, E. M. William Townsend Porter, 1862–1949: editor of the *American Journal of Physiology,* 1898–1914. *Am. J. Physiol.* 158: v–viii, 1949.

22. LONG, E. R. *History of the American Society for Experimental Pathology.* Bethesda, MD: Am. Soc. Exp. Pathol., 1972.

23. MATHEWS, A. P. A plan for the organization of the American Biological Society. *Biochem. Bull.* 2: 261–268, 1912–13.

24. MINOT, C. S. The relation of the American Society of Naturalists to other scientific societies. *Science* 15: 241–244, 1902.

25. PORTER, W. T. On the teaching of physiology in medical schools. *Boston Med. Surg. J.* 139: 647–652, 1898.

26. ROOT, E. H. Frances Emily White, M.D. *Women's Med. J.* 14(5): 97–99, 1904.

27. *Trans. Congr. Am. Physicians Surg.* 1–7, 1888–1907.

28. TUCKER, G. S. Ida Henrietta Hyde: the first woman member of the Society. *Physiologist* 24(6): 1–9, 1981.

_1887_

American Physiological Society. Meeting for Organization.

At a meeting held at the College of Physicians and Surgeons of New York on Dec. 30th 1887 to consider the advisability of instituting a society of American Physiologists the following persons were present — namely — H. P. Bowditch; R. H. Chittenden; J. G. Curtis; H. H. Donaldson; F. W. Ellis; G. Stanley Hall; W. H. Howell; Joseph Jastrow; W. P. Lombard; H. N. Martin; C. S. Minot; S. Weir Mitchell; Isaac Ott; W. T. Sedgwick; H. Sewall; J. W. Warren; H. C. Wood.

Dr. S. Weir Mitchell was called to the chair, and a proposed constitution for the Society was read clause by clause, and amended. After amendment it was unanimously adopted as follows ——

CONSTITUTION.

1. This Society shall be named "THE AMERICAN PHYSIOLOGICAL SOCIETY."

2. This Society is instituted to promote the advance of Physiology and to facilitate personal intercourse between American Physiologists.

3. The Society shall consist of ordinary and of honorary members.

4. Any person who has conducted and published an original research in Physiology or Histology (including Pathology and experimental Therapeutics and experimental research in Hygiene), or who has promoted and encouraged Physiological research, and who is a resident of North America, shall be eligible for election as an Ordinary member of the Society.

5. Distinguished men of science who have contributed to the advancement of Physiology shall be eligible for election as Honorary members of the Society. The number of honorary members resident in North America shall not exceed three. Honorary members shall pay no membership fees. They shall have the right of attending the meetings of the Society, and of taking part in its scientific discussions, but they shall have no vote.

6. The management of the Society shall be vested in a Council of five elected by ballot from the ordinary members. The Council shall appoint from its members a President of the Society, and a Secretary, who shall also act as Treasurer.

7. The members of the Council shall be elected at each annual meeting of the Society, to serve one year.

First page of minutes of the organizational meeting of APS, 30 December 1887.

COLLEGE OF PHYSICIANS AND SURGEONS.
59TH STREET FRONT.

Top: *College of Physicians and Surgeons of Columbia University, ca. 1887, location of the organizational meeting of APS. (From J. C. Dalton,* History of the College of Physicians and Surgeons in the City of New York: Medical Department of Columbia College, *New York, published by order of the college, 1888.)* **Bottom:** *John Green Curtis in the laboratory at the College of Physicians and Surgeons, 1887. Seated left to right: Curtis and Frederic S. Lee. Standing center: Warren P. Lombard. (Courtesy of H. W. Davenport.)*

Original members. **Top left:** *William Henry Howell, ca. 1891. (Courtesy of H. W. Davenport.)* **Top right:** *Edward T. Reichert.* **Center:** *Henry Newell Martin and the Department of Biology, Johns Hopkins University, 1880. Seated left to right: K. Mitsukura, a student; Martin; W. Keith Brooks, professor of morphology. Standing left to right: doctoral students E. B. Wilson, William T. Sedgwick, Samuel F. Clark, Henry Sewall, E. M. Hartwell. Sewall and Sedgwick were among the original members of APS.* **Bottom left:** *Henry Herbert Donaldson.* **Bottom right:** *Horatio C Wood.*

Original members. **Top left:** *Henry Gustav Beyer.* **Top right:** *Charles S. Minot.* **Center left:** *Isaac Ott.* **Center right:** *William Henry Welch.* **Bottom left:** *T. Wesley Mills.* **Bottom right:** *Victor C. Vaughan.*

Top left: *Frederick W. Ellis, original member.* **Top right:** *Joseph W. Warren, original member.* **Center left:** *Jacques Loeb. (National Library of Medicine.)* **Center right:** *Ida Henrietta Hyde, first woman member of APS. (Courtesy of G. S. Tucker.)* **Bottom left:** *John J. Abel. (Courtesy of the Alan Mason Chesney Medical Archives, Johns Hopkins University.)* **Bottom right:** *Reid Hunt.*

Program.

✦

Monday, December 30th,

Hull Physiological Laboratory, Room 26.

10 A. M.

General Business.   Reading of Papers.

G. T. KEMP—Relation of blood plates to the increase in number of red corpuscles at high altitudes.

G. T. KEMP and O. O. STANLEY—Some new observations on blood plates.

C. W. GREENE—Notes on the physiology of the circulatory system in the hagfish, Polistotrema stouti.

W. T. PORTER—The mechanism of fibrillar contraction of the heart.

W. B. CANNON—The movements of the intestines (studied by means of the Röntgen rays).

D. J. LINGLE (By invitation)—On further experiments on the importance of sodium for the heart beat.

J. LOEB (For W. ZOETHOUT)—Effects of calcium and potassium on striped muscle.

F. S. LEE and W. SALANT—The action of alcohol on muscle.

J. LOEB (For R. W. WEBSTER)—A contribution to the physiological analysis of absorption by muscle.

2 P. M.

A. R. CUSHNY (For C. A. GOOD)—The excretion of lithium.

P. A. LEVENE—Glycocoll in gelatoses.

G. LUSK—On the question whether dextrose is produced from cellulose in digestion.

L. BREISACHER—A contribution to the physiology of the thyroid gland.

W. S. CARTER—The relation of the parathyroid to the thyroid gland.

L. B. MENDEL and L. F. RETTGER—Experiments on the relation between the spleen and the pancreas.

W. JONES—On the nucleic acid of the suprarenal.

F. G. NOVY—On the surface action of metals.

W. J. GIES and P. B. HAWK—The composition and chemical qualities of the albumoid in bone.

W. J. GIES and L. D. MEAD—A comparative study of the reactions of various mucoids.

P. A. LEVENE—On gluco-phosphoric acid.

W. T. PORTER (For R. S. LILLIE)—The role of the cell-nucleus in oxidation and synthesis.

W. P. LOMBARD—Demonstration of apparatus.

W. T. PORTER—A moist chamber and other new physiological apparatus.

8.30 P. M.

Hotel del Prado,

Smoker.

✦

Tuesday, December 31st.

Hull Physiological Laboratory, Room 26.

9 A. M.

General Business.   Reading of Papers.

W. J. GIES and E. A. POSNER—Are proteids, prepared by the usual methods, combined with fat or fatty acid?

L. B. MENDEL—New experiments on allantoin excretion.

J. T. HALSEY—Studies on diuresis.

A. R. CUSHNY—An unrecognized feature in diuresis.

J. LOEB—On the influence of the electrical charge and valency of ions upon their toxic and antitoxic effects.

J. LOEB and MR. LEWIS—On the prolongation of life of unfertilized eggs of the sea urchin by potassium cyanide.

J. LOEB (For M. FISCHER)—Artificial parthenogenesis in annelids.

E. P. LYON (By invitation)—Effects of potassium cyanide and of lack of oxygen on the development of sea urchin eggs.

J. LOEB (For A. W. GREELEY)—Artificial parthenogenesis in star fish produced by lowering the temperature.

P. A. LEVENE—Embryochemical studies. II. The presence of mono-amino-acids in the developing egg.

R. HUNT—Experiments with Zygadenus venenosus.

2 P. M.

P. M. DAWSON—An attempt to obtain regeneration of the spinal cord.

H. H. DONALDSON—The formula for determining the weight of the central nervous system in frogs of different sizes.

A. P. MATHEWS—The nature of nerve stimulation and alterations of irritability.

W. KOCH (By invitation)—The chemical analysis of the brain.

S. I. FRANZ—The frontal lobes (cerebral) and the formation and retention of associations.

W. J. GIES and I. O. WOODRUFF—On the toxicology of selenium and its compounds.

Y. HENDERSON—A study of metabolism in a case of lymphatic leukaemia.

G. N. STEWART—The mode of action of certain substances on the colored blood-corpuscles, with special reference to the relation between so-called vital processes and the physico-chemical structure of cells.

W. S. HALL—Presentation of a new form of ergograph.

A. R. CUSHNY—Demonstration of the glands in the oviduct of the fowl.

J. ERLANGER (By invitation)—A new instrument for determining systolic and diastolic blood-pressures in man.

Program of Annual Meeting, University of Chicago, 30–31 December 1901. This was the first meeting held away from the East Coast.

American Physiological Society.

The American Physiological Society will hold its sixteenth annual meeting in Philadelphia, on Tuesday and Wednesday, December 29th and 30th, 1903. The usual smoker will be held at the Hotel Walton on Monday evening, December 28th. The sessions of Tuesday will be at the University of Pennsylvania, and those of Wednesday at the Jefferson Medical College. The headquarters of the Society will be at the Hotel Walton, Broad and Locust Streets. Members of the Society will please inform the Secretary at their earliest convenience, whether they intend to be present, and what communications they desire to make. Those who will require apparatus or other necessities for the making of demonstrations, may communicate with Professor E. T. Reichert, the University of Pennsylvania, or Professor A. P. Brubaker, Jefferson Medical College. Attention is called to the facts that the reading of papers is confined to the members of the Society and to guests specially invited by the President and Secretary jointly ; and that papers are limited to a length of fifteen minutes. The annual membership fee for 1903–04 will be two dollars.

FREDERICK S. LEE, Secretary.

437 West 59th Street, New York City.

November 23, 1903.

American Physiological Society.

Members desiring to vote by proxy are requested to return the attached ballot, properly prepared and enclosed in a signed envelope, to the Secretary, 25th and E Streets, N. W., Washington, D. C.

The members of the existing Council are:

> President—WILLIAM H. HOWELL
> Secretary—REID HUNT
> Treasurer—WALTER B. CANNON
> Additional Members—
> GRAHAM LUSK
> JOHN J. ABEL

American Physiological Society.

I hereby cast my ballot for

———————————————————For President
———————————————————For Secretary
———————————————————For Treasurer
———————————————————⎫ For Additional members
———————————————————⎭ of the Council

Dated,   190

Top left: *notice of Annual Meeting, 1903.* Top right: *mail proxy ballot for officers, 1908.*
Bottom left: *Lafayette B. Mendel.* Bottom right: *Graham Lusk.*

688 BOYLSTON STREET,

BOSTON, MASS., *January* 15 1903

Dr. W. P. Lombard

To The American Journal of Physiology, Dr.

The subscription price in the United States and Canada is five dollars per volume, postage free; in other countries, five dollars and twenty-five cents (£1 2s; marks, 22; francs, 27), payable in advance.	Each volume will contain about five hundred pages, divided into numbers, to be issued monthly.

Volume IX (will begin March 1) @ $5.00

2 copies $10.00

" VIII 2 " 10.00

$20.00

Received Payment

W. T. Porter

Top: *William T. Porter, founder and editor of the* American Journal of Physiology. *(Courtesy of the Francis A. Countway Library, Harvard Medical School.)* **Bottom**: *receipt for the* American Journal of Physiology *sent to Warren P. Lombard, 1903. (APS Archives, donated by H. W. Davenport.)*

CONTENTS.

First page of the table of contents of the American Journal of Physiology *1, 1898.*

Top: *honorary Sc.D. candidates at the IV International Physiological Congress and the International Zoological Congress, Cambridge University, 1898. Left to right: Etienne Marey (France), A. F. Dohrn (Germany), C. Golgi (Italy), Ernst Haeckel (Germany), A. A. W. Hubrecht (The Netherlands), Willy Kühne (Germany), Henry P. Bowditch (United States), Hugo Kronecker (Switzerland).* **Bottom:** *International Committee, VII International Physiological Congress, Heidelberg, 1907. Seated left to right: Hugo Kronecker, Albrecht Kossel (president of the congress), Sigmund Exner, P. Grutzner, J. L. Prevost, Charles S. Sherrington. Standing left to right: Paul Heger, R. Nicolaides, C. Bohr, A. Wedensky, John Newport Langley, Willem Einthoven, J. E. Johansson, N. Mislawsky, Charles Richet, N. Cybulski, William T. Porter. (IUPS Archives, American Philosophical Society.)*

CHAPTER 4

Second Quarter Century 1913–1937

TOBY A. APPEL

During the second quarter century of the American Physiological Society (APS, the Society), physiology in America came of age and became the equal of Old World physiology (10). This new stature was symbolized in 1929 by the first meeting of 600 foreign scientists in America at the International Physiological Congress in Boston. It was a period of remarkable growth and new discoveries. To many it represented the classic era of physiology.

Although periods of twenty-five years do not necessarily coincide with natural breaks in any organization's history, in the case of APS the beginning and ending of the second quarter century do at least roughly correspond to major turning points. In 1912–14, a number of propitious changes in the Society's governance and activities took place. The Federation of American Societies for Experimental Biology (FASEB, the Federation) was launched at the end of 1912 and in 1913 held its first meeting in Philadelphia. The following year the Federation successfully met on its own, apart from the American Association for the Advancement of Science (AAAS), the naturalists, and the zoologists. An especially momentous event in the history of the Society occurred early in 1914, when Porter resigned as editor of the *American Journal of Physiology* and the Society undertook ownership and management of its own journal. The new responsibilities for the journal precipitated a major overhaul of the constitution, which was approved at the annual meeting in December 1914 and affected not only publications policy, but terms of Council members, mode of election of officers, and membership criteria.

Through the twenty-five years the Society grew and prospered. From 230 members in 1913, membership increased nearly threefold to 661 by the time of the Semicentennial Celebration in 1938, though the requirements for membership became increasingly stricter. The number of papers presented at annual meetings grew much faster than membership, from 37 in 1913 to 331 in 1938. Annual meetings each year

63

were held in conjunction with the meetings of the Federation, which throughout the period consisted of APS, the American Society of Biological Chemists (ASBC), the American Society for Pharmacology and Experimental Therapeutics (ASPET), and the American Society for Experimental Pathology (ASEP). The Society's publications also continued to expand in pages published and number. The *American Journal of Physiology* was followed in 1921 by a second journal, the immensely successful *Physiological Reviews*. Also in 1921 the Porter Fellowship Program, the first of the Society's annual awards, was inaugurated. On 2 June 1923 the Society, which now had responsibility for two journals and fellowship funds, was incorporated in the state of Missouri, which was then the residence of both its secretary, Charles Wilson Greene (University of Missouri), and its treasurer, Joseph Erlanger (Washington University School of Medicine).

Before World War II, APS was a research society, devoted almost entirely to dissemination of the results of research through meetings and publications. It remained a volunteer organization, with dues of two dollars or less. The major work of managing the organization was carried out by the secretary and the treasurer. There were as yet no standing committees. As a society, APS was reluctant to involve itself in other issues, such as education, ethical issues concerning the profession of physiology, and the ever-present battle against the antivivisectionists, even though they might be of vital importance to physiology. The leaders of the Society generally felt that these issues were best handled by other organizations.

The second quarter century ended with a nostalgic Semicentennial Celebration, held in Baltimore in 1938 and presided over by William T. Porter as honorary president. From the vantage point of 1938, physiologists, including a few who were present at the founding of the Society in 1887, could rejoice in the tremendous progress in their field. If the beginning of the quarter century coincided with the beginning of an era in the Society's history, the end of the period roughly corresponded to the end of an era. World War II soon followed the Semicentennial Celebration and brought with it so great a transformation in the organization of the sciences that the Society was never to be the same again.

Officers and Organizational Changes

The second twenty-five years generated twelve presidents of APS: Meltzer, Cannon, Lee, Lombard, Macleod, Carlson, Erlanger, Meek, Luckhardt, Greene, Mann, and Garrey. They were elected for a year at a time, but with the general understanding that they would be reelected to serve two or three years. In 1935 the constitution was amended to limit the president to two years in office. There was never any such limitation on the terms of other officers, namely the secretary, who was to maintain correspondence, and the treasurer, who collected dues and managed the Society's finances. Those who served the longest included Greene as secretary for nine years (1915–23) and Erlanger as treasurer for eleven (1913–23).

After the constitutional revision of 1914, Council was to consist of seven members—the three officers and four others elected for four-year rotating terms. Election of officers and members of Council took place at business meetings, and the new officers assumed office at the close of the annual meeting. Before 1914, those members who were not going to attend the meeting could vote by mail proxy ballot. This worked well enough when there was only a small membership, but as the number of possible candidates grew, elections became more complicated. A series of ballots was sometimes necessary for one candidate to obtain a clear majority. There was discussion of sending two ballots (a nominating ballot and a final ballot) to members who were not attending the meeting. However, after considerable debate, the proxy ballot was eliminated altogether. Samuel J. Meltzer, who was among those who favored the change, offered the argument that he was "opposed to giving members who perhaps never attend a meeting of the Society, the right to decide who should be its officers."[1] After 1914, voting for officers from nomination to final ballot was left entirely to the minority of members who attended the business meetings (until mail ballots were instituted in 1975). The election procedure became increasingly cumbersome over the years and took up a great deal of time. Twice, in 1930–31 and in 1934–35, the creation of a nominating committee was seriously considered, but both times Council voted to retain the then-current method. Walter J. Meek wrote to Arno B. Luckhardt in 1930, "The nominating committee always arouses suspicions of clique ruling." APS remained (and still remains) one of very few societies in which the members at large select the candidates for office.

APS During World War I

World War I brought with it great disruption in the normal laboratory work of physiologists but also new opportunities for research on subjects of national interest. At the 1917 meeting of APS, Council decided to contribute to the war effort by investing $4,000 of the surplus fund of the journal in Liberty Bonds. When the National Research Council (NRC) of the National Academy of Sciences (NAS) was created in 1916 as a means of mobilizing scientific talent to aid the war effort, APS appointed members to serve on a Committee of Physiology of NRC, chaired by William Henry Howell, which directed investigations of shock, industrial fatigue, food and nutrition, and the effects of poisonous gases. In March 1917 APS President Frederic S. Lee appointed the three most recent presidents of APS—Walter B. Cannon (chairman), William Henry Howell, and himself—to an APS committee to confer with the Council of National Defense, headed by the Secretary of War, on the best ways to utilize the expertise of the members of the Society. In May 1917, for the use of the NRC committee, Lee sent a letter to all members of APS inquiring "what its members are already doing or are willing to do in the way of national service, whether their laboratories will be open during the coming summer, whether they desire to undertake research in case of need, and, if so, what general lines of research they are prepared to follow."[2]

After the war NRC was established on a permanent basis. APS was asked early in 1919 to name a representative to the Division of Medical Sciences. Lee acted as temporary representative until the annual meeting, when Howell was appointed. Howell was followed by A. J. Carlson, Yandell Henderson, and E. K. Marshall. In 1923 Carlson reported the establishment of NRC Research Fellowships, funded by the Rockefeller Foundation and especially favoring the preclinical sciences. Beginning in 1926, APS was also asked to appoint representatives to the Division of Biology and Agriculture of NRC. The Society's first delegate was Merkel H. Jacobs, a general physiologist.

Membership

Although membership in APS continued to increase as more workers entered physiology, at the same time it became more difficult to be elected a member of the Society. The 1914 revision of the constitution tightened requirements in several respects. "Original researches" were required instead of "an original research." The passage allowing for the election of those who had "promoted and encouraged physiological research" was finally eliminated. (It had never been used.) Instead the candidate was now required not only to have published research, but to be actively continuing research. Cannon explained that he had in mind when suggesting this change

> the instance of a man who was elected a member of the Society on the basis of work done some fifteen or twenty years previous to the time of his election, and who had done nothing since that time to justify his being a member of the Society. It seems to me that [the] Society should be composed, so far as possible, of active workers.[1]

The fields mentioned in addition to "Physiology" were modified (see chapt. 2); histology was dropped, and, at the suggestion of Carlson, experimental zoology was added.[1] The criteria, as adopted at the 1914 meeting, read:

> Any person who has conducted and published original researches in Physiology (including Experimental Zoology, Pathology, Pharmacology, Experimental Therapeutics, and Hygiene), or [sic] who is continuing researches in any of these fields, and who is a resident of North America, shall be eligible for election as an ordinary member of the Society.

Membership applications were scrutinized with increasing care after 1914. In December 1919, for example, out of twenty-three proposals, only twelve were approved. In 1920 twenty applied for membership and twelve were accepted. Each year a number of applications were "laid on the table," that is, deferred for a year, until more evidence of research and publication had accumulated. Candidates sponsored by the most eminent physiologists of the day, leaders of the next generation, and even future presidents were told to wait a year and try again. Among those deferred a year in the 1920s were such later-to-be distinguished names as Ralph Gerard, John Field II, William F. Hamilton, Laurence Irving, and Detlev Bronk. However, in contrast to a later period, the rank held by the candidate was not a bar to membership, so long as it allowed for the possibility of continued research; instructors, fellows, and research associates were commonly elected. If a candidate had published good research, he or she might well become a member at the next meeting after acquiring a Ph.D. degree.

In the 1930s physiology tended to become more narrowly defined than previously, in that candidates who were members or even prospective members of some other professional biological society were frequently rejected. Membership in more than one Federation society, formerly very common, was now discouraged. If the position the candidate held was not sufficiently physiological in nature, he or she was not welcomed. This notion that individuals should belong to only one society particularly hurt women, who, after receiving degrees in physiology, often turned to research in a related area because they were unlikely to find employment in a department of physiology. The percentage of women accepted for membership, which was over ten percent in the 1920s and early 1930s, decreased after 1935. In 1934 the first black member of the Society, Joseph L. Johnson of Howard University School of Medicine, was elected. He is now one of the Society's oldest living members.

In general, election policy in the second quarter century was not as liberal or as broad minded as it was in the early years of the Society or as it has again become in the past quarter century. The programs of the meetings and the title pages of the journals indicate that a great many prospective members were not selected or were discouraged from applying for membership.

Meetings

The first Federation meeting, held in Philadelphia at Jefferson Medical College and the University of Pennsylvania in December 1913 simultaneously with the meetings of the naturalists, zoologists, and anatomists, was a resounding success. A record 118 APS members attended (including eight newly elected members), as well as a number of nonmember guests. The format adopted served as a model for future meetings. As is still the custom, the councils of the various societies held meetings on the Sunday before the opening of the scientific sessions. The scientific meeting consisted of five half-day scientific sessions. (Later meetings had six sessions.) The first was a joint session of all the societies at which selected contributed papers were presented. It opened with an address by President Meltzer entitled "Theories of anesthesia." This was the only such presidential address given; the Society resolved at its business meeting "that a formal presidential address at the meeting of the Federation is not desirable." (Joint sessions at later meetings often began instead with memorials to deceased members of the Federation societies. In 1914, for example, S. Weir Mitchell and Charles S. Minot were eulogized.) Three sessions of the 1913 meeting were devoted to separate sessions for each society and to joint demonstrations held in the medical laboratories of the University of Pennsylvania. In all, ninety-three papers and twenty demonstrations were presented. In addition, social events were planned for every evening at local hotels and restaurants. The annual Federation banquet was inaugurated at this meeting.

In 1914, when the Federation began its tradition of meeting most years on its own, it accepted an invitation from the Washington University School of Medicine. The four Federation societies, as well as the anatomists, went west to St. Louis. (AAAS and the naturalists and zoologists met in Philadelphia.) To mark the occasion a panoramic photograph, over three feet long, of all those who attended was taken in front of the School of Medicine. Later meetings were generally held on university campuses and organized by local arrangements committees. The locations varied widely but were usually in the East or Midwest. They ranged as far north as Montreal (1931), as far west as Minneapolis and Rochester, Minnesota (1917), and as far south as Memphis (1937).

Because of World War I, the 1918 annual meeting was postponed from Christmas recess to the spring vacation period of 1919. Two meetings were thus held in 1919, the first in Baltimore in April and the second (originally planned for Toronto) in Cincinnati at the usual time in December. The programs of these meetings reflected some of the recent wartime activities of physiologists. As a special feature of the April meeting, there was an evening presentation of a film, "Efficiency and Sanitation in the Feeding of the United States Army," by John R. Murlin, who returned after the war to his position as professor of physiology and director of the Department of Vital Economics at the University of Rochester. In one of the joint sessions Lee presented a paper on "The Work of the U.S. Public Health Service in Industrial Physiology," and seven more papers were presented under the heading "Contributions from Officers and Former Officers in the Division of Food and Nutrition, Sanitary Corps, U.S. Army."

Occasional meetings were still held in conjunction with AAAS, the last of which was held in Washington, D.C., in 1924. APS sessions for this large gathering of societies were held at Central High School. By this time it had probably become clear that instead of competing with AAAS and other societies by holding a meeting at the same time but in a different city, it would be preferable to choose a separate meeting time. An unofficial ballot taken at the business meeting indicated a preference for a spring meeting. Therefore the 1925 meeting, held in December as previously planned, was the last of the December meetings. No meeting was held in 1926. In 1927 the Federation experimented with a spring meeting. Held in Rochester, New York, from 14 to 16 April 1927, the meeting proved sufficiently successful that the Federation has since met in late March or April.

With the increase in the number of members, it became necessary to accommodate an ever larger number of papers at meetings. The alternative of selecting papers to be presented or discussed on the basis of merit, although raised from time to time, was always discarded as being opposed to what were considered the democratic traditions of the Society. The ten-minute paper was already a feature of meetings by 1913, but according to the bylaws a member could still present two papers if there was room on the program. In 1925 the bylaws were changed so that members could present only a single paper before the Federation, even if they were members of more than one society. There was still no stated limit on introductions of nonmembers, as heads of departments often had several students they wished to sponsor. There was also no limit on the number of papers "read by title," which meant, in effect, that the title would appear on the program and the abstract would be published. In 1925 APS abstracts were for the first time required to be sent to the secretary simultaneously with the title of the paper instead of after the meeting. Beginning in 1932, a bound copy of APS abstracts in the form of page proof was distributed before the meeting.

In 1925 APS yielded to the inevitable and met in two simultaneous sections instead of a single session over which the president presided. The minutes reported, "This proved entirely satisfactory and may be considered a safe means of getting more papers on the program." Sessions were labeled "Program A" and "Program B" (in 1927) and were divided roughly by subdiscipline. "Program C" first appeared in 1932, and in that year the sessions were labeled by specialty as, for example, "Blood, Circulation and Respiration," "Muscle-Nerve," and "Gastrointestinal Activity." By 1937 the APS program could accommodate 247 papers in five simultaneous sessions, not including papers in the joint session and demonstrations.

There were no regular symposia until 1938, though some of the joint sessions, if there were several papers on a single topic, were similar to symposia and were sometimes even called symposia. One of the most memorable of these was the symposium on insulin at the 1922 Federation meeting at which F. G. Banting and Charles H. Best presented the results of their Nobel Prize-winning research. The joint sessions were generally placed at the beginning or end of the program. Although initially they consisted of specially selected contributed papers, by the 1930s the speakers were generally invited from among the most distinguished members of the Federation. In 1938 APS sponsored its own symposium and thus began a tradition that has continued and expanded. This pioneer symposium, held as one of five simultaneous sessions, was organized by L. J. Henderson on the subject of anoxia, its participants (each presenting a paper fifteen minutes in length) were E. S. Guzman Barron, Ralph W. Gerard, David Bruce Dill, Ernest Gellhorn, Ross A. McFarland, Dickinson W. Richards, Jr., and Harry G. Armstrong.

Before the 1930s the Federation was primarily a mechanism for organizing the annual meeting. It had no income of its own. Its expenses each year were prorated according to the number of members in each society. In 1916, for example, the annual budget for the Federation was $312.14, and APS's share was $138.08. The funds were spent mostly for stationery; postage; the annual Federation yearbook (which began in 1914) containing officers, constitutions, and membership lists of the societies; and the program of the meeting.[3] Not until the mid 1930s did the Federation begin to take a more active role. In 1933 a registration fee of one dollar was collected for the first time. In 1935 Hooker was appointed secretary of the Federation, and at about the same time, several standing committees, including one on international congresses and one on the defense of animal research, were organized.

To long-time members of APS, it seemed that the last few meetings of the 1930s had suddenly become very large. Beginning in 1933, when it was no longer possible to hold sessions in university facilities, meetings took place in hotels and meeting halls. Thus the 1938 Baltimore meeting was not held at Johns Hopkins, as all previous Baltimore meetings had been, but rather in the Fifth Regiment Armory. Ten meeting rooms were required for the Federation societies, as well as rooms for other societies, such as the American Institute of Nutrition, that were meeting with the Federation. By the time meetings resumed after World War II, Baltimore and most of the other cities in which the meetings had been held for the first fifty years could no longer accommodate the Federation.

Publications

As noted above, a turning point in the history of APS publications occurred in 1914, when the Society took on full ownership of and responsibility for the *American Journal of Physiology*. William T. Porter, under contract to the Society, had been editing the journal since it first appeared in 1898. It was published in monthly issues, which were gathered every three or four months into volumes that sold for $5.25 each in the United States and Canada and $5.75 abroad. By the end of 1913 the journal numbered thirty-two volumes. Porter's contract had been renewed in 1900 and again in 1905 and 1911.

Though the Society was grateful for Porter's generosity in running the journal, some degree of friction seems to have developed between Porter and the Council and its Editorial Committee. Porter, a man of high ideals and uncompromising standards, did not always work easily with others. At the 1913 meeting, the same meeting at which Porter's rival, Cannon, was elected president, members of the Society at a business meeting adopted the following resolution after "extensive" but unreported "discussion":

> Resolved that the Editorial Committee be appointed a special committee and instructed to consider the question as to the relation between the American Physiological Society and the American Journal of Physiology, and to recommend at the next annual meeting of the Society means by which more satisfactory conditions for publication by American physiologists may be obtained.

The Editorial Committee, which then consisted of Howell (chairman), Carlson, Erlanger, Lee, Graham Lusk, A. B. Macallum, and Porter (who was not in attendance), met immediately after the APS annual meeting and addressed to Porter a letter inviting him to discuss the matter. After receiving the letter, Porter wrote Carlson, secretary of the Society, a terse note announcing his decision to cease editing the journal after the current volume (i.e., volume 33) was completed, which would probably be April 1914. He set forth conditions for continued use of the title, which he owned: "(1) that the Society shall publish a journal that shall be their own property and (2) that such a journal shall be printed by a first-class house in a form little if at all inferior to that of the present journal."[4]

Porter's sudden decision precipitated a crisis. April was only three months away, but no official Society action could be taken until the following December. The

Editorial Committee had no authorization to take financial responsibility for a journal, and there was no one at hand to do the work. Quick measures had to be taken. A flurry of letters passed among the members of Council and the Editorial Committee. When it proved impossible to induce Porter to continue as editor until the next meeting of the Society, a meeting of the Editorial Committee was held in New York on 25 January; Howell, Lee, Lusk, Cannon, Meltzer, and Lafayette B. Mendel attended. The committee recommended that the Society own and manage the journal and continue to use the same title if possible, but, if Porter was unwilling, to adopt a new title. Lee and Cannon met with Porter in Cambridge on 5 February and succeeded in persuading him to waive the condition as to quality on the grounds that "there was no one to determine whether the standard was being maintained and no redress in case a difference of opinion arose and, furthermore, that no one in the Society wishes to have a journal of poor quality published." Thus both of Porter's conditions were satisfied. However, it was learned from Porter at this time that volume 33 would be completed in March rather than in April. Although Porter was willing to receive manuscripts for the April issue, he insisted that the Society assume responsibility for sending out notices and collecting bills by 1 March.[5]

Because there was no time for constitutional niceties, Cannon took matters in hand and, after conferring with Lee, decided to send members of the Society a printed notice requesting a vote on the question, "Shall the *American Journal of Physiology* be owned by the American Physiological Society and edited under its control?" The notice claimed that the journal as edited by Porter "results in little or no financial deficit," when in fact the journal had been running at a loss. Nevertheless, until other provisions were made, Council proposed to raise a "guarantee fund."[6] By 16 February, of the seventy-five members of the Society who responded, seventy-three were in favor, and ten men had pledged a total of over $1,000 for the emergency fund. A formal vote was taken by Council on the question of committing the Society to the publication of the journal, and the matter was decided.[7] The membership gave final approval at the 1914 meeting and altered the constitution to enable the Society to own the journal.

The wording of Article IX specifying the relationship of the journal to Council caused considerable disagreement among the members of Council. Cannon felt that the danger of subjecting the journal to "the temporary will of the members, perhaps a small minority, in attendance at any particular meeting" would be minimized by the creation of a separate editorial committee elected by Council for long terms. Howell worried about the problem of divided authority.[1] His view prevailed, with the result that the separate Editorial Committee was disbanded and Council itself acted as an editorial board.

By March, a candidate for managing editor was found.[8] Neither Lee nor Howell, the two most senior members of the committee, was willing to take on the work. At the 25 January meeting in New York, both had supported the idea of finding a younger man. There was some concern, however, that a young man would not have the necessary broad experience for editorial judgments. Donald Russell Hooker, first suggested by Lee, was an excellent choice, because, as a student of Howell and a member of the Department of Physiology at Johns Hopkins, he could readily avail himself of Howell's advice. Elected to APS in 1906, he was at this time thirty-seven years old. Howell, as chairman of the Editorial Committee, requested Hooker's appointment to the Editorial Committee, and, although Porter had not yet formally resigned and created a vacancy, Cannon approved. Hooker was to act as managing editor of the APS journals for the next thirty-two years.

With Hooker in place, journal procedures were regularized and some important changes were made. Porter had paid authors of accepted papers five dollars (although members of the Harvard department had for a long time not received the payment). This was eliminated as unnecessary. It was decided that authors should pay for reprints and for part of the cost of publishing articles that were long or contained many tables. Procedures for acceptance or rejection of manuscripts were spelled out. There were at this time no formal peer review procedures. It was decided that any member of Council would have the power to accept articles for the journal, but that two members of Council would be required to reject a manuscript. In fact, few articles were rejected. Ten rejections were reported in 1917, three in 1919, eight in 1920, and ten in 1921. In 1914 there were about 400 subscribers to the journal, for the most part libraries and institutions. Of 205 members at the beginning of 1914, 55 subscribed to volume 34; the number of subscribers increased to 71 as the result of a circular sent to all members. The price per volume was reduced in 1914 to $5.00 in the United States and Canada and $5.25 abroad.

Immediate attention was devoted to finding another press for printing the volumes, because the Plimpton Press used by Porter was considered too expensive. The chief competitors were the Journal of the American Medical Association Press, favored by Carlson and Meltzer, and Waverly Press, favored by Cannon and Lee. Among the advantages of Waverly Press were its location in Baltimore, close to the managing editor, and its extensive experience in handling similar journals. At the time, Waverly Press printed thirty-one journals, including the *Journal of Biological Chemistry* and the *Journal of Pharmacology and Experimental Therapeutics*.[8] With volume 35, beginning in August 1914, printing of the *American Journal of Physiology* was shifted to Baltimore, and Waverly Press has printed APS journals ever since. Erlanger, APS treasurer, handled subscriptions and finances for volume 34 (April–July). By mutual consent and the agreement of Council, Hooker assumed these duties for volume 35 and all subsequent volumes. Finances for the journal were thus from the beginning separated from finances of the Society. Hooker established an office for the journal in his home in Baltimore and in 1915 hired an assistant, Laura Campen, APS's first employee, who was responsible for all the day-to-day affairs of the APS journals throughout Hooker's long tenure as managing editor.

APS was one of the earliest societies of its size to assume full responsibility for managing its associated journal. By way of contrast, the *Journal of Biological Chemistry* and the *Journal of Pharmacology and Experimental Therapeutics* were still privately owned with only an informal relation to ASBC and ASPET. Early in its history, however, APS began a strong publications program that has led to a long period of steady growth and prosperity.

Biographical Sketch of Donald Russell Hooker

Born in 1876 in New Haven, Connecticut, Donald Russell Hooker received his B.A. and M.S. degrees from Yale and his M.D. degree in 1905 from Johns Hopkins. After spending a year in research at the University of Berlin, he joined the Department of Physiology under Howell at Johns Hopkins as an assistant in physiology. He was elected to membership in APS in 1906. In 1910 he became an associate professor but gave up the appointment in 1920 because of the pressures of editing the APS journals. In 1926 he accepted an appointment as lecturer in the School of Hygiene and Public Health. His research was of a high order. Between 1907 and 1935 he published over forty journal articles on the physiology of the circulatory system and is recognized as a pioneer in the study of venous pressure. Some of his most important work was done at Hopkins in the early 1930s as part of a historic team project, funded by Consolidated Edison, to investigate the effects of electricity on the human body. The collaborators were Hooker, O. R. Langworthy, a neurologist, and William Kouwenhoven, an electrical engineer. Their research, published in the *American Journal of Physiology* in 1933, demonstrated that electrical defibrillation was possible in animals and led to the eventual development of means to defibrillate the human heart (5; 9; 12, p. 98–100).

In addition to his duties as managing editor of the *American Journal of Physiology* and *Physiological Reviews* (begun in 1921), in 1935 he became secretary of the Federation; he was the founder and the first editor of *Federation Proceedings*. For most of his years as managing editor of APS journals, he worked without remuneration. He maintained the journal office at his home at 19 West Chase Street in Baltimore. Carlson said of him that he "was one of the ablest and most devoted servants of our science," and "the last man to claim the stature of a superman" (4). As an editor he was known for his wide experience, good judgment, high standards of accuracy, clarity and brevity, and conservative management, resulting in the accumulation of a substantial reserve fund for the protection of the journals. He was a man of strong social conscience, active in civic affairs and social reforms. In the volume of the *American Journal of Physiology* dedicated to him, was written, "No one of his generation has had a greater influence on American physiology" (1).

Founding of Physiological Reviews

Under the conservative management of APS Council and Hooker, the *American Journal of Physiology* began to accumulate a modest revenue. As early as 1915 Hooker reported a surplus of $2,565.76, an amount that grew steadily every year thereafter. Within a few years it was generally understood that the funds were to be used for publication purposes only. Among the uses for the money suggested in Hooker's annual report for 1915 was "to guarantee for one year the publication of a journal devoted to articles reviewing the various fields of biological science." At the time this was a quite novel suggestion, for there were few journals of this kind in existence.

The initial idea of *Physiological Reviews*, as well as its subsequent development and rapid success, was largely due to Hooker's mentor and colleague, Howell. In April 1919, Howell and Hooker brought the matter of the review journal before Council and the Society. Howell felt that there would be good demand for such a journal and that it would soon be self-supporting. Council appointed Howell and Hooker to a committee charged with presenting a detailed plan at the next meeting. The proposition was discussed informally by the Society. Its response was described in the minutes as "enthusiastic," though Howell recalled years later that the initial reaction was at best lukewarm.

In June 1919 Hooker and Howell sent Council a report proposing that a journal, called either *Physiological Reviews* or *Quarterly Reviews of the Physiological Sciences*, be published quarterly in a single volume per year. The journal would "cover the subject of physiology, physiological chemistry, pharmacology, experimental pathology and such other subjects as may from time to time appeal to those interested in the biological sciences." Editorial responsibility would rest with a Board of Editors representing various branches of physiology, who would appoint a managing editor, select the subjects to be reviewed, and assign them to authors. "The ultimate subscription list is estimated as over a thousand," they wrote. Appended to the report was a list of suggested contents for volume 1. In the plan presented to and approved by the Society in December 1919, Council was to appoint the managing editor and deal with policy matters other than selection of authors and articles. Council appointed Hooker managing editor and a board of seven editors, four from APS and one each from ASBC, ASPET, and ASEP. They were Howell (Johns Hopkins), Macleod (Toronto), Lee (Columbia), and Hooker (Johns Hopkins) representing APS, Mendel (Yale) representing ASBC, Reid Hunt (Harvard) representing ASPET, and H. Gideon Wells (Chicago) representing ASEP. The board chose Howell to be its chairman, and he was to remain in that position, despite several offers to resign, until 1932, when he was replaced by Carlson.

By 1920 a prospectus was drawn up and widely distributed. It described the aims of the journal in the following terms:

> The main purpose of the PHYSIOLOGICAL REVIEWS is to furnish a means whereby those interested in the physiological sciences may keep in touch with contemporary research. The literature, as every worker knows, is so extensive and scattered that even the specialist may fail to maintain contact with the advance along different lines of his subject. The obvious method of meeting such a situation is to provide articles from time to time in which the more recent literature is compared and summarized. The abstract journals render valuable assistance by condensing and classifying the literature of individual papers, but their function does not extend to a comparative analysis of results and methods. Publications such as the Ergebnisse der Physiologie, the Harvey Lectures, etc., that attempt this latter task, have been so helpful as to encourage the belief that a further enlargement of such agencies would be welcomed by all workers. It is proposed, therefore, to establish a journal in which there will be

published a series of short but comprehensive articles dealing with the recent literature in Physiology, using this term in a broad sense to include Bio-chemistry, Bio-physics, Experimental Pharmacology and Experimental Pathology.

It was hoped that the journal would appeal to teachers and clinicians, as well as to physiological researchers. Although the first issue had yet to appear, cash subscriptions in advance at $6 per year had been adequate to meet initial expenses, and the reserve fund of $3,000 did not have to be used.

The first volume, which appeared in 1921, contained nineteen contributions, including articles on the conduction of the heartbeat by J. A. E. Eyster and Meek, functions of the capillaries and venules by Hooker, blood volume and its regulation by Erlanger, the sugar of the blood by Macleod, the regulation of pulmonary circulation by Carl J. Wiggers, afferent paths for visceral reflexes by S. W. Ranson, the physiology of undernutrition by Lusk, and physiological effects of altitude by E. C. Schneider. The second volume in 1922 contained papers by such notable biomedical scientists as Ralph Stayner Lillie, Florence Sabin, Janet Howell Clark, A. V. Hill, Alexander Forbes, George H. Whipple, Otto Folin, and Torald Sollmann. *Physiological Reviews* proved an immediate and overwhelming success. The first issue was exhausted by July 1921. By the end of the first year of operations the journal had 838 subscribers "and new subscriptions constantly coming in." It was not long before the "ultimate subscription list" of 1,000 was reached and surpassed.

Establishment of the Board of Publication Trustees and Review Procedures

The major concerns of the APS journals in the 1920s were the products of their success: first, too many manuscripts, and second, how to handle a rapidly growing reserve. Hooker and the APS Council took pride in the rapid publication of research results; they had been able to publish articles two months after they were submitted. The mounting accumulation of manuscripts was at first handled by increasing the number of pages in the volumes and by printing in smaller type. In 1921 Council authorized a free volume to all subscribers to eliminate the backlog, but this was only a temporary solution. At this time the rejection rate for manuscripts was low; only the obviously unfit were refused publication. Hooker and Council were reluctant to select articles on the basis of merit. Instead they decided to restrict the field covered by the journal, by eliminating almost all articles that would be more appropriately submitted to some other journal. Thus the journal became far less hospitable to articles in such areas as general physiology, comparative physiology, nutrition, clinical physiology, and industrial physiology, because these fields were covered by professional journals. Articles describing apparatus, those containing no new research (i.e., theoretical articles), and those coming from a foreign laboratory were also routinely rejected. In addition, Hooker urged authors to avoid historical reviews of the literature, to state their results with clarity and brevity, and to include only the most important tables and group them together whenever possible. An increasing number of articles was returned for condensation. These measures only staved off for a time the need for closer scrutiny of the manuscripts.

Nearly everyone agreed that Hooker was the ideal managing editor and that the Society was extraordinarily fortunate in having his services, but there was also an increasing concern over placing too much responsibility on a single individual. Manuscripts were sent directly to Hooker in Baltimore, who approved them on his sole authority, returned them to the author for condensation, or rejected them if they fell into various categories of unsuitable manuscripts as previously defined by Council. The relatively small number of manuscripts on which there was some question was sent by Hooker with his recommendation to members of Council. No one man, it was felt, could possibly have the expertise to render judgment on papers in all branches of physiology. Occasionally members of the Society would bring to Hooker's attention examples of seriously flawed papers that had been published. By the 1930s the problem of too many manuscripts had reached a critical point. In some years, five volumes of the journal had appeared at a high total cost to subscribers of $37.50 for nonmembers and institutions and $18.75 for members. (Member rates had been introduced in 1922.) As billing for journals was still done by volume, rather than by year, there was pressure from subscribers, especially in a time of depression, not to publish more than four volumes a year. Although the journal was recognized as excellent, there was a growing feeling that manuscripts should be more carefully selected.

In his 1929 annual report, Hooker, concerned with the high cost of subscriptions and with the accumulation of manuscripts, suggested that the Society should "create a continuously functioning editorial board with specified jurisdiction for each member, to which all manuscripts will be referred." In 1930 he requested and received permission to consult nonmembers of Council on articles with the understanding that appeals could still be made to Council. That year an effort was made to reduce the cost of subscription by including the cost of the journal in the Society dues. Council voted to set dues at fifteen dollars to include a subscription to the journal, but this was rejected at the business meeting.

The above problems were brought to a head by the loss of some of the Society's funds, then held in a Boston bank, through bank closures in early 1932. It was imperative to protect the publications reserves, which by 1932 amounted to nearly $100,000. Hooker had the major responsibility for the supervision of the publication reserves and for deciding whether to purchase bonds or other securities or keep them in a bank account. The choice of bank was no less important than the choice of securities, for accounts were not federally insured. Council felt that Hooker should not bear this responsibility alone. Thus in 1932 a committee, consisting of Meek (chairman), Cecil K. Drinker, and A. C. Ivy, was appointed to examine editorial and financial policies of the Society's publications and to report back at the next meeting. By putting Drinker on the committee, the Council knew what to expect, as Drinker had already proposed a trustee arrangement to APS Council in the 1920s.

The committee's report, presented to Council in 1933 recommended:

1. To establish a Board of Publications Trustees of the Council which shall devote special attention to the fiscal and editorial policies of the Journal.
2. To appoint a Board of Editors for the American Journal of Physiology.
3. To pay a salary to the Managing Editor.

The Board of Publication Trustees (BPT) was to consist of three members appointed by the president for three-year rotating terms and subject to the approval of the Society. It was to advise Council on journal finances and editorial policy and recommend members to serve terms on the Editorial Boards of the *American Journal of Physiology* and *Physiological Reviews*. After consultation with Council, President Luckhardt selected the first three BPT members: Meek (chairman), Ivy, and Wallace O. Fenn.

Beginning in 1933, the BPT voted for Hooker each year an honorarium in lieu of a salary, which increased from $1,500 in 1933 to $4,500 in 1946. It was realized that when Hooker retired, his successor would have to be salaried. A ten percent reduction in journal prices was enacted for prompt cash payment of subscriptions. Despite the additional costs of meetings of the board, the honorarium, and the reduction in subscription prices, the journals under the careful and conservative management of the BPT continued to prosper and the reserves to rise.

The first Board of Editors of the *American Journal of Physiology* was soon appointed and took up its duties of reviewing all submitted manuscripts in the summer of 1933. Its members were Cecil K. Drinker, Carl J. Wiggers, Herbert S. Gasser, Andrew C. Ivy, Walter M. Boothby, Roy G. Hoskins, J. G. Dusser De Barenne, and Alfred Newton Richards. Hooker continued to read all manuscripts as they came in and sent them to the appropriate member of the board with his own opinion attached. The number of manuscripts rejected or returned for revision increased dramatically after 1933. In 1935, for example, less than half the manuscripts were accepted as submitted. The new editorial review procedures proved to be not as cumbersome as some had feared and were effective in limiting the size and improving the quality of the journal.

Though the BPT was to act in an advisory capacity, it soon assumed, with the tacit approval of Council, much stronger powers. It appointed the managing editor and the boards and made financial and policy decisions. It submitted these decisions in the form of an annual report, which Council always approved. Though the members of the BPT were to rotate, in fact the same members tended to be reappointed for long periods, because they were known to have the dedication and experience required for the job. Meek served as chairman of the BPT until 1946. Ivy served for a total of nine years, H. C. Bazett for eight, and Fenn for fourteen. Except for Homer W. Smith, who served in 1944–48, all fifteen members of the BPT from 1933 until its dissolution in 1961 were past or future presidents of APS. The creation of the BPT led to a more efficient and regularized operation of the journals, an increase in quality, and a more secure control over reserve funds. It also meant that journal affairs were increasingly separated from those of Council and the Society. The separation posed no difficulty in this period, even though the disparity between journal funds and Society funds was far greater than in 1961. It was not until the Society became involved in a broader range of activities that conflicts began to arise (see chapt. 5).

In the second quarter century, APS was involved in three other publication ventures besides its own journals: *Physiological Abstracts, Biological Abstracts,* and the *Annual Review of Physiology*. In 1915 the Physiological Society requested collaboration of APS in the publication of an abstract journal. The president appointed Macleod and Percy G. Stiles as reviewers for APS. From 1919 to 1923, the Society voted annual grants from Society funds of $125 in support of the publication with the hope that it would eventually become self-supporting. *Physiological Abstracts* was published by the Physiological Society until 1937.

In 1922 delegates from about thirty biological societies met in the offices of NRC in Washington and agreed to form the Union of American Biological Societies. A constitution was adopted the following year. APS was represented then and for many years thereafter by C. W. Greene. Although some hoped the new federation of biological societies would carry on a broad range of activities, in practice the union concentrated on one major project, the publication of *Biological Abstracts,* begun in 1926 and funded initially by a major grant from the Rockefeller Foundation. In 1924 $1 for each member of APS, or $347 in all, was donated by APS from Society funds in support of the union. APS continued to cooperate with the union and *Biological Abstracts* and made frequent contributions. The union remained in existence until the formation of the American Institute of Biological Sciences in 1947.

Annual Review of Physiology was initiated in 1937 by J. Murray Luck of Stanford University, who had previously founded *Annual Review of Biochemistry*. The APS BPT was not at all happy to see a physiological journal begun outside the control of the Society, but when it became clear that the venture would go ahead regardless, they agreed to Luck's offer to cosponsor the journal. A five-year contract to share in ownership and management was signed. Luck was appointed editor, and the Society helped name an editorial board. The first volume appeared in 1939. The contract arrangement with Annual Review of Biochemistry, Inc. (later Annual Reviews, Inc.), was maintained until 1962.

Porter Fellowship

The Porter Fellowship, founded by William T. Porter and continued today as the Porter Development Program, was the first of the Society's regular awards programs. In 1900 Porter had founded the Harvard Apparatus Company to provide high-quality physiological apparatus at low cost for student laboratories. Originally set up within Harvard, because of tax considerations it had to be reorganized as a separate company. By 1920 the company had accumulated surplus funds that Porter decided to use to benefit physiological research. That summer, when APS President Warren

P. Lombard, professor of physiology at the University of Michigan, was traveling in the East, Porter arranged to meet with him. As Lombard reported to the secretary of APS, C. W. Greene, on 17 November 1920:

> Dr. Porter has told me that he has always carried on this business, as it were, in trust for American physiology. He is getting on in life, and hopes that in some way the Company may continue to be of service to physiological research after his management has ceased.

Through Lombard, Porter made two proposals to the Society. First, he wished to found a fellowship for physiological research to be funded by profits of the Harvard Apparatus Company and to be awarded by APS. He offered to provide a stipend of $1,200. Secondly, he suggested that he would like the Harvard Apparatus Company to become the property of the Society on his death.[9]

The offer to found a fellowship was gratefully accepted by the Society. It was decided that Council would select the fellow each year from nominations made by members of the Society. The first announcement, sent to all members on 4 June 1921, resulted in two applications. John Hepburn, the Society's first Porter Fellow began his fellowship in October 1921; he worked with Macleod at the University of Toronto, at first on a study of oxygen saturation of arterial and venous blood by using the Barcroft method, and then he joined the group of physiologists at Toronto who were working on insulin. It is interesting to note that several of the Porter Fellows were women. The second fellow was Florence B. Seibert, who had a distinguished career as a biochemist on the faculty of the Henry Phipps Institute of the University of Pennsylvania. She held the fellowship for two years; the first year she worked with Lafayette B. Mendel at Yale as a predoctoral student and the second with H. Gideon Wells at the University of Chicago as a postdoctoral fellow. Among early Porter Fellows who became long-time APS members were Dea B. Calvin, Donald Eaton Gregg, Herbert Silvette, Harold C. Wiggers, Earl R. Loew, Gordon K. Moe, J. Henry Wills, and Jane A. Russell. (Another early fellow, Ellen Robinson, married Albert Grass, and together they founded the Grass Instrument Company. The Grass Foundation, an outgrowth of the company, now sponsors the Society's Walter B. Cannon Lecture, given each year by an outstanding physiological scientist.) The Porter Fellowship was awarded annually through 1968, by which time its goals were changed because it had become overwhelmed by the influx of federally funded fellowships (6, p. 132–135).

Though Council appreciated the generous motives behind Porter's second proposition, they were of the opinion that it was not possible for a scientific society to manage effectively a business establishment.[10] Not long before his death in 1949 at the age of eighty-six, Porter tried once more to offer the Harvard Apparatus Company to the Society, but once again the Society regretfully declined. In 1948 Porter was made an honorary member of APS for his many services to the Society. The Harvard Apparatus Company, through the Harvard Apparatus Foundation, remains a steadfast benefactor of APS and generously funds the Porter Development Program, the successor of the Porter Fellowship awards (see chapt. 20).

Animal Care and Experimentation

Though the defense of animal experimentation was a continuing problem for physiologists, there are surprisingly few references to it in Society minutes from 1913 to 1938. In response to protests in 1919, the Society investigated a single case of possible mistreatment of animals in experiments on shock reported at the annual meeting. The author of the paper was able to exonerate himself, though it was admitted that he was careless in the manner in which he presented his results and therefore gave the audience the wrong impression. Cannon, in the name of the Committee on Defense of Medical Research of the American Medical Association (AMA), commended the Society for undertaking the investigation, but no further cases were considered. Also in 1919 Lombard, as president of APS, at the suggestion of AMA, wrote to the chairman of a U.S. Senate committee in opposition to an antivivisection bill, then before the U.S. Congress, that would outlaw experiments on dogs in the District of Columbia.[11] In 1925 the leaders of the Society were disturbed to find examples of "cruel experiments" quoted from the *American Journal of Physiology* in a pamphlet put out by the Vivisection Investigation League. Hooker was requested to make every effort to keep injudicious statements out of the journal and not to publish research based on morally questionable procedures.

In the 1930s the Federation began to take a more active role in the defense of medical research. At the 1932 Federation meeting, on the initiative of Carlson and Luckhardt, both of the University of Chicago, over 500 Federation members signed a petition thanking the mayor of Chicago for his support in the passage of a city ordinance restoring the use of dogs from the pound in experimentation. Three years later, the Federation instituted a Committee on Antivivisection Activities consisting of Elliott C. Cutler (chairman), who was also chairman of the Committee on Defense of Medical Research of AMA, Carlos I. Reed, and George H. Whipple. APS voted $250 for the support of the committee in 1935. Despite isolated instances of concern with the problem posed by opponents to medical research, APS relied on other organizations to take action rather than initiating action itself.

Education

The teaching of physiology was at this period considered outside the scope of concerns of the Society. When a suggestion was made to Council in 1919 to hold a symposium on teaching of physiology at the annual meeting, it was turned down. The minutes read, "It was the sense of the Council that such would be an unwise program step, since the settled policy was to limit the program to the presentation and discussion of scientific research." Similarly, when in 1927 the Association of American Medical Colleges suggested that the Society appoint a committee to study the relation of physiology to the medical curriculum, Council "decided that it would not enter into such an undertaking unless questions of research were concerned, since the society is essentially a research organization." This attitude was not shared by other Federation societies, for at this time the biochemists were holding an

Annual Conference of Biological Chemists, in effect an annual symposium on the teaching of biochemistry. Not until after World War II did APS take an active role in the improvement of physiological teaching.

International Physiology and the XIII International Physiological Congress, Boston, 1929

A milestone in the history of American physiology was the first International Physiological Congress held in America in 1929. It was the first meeting of the congress to be held outside Europe in its forty-year history. The invitation to the International Committee was made at the initiative and in the name of APS. At the request of APS, the Federation assumed management of the congress and acted as official host for a gathering of 1,606 physiologists representing forty-one countries.

As early as 1897 the question of inviting the congress to meet in America had been brought before the Society. Bowditch planned at that time to extend an invitation at the fourth congress in Cambridge in 1898 for a meeting at Harvard in 1901, but at the April 1898 meeting, he reported that it seemed unwise to press the invitation. Yet the hope of an American meeting was not forgotten; when the Federation was formed in 1912, representatives of the three societies agreed that the Federation should invite the congress to meet in America in 1916. In one of the earliest extant letters written on Federation stationery, Meltzer, first chairman of the Executive Committee of the Federation, wrote to Simon Flexner about the proposed invitation to be made at the ninth congress in Groningen in 1913.[12] Meltzer presented the invitation, but Paris rather than New York was chosen as the site of the next congress. The stumbling block was the great expense of travel to America.[13] World War I soon intervened, and no congress was held in 1916.

After the war it was proposed that an Interallied Physiological Congress be held in Paris in 1920. Frederic S. Lee was elected APS representative. Several of the leaders of American physiology, particularly Meltzer, were unhappy at the precedent established by a congress that was not truly international, since the Germans and Austrians were excluded. Because of the political situation, Council decided in 1919 that it was "unwise to extend an invitation at this time."

By the eleventh congress in Edinburgh in 1923, the international character of the congresses was resumed and the Americans could again press their invitation. On behalf of APS, Carlson invited the congress to meet in America in 1926. At a preliminary meeting of the International Committee, it was decided that the invitation could not be accepted unless a subvention could help pay expenses of European visitors, especially the German and Central European physiologists. Because the leaders of APS were unenthusiastic about obtaining the congress in this way, the decision was made to hold the twelfth congress in Stockholm. At the 1924 meeting of APS, Council reluctantly accepted the necessity of fund-raising if a congress was to be held in America, and a committee, chaired by Howell, was appointed to consider the problem.

The following year, Council and the Society officially voted to extend an invitation for 1929. The invitation was formally presented by APS President Joseph Erlanger at the twelfth congress at Stockholm in August 1926, put to a vote, and accepted. The Americans now had an enormous amount of work ahead of them to choose the congress site; extend invitations; organize the program; provide information, housing, and entertainment for visitors; organize postcongress trips to laboratories in major centers of physiology; and above all, raise money. A first informal meeting of APS Council on organization of the congress, in which Erlanger, Forbes, Murlin, Wiggers, and Meek participated, took place on the train from Uppsala to Stockholm on 6 August.

On their return to America, APS Council discussed how to set up a national committee to manage the congress. By October, they realized that APS should not act alone, but that the physiologists should work with the biochemists, pharmacologists, and pathologists through the Federation. To gain a broader basis of support for the congress, in November Council invited the Federation to take charge. The Executive Committee of the Federation recommended the appointment of a national committee consisting of four physiologists, three biochemists, two pharmacologists, and two pathologists. APS Council chose as its four representatives President Erlanger, Howell, Lusk, and Macleod. The other members of the committee were Lafayette B. Mendel, Donald van Slyke, and Philip A. Shaffer for ASBC; J. J. Abel and Torald Sollmann for ASPET; and Aldred Scott Warthin and Wade H. Brown for ASEP.

At the first meeting of the National Committee held on 15 April 1927, at the time of the Federation meeting in Rochester, several of the major decisions were hammered out. Boston was chosen as the congress site. Howell was elected chairman; Lusk, treasurer; and Shaffer, secretary; and Cannon was chosen chairman of the Local Committee with power to select other members. Cannon chose Alfred C. Redfield and Edwin J. Cohn as secretaries. To help pay for the congress all members of the Federation would be assessed five dollars a year for two years, a not inconsiderable sum, since Society dues were still no more than two dollars. Howell was nominated by the Federation for the high honor of president of the congress. This choice was ratified by the members of APS at the 1928 meeting in Ann Arbor. The motion was made by Cannon, who "stated that he was glad to make this motion particularly since Dr. Howell was a charter member of our Society, our oldest active past president, and a contributor of important researches and men for over forty years."

The formal sessions of the congress were held at the Harvard Medical School in Boston from 19 to 23 August 1929. The scientific communications numbered 495 and were presented in six parallel sessions. Official languages were English, French, German, and Italian. The proceedings were published (with the help of the publications reserve) in volume 110 of the *American Journal of Physiology.* At the opening session, a welcoming address was presented by President Howell, and August Krogh of the University of Copenhagen spoke "On the Progress of Physiology." At the final session, Léon Fredericq, emeritus professor of the University of Liége, presented "Reminiscences of the Early Days of the Physiological Congresses," a history of the first nine congresses (7, 8, 13). Also at the closing plenary session, H. H. Dale presented APS, through the congress, a photostatic copy of the first bound minutes of the Physiological Society. This volume, covering the years 1876–92 and including material on the organization of the first congress, is now in the APS Archives in the same fireproof filing case as the volumes of APS bound minutes.

Meals and socializing took place outdoors, where the courtyard of Harvard Medical School was festively set up with tables and umbrellas. Among the entertainments was a concert in the courtyard by the Boston Pops, with Arthur Fiedler conducting. (It was the first year of Fiedler's fifty-year tenure.) To commemorate the congress, a panoramic photo was taken of all members before the Medical School building. Many of the illustrious participants were even captured in movie footage, interspersed with captions, a copy of which is now at the National Library of Medicine. Continuing the tradition begun at Torino in 1901, a congress medal was given to all members of the congress. By custom the congress medal featured a physiologist from the host country. For the first congress in America, the obvious choice was the pioneer physiologist William Beaumont. His likeness was sculpted by Baltimore sculptor Joseph Maxwell Miller, and the medal was produced by the Medallic Art Company (2). To accompany the medal was a facsimile reprint of Beaumont's classic work, *Experiments and Observations on the Gastric Juice and the Physiology of Digestion*, originally published in Plattsburgh, New York, in 1833 (3). The reverse of the medal showed two allegorical male figures together holding a single torch. Howell explained to Redfield that the design "symbolizes the old and the new worlds uniting in their effort to hold aloft the torch of learning to illuminate the world."[14] (These and other congress memorabilia can be found in the APS Archives.)

The foreign delegates enjoyed elaborate postcongress tours of American laboratories, which were financed completely by the funds raised. They visited the U.S. Fish Commission Laboratories and Marine Biological Laboratory at Woods Hole on 24 August, toured Yale University on 25 August, and spent 26–30 August in New York visiting the College of Physicians and Surgeons of Columbia University, Cornell Medical School, The Rockfeller Institute for Medical Research, the Russell Sage Institute of Pathology, and the Department of Genetics of the Carnegie Institution at Cold Spring Harbor. Some went on to the University of Rochester, the University of Toronto, and McGill University.[15]

The fund-raising effort had been a spectacular success. In all, about $80,000 were raised from public and private sources. The Carnegie Endowment for International Peace contributed $10,000 to assist the trans-Atlantic travel of sixty-six foreign scientists from seventeen countries chosen by a European committee on the basis of both scientific attainments and need. Perhaps the most celebrated physiologist assisted in this way was Pavlov, then nearly eighty years old. Once they arrived in America, all foreign delegates became the guests of the Americans, who provided all their living expenses in America. They were housed in the dormitories of Harvard University in Boston and in the dormitories at Columbia when they visited New York. Many more foreigners—540 foreign delegates in all—made the journey than anyone ever expected.[16] They booked ships together and turned the voyage into a sort of minicongress. British visitors arrived on the SS *Minnekahda*, whose voyage was so well described by Yngve Zotterman in Fenn's history of the International Congresses (15). German and Swiss visitors came on the SS *Stuttgart*. With funds remaining from the congress, the Federation instituted awards to young scientists to attend future congresses and to present the results of their research. In 1932, for the fourteenth congress in Rome, the first of these awards was given, one for each of the four Federation societies. The first APS awardee was Francis O. Schmitt.

Howell's words of welcome expressed what the thirteenth congress represented for American physiology:

> May I express to you first the gratification and pride that we physiologists of the United States feel in welcoming the Thirteenth International Physiological Congress to our own country. It was only about fifty years ago, barely the space of a single lifetime, that physiology in this country began its career as an independent science with special facilities for instruction and research and for the training of its own group of workers.
>
> Before that period individuals here and there had made notable contributions to the subject. Beaumont, Dalton, Flint, Weir Mitchell and Wood are names especially to be remembered. But the real birth of physiology with us began with the establishment of the laboratories of Bowditch at Boston, of Newell Martin at Baltimore and of Chittenden at New Haven. Into these laboratories were transplanted from Europe the spirit and the methods of modern physiological research. . . .
>
> From these modest beginnings have sprung the splendid laboratories in experimental physiology and medicine which we now possess. Our workers are numbered by the hundreds and our contributions to the advancement of the subject increase constantly in importance. We look upon this meeting of the Congress at Boston as a recognition that American physiology has come of age, and has been admitted to full membership in that old world group to whom we have so long looked for guidance and inspiration (11; 12, p. 138).

Semicentennial Celebration, 1938

In 1936, at the suggestion of Charles Wilson Greene, President Frank C. Mann appointed a Committee on the Semicentennial consisting of Meek (chairman), Howell, and Greene. It was decided to celebrate the event in 1938, the fiftieth anniversary of the first scientific program, and to hold the Federation meeting that year in Baltimore for its historic association with Martin and Howell. The fifty-year anniversary of the Society was celebrated in two ways: first, by the publication of the Society's history; and second, by a memorable banquet over which Porter would preside.

The preparation of the Society's history (12) was a major undertaking. Howell, who had been present at the organization of the Society, agreed to write the history of the first twenty-five years. For his section of the volume, he sought out photographs (still in the APS Archives) and biographical information on each of the Society's twenty-eight original members, some of whom were long dead and relatively obscure. To his imaginatively written biographies and lively account of the early days of the Society, Howell added personal reminiscences of considerable historical value. Greene, secretary from 1915 to 1923, given responsibility for the second quarter century, organized his section by annual meeting and relied heavily on the volumes of minutes. To safeguard the original volumes of minutes, photostatic copies of the earliest volumes were made for the use of authors of the history and the later volumes of minutes were microfilmed, then a very new technology. After the history was written, one of the photostatic copies of the first volume of the minutes of the Society would be presented to the Physiological Society in return for the first volume of their minutes that APS had received in 1929. Howell and Greene's sections were combined with a verbatim transcript of the banquet speeches, and *History of the American Physiological Society Semicentennial, 1887–1937*, published by the Society, was presented in 1938 to each member of APS.

The highlight of the celebration was the semicentennial banquet held at the Federation meeting in Baltimore in 1938. At the 1937 meeting, Greene, Luckhardt, and others suggested to President Mann that the Society honor its benefactor, William T. Porter, by naming him "Honorary President of the Society for the Semi-Centennial Celebration." Cannon was among those who warmly supported the idea. Porter, who had not attended a meeting in many years, was taken completely by surprise and was deeply touched by the honor. The banquet over which Porter presided took place on the evening of 1 April 1938 at the Lord Baltimore Hotel. Everyone who attended received a souvenir placemat with photographs of the three main founders of the Society, Mitchell, Bowditch, and Martin, and the text of the letter of invitation to attend the organizational meeting. Four of the five still-living original members were present as guests of honor—Russell H. Chittenden, Howell, Joseph Jastrow, and Warren P. Lombard. J. J. Abel was also a distinguished guest. Greetings were presented from the Physiological Society by W. H. Newton and from the Canadian Physiological Society by Charles H. Best.

The program began with the roll call of the twenty-eight original members by President Walter E. Garrey and the introduction of Porter. Porter made a delightful toastmaster. His remarks were a blend of whimsical humor, anecdote, and philosophy spiced with literary and historical allusions. He introduced the three main speakers of the evening, three great men of physiology who each presented a eulogy of one of the founders. Cannon spoke of his teacher, Bowditch; Howell spoke of his mentor, Martin; and Carlson ended with a tribute to S. Weir Mitchell and remarks on the present and future state of physiology. The whole was recorded by dictaphone and transcribed (including the laughter) for inclusion in the history (12, p. 191–221).

Meek and Ivy reported:

> It is safe to say that never in the history of the society had the membership ever listened to such eloquent and finished speeches. From the literary remarks of Dr. Porter which were filled with humor to the vigorous words of Dr. Carlson, who closed the program, there was not an instant when the audience did not give rapt attention. Not only was the applause in honor of the society but it expressed the affection and admiration of all physiologists for four of the greatest scientists and characters that America has produced: Howell, Porter, Cannon and Carlson. . . . At the end of fifty

years the American Physiological Society is more active and prosperous than ever in its history. Its achievements have been due to the spirit of its founders and the true devotion to physiological investigation as exemplified by its members (14).

NOTES

[1] W. B. Cannon, letter to J. Erlanger, 16 November 1914. Joseph Erlanger Papers, Washington Univ. School of Medicine Archives, St. Louis, MO. (Opinions of APS Council members on proposed constitutional changes.)

[2] Forms, completed by members of APS, are in Frederic S. Lee Papers, Rare Book and Manuscript Library, Butler Library, Columbia Univ., New York.

[3] Federation of American Societies for Experimental Biology, "Statement of Federation Expenses for the Year 1916," 19 October 1917. Joseph Erlanger Papers, Washington Univ. School of Medicine Archives, St. Louis, MO.

[4] A. J. Carlson, letter to APS Council, n.d. (ca. 25 January 1914), including copy of W. T. Porter to A. J. Carlson, 10 January 1914. Joseph Erlanger Papers, Washington Univ. School of Medicine Archives, St. Louis, MO. (Another copy in Frederic S. Lee Papers, Rare Book and Manuscript Library, Butler Library, Columbia Univ., New York.)

[5] W. B. Cannon, letter to J. Erlanger, 6 February 1914. Joseph Erlanger Papers, Washington Univ. School of Medicine Archives, St. Louis, MO. (See correspondence between Lee and Howell in Frederic S. Lee Papers, Rare Book and Manuscript Library, Butler Library, Columbia Univ., New York; and between Erlanger and Cannon in the Erlanger Papers.)

[6] APS Council, letter to APS members, n.d. [1914]. Joseph Erlanger Papers, Washington Univ. School of Medicine Archives, St. Louis, MO.

[7] W. B. Cannon, letter to J. Erlanger, 16 February 1914. Joseph Erlanger Papers, Washington Univ. School of Medicine Archives, St. Louis, MO.

[8] W. B. Cannon, letter to APS Council, 6 March 1914. Joseph Erlanger Papers, box 1, folder 1a, Washington Univ. School of Medicine Archives, St. Louis, MO.

[9] Initial reactions of members of APS Council to proposals are in Warren P. Lombard Papers, Michigan Historical Collections, Bentley Historical Library, Univ. of Michigan, Ann Arbor, MI.

[10] W. P. Lombard, letter to W. T. Porter (copy), 5 January 1921. Warren P. Lombard Papers, Michigan Historical Collections, Bentley Historical Library, Univ. of Michigan, Ann Arbor, MI.

[11] W. P. Lombard, letter to G. W. Norris (copy), 17 July 1919. Warren P. Lombard Papers, Michigan Historical Collections, Bentley Historical Library, Univ. of Michigan, Ann Arbor, MI.

[12] S. J. Meltzer, letter to S. Flexner, 31 March 1913. Simon Flexner Papers, American Philosophical Society, Philadelphia, PA.

[13] W. P. Lombard, letter to J. Erlanger, 14 May 1920. Joseph Erlanger Papers, box 12, folder 129, Washington Univ. School of Medicine Archives, St. Louis, MO.

[14] W. H. Howell, letter to A. C. Redfield, 24 July 1928. XIII International Congress Papers, Francis A. Countway Library, Harvard Medical School, Boston, MA.

[15] E. J. Cohn and A. C. Redfield, letter to F. S. Lee, 27 November 1929 (draft report to the Carnegie Endowment for International Peace). Frederic S. Lee Papers, Rare Book and Manuscript Library, Butler Library, Columbia Univ., New York.

REFERENCES

This chapter is based on APS Council and Society minutes, annual reports of the managing editor, programs of annual Federation meetings, and annual membership directories of the Federation. Only additional sources are cited in footnotes.

1. ANONYMOUS. Donald Russell Hooker, 1876–1946. *Am. J. Physiol.* 148: i, 1947.

2. APPEL, T. A. The medal [William Beaumont]. *Federation Proc.* 49: 1, 1985.

3. BEAUMONT, W. *Experiments and Observations on the Gastric Juice and the Physiology of Digestion.* Boston, MA: Federation Am. Soc. Exp. Biol., 1929.

4. CARLSON, A. J. Donald Russell Hooker, 1876–1946. *Federation Proc.* 5: 439–440, 1946.

5. CORNER, G. W. Donald Russell Hooker. In: *Dictionary of American Biography.* New York: Scribner, 1974, suppl. 4, p. 390–391.

6. FENN, W. O. *History of the American Physiological Society: The Third Quarter Century, 1937–1962.* Washington, DC: Am. Physiol. Soc., 1963.

7. FRANKLIN, K. J. A short history of the International Congresses of Physiologists, 1889–1938. In: *History of the International Congresses of Physiological Sciences, 1889–1968,* edited by W. O. Fenn. Washington, DC: Am. Physiol. Soc., 1968. (Reprinted with original pagination from *Ann. Sci.* 3: 241–335, 1938.)

8. FREDERICQ, L. Reminiscences of the early days of the physiological congresses. *Science* 70: 205–207, 1929.

9. FYE, W. B. Ventricular fibrillation and defibrillation: historical perspectives with emphasis on the contributions of John MacWilliam, Carl Wiggers, and William Kouwenhoven. *Circulation* 71: 858–865, 1985.

10. GEISON, G. L. International relations and domestic elites in American physiology, 1900–1940. In: *Physiology in the American Context, 1850–1940,* edited by G. L. Geison. Bethesda, MD: Am. Physiol. Soc., 1987, p. 115–153.

11. HOWELL, W. H. Address of welcome of the President of the Thirteenth International Physiological Congress. *Science* 70: 199–200, 1929.

12. HOWELL, W. H., AND C. W. GREENE. *History of the American Physiological Society Semicentennial, 1887–1937.* Baltimore, MD: Am. Physiol. Soc., 1938.

13. KROGH, A. The progress of physiology. *Science* 70: 200–204, 1929.

14. MEEK, W. J., AND A. C. IVY. The fiftieth anniversary of the American Physiological Society. *Science* 87: 361–362, 1938.

15. ZOTTERMAN, Y. The Minnekahda voyage. In: *History of the International Congresses of Physiological Sciences, 1889–1968,* edited by W. O. Fenn. Washington, DC: Am. Physiol. Soc., 1968, p. 3–14.

The Federation of American Societies for Experimental Biology

Comprising
THE PHYSIOLOGICAL SOCIETY
THE SOCIETY OF BIOLOGICAL CHEMISTS
and THE SOCIETY FOR PHARMACOLOGY AND
EXPERIMENTAL THERAPEUTICS

Headquarters: HOTEL WALTON

Annual Meeting, Philadelphia, Pa.
December 28–31, 1913

Sunday, December 28   *Hotel Walton*
  8:00 P.M. Physiological Society . . . Meeting of Council
              Biochemical Society . . . . Meeting of Council
              Pharmacological Society . Meeting of Council

Meetings at the Jefferson Medical College
Monday, December 29
  9:00–12:00 A.M. Joint session of the three societies
  12:15–1:00 P.M. Business Session } Physiological Society
  2:00–5:00 P.M. Scientific Session }
  2:00–2:30 P.M. Business Session } Biochemical Society
  2:30–5:00 P.M. Scientific Session }
  2:00–4:30 P.M. Scientific Session } Pharmacological Society
  4:30–5:00 P.M. Business Session }
  7:00 P.M. Dinner and Smoker: Hotel Walton[1]

Meetings at the University of Pennsylvania
Tuesday, December 30   *Engineering Building*
  9:00–12:00 A.M. Physiological Society, Room 314
                  Biochemical Society, Room 311
                  Pharmacological Society, Room 315
  12:00–12:30 P.M. Business Sessions { Biochemical Society
                              { Pharmacological Society
  1:00 P.M. Luncheon, Houston Club
  2:00–5:00 P.M. Medical Laboratories—Joint session of the
                  three societies—*Demonstrations*
  7:00 P.M. Dinner and Smoker, Kugler's Restaurant (1412
                  Chestnut Street)[1]

Wednesday, December 31   *Engineering Building*
  9:00–12:00 A.M. Physiological Society, Room 314
                  Biochemical Society, Room 311
  12:00–12:30 P.M. Physiological Society, Business Meeting
  7:00 P.M. Naturalists' Dinner and Smoker, Hotel Walton[2]

---

[1] The Anatomists, Naturalists, Zoölogists, and Pathologists are invited
to these dinners and smokers.

[2] The Physiologists, Biochemists, and Pharmacologists are invited to
this dinner.

Top: *first page of the program from the First Federation Meeting, Philadelphia, PA, December 1913.* **Bottom:** *detail from panoramic photograph taken at the Second Federation Meeting, St. Louis, MO, December 1914. Seated left to right: Graham Lusk, J. J. R. Macleod, Frederic S. Lee, Anton J. Carlson, Charles W. Greene.*

CONTENTS

Top: *Donald Russell Hooker, managing editor of APS publications (1914–46). (Courtesy of the Alan Mason Chesney Medical Archives, Johns Hopkins University.)* **Bottom:** *table of contents of* Physiological Reviews *1, 1921.*

Friday Afternoon 2:00 to 5:00

JOINT SESSION OF THE PHYSIOLOGICAL, BIOCHEMICAL, PHARMACOLOGICAL AND PATHOLOGICAL SOCIETIES

1. J. J. R. MACLEOD, N. A. McCORMICK (by invitation), E. C. NOBLE (by invitation) and K. O'BRIEN (by invitation), University of Toronto.
   Experiments on the Mechanism of Action of Insulin.

2. F. H. BANTING (by invitation), C. H. BEST (by invitation), G. M. DOBBIN (by invitation) and J. A. GILCHRIST (by invitation), University of Toronto.
   Quantitative Parallelism of Effect of Insulin in Man, Dog, Rabbit.

3. J. B. COLLIP, University of Alberta.
   Delayed Manifestation of the Physiological Effects of Insulin Following the Administration of Certain Pancreatic Extracts.

4. F. G. BANTING (by invitation), G. M. DOBBIN (by invitation) and MISS S. CAIRNS (by invitation), University of Toronto.
   Methods of Administration of Insulin.

5. C. CLYDE SUTTER (by invitation), C. B. F. GIBBS (by invitation) and JOHN R. MURLIN, University of Rochester.
   Methods of Administration of Pancreatic Extracts to Diabetic Animals and Man.

6. E. A. DOISY, MICHAEL SOMAGYI (by invitation) and P. A. SHAFFER, Washington University.
   Some Properties of an Active Constituent of Pancreas (Insulin).

7. C. H. BEST (by invitation) and J. J. R. MACLEOD, University of Toronto.
   Some Chemical Reactions of Insulin.

8. G. S. EADIE (by invitation), N. A. McCORMICK (by invitation), MISS K. O'BRIEN (by invitation) and J. J. R. MACLEOD, University of Toronto.
   The Physiological Assay of Insulin.

9. SIDNEY R. BLISS (by invitation), Physiatric Institute.
   Insulin Treatment of Diabetic Dogs.

10. HARRY D. CLOUGH (by invitation), ARTHUR M. STOKES (by invitation), C. B. F. GIBBS (by invitation) NEIL C. STONE (by invitation) and JOHN R. MURLIN, University of Rochester.
    The Influence of Pancreatic Perfusates on the Blood Sugar, D:N Ratio and Respiratory Quotient of Depancreatized Animals. 20 min.

11. A. I. RINGER, S. BLOOM (by invitation) and S. JACOBSON (by invitation), New York.
    Studies in Diabetes.
    1. The Protein Requirements in Diabetes.
    2. The Influence of Iletin (Insulin) on Glycosuria, Blood Sugar Concentration, Nitrogen and Ammonia Excretion in the Urine, Acidosis and Respiratory Quotient.

18

THIRD SCIENTIFIC SESSION

Friday Morning 9.30 to 12.30

Joint Session of the Physiological, Biochemical, Pharmacological and Pathological Societies.

1. JOHN J. ABEL
   On the Presence of Albumoses and of a Biuret Peptide in the Blood, and the Bearing of these Facts Upon Our Theories of Protein Metabolism

2. LAWRENCE J. HENDERSON
   The Equilibrium Between Oxygen and Carbon Dioxide in the Blood

3. FREDERIC S. LEE
   The Work of the United States Public Health Service in Industrial Physiology

4. A. H. RYAN
   Rythm in Industry

Contributions from Officers and Former Officers in the Division of Food and Nutrition, Sanitary Corps, U. S. A.

1. R. J. ANDERSON
   Applications of the Principles of Nutrition in an Army Camp

2. JOHN R. MURLIN
   Average Food Consumption in Training Camps of the United States Army. (With lantern slides)

3. PHILIP A. SHAFFER
   The Army Ration in France

4. PAUL E. HOWE
   Fluctuations in Food Consumption in the Same Organization Under Similar Conditions. (With lantern slides)

5. R. G. HOSKINS
   Military Hospital Dietaries. (With lantern slides)

6. N. R. BLATHERWICK
   The Acid-Base Balance of Food Consumed in Army Camps

7. K. GEORGE FALK
   The Work of the Harriman Research Laboratory in Affiliation with the Division of Food and Nutrition

9

Left: joint session on insulin, Federation Meeting, Toronto, December 1922. Right: joint session on war research, Federation Meeting, Baltimore, MD, April 1919.

Friday Morning 9:00 to 12:00
Business Session 12:00 to 12:30

PHYSIOLOGICAL SOCIETY

Presentation of papers limited to ten minutes.

1. R. S. LILLIE and S. E. POND (by invitation), Nela Research Laboratory.
   Chemical Effects Produced by Passing Electric Curents through Thin Artificial Membranes of High Electrical Resistance.

2. A. J. CARLSON, University of Chicago.
   Some Physiological Effects of Saccharin.

3. ALEXANDER FORBES, F. R. GRIFFITH (by invitation) and L. H. RAY (by invitation), Harvard Medical School.
   Delay in the Response to the Second of Two Stimuli in Nerve, and in the Nerve-Muscle Preparation.

4. HERBERT S. GASSER and JOSEPH ERLANGER, Washington University, St. Louis.
   Some Characteristics of the Action Current in Nerve.

5. E. J. COHN, Harvard University.
   On the Concentration of Proteins in Tissues.

6. C. A. MILLS, University of Cincinnati.
   Rapid Intestinal Absorption, without Digestion, of a Complex Protein Substance,—Tissue Fibrinogen.

7. A. M. BLEILE and R. J. SEYMOUR (by invitation), Ohio State University.
   The Effect of Formaldehyde upon the Vitamin Content of Milk.

8. A. C. IVY and G. B. McILVAIN (by invitation), Loyola University.
   The Excitation of Gastric Secretion by Application of Substances to the Duodenal Mucosa.

9. T. L. PATTERSON, Detroit College of Medicine and Surgery.
   The Behavior of the Empty Stomach in the Mullusca.

10. O. O. STOLAND, University of Kansas.
    The Site of the Anaphylactic Reaction of the Uterus.

11. STUART MUDD, Harvard University.
    Certain Factors Affecting Filtration through Berkefeld Candles.

12. E. L. SCOTT and F. B. FLINN (by invitation), Columbia University.
    Some Effects of Various Environmental Temperatures Upon the Blood of Dogs.

14

13. H. C. BAZETT, University of Pennsylvania.
    Causation of the Differential Blood Pressure in Aortic Regurgitation.

14. JANE SANDS (by invitation), University of Pennsylvania.
    Study of Pulse Wave Velocity in Arteriosclerosis.

15. E. L. PORTER, Western Reserve University.
    The All-or-None Character of Minimal Flexion Reflex Contractions in the Spinal Cat.

16. MARY A. M. HAUPT (by invitation) and N. KLEITMAN (by invitation).
    The Influence of Insomnia, Fasting, etc., on the Kneejerk.

PHYSIOLOGY A

Friday, 2:00 p.m.

See Directory at Armory

———

Symposium on Anoxia

L. J. HENDERSON, Presiding

(Papers are limited to 15 minutes)

1. E. S. GUZMAN BARRON
   Anoxia and cellular oxidations—I. Enzymatic processes.

2. R. W. GERARD
   Anoxia and cellular oxidations—II. Metabolism of nervous tissues.
   Discussion by A. Baird Hastings

3. DAVID BRUCE DILL
   Respiratory adaptation to low oxygen pressure.
   Discussion by Donald D. Van Slyke

4. ERNEST GELLHORN
   Cardio-vascular changes in anoxia.
   Discussion by Arthur Grollman

5. ROSS A. McFARLAND
   Psychological effects of anoxia.
   Discussion by Edward C. Schneider

6. D. W. RICHARDS, JR.
   Anoxia in clinical medicine.
   Discussion by Wm. S. McCann

7. CAPT. H. G. ARMSTRONG
   Anoxia in aviation.
   Discussion by A. B. Luckhardt

Session of APS contributed papers, Federation Meeting, Toronto, December 1922. At this time, APS still held only a single session at a time. **Inset at lower right:** *program of the first invited symposium sponsored by APS, Federation Meeting, Baltimore, MD, 1938.*

Top left: *Herbert S. Gasser. (Photo by H. W. Davenport, 1954.)* **Top right:** *Alexander Forbes. (Photo by F. A. Hitchcock, 1954.)* **Center left:** *John R. Murlin.* **Center right:** *Lawrence J. Henderson.* **Lower left:** *Joseph L. Johnson, first black member of APS.* **Lower right:** *Cecil K. Drinker. (Courtesy of the Francis A. Countway Library, Harvard Medical School.)*

XIII International Physiological Congress, Boston, MA, 1929. **Top left:** *Reid Hunt, Ross G. Harrison, William Henry Howell (president of the Congress). (Courtesy of the Alan M. Chesney Medical Archives, Johns Hopkins University.)* **Top right:** *A. B. MacCallum.* **Bottom:** *Thorne M. Carpenter, Ivan P. Pavlov, Walter B. Cannon.*

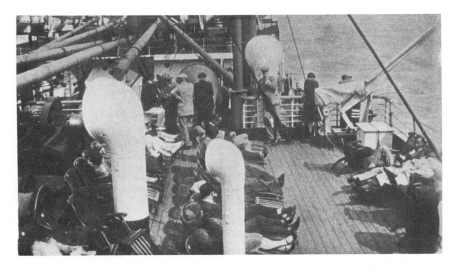

XIII International Physiological Congress, Boston, MA, 1929. **Top:** *deck of S. S.* Minnekahda.
Center left: *A. V. Hill. (Donated by Oscar W. Richards.)* **Center right:** *E. F. Adolph, K.
Gollwitzer-Meier (Frankfurt), A. C. Redfield, and Gustav Endres (Würzburg) on the beach
at Cohasset, 20 August 1929. (Donated by E. F. Adolph.)* **Bottom:** *William Beaumont
medallion given to all members of the XIII Congress.*

CELEBRATION OF THE FIFTIETH ANNIVERSARY
OF THE FOUNDING OF
THE AMERICAN PHYSIOLOGICAL SOCIETY

ANNOUNCEMENTS.....................By Dr. WALTER E. GARREY
President of the Physiological Society

TOASTMASTER....................................Dr. WILLIAM T. PORTER
Honorary President of the Physiological Society

INTRODUCTION OF ORIGINAL MEMBERS

INTRODUCTION OF GUESTS

EULOGIES OF THE FOUNDERS OF
THE AMERICAN PHYSIOLOGICAL SOCIETY

HENRY PICKERING BOWDITCH
By Dr. WALTER B. CANNON

SILAS WIER MITCHELL
By Dr. A. J. CARLSON

HENRY NEWELL MARTIN
By Dr. WILLIAM H. HOWELL

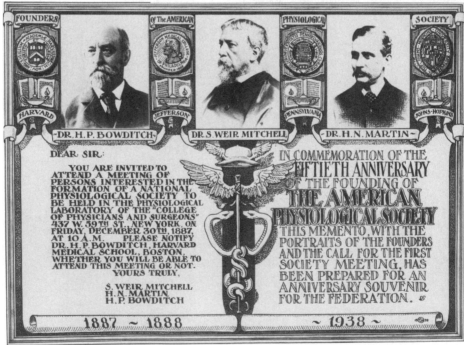

Top left: *program of the APS Semicentennial Celebration, Lord Baltimore Hotel, Balti-more, MD, 1 April 1938.* Top right: *William T. Porter, honorary president of APS and master of ceremonies at the Semicentennial Celebration.* Bottom: *souvenir from the APS Semicentennial Celebration, 1938.*

CHAPTER 5

Third Quarter Century 1938–1962

TOBY A. APPEL

The period from 1938 to 1962 was one of rapid change in the American Physiological Society (APS, the Society). In a relatively short period of time the Society was transformed from an organization with a strictly research orientation to one with a wide range of professional activities. In 1938 the Society's activities were limited almost entirely to holding annual meetings and publishing journals. There were as yet no standing committees, and because the Society did not undertake much, it needed little in the way of funds. Dues in 1938 were set at the same two dollars as in 1887. In 1939 there was an accumulation sufficient that dues were set at only one dollar!

During the course of the third quarter century, APS acquired its first fully paid executive secretary-treasurer and managing editor, set up a central office, and not long thereafter moved to property it helped purchase for the Federation of American Societies for Experimental Biology (FASEB, the Federation) in Bethesda, Maryland. Its traditional activities as a society of researchers were expanded and a host of new activities begun. A fall meeting was instituted, and several new publications projects were undertaken. In the years following World War II it also began to involve itself in social issues beyond simple monetary contributions or signing of resolutions: it sought to help physiologists in war-stricken countries, it worked to defend physiologists against loyalty tests, and for the first time it took an active role in the defense of animal research through formation of its own Committee on Animal Care and Experimentation. The system of standing committees, established in the early 1950s, allowed for wider membership participation in such Society activities as determining membership and organizing scientific programs. Two broad surveys of physiology conducted by the Society in 1945 and in 1952–53 revealed that there were a great many physiologists in colleges and universities whose needs APS was not serving. As a result, for the first time the Society became actively concerned with the teaching of physiology, not just in medical schools, but in colleges as well, and with the recruitment of physiologists in colleges and high schools. With the assistance of

97

federal grants, the Education Committee oversaw a range of projects—it made available refresher courses at meetings, summer workshops and summer traineeships for college teachers, and visiting lecturers; it published collections of college experiments; and it sent out career brochures. Efforts were made to bring college teachers of physiology into the Society by the creation of a category of associate membership. Industrial organizations were brought into closer relationship with the Society through the category of sustaining associates. In addition the Senior Physiologists Committee concerned itself with finding employment for "retired" physiologists. APS was transformed from a society for research in physiology to a society for physiology and physiologists.

War Years and the First Survey of Physiology

Even before the war, there were indications of change in the increased activity of the Federation, which had set up a number of committees in the 1930s with representatives from each of the societies. Of special importance was the FASEB Committee on Antivivisection Activities. In 1940 APS voted to donate $250 a year for five years to support the committee's activities. Not long after Hooker was appointed secretary of the Federation in 1935, he began to promote the establishment of *Federation Proceedings* as a way of removing the abstracts of the annual meeting from the *American Journal of Physiology* (*AJP*), which he had edited since 1914. He felt that the abstracts, which each year continued to grow in number, were for the most part of only transient value and that the subscribers to the journal should not have to pay for them. *Federation Proceedings*, which first appeared in 1942, was initially financed by a loan of $1,000 from APS funds, but its publication was soon made self-sufficient by an annual assessment of one dollar per member of the Federation. The early volumes contained four issues a year: one issue consisted of the program and abstracts of the annual meeting, another was the annual directory of members, and the other two contained symposia from the annual meeting (27). The existence of *Federation Proceedings* encouraged the inclusion of symposia in the Federation program. It was because of such undertakings that APS dues finally rose above two dollars in the early 1940s.

The great divide in the Society's history came with World War II. Wallace Fenn wrote in his memorable 1949 past president's address, "Physiology on horseback," that with the war "physiology has now taken to horseback like its elder sisters, physics and chemistry." While World War I had little effect in removing physiology from the ivory tower, World War II "has changed all that."

> One sign of the times is the founding of the *Journal of Applied Physiology*. Industry beckons with a strong arm. The persuasive call of the Almighty Dollar has penetrated to the once quiet recesses of the laboratory. The portentous international situation has cast its spell over us and the urgency of National Defense problems has diverted our former interests into new channels of military significance which now claim varying proportions of our time. Suddenly we have found ourselves projected from our Ivory Towers into the Marts of Men. The "long-haired boys" have left their laboratories and are out with the Men on Horseback in the military camps, on the sea

and in the air, and in all the unfriendly environments from the pole to the equator (19, p. 552).

Among other activities, the war affected scheduling of meetings. Thus the meeting in Boston in 1942 was the last before a three-year hiatus. The 1943, 1944, and 1945 annual meetings were canceled when the Office of Defense Transportation strongly discouraged scientific meetings to spare the railroads the burden of nonessential wartime transport. The officers and Council of APS elected in 1942—Philip Bard, president; Wallace Fenn, secretary; Hallowell Davis, treasurer; and Maurice B. Visscher, Charles H. Best, Hiram E. Essex, and William F. Hamilton, members of Council—remained of necessity in office until a new election could be held in 1946 (20, p. 9–12).

Nevertheless Society activity by no means ceased. The Council communicated regularly by mail, appointed ad hoc committees, and voted on issues. A revision of the constitution was drafted (to be submitted to the membership at the next meeting) to accommodate any future emergency cancellations of annual meetings, and new guidelines for election of members were hammered out. Membership applications were accepted, and new members were nominated and temporarily granted the privileges of membership until their election could be ratified by the Society. APS journals continued publication, although only two volumes per year of *AJP* were published because of the reduction in the number of manuscripts submitted. Some difficulty was also found in obtaining articles for *Physiological Reviews*, because so many physiologists were called into war service and much of the research was classified. Though no meetings were held, abstracts were collected and published each year in *Federation Proceedings*. Even symposia were organized for publication in *Federation Proceedings*. In 1944 APS sponsored two symposia: "Cerebral Blood Flow" was chaired by C. F. Schmidt, and "Physiological Aspects of Convalescence and Rehabilitation" was organized by Ancel Keys. In 1945 Andre Cournand organized a symposium on "Cardiac Output," which, when the meeting was canceled, was presented at the Philadelphia Physiological Society. One regional meeting of APS was held in San Francisco, organized by F. Weymouth in 1943.

The pioneer first Survey of Physiology grew out of discussions by APS Council in 1945 of the shape of "postwar physiology" and long-range plans for the Society. When the 1945 meeting scheduled for Cleveland had to be canceled in February 1945, APS Council arranged to meet in Rochester, New York, on 18–19 May 1945. All members of Council attended except Davis, who was ill. In response to a request for suggestions for the annual Society-sponsored symposium, Edward F. Adolph proposed a symposium on "Future Physiology." This was rejected in favor of a symposium on "Physiological Contributions to War Problems" (12). When Council discussed postwar physiology, however, Adolph's symposium suggestion was again raised. Council decided that a fact-finding committee was needed to survey the status of physiology and to make a report and recommendations to the Society. Adolph, T. E. Boyd, and Philip Dow were suggested as a committee to undertake the survey. At Adolph's suggestion, Julius Comroe, then a young assistant professor, was added.

The committee drew up a questionnaire addressed "to physiological scientists of North America." The survey, sent to nearly 2,000 "physiologists," including many who were not members of APS, was intended to assess the status of physiology "at the post-war crossroads." Information was solicited on training, employment during and after the war, shifts between physiology and related disciplines, problems of research and teaching, and income and research funding. The results were presented in a special evening session at the 1946 Federation Meeting in Atlantic City and published in *Federation Proceedings* (2). One conclusion was the need for greater stimulation of new entrants into physiology. Also of importance for the history of the Society was the awareness of the number of physiologists who existed outside the Society. Adolph determined that "probably less than half of those who by first choice identify themselves with this science belong to the American Physiological Society." An unpublished "Supplementary Report of the Survey Committee," written by Philip Dow in 1946, recommended that the Society "undertake a thorough study of the teaching of physiology."

One immediate result of the survey was the decision to organize a symposium at the 1947 meeting on the training of physiologists. Ralph Gerard headed the symposium committee whose members were Eugene M. Landis, Paul Weiss, and Laurence Irving. The symposium, "Perspectives in Physiological Education," was published in *Federation Proceedings* (22). Gerard recommended to Council in 1946 a blueprint for change in the Society. He argued that APS should "strive to be inclusive rather than exclusive" by the admission of associate members.

> The A.P.S. should face formally and actively the extra-technical needs of physiology and physiologists. It should undertake and support activities dealing with educational standards, economic problems, public relations, and cooperation with other scientific groups. Appropriate continuing committees should be created to these ends.

The first survey of 1945–46 and the 1947 symposium eventually led to the much larger and better-funded Survey of Physiological Science of 1952–53 chaired by Gerard (see below).

Changes in the Governance of the Society, 1946–62

At the first meeting of the Society after the war years, held in Atlantic City in 1946, changes in the constitution and bylaws were adopted as proposed by a committee consisting of Hiram E. Essex, F. C. Mann, and Maurice B. Visscher that had been appointed in 1943. Its initial purpose was to amend the constitution so that if the annual meeting ever had to be canceled for emergency reasons again, the Society could still operate legally.

The Board of Publication Trustees (BPT) was included in the constitution at this time, and it was granted full power to manage the Society's publications, a power it had always exercised de facto. The change was made in response to a crisis that had arisen in 1941 when John F. Fulton and others, in response to a plea for aid for the publications of the war-torn Physiological Society, had induced the members at the business meeting to vote to send up to $20,000 to the Royal Society of Great Britain.

Because the Society had no reserves to speak of, the money could come only from publications funds, which, it had long been assumed, were to be used for publications purposes only. Council took the liberty of interpreting "up to $20,000," and decided to send $5,000. A special canvas of the membership raised over $1,000, but the remainder had to be taken from publications income. Some members were angered that the full $20,000 was not sent. BPT members, however, were horrified by this unauthorized "raid" on the publications funds and sought to have the constitution modified to prevent any future impulsive raids by the membership. The $5,000 sent to Britain sufficed for the needs of the Physiological Society, and in fact there was a surplus of $933, which was used to supply physiological publications to continental laboratories (11; 20, p. 9, 71–72).

Although the BPT had been created in 1933, it did not become part of the constitution until the revision of 1946. *Section 3* of the new Article X read:

> The Board of Publication Trustees shall be vested with full power of the Society to control and manage, both editorially and financially, all of the publications owned in whole or in part by the Society; to appoint editorial boards; to appoint and compensate a Managing Editor; and to control all publication funds, none of which, however, may be diverted from support of publications of the Society except by consent of the Council.

This consolidation of the power and autonomy of the BPT led to increasing resentment among members of Council in the following decade.

One of the most important events of 1946 was the death of Donald Russell Hooker, who, with his assistant, Laura Campen, as secretary of publications, had managed the Society's publications since 1914. Hooker had already stepped down as Federation secretary at the 1946 meeting and was replaced by William H. Chambers. Campen carried on the publications alone under the supervision of Bard while the BPT searched for a new managing editor. The publications office was moved in May 1947 from Hooker's home at 19 West Chase Street in Baltimore to the Medical and Chirurgical Faculty of Maryland, also in Baltimore. Ivy, chairman of the BPT, proposed at the 1947 spring meeting the appointment of Milton O. Lee. APS Council appropriated $1,000 so that Lee might also act as executive secretary of the Society. The next problem was where to house the publications office. Through the generosity of Detlev Bronk, a member of APS and president of the National Academy of Sciences, space in the National Academy of Sciences Building at 2101 Constitution Avenue was made available at no cost for the work of the Society, as well as of the Federation. The Federation appointed Lee executive secretary as well; thus Lee had three jobs and three sometimes conflicting masters to serve.

Biographical Sketch of Milton O. Lee

Milton O. Lee was born in Conneaut, Ohio, in 1901. He received his A.B. and A.M. degrees in 1923 and Ph.D. degree in physiology in 1926 from Ohio State University. He was elected a member of APS in 1927. Through the course of his career he had acquired unusual administrative and editorial experience. From 1927 to 1942 he was research physiologist and deputy director of the Neuro-Endocrine Research Foundation at the Harvard Medical School. During the war he served in the Army as nutrition consultant at the Fourth Service Command Headquarters in Atlanta, Georgia. Just before coming to APS and the Federation, he was Senior Military Government Officer for the island of Hokkaido, Japan. He had served since 1936 as managing editor of *Endocrinology* and the *Journal of Clinical Endocrinology* and had been a member of the Editorial Board of AJP since 1941.

Lee's obituary in *The Physiologist* (9) states:

> The American Physiological Society owes a great debt of gratitude to Milton O. Lee. Its successful development during the turbulent post-war years was assisted and stimulated by the quiet and persuasive work of one who was not only a good scientist but one who believed that Physiology had more to offer than its own advancement as a strong and vital discipline of the basic sciences. Dr. Lee gave us perspective and vision—he encouraged the Society to become an essential element in the broad area of active involvement in society as a whole.

It was largely Lee's vision and enterprise as executive secretary of FASEB that brought about the Federation's rapid growth and rise to eminence. In honor of Lee, the main office building on the Beaumont campus, since 1962 the location of APS headquarters, was in 1965 named the Milton O. Lee Building.

Washington, D.C., Years, 1947–54

Lee, who assumed his multiple duties in June 1947, worked to assemble a new staff. In the summer of 1947, he appointed Sara Leslie, a recent employee of the U.S. State Department, to be senior assistant; she became secretary of publications in 1948 and eventually succeeded Lee as publications manager of APS. Dorothy E. Briggs was hired to take care of the billing, accounting, and other business affairs; and a recent college graduate was hired as editorial assistant, although Laura Campen continued her editorial duties until her retirement in September 1947. In October the APS office was moved to the National Academy of Sciences.

Sara Leslie recalled:

> Our first year there was far from easy. Office space was small, inconvenient and poorly lighted; supplies were limited; the staff was not adequate for increasing activities, although new people were hired from time to time; there was little furniture—we had brought from Baltimore all APS owned: an ancient safe, two wooden cabinets, a bookcase, later supplemented by a temperamental adding machine from a Government warehouse. We borrowed desks and chairs from the Academy. For file cabinets, we made do with fruit boxes stained walnut to disguise their origin. . . . We were a congenial good-natured group who worked often far into the late evening and on weekends, though salaries were low and there was no overtime pay. We all knew that once a budget was approved by APS and FASEB for our office, conditions would improve. And improve they did.[1]

One of Lee's first achievements was to modernize the antiquated system of record keeping. Campen had kept books by hand and had no organized system of bookkeeping. She did not even have a separate subscription list. Lee, who was said to have been ahead of his time in his appreciation of automation and of the future importance of computers, contracted with Remington-Rand to set up an automated accounting system. With this new system in place, one could calculate balances at any time.[1] In 1947, for the first time the journal was issued on a calendar-year basis with four volumes a year (changed to two much larger volumes per year in 1959).

While the Society offices were located at the National Academy of Sciences, the American Institute of Biological Sciences (AIBS) was founded in 1947 with help from the National Academy of Sciences. APS became the first charter member of AIBS and each year paid dues of $1,000 (later $1 per member). From 1947 to 1963, Wallace Fenn served as APS representative (20, p. 124–126). Lee for a time added to his three positions two more, those of executive director of AIBS and secretary of the Division of Biology and Agriculture of the National Research Council! These duties ceased by 1951 when APS moved from the National Academy of Sciences Building on Constitution Avenue to property rented by the academy at Dupont Circle. Dividing the services of Lee and the staff among the Federation, APS publications, and the Society was a tricky business. For 1950 costs were allocated according to the ratio 5:5:1. For 1951 it was decided that the Federation and the Society would contract for services from the BPT, the principal employer. In 1951 APS dues were raised to $7.00, of which $3.00 went to the Federation, $0.85 to AIBS ($1,000 a year), and $3.15 to the Society office. Dues continued to rise from six dollars in 1950, to seven dollars in 1951, and to eight dollars in 1952.

With an executive secretary in place to handle correspondence and collection of dues, the volunteer offices of secretary and treasurer were no longer needed. In 1948 these offices were abolished, and APS adopted the current system of electing a president elect who would take on some of the duties of secretary and would assume the office of president the following year. Carl J. Wiggers became the first president elect, succeeding Maurice B. Visscher as president in 1949. In 1951 the constitution was amended again so that officers would take office on 1 July rather than at the close of the meeting.

With the many changes in the organization of the Society, it seemed as though the constitution was always in the process of modification. At least a year was required to adopt amendments, because they had to be approved at two meetings. A committee composed of Gerard (chairman), E. M. Landis, and Lee was appointed in 1951 to draft an entirely new constitution for APS that would relegate detail to the bylaws, which could be more easily amended. When the new constitution was then adopted in 1953, it consisted of only two articles, as follows:

Article I. Name. The name of this organization is THE AMERICAN PHYSIOLOG-ICAL SOCIETY.

Article II. Purpose. The purpose of the Society is to promote the increase of physiological knowledge and its utilization.

The remaining information, now amounting to thirteen articles, is found in the bylaws, which can be amended at any business meeting of the Society (originally) by a three-fourths majority, changed to two-thirds in 1966. Because of this new constitution it was possible to dissolve the BPT in 1961 through the action of a single business meeting.

In other ways also the late 1940s and the early 1950s were active years for the Society. For example, the Society took a leading role in founding an International Union of Physiological Sciences (see chapter on international physiology). The initiative to form a union was especially due to Henry Cuthbert Bazett who first brought the project before the Society in 1947. The Society formed a committee, with Bazett as chairman, to correspond with European leaders and formulate plans. In 1953 at the International Physiological Congress in Montreal, the union was formally established. Bazett also founded and chaired a Committee for Scientific Aid to provide equipment, books, and journals to physiological laboratories in war-stricken foreign countries. After Bazett's death in 1950 this committee was chaired by Walter Root, and the fund became known as the Bazett Memorial Fund.

In the postwar years, questions of loyalty and security took on new importance. In 1951, on the initiative of Ralph Gerard, the Society sent the governor of California, in his role as chairman of the Board of Regents of the University of California, a resolution (3) congratulating the California court system for upholding academic freedom against attempts to institute loyalty tests for university personnel. Also in the fall of 1951 the Society sent the National Academy of Sciences and the American Association for the Advancement of Science letters encouraging them to "give continued formal attention to security and loyalty matters as they affect scientific disciplines and individuals."

A major change in the Society came about when, during Gerard's presidency, the Society instituted standing committees. Among those who played an important role was Orr Reynolds. Gerard and Reynolds had discussed the problem of how to increase membership involvement, especially of younger members, in APS. The minutes of the 1951 fall meeting in Salt Lake City read, "at this session there was a discussion with regard to membership participation in the committee structure of the Council which grew out of suggestions made by Dr. Orr E. Reynolds in a letter to Dr. Milton O. Lee." As a result, the decision was made to increase the committee structure of the Society and to place more non-Council members on committees.

The first committee to be established was a Membership Committee to consist of two members from Council and four from the Society, with "thought being given to the desirability of there being wide distribution as regards geography and discipline, with young men to be included." Among the young men placed on the first Membership Committee was Reynolds. At the same meeting, finally, a Program Advisory Committee was established under the chairmanship of A. B. Hertzman to suggest symposium topics, consider problems of overcrowding at the spring meeting, and, in general, seek ways to improve the program. These committees were soon joined by the Senior Physiologists Committee (1952) under the chairmanship of David Bruce Dill, the Porter Fellowship Committee (1952) under the chairmanship of Horace Davenport, and the Education Committee (1953) under the chairmanship of Adolph. Dill had already begun in 1951 to survey APS members to provide assistance to senior physiologists desiring employment. During the decade the main role of the committee evolved into providing a means of maintaining friendly contact with senior members of the Society (17, 18).

One of the most ambitious undertakings of the Society in the 1950s was the Survey of Physiological Science of 1952–54 (see above). This project was initiated by Gerard, who requested permission of Council in 1951 to investigate means to carry it out. As a result of an informal conversation with John Field, Assistant Director for Biology of the new National Science Foundation (NSF), in November 1951 NSF took an interest in the project. Through several more meetings with NSF, a proposal was worked out and submitted in March 1952. This was one of NSF's earliest grants. The proposal, with a budget of $117,500, called for "a survey and inventory of physiological science, with special reference to the future welfare of the United States as served by science." The project was to be overseen by a Central Committee chaired by Gerard, with Fenn as deputy chairman, and was to be undertaken in two stages, a nine-month pilot phase and an eighteen-month operational stage. Reynolds took leave from his position at the Office of Naval Research to act as executive secretary for the first stage. L. M. N. (Matt) Bach headed the survey staff, housed near the APS office, during the second stage. To these phases was later added a report phase, funded by additional grants.[2]

From the beginning it was decided to study physiology in its broadest sense as functional biology and therefore to include plant and bacterial, as well as animal, physiology. An extensive network of committees gathered massive amounts of data on the background and training of physiologists; their activities, employment, and income; occupational motivations, satisfactions, and mobility; the teaching of physiology in high schools, colleges, and medical schools; research activities, funding, and support; forms of professional communication (societies and publications); and the public perception of physiology. The chief instrument of the Survey of Physiological Science was a questionnaire sent to some 7,000 individuals in the United States and Canada. It was estimated from the survey that there were some 1,900 "central" or primary physiologists (those who identified themselves with some branch of physiology—animal, plant, or bacteria) and about double this number of "peripheral" physiologists.

The Survey of Physiological Science, as a pioneer study of a scientific discipline, became the model for studies by several other disciplines and catalyzed interest in the Society in improving education in physiology and in attracting more young men and women into the field. Survey results were presented in symposia at the APS Fall Meeting in Madison, Wisconsin, in 1954 and published in 1958 (21).

The international aspects of physiology had been discussed by leading physiologists of various nationalities in a series of conferences held during and just after the XIX International Congress in Montreal in 1953. These resulted in a volume edited by the historian of medicine, Ilza Veith, and published by APS in 1954 (31).

Move to Beaumont

A matter of great importance occupying the Society during the time of its survey was that of finding a home for APS and the Federation. Soon after the Society and

Federation offices moved to the National Academy of Sciences Building on Dupont Circle, APS and the Federation were asked to pay rent at a cost of $12,000 a year. The BPT was concerned with this expense and began considering the possibility of purchasing property for offices. At the same time, Lee saw the opportunity to provide a home for the Federation with space sufficient to permit all the member societies eventually to establish executive offices. An APS Committee on Society Headquarters was formed in 1953; it consisted of Hiram Essex, J. P. Quigley, William F. Hamilton, Visscher, Fenn, and Dill. Although the initial idea was to find space downtown, when Lee and members of the committee were shown the Hawley estate, a thirty-eight-acre property with a mansion house and outlying buildings, just north of the National Institutes of Health (NIH) in Bethesda, most realized that the property was an excellent bargain. The sale dragged on, first over negotiations with the estate and then with the zoning board and Montgomery County authorities, since a county ordinance had to be changed to allow occupancy of the property by nonprofit educational institutions. When another offer for the property was made by the Catholic Diocese with the intention of using it for a parochial school, APS had to agree to purchase the property in early 1954 before legal occupancy could be assured. The purchase price for the property and buildings was $212,500 paid in full, mostly from the APS publications reserves. (The Federation made a short-term loan of its savings as a matter of good faith.) Portions of the property were then sold for something over $100,000 to the State Highway Commission to widen Rockville Pike (which was then only two lanes wide) and to a developer to be used for single-family homes. At the request of the Federation, strongly supported by Lee and K. K. Chen, chairman of the Federation Board, the APS BPT agreed to transfer title of the remaining eleven acres to the Federation for $100,000 and loan the Federation part of the funds needed to pay for it. Thus the Federation obtained in 1954, through APS and in particular through APS publications, the valuable property it now owns for what was even then considered an exceedingly good price (20, p. 79–84; 24).

Sara Leslie, in her account for the APS Archives, has written of the move to Beaumont:

> Moving day was Thursday, August 12, 1954. The Washington movers were not familiar with the area beyond Bethesda and I volunteered to ride with them and lead the way. It was late in the evening when the moving van pulled away [from Beaumont], and the small group of weary staff members gathered in Dr. Lee's new office, wishing aloud that we had champagne to toast this special moment in APS and FASEB history. Instead we welcomed the hot coffee, thoughtfully provided by our new caretakers, Paul and Madeleine Schroeter, which strengthened us while we discussed plans for the following day. Friday was a work day, said the Managing Editor, and there was much to be done before Monday when the full staff [then about 15] would be in to work "as usual."[1]

The estate was renamed Beaumont and the mansion house Beaumont House. William F. Hamilton's copy of the Beaumont medallion from the 1929 congress was embedded in the front door.

Appointment of Ray G. Daggs

Lee's increased duties soon brought about the need for a separation of the work of the executive secretary-treasurer of the Society from that of the managing editor of the journals. The 1950s was a busy time for the Education Committee, which undertook several major projects funded by outside grants (see below). Lee had little time left for working with the expanded committee structure of APS and administering the grants. Ray Gilbert Daggs, previously Director of Research at the Army Medical Research Laboratory, Fort Knox, Kentucky, became executive secretary-treasurer of APS in 1956, while Lee remained as managing editor of the journals and executive secretary of the Federation until 1965. Daggs' salary was to come in part from the overhead of grants with the understanding that the BPT would fill in if necessary in return for duties to publications. In 1957 he thus became associate editor of *Physiological Reviews*. His biography and his many contributions to the Society appear elsewhere in this volume. Daggs' active involvement in the work of Council and the committees encouraged new projects and activities but also led to frustration on the part of Council over the limited financial resources of the Society and to a deepening conflict with the BPT. Society operations depended almost entirely on dues and grant support. President Robert F. Pitts pointed out in 1959 that of $10 dues (not including compulsory subscription to an APS journal) $4 went to the Federation, $1 went to AIBS, and about $0.50 went to the Federation in payment for business service; only $4.50 per member or under $8,000 remained for Society operations (29). When certain grants came to an end in 1960, the Society reached a point of financial crisis (5).

The last few years of the third quarter century, a period of severe upheaval, culminated in a major reorganization of the Society in 1961. The BPT was abolished, and publications and finances of the Society were restructured in the aftermath of the creation of a separate Publications Committee and a Finance Committee (see below). The Society was still undergoing the process of readjustment when it moved its offices in July 1962 from Beaumont House to the new Milton O. Lee Building. Philosophic calm had finally begun to prevail by the time of the Society's Seventy-fifth Anniversary Celebration at the 1963 fall meeting in Coral Gables, Florida.

Membership

At the commencement of the third quarter century, there were still only two membership categories, ordinary and honorary, and there were very few left of the latter, since no honorary members had been elected since 1912! From 1938 to 1962 there were a number of major changes in membership: the continued growth of the Society, from 656 members in 1938 to 2,188 in 1962, simultaneously with a tightening of membership qualifications; the institution of the category of retired member; changes in the constitution to facilitate election of honorary members; the establishment of a Membership Committee to make recommendations on membership to

Council; and, after long and heated debate, the decision to establish the category of associate member (20, p. 56–69).

The present category of emeritus member goes back to an amendment to the constitution proposed by Council in 1938 and approved in 1939. Council had become concerned that long-time and distinguished members were resigning when they ceased active research so that they could be relieved of annual dues. At first, no separate formal category of membership was envisioned. The amendment read:

> Any member who has paid the annual assessment for thirty years, or who has attained the age of sixty-five years, or who has retired because of illness, may be relieved from the payment of the annual assessment.

The first four members to avail themselves of the new amendment were Carl L. Alsberg, Francis Gano Benedict, E. D. Brown, and William J. Gies. The treasurer, Fenn, determined that fifty-nine members were then eligible. Because it was unclear whether members were automatically relieved from dues or must request relief, the provision was modified a few years later to read, "Any member who has retired because of illness or age may, upon application to the Council be relieved from payment of the annual assessment." In 1948, the first year "retired members" were indicated in the Federation directory of members, there were thirty-seven such members. The category of retired member first appeared in the bylaws in 1966 and was changed to "emeritus member" in 1982.

Throughout the period, it remained difficult to become a member of APS. In 1939 at the annual business meeting, Council nominated to the Society forty-three candidates for membership, deferred thirty-four, and rejected eight. The primary reason for rejection was "insufficient research," but many were also deferred or rejected because it was thought that they "belonged" in some other Federation society, whether or not they were yet a member of that society. Attendance or nonattendance at meetings was also becoming a factor of increasing importance. It was pointed out by an angry member that rejecting candidates because they belonged in some other scientific society was unconstitutional. Indeed it was, because the constitution still defined the scope of ordinary membership by a hodgepodge of fields, which had not changed since 1914. According to the constitution, any person was eligible

> who has conducted and published original researches in Physiology (including Experimental Zoology, Pathology, Pharmacology, Experimental Therapeutics, and Hygiene), or who is continuing researches in any of these fields, and who is a resident of North America.

The constitutional provision was therefore streamlined in 1942 to read:

> Any person who has conducted and published original researches in Physiology, and who is a resident of North America, shall be eligible for election as an ordinary member of the Society.

The increasing tendency to regard membership as a special honor was reflected in the decision of the Federation to include the year of election under each member's name in the annual Federation directory of members. This practice began with the 1940–41 directory and has continued.

During the war, the officers, Bard, Fenn, and Davis, had been constituted a committee to consider revisions in the qualifications for membership. Fenn drew up guidelines that were discussed among Council members and revised and approved in 1944. According to this document, successful candidates would ordinarily have been three or four years beyond the completion of the M.D. or Ph.D. degree and have published four or five substantial papers (not abstracts). Applicants would not be refused because they belonged to another society or because they did not have a permanent position, as long as their publications were "clearly physiological." The guidelines concluded, "In general, the effort will be made to eliminate or avoid disinterested members by increasing the obligations or professional standards of membership, rather than by exclusiveness for non-professional reasons." Fenn wanted to require subscription to at least one of an approved list of physiological journals (including but not limited to those published by APS) as a means of weeding out deadwood from the membership rolls, but this provision was not agreed to. In 1944 a new nomination form was introduced, and separate confidential letters of recommendation were required from the sponsors. At this time qualifications of candidates were still posted at the meetings.

At Fenn's suggestion, the changes in the constitution that were voted on at the 1946 meeting included the addition of the word "meritorious" to the qualifications for membership. The constitution then required the prospective member to have conducted and published not simply "original researches in Physiology" but "meritorious original researches," with the judgment of merit left to the discretion of Council.

In 1952 the initial screening of applications was taken over by the Membership Committee, created in 1951, under the chairmanship of Fred A. Hitchcock. The deliberate inclusion of nonmembers of Council on the committee was something of an innovation. However, the committee complained at the 1952 meeting that its work was hampered because only those members of the committee who were on Council were allowed to read the confidential letters from sponsors! Council agreed that from then on the whole committee could have access to the letters.

The Council and committee discussed the membership structure at length in 1952. It was the general consensus that the Society was bringing in too many "clinical people." They pointed to the "tremendous tide of clinical investigations" in recent years. Physiology had been so successful that it was being taken over by clinicians in clinical departments where it was easier to get money. Some found it unjust that it was easier for clinicians to be accepted for membership in the Society than for doctorates in physiology who were teaching the subject. At the Council meeting the previous year, Adolph had decried the fact that

> college teachers of physiology, upon whom the future of the science depends, do not qualify for membership, do not apply for it. . . . The view can be put forward that [the] American Physiological Society is concentrating on *research only* at a time when the need for that emphasis has passed.

Despite the prevalent concern that the Society represented only medical physiology and only research interests, the initial effect of the Membership Committee was to examine applications more critically than before.

Fenn, in his history, noted a marked slump in the membership growth rate from 1952 to 1956 (20, p. 58). Perhaps it was the negotiations in 1955 and 1956 to found a biophysical society that reversed the trend. The hasty revision of the membership definition in the bylaws whereby "physiology" was replaced by "physiology and/or biophysics" did nothing to prevent the formation of the Biophysical Society in 1957. It became clear by then that too restricted a definition of physiology encouraged splinter groups. After 1956 the rate of growth of the Society increased.

The issue of associate members, mentioned in the minutes as early as 1946, was one that deeply divided members of the Society and of Council. The new constitution and bylaws then proposed were to include provision for "associates." (It had been voted not to call them "associate members.") By the time the constitution and bylaws were circulated to the membership in December 1951, the portions relating to associates were placed in brackets to indicate that there was a great deal of misgiving over the wisdom of the step. The provision defined associates in a much broader manner than the form that was later adopted. It provided that

> any individual who is engaged in biological research or teaching, or who has pursued graduate studies in a recognized institution, or who is an undergraduate exhibiting unusual promise and interest in physiological science, shall be eligible for proposal as an associate of the Society.

After discussion of the matter in 1952, Council agreed that "opinion had crystallized against the creation of such a class of membership." It was taken out of the proposed new constitution and bylaws, but the debate continued. In 1956 Julius Comroe was appointed chairman of an ad hoc committee to make recommendations to broaden the membership of the Society. The committee's report presented to Council in 1957 recommended associate membership as a means of attracting into the Society everyone with an interest in physiology while at the same time retaining the prestige of regular membership (14). The bylaw change finally approved by the Society at the 1958 spring meeting defined associate members as "advanced graduate students in physiology at a pre-doctoral level, teachers of physiology, and investigators who have not yet had the opportunity or time to satisfy the requirements for full membership." The first sixteen associate members were elected at the 1959 spring meeting. Dues were low (five dollars) and privileges were limited. Associate members received *The Physiologist* and obtained reduced rates for APS journals, but they were not members of the Federation, could not vote, and could present papers without member sponsorship only at fall meetings. The new category was only moderately successful in broadening the membership; by the end of the third quarter century there were 136 associate members. (In 1977 the definition of associate members was greatly modified.)

At the 1960 spring meeting the category of sustaining associates was approved. These were to be "individuals and organizations who have an interest in the advancement of biological or biophysical investigation" and who were invited by the president with the approval of Council. The purpose of the new category, established at the initiative of Comroe, was to raise much needed funds for Society operations. It also reflected the Society's closer ties to industry after World War II. Sustaining associates were encouraged to pay a minimum of one hundred dollars in annual dues. By 1962 twenty-six industrial organizations had become sustaining associates (20, p. 63–64).

In 1938 election to honorary membership still required nomination by Council and the unanimous approval of all members present at a business meeting. This created a difficult situation, because unanimity could never be counted on. Council would be placed in an awkward situation if it should nominate honorary members only to have them rejected in a public meeting on the whim of a single member. Consequently the hope of electing honorary members in time for the Semicentennial Celebration had to be abandoned. The solution to the dilemma was to loosen the restrictions on approval of honorary members. In 1939 the constitution was changed to require one vote in twenty to exclude, and Bernardo A. Houssay was elected an honorary member in 1942. In 1946 the constitution was changed once more to require only a majority ballot of the membership at a meeting. The way was then clear to elect a number of honorary members, including E. D. Adrian, Joseph Barcroft, A. V. Hill, August Krogh, Leon Orbeli, and Louis Lapique. Records of these actions were so poorly kept that Hill was elected an honorary member again in 1950. On receipt of the second notice of honorary membership, Hill humorously pointed out the error (20, p. 62–63). From 1946 honorary members were elected from time to time, but only in the past few years have they been elected with any regularity (see chapter on membership).

Meetings

APS spring meetings continued to be held each year with FASEB. Early in the third quarter century, the Federation was enlarged, for the first time since 1913, by the addition of two other societies, the American Institute of Nutrition (1940) and the American Association of Immunologists (1942). After 1948, in addition to the spring meetings, APS began holding an annual fall meeting on its own.

Spring Meetings

The postwar years began with a radical departure, the first of many meetings to come in the nonuniversity resort town of Atlantic City. The story of "How the Federation Came to Atlantic City" is related by Carl F. Schmidt in Fenn's history (20, p. 13–14). When it proved impossible to hold the 1946 meeting in Cleveland, the planned site of the canceled 1945 meeting, or in Chicago, the Federation committee turned to the University of Pennsylvania, which had offered an invitation five years previously. The Philadelphia group, familiar with meetings of the Society for Clinical Investigation and the Association of American Physicians in Atlantic City, investigated the possibility and received unusually favorable terms. Space in Convention

Hall, about to be returned to civilian use, was made available without charge. The decision to meet in Atlantic City rather than Philadelphia was made in early 1946 for an early March meeting. Many found the Atlantic City meetings over the next thirty years to be enjoyable, despite their size, because of the large variety of accommodations and because one could see all one's colleagues while strolling the boardwalk.

Year after year Federation meetings continued to grow. By 1962 the total attendance was 14,549. Thirty-one simultaneous sessions were required to program 2,986 papers. APS was responsible that year for programming 75 of 285 sessions (7). Only Atlantic City and Chicago were found to have the facilities for meetings of this size. Between 1946 and 1959 there were five Atlantic City meetings and two Chicago meetings. From 1959 through 1975, all Federation meetings were held in one of these two locations, usually three years in Atlantic City followed by one in Chicago. In part because space was at such a premium for ten-minute papers, there were few symposia. In 1962 APS was still sponsoring only three symposia and a teaching session.

The large size and limited locations brought several changes in the character of the meetings. In large hotels and convention centers, the traditional experimental demonstrations were no longer possible. Films were encouraged for a time as a substitute. Soon after his arrival, Lee arranged for commercial exhibits, and these soon provided a valuable source of income. Because meetings could be held in only a few places, it was no longer possible to organize them by a local committee. The work of planning the meeting had to be done entirely by the central office. Social events also changed in character. The meetings had become so large that it was no longer possible to hold a Federation banquet or mixer. Instead numerous smaller banquets were organized by various informal special interest groups, several of which have since evolved into sections of APS (20, p. 92–112; see chapter on sectionalization).

A perennial topic for discussion at APS Council and business meetings, especially toward the end of the third quarter century, was the relation of APS to FASEB. Many felt that the meetings had become so unwieldy that APS should leave the Federation altogether. In April 1960 the Society requested that Council study the costs and benefits of remaining in the Federation (6). At the 1961 meeting a survey of the amount of "crossover" from one Federation society to another was conducted. It was determined that fifty-six percent of those in attendance crossed over from sessions organized by one society to sessions organized by another. These and other statistics on the role of the Federation were discussed at length at the fall meeting in Bloomington in 1961 (20, p. 121–122). The value of remaining in the Federation was established, but the problem of how to limit the ten-minute paper remained. In 1961 Davenport addressed this problem, which had reached critical proportions, in his president's message, "Somebody must do something!" (16). It was during the fourth quarter century that various remedies were tried, with mixed results (see chapter on meetings).

Fall Meetings

The fall meeting was begun in 1948, primarily as a means of trying to relieve the spring meeting of its surfeit of contributed papers (20, p. 85–91). It was also hoped that the fall meeting might recreate some of the intimacy of former spring meetings. The first hope was not realized; the fall meetings did little to relieve the congestion of the spring meeting. The second hope, however, was amply rewarded; the early fall meetings were delightful affairs. Because the fall semester at colleges did not begin until late September, the meetings were generally held in September on university campuses and attendees were housed in dormitories. The first fall meeting was held 15–18 September 1948 at the University of Minnesota with a side trip to the Mayo Clinic. The meeting was organized by a large local committee chaired by Ancel Keys. Visscher, president of the Society and chairman of the Department of Physiology at the University of Minnesota, hosted the meeting.

The entire work of organizing the fall meetings—the programming, the symposia, the demonstrations, the meals and housing, the barbecues and excursions, and the ladies' programs—was left to the local committees. They received an advance from headquarters and set a registration fee to cover costs. Each local committee wrote an account of its experiences, and this and copies of all the announcements, forms, tickets, badges, and programs were entered into the "brown" and the "black" notebooks that were handed down from one local committee to the next. In this way, committees profited from the experience of their predecessors. Meetings could be held in locations all over the country, many where the Society had never met before and where a Federation meeting would have been impossible—the University of Minnesota (1948), the University of Georgia (1949), Ohio State University (1950), the University of Utah (1951), Tulane University (1952), the University of Wisconsin (1954), Tufts University and Woods Hole (1955), and so on. The pleasant informality of the campus settings has been admirably captured on film by Davenport, who took numerous candid photographs at meetings in the early 1950s.

One problem that the fall meetings organizers soon had to face was the matter of discrimination against the black members of APS. Some Council members questioned whether the Society should hold the 1949 meeting in Augusta, Georgia, but by then it was too late to change. None of the small number of black members of APS tried to attend. At the 1952 meeting in New Orleans the issue came to a head when Tulane University went back on its informal promises and denied black members use of the dormitories and cafeteria. Walter Booker and Joseph L. Johnson's letter of protest was read before the business meeting. It was thereupon voted that for future meetings the Society require a firm agreement from the host institution that the same facilities would be offered to all members of the Society without exception.

Several APS traditions had their origin in connection with the fall meetings. The first, begun in 1948, was the APS banquet and past president's address. As Fenn explained, the tradition of the past president's address came about by accident when the distinguished afterdinner speaker selected by Visscher belatedly declined the invitation and Visscher called on Fenn, the immediate past president, to fill in. Fenn's well-known essay, "Physiology on horseback," set a high standard for future addresses (19). The refresher course organized by the Education Committee to precede the regular sessions became a feature of fall meetings in 1954. In 1956 John R. Pappenheimer, at the invitation of President William F. Hamilton, delivered the first in the series of Henry Pickering Bowditch Lectures (see chapter on honors and awards).

Publications

Through the 1950s the publications of APS continued to expand. A new journal, the *Journal of Applied Physiology* (*JAP*), was inaugurated in 1948, *The Physiologist* began publication in late 1957, and the first volume of the *Handbook of Physiology* appeared in 1959. In addition, in the period 1937–62, APS published twelve special volumes, five supplements to *Physiological Reviews*, and one supplement to *JAP*. The end of the third quarter century was marked by the complete reorganization of APS publications, the purchase of the *Journal of Neurophysiology*, the initiation of page charges, and the appointment of the first section editors for *AJP* (20, p. 70–84).

Journal of Applied Physiology

The *Journal of Applied Physiology* originated in a proposal to Council made at the 1947 annual meeting from a group of physiologists. This group, consisting of Norman R. Alpert, William B. Bean, Henry Borsook, Ernest William Brown, J. Brozek, Robert S. Goodhart, Robert M. Kark, Geoffrey L. Keighley, Keys, Ernst Simonson, and Visscher, had met at the University of Minnesota in January 1947 to discuss

> the need for a journal of high standard for the broad field of human physiology, with emphasis on the relationship between man and environment. Specifically, the journal should be concerned with experimental investigations in the fields of industrial physiology, physiology of exercise and athletics, climate physiology and military physiology.

Under the leadership of Simonson, the group conducted a poll of workers in the area and collected statistics from recent review articles on the number and location of articles being produced in the field. They felt that sponsorship by the Society would be the most desirable means of publication.

The BPT agreed to take up the project in a modified form and chose as a first editorial board R. A. Cleghorn, Dill, Chalmers L. Gemmill, A. H. Steinhaus, Adolph, Frances A. Hellebrandt, Simonson, Landis, Fenn, Robert E. Johnson, Hudson Hoagland, and E. J. Van Liere. Six issues of *JAP* were published between 1 July 1948 and 1 January 1949. Unlike other APS publications, *JAP* was somewhat slow in getting off the ground financially. The journal had an initial problem of identity, and it took some time before it acquired enough subscribers to be self-supporting. According to the annual report of the BPT for 1948, *JAP* had 604 subscribers, compared with

1,408 subscribers for *AJP* and 2,423 for *Physiological Reviews.* By 1961, however, *JAP* had 2,152 subscribers and had become the journal of choice for papers in respiratory physiology.

Handbook of Physiology

The *Handbook of Physiology* was initiated in 1953 by Visscher, a member of the BPT and chairman from 1954 to 1959, as a means of making constructive use of the mounting publications reserves. The early history of this ambitious and highly successful project has recently been related by Frances O'Malley and Horace W. Magoun (28). Visscher proposed in 1953 that the BPT "undertake to edit and publish over the next 5 or 10 years a multi-volume *Handbook of Physiology* which would be a somewhat streamlined version of the Bethe *Handbook* series developed before the war in Germany." After questionnaires were sent to members of Council and prominent physiologists, an ad hoc committee was formed and recommended proceeding with the project on an experimental basis. John Field, appointed editor-in-chief of the handbook project, was authorized in 1955 to begin sending invitations and collecting manuscripts for the first two or three volumes on neurophysiology. The entire work of selecting chapters and authors and assembling the first three volumes was carried out by Field, Victor Hall (executive editor), and Horace Magoun (section editor in neurophysiology), aided by Sally Field. It was not until 1958 that the BPT decided to issue the volumes as *Handbook of Physiology, Section 1. Neurophysiology* rather than as supplements to *Physiological Reviews.*

In the introduction to the first volume of section 1, Field explained the purpose of the series:

> The *Handbook of Physiology*, like its predecessors from von Haller on, is designed to constitute a repository for the body of present physiological knowledge, systematically organized and presented. It is addressed primarily to professional physiologists and advanced students in physiology and related fields. Its purpose is to enable such readers, by perusal of any Section, to obtain a working grasp of the concepts of that field and of their experimental background sufficient for initial planning of research projects or preparation for teaching.

The three volumes of section 1 appeared in 1959–60. After 1960 the post of editor-in-chief was discontinued. Instead section editors were appointed for each section, and the entire project was overseen by a Handbook Committee. By the end of the third quarter century, the first volume of *Section 2. Circulation* had appeared with William F. Hamilton as section editor and Philip Dow as executive editor.

The Physiologist

The founding of the Society's house organ, *The Physiologist*, in 1957 reflected the increased activity of the Society and the need for an appropriate outlet for communication. The journal was identified in the masthead as "A Publication for Physiologists and Physiology." It continued the biannual "President's News Letter" begun in 1952. Edited by the executive secretary-treasurer, Ray G. Daggs, early issues of *The*

Physiologist contained Society news; the annual Bowditch Lecture; the past presidents' addresses; news of senior physiologists; announcements of events; and articles on the teaching of physiology, history, and other topics of timely interest to physiologists but not suitable for publication in a research journal. One issue a year contained the abstracts of the fall meeting (previously published in *AJP*). The first issue of *The Physiologist*, which appeared in November 1957, carried for the first time the now familiar seal of the Society designed by Hamilton.

The founding of the Society's house organ was part of a controversial agreement with the BPT whereby the board would fund the new publication and distribute it to members provided that members agree to a compulsory subscription to *Physiological Reviews* at eight dollars or a subscription to *AJP* or *JAP* at a fifty percent member rate. Dues, including the required subscription, were increased to a minimum of eighteen dollars. This proposition was put to a mail vote of the membership with the stipulation that it would be passed unless a clear majority voted against it. Although the measure went through, it proved unpopular, and there were a number of resignations in protest. In 1961, after publications were reorganized, compulsory subscription was dropped, but *The Physiologist* was continued. Dues (now omitting the journal subscription) were immediately raised from ten to fifteen dollars, and eventually they were raised sufficiently to cover the costs.

Journal of Neurophysiology

The *Journal of Neurophysiology* (*JN*) was founded in 1937 by John F. Fulton and the Charles C Thomas family. When Fulton, professor of physiology at Yale University, died in 1960, he gave his share of the journal to Yale. APS's interest in purchasing the journal was motivated in part by the desire to prevent the formation of a splinter group of neurophysiologists and in part by the expectation that *AJP* would eventually be sectionalized. APS acquired full ownership of *JN* in November 1961 through purchase from Charles C Thomas publisher ($35,000) and Yale University ($25,000). The negotiations with Yale were delicate, because Yale was willing to sell only on approval of the journal's editorial board. APS took over publication of the journal in January 1962. Vernon B. Mountcastle, who had edited the journal in 1961, became chief editor of the new Editorial Board. At this same time (January 1962), a system of section editors for *AJP* and *JAP* was initiated and the articles in *AJP* began to be organized according to section (10).

Reorganization of Publications

Throughout the 1950s, as APS became involved in an expanded range of activities, tension between APS Council and the BPT grew. There was a great disparity between the Society's reserves (about $30,000 in 1961) and the publications reserves (about $600,000) (15). The estimated budget for 1961, as published in *The Physiologist*, showed a Society income of $20,500, mostly from dues, and expenses of $28,000, leaving a deficit of $7,500, while the net income from publications operations was

$25,000 and the income from sale of securities was $32,500 (5). Under the bylaws, BPT had been given full control over the income of the publications funds, which could be used only for publications purposes.

Over this decade the frustrations of Council mounted while the BPT grew increasingly defensive. The matter was passionately debated at several Council meetings before it came to a head in 1960–61. A number of factors seem to have contributed to the "Comroe revolution" (13). The demise of the BPT at the APS Spring Meeting in Atlantic City in April 1961 was precipitated by the lack of money available for Society activities; the desire for publications, such as *Physiology for Physicians* (begun in 1963), that the board did not wish to support; distrust on the part of some members of Milton Lee's management of publications; and the general belief that Council should be in full control of all the activities of the Society. Those who supported the BPT, such as Lee, Visscher, and Hamilton, maintained that the publications reserves had come from subscribers, only a small proportion of whom were Society members, and therefore should be used for publications purposes and not to benefit members of the Society; that the BPT had supported the Society in many ways; that BPT members, who held office for a long period, were better able to manage publications and enter into long-term agreements than a constantly changing Council; and that the plan to separate finances from editorial policy was schizophrenic and unworkable (24, 26, 32).

At the first session of the business meeting of the 1961 spring meeting in Atlantic City, Julius H. Comroe, Jr. (president, 1960–61), offered for a vote an amendment to the bylaws that had previously been circulated to the membership. The amendment, formulated by Robert Pitts and Comroe at the previous fall meeting, created a Publications Committee to oversee editorial policy of the publications and a separate Finance Committee to be responsible for all the financial affairs of the Society. Both committees were to be under the direct control of Council. The capital fund of the BPT was to remain as a reserve fund for publications, but annual income from investment of the fund "may be used for any of the activities of the Society including publications" (15). When the vote was taken by written ballot at this session, the seventy-four percent in favor of the measure was just short of the required three-fourths majority. However, a member who had voted against the motion was prevailed on to bring the motion up for reconsideration at the second business session two days later. This time the motion was passed by a small margin (seventy-eight percent). At a stroke, the BPT, in existence since 1933, was abolished (20, p. 77–79). This was the second of the "limbic elements" in the publications policies of APS described by Philip Bard at the seventy-fifth anniversary in 1963. [The first was the "raid" initiated by Fulton in 1941 (11).] Fenn observed in his history, "As society disputes go, this was a rather heated controversy" (20, p. 78).

In charge of the difficult transition period were Bard as the first chairman of the Publications Committee (the one holdover from the BPT), Landis, first chairman of the Finance Committee, and Davenport, president of APS in 1961–62. A five-year contract was arranged with Lee and Sara Leslie to serve as managing editor and executive editor, respectively. In 1962, with the aid of a grant from NIH, Lee established the FASEB Publications Service Center, which handled publication of *AJP* and *JAP*. (The continued progress of APS publications is described in the chapter on publications.)

Education Committee

The Committee on Educational Matters, one of the most visible and active committees of the 1950s, made full use of the new opportunities afforded by federal grants. Chaired by Adolph (1953–58), Davenport (1958–59), and John R. Brobeck (1960–62), the committee undertook a wide range of projects involving hundreds of people across the country (1). One of its first duties was to take over the organization of the teaching session at the Federation meetings, which had been started in 1951. This was intended to be a forum for discussion of teaching problems, methods, and materials. The tradition of holding a refresher course just preceding the annual fall meeting in some area of physiological teaching was begun at the 1954 fall meeting in Madison, Wisconsin. The first such course was organized by Julius Comroe on pulmonary physiology.

The committee soon cast its horizons beyond APS members and APS meetings. In 1954 the committee's first career brochure appeared for distribution to high school and college students and vocational counselors. With the aid of a grant, a much larger brochure was published in 1960, and over the next few years 100,000 copies were distributed. Many of the projects of the committee were aimed at bridging the gap between physiologists at major research centers and isolated college teachers of physiology. Funded by NSF, two-week workshops on teaching problems for college teachers of physiology were begun in 1955. The first, held at the University of Connecticut, was organized by C. L. Prosser. For a period of five years beginning in 1956, NSF and NIH funded summer traineeships that enabled college teachers of physiology to spend the summer in a research laboratory. With the Education Committee acting as a selection committee, 171 awards were made to 136 teachers from 122 colleges in 41 states. For a time the committee acted as mediator so that trainees might obtain, through NIH, small start-up grants to continue research in their home institutions. Another major project funded by grants from NSF and NIH was the visiting lecturers. In a restricted geographical area, physiologists were made available on request to colleges to confer with students and faculty. This project was carried out in Louisiana under the direction of H. S. Mayerson and in Washington state under the direction of A. W. Martin.

Other projects initiated by 1962 included courses for physicians in connection with the College of Physicians (begun in 1960), the compilation and distribution of mimeographed laboratory manuals for college courses in general physiology and in human physiology, the publication of articles on teaching in APS journals, the maintenance of a roster of college teachers, and exhibits of teaching materials at APS meetings and workshops. Apart from the projects mentioned previously, a general effort was made to reach out to physiologists in settings other than medical schools. Many of the president-elect tours, begun by Landis in 1951, were to physiologists in out-of-the-way locations and were invaluable in bringing problems of training and recruitment to the attention of Council and the Education Committee. Articles appeared in *The Physiologist* on physiologists in dental and veterinary schools (23, 30). After 1959 the Society of General Physiologists was invited to appoint representatives to the Education Committee. Many of the projects begun by the Education Committee in the 1950s were continued in the 1960s and are described in further detail in the chapter on education.

Animal Care and Experimentation

APS involvement in matters related to animal care and experimentation stepped up considerably after World War II. Just after the war, in 1946, the National Society for Medical Research (NSMR), an organization to educate the public on the benefits of animal experimentation, was founded by A. J. Carlson and Ivy. Carlson was named first president. In response to Ivy's appeal at the 1946 business meeting, APS donated to NSMR over $300 raised from voluntary contributions added to Society dues. In the postwar years, NSMR became active in promoting state legislation to release unclaimed pound animals to laboratories. It was in part the success of this type of legislation that led to the creation of the Animal Welfare Institute by Christine Stevens, the daughter of APS member Robert Gesell, professor of physiology at the University of Michigan. At the APS Spring Meeting in 1952, Gesell shocked the Society at the first business session by presenting a prepared speech in which he accused physiologists of inhumanity in their treatment of laboratory animals and censured the NSMR. A committee, consisting of Landis and Visscher, was appointed on the spot to formulate a response at the second business session. After conferring with Gesell, the committee stated that "the American Physiological Society rejects unequivocally the inference that its members are insensitive to the moral responsibilities which they have in protecting the welfare of man and animals."

At the 1952 Council meeting preceding Gesell's speech, there had been a discussion of the ethics of animal experimentation and of the possible need for a Society disciplinary committee to deal with infringements of ethics. As a result of the discussion, a committee consisting of Essex (chairman) and Davenport was appointed to present recommendations. The Committee on the Use and Care of Animals, enlarged by the addition of John Haldi and Hamilton, conferred with Gesell over the next year by mail and in person to clarify his stand.

In part because of the Gesell episode, the committee decided to direct its efforts toward a statement of experimental ethics that might be posted in laboratories and published in the APS journals. The APS "Guiding Principles in the Care and Use of Animals" was first approved by Council in 1953 and continues in use in slightly modified form. Adherence was considered an obligation of membership, and articles in violation are refused publication in the Society's journals (4). Special attention was given to assuring that physiologists used proper anesthetization in experiments. The "Guiding Principles" were in part based on guidelines drawn up by Walter B.

Cannon while he was chairman of the American Medical Association's Committee on Defense of Medical Research. In turn they have served as the basis for guidelines drawn up by several other organizations including NIH.

Seventy-fifth Anniversary

The seventy-fifth birthday party of APS was held in conjunction with the fall meeting of the Society in Coral Gables, Florida, 27–30 August 1963. President Hermann Rahn arranged a special celebratory convocation on Tuesday evening to which all living presidents of the past twenty-five years were invited. The evening began with a resume of the third quarter century by Fenn followed by short addresses, later published in *The Physiologist*, by Ivy, Bard, Visscher, Dill, Gerard, Landis, Adolph, Essex, Hamilton, Burton, Katz, and Davis (8). Each president was adorned with a tinseled halo held aloft by wires. Mayerson, invited to join the other presidents, chose instead to give the customary presidential address, "Physiology and physiologists of the gay nineties." On display was a series of photographs of all APS presidents from Bowditch to Rahn. Fenn's excellent *History of the American Physiological Society: The Third Quarter Century, 1937–1962* was distributed gratis to all the members of the Society who attended the Coral Gables meeting and was later mailed to all other members. In the forward to the volume, Fenn wrote, "I shall not enjoy the privilege of reading the history of the next twenty-five years which may prove to be even more revolutionary and exciting than the last. I might express the pious hope, however, that it will be rather a time for more or less calm adjustment to the many dramatic changes which have occurred in the last quarter century" (20, p. vii–viii).

NOTES

[1] S. F. Leslie. "On the way to Beaumont." July 1986, APS Archives, Bethesda, MD.
[2] Survey of Physiological Science papers, 1952–58. APS Archives, Bethesda, MD.

REFERENCES

This chapter is based on the minutes of APS Council and business meetings in the APS Archives, programs of spring and fall meetings, and the annual membership directories published by the Federation. Only additional sources are cited here.

1. ADOLPH, E. F. Educational activities in the Society. In: *History of the American Physiological Society: The Third Quarter Century, 1937–1962*, edited by W. O. Fenn. Washington, DC: Am. Physiol. Soc., 1963, p. 146–154.
2. ADOLPH, E. F., T. E. BOYD, J. H. COMROE, JR., AND P. DOW. Physiology in North America, 1945: survey by a committee of the American Physiological Society. *Federation Proc.* 5: 407–436, 1946.
3. ANONYMOUS. The California Oath. *Science* 114: 1–4, 1951.
4. ANONYMOUS. The care and use of animals. *Physiologist* 2(4): 47, 1959.
5. ANONYMOUS. Finances of the American Physiological Society. *Physiologist* 3(4): 33–39, 1960.
6. ANONYMOUS. The American Physiological Society and the Federation. *Physiologist* 4(1): 12–20, 1961.
7. ANONYMOUS. Spring meeting statistics. *Physiologist* 5: 46, 1962.
8. ANONYMOUS. Seventy-fifth anniversary celebration. *Physiologist* 6: 316–317, 1963.
9. ANONYMOUS. Milton O. Lee, 1901–1978. *Physiologist* 22(1): 7–8, 1979.
10. BARD, P. Section editors for APS journals. *Physiologist* 5: 8–9, 1962.

11. BARD, P. Limbic elements in the publication policies of the APS. *Physiologist* 6: 324–325, 1963.

12. BAZETT, H. C., ET AL. Symposium on physiological contributions to war problems. *Federation Proc.* 5: 318–361, 1946.

13. BERLINER, R. W. Julius Hiram Comroe, Jr. (1911–1984). *Physiologist* 28: 3–4, 1985.

14. COMROE, J. H. Associate member proposal. *Physiologist* 1(2): 21–23, 1958.

15. COMROE, J. H. President's letter. *Physiologist* 3(4): 10–13, 1960.

16. DAVENPORT, H. W. President's message: somebody must do something! *Physiologist* 4(4): 27–28, 1961.

17. DILL, D. B. Roles for senior physiologists. *Physiologist* 1(2): 28–30, 1958.

18. DILL, D. B., W. O. FENN, E. M. LANDIS, AND E. F. ADOLPH. Status and views of senior physiologists. *Physiologist* 4(1): 27–34, 1961.

19. FENN, W. O. Physiology on horseback. *Am. J. Physiol.* 159: 551–555, 1949.

20. FENN, W. O. *History of the American Physiological Society: The Third Quarter Century, 1937–1962.* Washington, DC: Am. Physiol. Soc., 1963.

21. GERARD, R. W. *Mirror to Physiology: A Self-Survey of Physiological Science.* Washington, DC: Am. Physiol. Soc., 1958.

22. GERARD, R. W., P. WEISS, E. M. LANDIS, AND L. IRVING. Report of the Committee on Teaching Problems in Physiology. perspectives in physiological education. *Federation Proc.* 6: 522–537, 1947.

23. GREEP, R. O., J. R. BROBECK, J. HALDI, AND N. R. ALPERT. Some basic problems pertaining to the teaching of physiology to dental students. *Physiologist* 2(1): 42–68, 1959.

24. HAMILTON, W. F. Personal recollections as to the affairs of the Society from 1921 to date. *Physiologist* 7: 38–41, 1964.

25. LEE, M. O. A home for the Federation. *Federation Proc.* 13: 821–824, 1954.

26. LEE, M. O. American Physiological Society journals: editorial procedures and practices. *Physiologist* 3(2): 23–30, 1960.

27. MCMANUS, J. F. A., K. F. HEUMANN, AND M. L. WESTFALL. *Federation Proceedings*: the first quarter century, 1942–1966. *Federation Proc.* 27: 845–855, 1968.

28. O'MALLEY, F. K., AND H. W. MAGOUN. The first American-based *Handbook of Physiology. Physiologist* 28: 35–39, 1985.

29. PITTS, R. F. President's message: problems of democracy in a republican Society. *Physiologist* 2(4): 55–59, 1959.

30. RAHN, H., AND J. R. PAPPENHEIMER. Physiology in schools of veterinary medicine. *Physiologist* 6: 89–93, 1963.

31. VEITH, I. (Editor). *Perspectives in Physiology: An International Symposium, 1953.* Washington, DC: Am. Physiol. Soc., 1954.

32. VISSCHER, M. B. American Physiological Society publications: history and policy. *Physiologist* 3(2): 15–22, 1960.

Top: *Milton O. Lee.* **Center:** *APS Council, Spring Meeting, Atlantic City, NJ, 1954. Left to right: Louis N. Katz, Horace W. Davenport, Hiram E. Essex, Edward F. Adolph (president), Alan C. Burton, Eugene M. Landis, Fred A. Hitchcock. (Photo by H. W. Davenport.)*
Bottom: *APS Council, Fall Meeting, San Francisco, CA, 1960. Left to right: John M. Brookhart, Julius H. Comroe, Jr. (president), Ray G. Daggs, Philip Bard (Board of Publication Trustees), Hymen S. Mayerson, Theodore C. Ruch, Hermann Rahn, Horace W. Davenport. (Photo by H. W. Davenport.)*

Top: *Beaumont House, Bethesda, MD.* **Bottom**: *Milton O. Lee, K. K. Chen, and F. S. Cheever (president of FASEB) at the dedication of the Milton O. Lee Building on the Beaumont campus, Bethesda, MD, 1962.*

Top left: *E. B. Brown, Jr., Fall Meeting, San Francisco, CA, 1960. (Photo by H. W. Davenport.)* **Top right:** *Ellen Brown and Charlotte Haywood, Fall Meeting, Salt Lake City, UT, 1951. (Photo by H. W. Davenport.)* **Center left:** *W. D. McElroy and E. Newton Harvey, Fall Meeting, Medford and Woods Hole, MA, 1955. (Photo by H. W. Davenport.)* **Center right:** *Chandler McC. Brooks, Spring Meeting, San Francisco, CA, 1955. (Photo by H. W. Davenport.)* **Bottom left:** *Homer W. Smith, Salt Lake City, UT, 1953. (Photo by H. W. Davenport.)* **Bottom center:** *Bodil Schmidt-Nielsen, Spring Meeting, Atlantic City, NJ, 1954. (Photo by H. W. Davenport.)* **Bottom right:** *Charles Code, Fall Meeting, Madison, WI, 1954. (Photo by H. W. Davenport.)*

Top left: *Wallace O. Fenn, C. F. Schmidt, Fred A. Hitchcock, and Chauncey D. Leake by the entrance of one of the buildings of the Institute of Experimental Medicine, Leningrad, 8 August 1956. This visit by prominent American physiologists to physiological laboratories in the USSR was organized by APS.* **Top right**: *Jere Mead, Fall Meeting, Madison, WI, 1954. (Photo by H. W. Davenport.)* **Center left**: *L. M. N. Bach, Fall Meeting, Medford and Woods Hole, MA, 1955. (Photo by H. W. Davenport.)* **Center right**: *Charles H. Best and F. C. MacIntosh, XIX International Physiological Congress, Montreal, 1953. (Courtesy of F. C. MacIntosh.)* **Bottom left**: *Ancel Keys, Fall Meeting, Salt Lake City, UT, 1951. (Photo by H. W. Davenport.)* **Bottom center**: *Louise H. Marshall with copy of APS career brochure, ca. 1962.* **Bottom right**: *Orr E. Reynolds, Fall Meeting, Madison, WI, 1954. (Photo by F. A. Hitchcock.)*

JOURNAL OF
NEUROPHYSIOLOGY

J. NEUROPHYSIOL.

Founded by John F. Fulton and the Charles C Thomas Family

EDITORIAL BOARD

VERNON B. MOUNTCASTLE, *Chief Editor*

JOHN M. BROOKHART   BERNARD KATZ
CHANDLER McC. BROOKS   STEPHEN W. KUFFLER
THEODORE H. BULLOCK   JAMES L. O'LEARY
RAGNAR GRANIT   W. DEWEY NEFF

HARRY D. PATTON

ADVISORY EDITORS

LORD ADRIAN (Cambridge)
PERCIVAL BAILEY (Chicago)
PHILIP BARD (Baltimore)
G. H. BISHOP (St. Louis)
F. BREMER (Brussels)
DETLEV W. BRONK (New York)
FRITZ BUCHTHAL (Copenhagen)
HSIANG TUNG CHANG (Shanghai)
STANLEY COBB (Boston)
HALLOWELL DAVIS (St. Louis)
SIR JOHN ECCLES (Canberra)
JOSEPH ERLANGER (St. Louis)
WALLACE O. FENN (Rochester, N. Y.)

A. FESSARD (Paris)
ALEXANDER FORBES (Boston)
H. S. GASSER (New York)
W. R. HESS (Zürich)
MARION HINES (Atlanta)
DAVENPORT HOOKER (New Haven)
YASUJI KATSUKI (Tokyo)
ARISTIDES A. P. LEÃO (Rio de Janeiro)
ROBERT B. LIVINGSTON (Bethesda)
DAVID P. C. LLOYD (New York)
GIUSEPPE MORUZZI (Pisa)
WILDER PENFIELD (Montreal)
H. PIÉRON (Paris)

VOL. XXV   **JANUARY 1962**   NO. 1

PUBLISHED BIMONTHLY BY THE AMERICAN
PHYSIOLOGICAL SOCIETY

Cover of the first issue of Journal of Neurophysiology *to be published by APS signed by a number of people who took part in the negotiations: Charles C Thomas, Vernon Mountcastle, Philip Bard, Eugene M. Landis, Horace Davenport, and Ray G. Daggs. (Courtesy of H. W. Davenport.)*

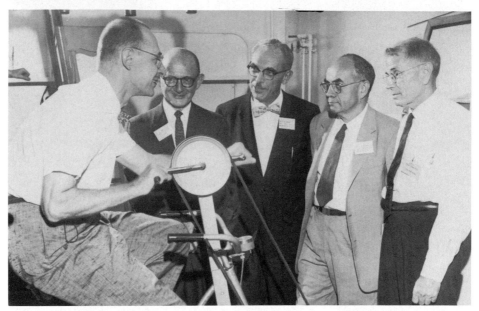

Top left: *Victor S. Hall, executive editor of the first volumes of the* Handbook of Physiology.
Top center: *John Field II, editor-in-chief through 1960 of the* Handbook of Physiology.
Top right: *H. W. Magoun, section editor of the* Handbook of Physiology, The Nervous
System, *section 1.* **Center**: *demonstration at Fall Meeting, Bloomington, IN, 1961. Left to
right: Horace W. Davenport, Hymen S. Mayerson, Ray G. Daggs, Julius H. Comroe, Jr., Sid
Robinson.* **Bottom left**: *Clara Hamilton and Ruth Conklin, Fall Meeting, Madison, WI,
1954. (Photo by H. W. Davenport.)* **Bottom right**: *John W. Bean, Spring Meeting, Atlantic
City, NJ, 1956. (Photo by H. W. Davenport.)*

Seventy-fifth Anniversary Celebration of APS, Fall Meeting, Coral Gables, FL, 1963. **Top:** *Arthur B. Otis, Hermann Rahn, Wallace O. Fenn.* **Center left:** *H. S. Mayerson delivering the past president's address.* **Center right:** *Andrew C. Ivy in halo presented to past presidents.* **Bottom left:** *Alan C. Burton and Norman Staub.* **Bottom right:** *Bruce Dill.*

Top: *participants in a series of conferences on the future of physiological research held in conjunction with the XIX International Physiological Congress, Montreal, 1953, resulting in the publication of* Perspectives in Physiology *by APS in 1954. Left to right: Giulio Stella (Italy), E. Lundsgaard (Denmark), Hans F. Hausler (Austria), Goran Liljestrand (Sweden), Homer W. Smith (United States), Charles H. Best (Canada), Ilza Veith (United States, historian of medicine and editor of the volume), P. Hoffman (Germany), Bernardo A. Houssay (Argentina), Alexander von Muralt (Switzerland), Maurice B. Visscher (United States), E. D. Adrian (United Kingdom), R. A. Peters (United Kingdom), Arturo Rosenblueth (Mexico).* **Bottom:** *past presidents at Seventy-fifth Anniversary Celebration of APS, Fall Meeting, Coral Gables, FL, 1963. Seated left to right: Edward F. Adolph, Eugene M. Landis, Ralph W. Gerard, David B. Dill, Maurice B. Visscher, Wallace O. Fenn, Philip Bard. Standing left to right: Ray G. Daggs, Hymen S. Mayerson, Hallowell Davis, Louis N. Katz, Alan C. Burton, William F. Hamilton, Hiram E. Essex, Hermann Rahn.*

CHAPTER 6

Presidents, 1888–1962

TOBY A. APPEL AND ORR E. REYNOLDS

In keeping with the tradition of earlier histories of the American Physiological Society (APS, the Society), we devote this chapter to biographies of the thirty-five eminent men who served as president of the Society during its first seventy-five years. Each brief biographical sketch is followed by a short bibliography for further information. Note that before 1927, presidents served for calendar years; after 1927 they assumed office following the annual or spring meeting or, after 1951, on 1 July. Biographies of presidents from 1888 to 1938 were written by Toby A. Appel and those from 1939 to 1962 by Orr E. Reynolds.

1 (1888, 1891–95)

Henry Pickering Bowditch
(1840–1911)

See biography in chapter 2.

2 (1889–90)

Silas Weir Mitchell
(1829–1914)

See biography in chapter 2.

3 (1896–1904)

Russell H. Chittenden
(1856–1943)

See biography in chapter 2.

4 (1905–10)

William H. Howell
(1860–1945)

Born in Baltimore, William Henry Howell spent most of his career at Johns Hopkins. He entered the third undergraduate class in 1878 and pursued graduate studies under Henry Newell Martin; he received his doctorate in 1884 and joined the department as a faculty member. After three years at the University of Michigan (1889–91) and a year under Bowditch at Harvard, Howell returned to Baltimore in 1893 as first professor of physiology in the Johns Hopkins Medical School, a position he held until 1916. He was dean of the Medical School from 1899 to 1911. With William Henry Welch, Howell organized the Johns Hopkins School of Hygiene and Public Health and served as assistant director under Welch from 1916 to 1925 and as director from 1925 to 1931, while simultaneously continuing his teaching and research as professor of physiology in the Department of Physiological Hygiene.

After 1893 most of Howell's publications dealt with the physiology and pathology of blood, and especially with the process of coagulation. His best-known achievements were the isolation of thrombin (1910), the discovery and naming of the anticoagulant heparin (1918), and during the last years of his life, the isolation of thromboplastin. In addition to his reputation as an authority in his field, Howell was known as a superb teacher. In 1896 he was editor of *An American Textbook of Physiology*, with chapters written by several members of the Society, and in 1905 published his own textbook for medical students, *Text-book of Physiology*, which went through fourteen editions in his lifetime.

A charter member of APS, he was a mainstay of the Society during its first half century. He holds the record for tenure on Council—a total of twenty-three years! He was a member of the Publications Committee from 1897 to 1914, all through the time that Porter was editing the journal, and was always ready with advice and assistance when his student, Hooker, took over as managing editor in 1914. He was the initiator of the highly successful *Physiological Reviews* and served as first chairman of the journal's Editorial Board from 1920 to 1932. When the XIII International Physiological Congress was held in America in 1929 at the invitation of APS, Howell was selected president of the congress. Finally, for the semicentennial of the Society in 1938, he authored the excellent history of the first twenty-five years and at the celebratory banquet paid tribute to his teacher, Martin, one of the founders of the Society. Howell's student, Joseph Erlanger, wrote of him, "Howell was one of the best loved of American physiologists. A kindly disposition and unpretentiousness of manner endeared him to all who knew him well."

BIBLIOGRAPHY

1. ANONYMOUS. William Henry Howell. *Physiologist* 4(1): 5–11, 1961.
2. CORNER, G. W. William Henry Howell. In: *Dictionary of American Biography*. New York: Scribner, 1973, suppl. 3, p. 369–371.
3. ERLANGER, J. William Henry Howell. *Biogr. Mem. Natl. Acad. Sci.* 26: 153–180, 1951.
4. FEE, E. William Henry Howell: physiologist and philosopher of health. *Am. J. Epidemiol.* 119: 293–300, 1984.
5. RODMAN, A. C. William Henry Howell. In: *Dictionary of Scientific Biography*. New York: Scribner, 1972, vol. 6, p. 525–527.

5 (1911–13)

Samuel James Meltzer
(1851–1920)

Samuel J. Meltzer was one of the three members of the APS delegation to the Conference Committee that founded the Federation of American Societies for Experimental Biology (FASEB, the Federation) on the evening of 31 December 1912. Because the chairmanship of the Executive Committee was rotated among the societies making up the Federation in order of seniority, Meltzer, as president of APS in 1913, became the first chairman of the Executive Committee.

Meltzer was born in Ponevyezh, Russia, into an orthodox Jewish family. He studied philosophy and medicine at the University of Berlin, where he pursued experimental research under the direction of Hugo Kronecker on the mechanism of swallowing. Soon after receiving his medical degree in 1882, he came to America, where he established himself in medical practice in New York City and continued research in his spare time in W. H. Welch's laboratory at Bellevue and J. G. Curtis's laboratory at the College of Physicians and Surgeons. He was elected to APS at its first annual meeting in 1888. In 1904 his devotion to physiological research was rewarded when he was invited to head the Department of Physiology and Pharmacology of the newly formed Rockefeller Institute for Medical Research. He retired from this position in 1919.

Meltzer's experimental research covered a wide spectrum of subjects in physiology, pharmacology, pathology, and clinical medicine. He is especially remembered for the Meltzer-Kronecker theory of deglutition, generalized into a broad theory of inhibition (i.e., every excitation or stimulation of a tissue was accompanied by a corresponding inhibitory action) that acted as a stimulus to much of his work. Other important contributions included work on the anesthetic effects of magnesium salts, artificial respiration through the technique of intratracheal insufflation, and the action of epinephrine on blood vessels and the muscles of the iris.

Meltzer played a major role in his day as a liaison between laboratory scientists and clinicians and as an ardent proponent of the concept of clinical research. He was a founder and officer of several medical and biomedical societies, including the Society for Experimental Biology and Medicine (1903), popularly known as the Meltzer Verein, and the American Society for Clinical Investigation (1908). Howell recalled that from his election to APS until his death, Meltzer was perhaps the Society's most faithful attendant at meetings, where he gave frank but friendly criticisms of papers and took a leading part in discussions of Society policy. He was said to have been something of a "kingmaker" when it came to selecting officers of the Society. F. C. Mann, who first met Meltzer in 1916 on the occasion of giving his first paper at an APS meeting, wrote of him (*Annu. Rev. Physiol.* 17: 1–16, 1955):

> Dr. Meltzer was a sincere idealist, a rugged champion of the experimental method in research, an enthusiastic physiologist, a physician who diligently attempted to use his knowledge of physiology to aid his patients, a teacher of all who had an interest in science.

BIBLIOGRAPHY

1. Anonymous. Samuel James Meltzer. *Physiologist* 5: 1–7, 1962.
2. Chittenden, R. H. Samuel James Meltzer. In: *Dictionary of American Biography.* New York: Scribner, 1933, vol. 12, p. 519–520.
3. Harvey, A. M. Samuel J. Meltzer: pioneer catalyst in the evolution of clinical science in America. *Perspect. Biol. Med.* 21: 431–440, 1978.
4. Howell, W. H. Samuel James Meltzer. *Science* 53: 99–106, 1921.
5. Howell, W. H. Biographical memoir, Samuel James Meltzer, 1851–1920. *Biogr. Mem. Natl. Acad. Sci.* 21: 15–23, 1926.
6. Parascandola, J. Samuel James Meltzer. In: *Dictionary of Scientific Biography.* New York: Scribner, 1974, vol. 9, p. 265–266.

6 (1914–16)

Walter Bradford Cannon
(1871–1945)

Walter B. Cannon's three years as president of APS coincided with the early years of the Federation. Perhaps the most important event of his presidency, one that required extremely delicate handling on his part, occurred in his first two months in office, when Porter resigned as editor of the *American Journal of Physiology* and the Society assumed ownership and management of the journal.

Cannon was born in Prairie du Chien, Wisconsin, where, at Fort Crawford in the 1820s, Beaumont had carried out some of his celebrated experiments on St. Martin.

Attracted to the biological sciences as an undergraduate at Harvard, Cannon began working in Bowditch's laboratory as a first-year student at Harvard Medical School in 1896. That year he began an innovative investigation in which he used the newly discovered X rays to study the mechanism of swallowing and the motility of the stomach. He demonstrated deglutition in a goose at the APS meeting in December 1896 and published his first paper on this research in the first issue of the *American Journal of Physiology* in January 1898. In 1900 he received his medical degree and became a member of APS.

Cannon became an instructor in the Department of Physiology at Harvard in 1900 and was promoted to assistant professor in 1902. When Bowditch retired in 1906, Cannon succeeded him as Higginson Professor and chairman of the department, a position he retained until 1942. He headed one of the most active departments in the country, where students from around the world were trained. Cannon's early research on gastrointestinal motility led to pioneering research on the physiological basis of the emotions and to the development of the concept of the emergency function of the sympathetic nervous system. During World War I he studied problems of traumatic shock. His later research on the sympathetic nervous system and neurochemical transmission of nerve impulses culminated in his enunciation and development of the key physiological concept of homeostasis. He was the author of *A Laboratory Course in Physiology* (1910 and subsequent editions), *The Mechanical Factors of Digestion* (1911), *Bodily Changes in Pain, Hunger, Fear and Rage* (1915, 2nd ed. 1929), *Traumatic Shock* (1923), *The Wisdom of the Body* (1932), *Digestion and Health* (1936), *Autonomic Neuro-effector Systems* (1937, with Arturo Rosenblueth), and *The Way of an Investigator* (1945), which was reprinted and distributed as a souvenir volume at the XXIV IUPS Congress held in Washington, D.C., in 1968.

Cannon's service to APS extended over a period of nearly forty years. He scarcely missed a meeting between his election to APS and his death in 1945. He was a member of Council from 1905 to 1920; he served as treasurer from 1905 to 1913 and was a member of the Conference Committee that established the Federation in 1912. From 1908 to 1926, as chairman of the Committee on Defense of Medical Research of the American Medical Association, he directed the defense of animal experimentation against the antivivisectionists and kept the Society informed of the state of the problem. When the XIII International Physiological Congress was held at Boston in 1929, Cannon was host in charge of local arrangements. In the 1930s he was the American representative to the International Committee that organized the congresses. At the fiftieth anniversary of the Society in 1938, Cannon presented a tribute to his teacher, Bowditch, at the celebratory banquet. He is commemorated by the Society through the Walter B. Cannon Memorial Lecture, a plenary lecture given at the spring meeting of the Society and sponsored by the Grass Foundation. Coinciding with the centennial of APS in 1987, the first volume of Cannon's life and letters is expected to appear, authored by APS members A. Clifford Barger and Saul Benison as well as by Elin Wolfe.

BIBLIOGRAPHY

1. ANONYMOUS. Walter Bradford Cannon. *Physiologist* 6: 4–5, 1963.
2. BARGER, A. C. New technology for a new century: Walter B. Cannon and the invisible rays. *Physiologist* 24(5): 6–14, 1981.
3. BENISON, S., AND A. C. BARGER. Walter Bradford Cannon. In: *Dictionary of Scientific Biography.* New York: Scribner, 1978, vol. 15, p. 71–77.
4. FLEMING, D. Walter Bradford Cannon. In: *Dictionary of American Biography.* New York: Scribner, 1973, suppl. 3, p. 133–137.
5. HOWELL, W. H., AND C. W. GREENE. *History of the American Physiological Society Semicentennial, 1887–1937.* Baltimore, MD: Am. Physiol. Soc., 1938, p. 94–96.
6. RING, G. C. Walter Bradford Cannon, born October 19, 1871, died October 1, 1945. *Physiologist* 1(5): 37–42, 1958.

7 (1917–18)

Frederic Schiller Lee
(1859–1939)

It is appropriate that Frederic Schiller Lee was president of APS during the war years, for he was one of the most active physiologists in the country in the application of physiology to war-related problems. A student of Martin at Johns Hopkins, Lee received his Ph.D. degree in 1885 with a dissertation on the subject of arterial tonicity. The following year was spent in Ludwig's laboratory at Leipzig, where he worked on electrical phenomena of muscular contraction. He was an associate in physiology at Bryn Mawr, when in 1888 he was elected to APS at the Society's first annual meeting. Most of his career was spent at the College of Physicians and Surgeons of Columbia University. He began as Curtis's demonstrator in 1891 and succeeded him as Dalton Professor of Physiology in 1904. He served as executive officer of the department from 1911 to 1920; in 1920 he became research professor. Like his teacher, Martin, Lee took a broadly biological view of physiology and was instrumental in opening courses in the medical department to students outside the medical school.

Lee's early work dealt with the semicircular canals, vestibular sacs, and lateral lines in fishes with respect to the body equilibrium function of the inner ear. Much of his later research focused on muscular fatigue. His studies on fatigue in isolated muscle led to the practical application of fatigue to workers in factories, and eventually he became America's leading scientific authority on the subject of industrial fatigue. During the war years he carried out studies for the Public Health Service, on which he reported to APS in 1919. He was the author of several books, including *Scientific Features of Modern Medicine* (1911) and *The Human Machine and Industrial Efficiency* (1918).

Lee quickly took an active role in Society affairs; he served as secretary from 1895 to 1903, as a member of Council for a total of seventeen years, and as a member of the Publications Committee from 1897 to 1914. He played a leading part in the negotiations resulting in the Society's acquisition of full ownership of the *American Journal of Physiology* in 1914. As president in 1917 he conducted a survey of the war-related activities of all APS members for the National Research Council. Active in international physiology, he was the APS delegate to the International Physiological Congress (Interallied) held in Paris in 1920 and was instrumental in raising funds to hold the XIII International Congress in America in 1929. He was remembered by contemporaries for his unfailing courtesy and his gentle manner.

BIBLIOGRAPHY

1. HOPKINS, J. G. A tribute to the work of a leader in modern physiology: Frederic Schiller Lee. *Columbia Univ. Q.* 47–51, Feb. 1911.
2. HOWELL, W. H., and C. W. GREENE. *History of the American Physiological Society Semicentennial, 1887–1937.* Baltimore, MD: Am. Physiol. Soc., 1938, p. 52–53. (Reprinted in *Physiologist* 7: 1–2, 1964.)
3. ROOT, W. S. Frederic Schiller Lee. In: *Dictionary of American Biography.* New York: Scribner, 1958, suppl. 2, p. 373–374.

8 (1919–20)

Warren Plimpton Lombard
(1855–1939)

Lombard was the last of the original members of APS to serve as president. His presidency coincided with the immediate aftermath of World War I. Because there were no meetings in 1918, two were held in 1919, at which war-related research was presented. In 1920 the first international congress since 1913 was held in Paris, but, to the disapproval of many Americans, German and Austrian physiologists were excluded. Lombard took part in the negotiations concerning the establishment of *Physiological Reviews*, and it was also during his presidency that Porter, through personal conversation with his old friend, offered the funds to establish the Porter Fellowship.

After receiving A.B. (1878) and M.D. (1881) degrees from Harvard, Lombard spent three years in Leipzig, where in Ludwig's institute he studied spinal reflexes in the frog by means of an ingenious apparatus that he designed to record simultaneously the contractions of as many as fifteen muscles. On his return, Lombard carried out research in the physiological laboratories at Harvard, Johns Hopkins, and Columbia. When APS was founded, Lombard was working as an assistant in Curtis's new laboratory at the College of Physicians and Surgeons. In 1889 he became assistant professor of physiology at the newly founded Clark University. The remainder of his career was spent at the University of Michigan, where he served as professor of physiology from 1892 until his retirement in 1923. Lombard's research included studies on the knee jerk, muscular fatigue, blood pressure, and metabolism. He was especially noted for his ability to devise new techniques and apparatus.

One of the seventeen men who attended the organizational meeting of APS in 1887, Lombard presented a paper on the knee jerk at the first special meeting of the Society in September 1888. For many years he was a constant attendant at meetings and a frequent contributor to the program. He served as second secretary-treasurer of the Society in 1893 and 1894, as a member of Council for a total of thirteen years, and as a member of the Publications Committee (1897–1911). He lived to attend the Society's Semicentennial Celebration in 1938, where he recalled the early days of physiology in Curtis's laboratory and concluded, "I have had more pleasure in research and the associations which were given me with the American Physiological Society, than perhaps anything else in my life,—except my wife." John W. Bean wrote of him, "His dignity of manner, gracious poise, and instinctive politeness marked him as a professor of the old school."

BIBLIOGRAPHY

1. ANONYMOUS. Warren Plimpton Lombard, 1855–1939. *Physiologist* 8: 1–2, 1965.
2. BEAN, J. W. Warren Plimpton Lombard. In: *Dictionary of American Biography*. New York: Scribner, 1958, suppl. 2, p. 390–391.
3. DAVENPORT, H. W. Physiology, 1850–1923: the view from Michigan. *Physiologist Suppl.* 24(1): 50–76, 1982.
4. HOWELL, W. H., AND C. W. GREENE. *History of the American Physiological Society Semicentennial, 1887–1937.* Baltimore, MD: Am. Physiol. Soc., 1938, p. 34–35, 107, and 197–198.

9 (1921–22)

John James Rickard Macleod
(1876–1935)

At the APS Meeting in New Haven in 1921, the first year of his presidency, Macleod introduced a paper by F. G. Banting and C. H. Best; it was their initial announcement of their celebrated research on insulin carried out in Macleod's laboratory. The following year, at a joint session of the Federation in Toronto, Banting and Best, again introduced by Macleod, reported the isolation and purification of insulin. In 1923, for the discovery of insulin, Macleod and Banting were awarded the Nobel Prize, which they divided with their co-workers, Best and J. B. Collip.

Macleod was born in Scotland and received his medical training at the University of Aberdeen. He was a demonstrator of physiology and a lecturer in biochemistry at the London Hospital School before being offered the chair of physiology at Western Reserve Medical School in 1903. He was immediately elected a member of APS. In 1918 he became professor of physiology at the University of Toronto. In 1928 he returned to his alma mater, the University of Aberdeen, as Regius Professor of Physiology. Macleod's publications dealt with a wide range of physiological and biochemical topics, including carbamates, purine metabolism, the breakdown of liver glycogen, intracranial circulation, ventilation, and surgical shock, as well as diabetes, on which he published a book as early as 1913. Macleod's textbook, *Physiology and Biochemistry in Modern Medicine* (1918), which went through seven editions during his lifetime, was unique in its emphasis on the important role of chemistry in physiology.

Macleod was first elected to APS Council in 1915. During his presidency, the first APS Porter Fellow, John Hepburn, spent his fellowship year (1921) with Macleod in Toronto working with the insulin group. Macleod was named to the initial Board of Editors of *Physiological Reviews* established in 1920. When he left North America in 1928 he was asked to remain on the board, because APS Council decided it would be valuable to have a British representative. This was the origin of the present European Editorial Committee.

BIBLIOGRAPHY

1. ANONYMOUS. John James Richard Macleod, 1876–1935. *Physiologist* 9: 1, 1966.

2. HOWELL, W. H., AND C. W. GREENE. *History of the American Physiological Society Semicentennial, 1887-1937.* Baltimore, MD: Am. Physiol. Soc., 1938, p. 113-115.

3. STEVENSON, L. G. John James Rickard Macleod. In: *Dictionary of Scientific Biography.* New York: Scribner, 1973, vol. 8, p. 614-615.

10 (1923–25)

Anton Julius Carlson
(1875–1956)

Carlson was president of APS during exciting years for American physiology. In the first year of his term, papers on newly discovered insulin were featured at the Federation meeting. That year Carlson presented an invitation at Edinburgh on behalf of APS to hold the next International Physiological Congress in America. The invitation, reissued in 1926, was accepted for the 1929 congress.

One of the most colorful characters in the history of American physiology, Carlson was born in Sweden and came to America alone in 1891, knowing scarcely any English. Abandoning his initial plan of entering the ministry, he received a Ph.D. degree in physiology at Stanford in 1902 under O. P. Jenkins, with a dissertation on the rate of the nerve impulse in mollusks. At Woods Hole in 1904 he acquired a reputation by his studies on the heart of the horseshoe crab *Limulus*, which showed that the cardiac nerves controlled the heartbeat. In 1904 he joined the Department of Physiology at the University of Chicago, where he became professor in 1914 and chairman in 1916, positions he held until 1940. Known by his students with affection and awe as "Ajax," he presided over the most prolific department in the country for the training of physiologists.

After 1909 Carlson turned from comparative to mammalian physiology. He worked, often in association with his colleague at Chicago, Arno B. Luckhardt, on the hunger mechanism, the physiology of the thyroid and parathyroid, the pancreas, and the visceral sensory nervous system. He was the author of *The Control of Hunger in Health and Disease* (1916) and of the popular textbook, *The Machinery of the Body* (1941), written with Victor Johnson.

Elected to APS in 1904, Carlson served on Council for a total of thirteen years. As secretary of APS from 1909 to 1914, during the period of the founding of the Federation, he became the first secretary of the Executive Committee of the Federation. He long took an active role in APS publications, first as a member of the Publications Committee (1912–14), when Porter was editing the journal, and later as chairman of the Board of Editors of *Physiological Reviews* (1932–50). At APS meetings he was well known and at times feared for his aggressive and pungent criticism of papers. Long active in civic affairs, in 1946 with his former student, A. C. Ivy, he founded the National Society for Medical Research to educate the public on the dangers of antivivisectionist legislation. Ivy wrote of him, "The influence of a zeal for the truth, a critical judgment, a colorful personality and dynamic teaching has nowhere been better exemplified than in the life of Dr. Carlson."

BIBLIOGRAPHY

1. DRAGSTEDT, L. R. Anton Julius Carlson, January 29, 1875–September 2, 1956. *Biogr. Mem. Natl. Acad. Sci.* 35: 1–32, 1961.
2. DRAGSTEDT, L. An American by choice: a story about Dr. A. J. Carlson. *Perspect. Biol. Med.* 7: 145–158, 1963–64.
3. GARRETT, C. G. B. Anton Julius Carlson. In: *Dictionary of American Biography.* New York: Scribner, 1980, suppl. 6, p. 99–100.
4. HOWELL, W. H., AND C. W. GREENE. *History of the American Physiological Society Semicentennial, 1887–1937.* Baltimore, MD: Am. Physiol. Soc., 1938, p. 122–124.
5. INGLE, D. J. Anton J. Carlson: a biographical sketch. *Perspect. Biol. Med.* 22(2), part 2: S114–S136, 1979.
6. IVY, A. C. Anton Julius Carlson. *Physiologist* 2(2): 33–39, 1959. (Reprinted in *Physiologist* 10: 1–6, 1967.)
7. VISSCHER, M. B. Anton Julius Carlson. In: *Dictionary of Scientific Biography.* New York: Scribner, 1971, vol. 3, p. 68–70.

11 (1926–29)

Joseph Erlanger
(1874–1965)

As president of APS Erlanger traveled to Stockholm to repeat the invitation on behalf of APS to hold the XIII International Physiological Congress in America. He remained president through the congress held in Boston in August 1929. Born in San Francisco, Erlanger attended Berkeley and then went east to earn his medical degree at Johns Hopkins, where he worked during the summers with Lewellys Barker. Shortly after Erlanger received his M.D. degree in 1899, he was offered an assistant professorship at Johns Hopkins under William Henry Howell. He was elected to APS in 1901.

After 1904 Erlanger's research concerned the conduction of excitation in the heart; he showed that Stokes-Adams syndrome resulted from blockage of conduction between the auricles and ventricles. After four years at the University of Wisconsin, in 1910 he accepted the chair of physiology at Washington University in St. Louis, which he held until his retirement in 1946. His department became one of the major research centers in physiology in America. At Washington University, he continued his work on cardiovascular physiology and, during the war, carried out research on the problem of shock. In 1921 he shifted his interests to neurophysiology, and he and his colleague, Herbert Gasser, began their celebrated joint work on the amplification and recording of nerve action potentials with the cathode ray oscilloscope, for which they were awarded the Nobel Prize in 1944. Their major discoveries in neurophysiology were published beginning in 1922 in the *American Journal of Physiology*.

Elected to Council in 1910, Erlanger remained on Council, with the exception of two years, until 1929. As treasurer from 1913 to 1923, he helped shape the financial policies of the Society and played an important part in the financial aspects of the transfer of the *American Journal of Physiology* to the Society. He served on the Publications Committee in 1913–14 (and also through the remainder of his time on Council, since Council served as Publications Committee after 1914) and was a member of the Board of Editors of the *American Journal of Physiology* from 1936 to 1938. C. W. Greene wrote, "He has always been characterized by his conservative leadership in the business and organization work of the Society, of which he has carried a large share of responsibility."

BIBLIOGRAPHY

1. Anonymous. Joseph Erlanger, 1874–1965. *Physiologist* 11: 1–2, 1968.
2. Davis, H. Joseph Erlanger, January 5, 1874–December 5, 1965. *Biogr. Mem. Natl. Acad. Sci.* 41: 111–139, 1970.
3. Erlanger, J. Prefatory chapter: a physiologist reminisces. *Annu. Rev. Physiol.* 26: 1–14, 1964.
4. Howell, W. H., and C. W. Greene. *History of the American Physiological Society Semicentennial, 1887–1937.* Baltimore, MD: Am. Physiol. Soc., 1938, p. 131–133.
5. Ludmerer, K. M. Joseph Erlanger. In: *Dictionary of American Biography.* New York: Scribner, 1981, suppl. 7, p. 225–227.
6. Monnier, A. M. Joseph Erlanger. In: *Dictionary of Scientific Biography.* New York: Scribner, 1971, vol. 4, p. 397–399.

12 (1929–32)

Walter Joseph Meek
(1878–1963)

During Walter J. Meek's term in office a committee was created to examine fully the policies and finances of the Society's publications. Its report led to the establishment in 1933 of a Board of Publication Trustees to oversee all aspects of the publications. Meek was appointed first chairman of the Board of Publication Trustees and served in this position of great responsibility and authority from 1933 to 1946.

Meek received his A.B. degree from the University of Kansas in 1902, his A.M. degree from Penn College (Iowa), and his Ph.D. degree from the University of Chicago in 1909. Appointed an instructor in physiology under Joseph Erlanger at the University of Wisconsin in 1908, Meek rose through the ranks to become professor of physiology in 1918. In 1919 he became chairman of the department, a position he retained until 1948. He also served as assistant dean of the Medical School from 1920 to 1942 and acting dean from 1945 to 1949.

Meek's research dealt mainly with the cardiovascular system, gastrointestinal tract, and autonomic nervous system. His work on the heart included studies with A. J. Carlson on the heart of the *Limulus* and a series of papers with J. A. E. Eyster on the origin and conduction of the heartbeat in mammals. On this subject, Eyster and Meek published the first article in the first volume of *Physiological Reviews* (1921). In other studies, Meek investigated the effects of anesthetic agents, in particular cyclopropane, on cardiac irritability and rhythm, studied the origin of fibrin in the liver, and investigated distension as a factor in intestinal obstruction.

Meek was elected a member of APS in 1908 and soon became a stalwart of the Society. He was a member of Council from 1915 to 1919 and again from 1924 to 1936. He served as secretary from 1923 until 1929 when he was elected to the presidency. Perhaps his greatest service to the Society was as chairman of the Board of Publication Trustees. Under his guidance the publications of the Society were reorganized and new procedures and regulations instituted, including for the first time peer review of all papers submitted. As chairman of the Semicentennial Committee, Meek arranged the nostalgic banquet program, presided over by Porter, which was published verbatim in the semicentennial history. An avid historian of medicine, Meek assisted Howell and Greene with the writing of the history and held the office of Society historian from 1938 to 1954.

The memorial notice in *The Physiologist* said of him that he was a man of a modest, unassuming manner and was easily approachable by students and colleagues alike. He had keen critical faculties based on sound logic and was quickly able to get to the meat of a problem. His scientific work was characterized by energy, experimental skill, originality and critical ability. These characteristics were conveyed to the many co-workers and students who were associated with him.

BIBLIOGRAPHY

1. ANONYMOUS. Walter Joseph Meek, 1878–1963. *Physiologist* 12: 1–2, 1969.
2. BROOKS, C. McC. Walter Joseph Meek, August 15, 1878–February 15, 1963. *Biogr. Mem. Natl. Acad. Sci.* 54: 251–268, 1983.
3. HOWELL, W. H., AND C. W. GREENE. *History of the American Physiological Society Semicentennial, 1887–1937.* Baltimore, MD: Am. Physiol. Soc., 1938, p. 135–137.

13 (1932–34)

Arno B. Luckhardt
(1885–1957)

The major events of Arno B. Luckhardt's term as president of APS were the creation in 1933 of the Board of Publication Trustees to oversee the management of the Society's journals and the initiation of formal review procedures for manuscripts submitted for publication in the *American Journal of Physiology.* Previously the managing editor, Hooker accepted articles on his sole authority; in 1933 for the first time all articles were to be submitted for review by a Board of Editors.

Born in Chicago, Luckhardt spent his entire professional career at the University of Chicago. He received his B.S. (1906) and Ph.D. (1911) degrees from the University of Chicago and an M.D. degree from the associated Rush Medical College (1912). He joined the Department of Physiology in 1909 and eventually became William Beaumont Distinguished Service Professor. He retired in 1950. As an investigator he is most remembered for his demonstration of the anesthetic value of ethylene gas and his perfection of the technique in practical surgery. Also of clinical importance was his demonstration that after complete parathyroidectomy animals could be maintained alive by administration of organic calcium salts. Other research themes included studies of gastric motility, gastric and pancreatic secretion, parathyroid physiology, general and local anesthetics, the visceral sensory nervous system, and respiratory physiology.

Luckhardt was elected to APS in 1911 and served as secretary from 1930 to 1932 just before his term as president. He was a long-time member of Council (1923–24, 1927–29, and 1930–42). In addition to his scientific work and his excellent reputation as a teacher, Luckhardt was an avid book collector and an enthusiastic historian of physiology. A recognized expert on the subject of William Beaumont and a collector of Beaumontiana, he presented to the Society an oil sketch by Dean Cornwell of his well-known painting of Beaumont and St. Martin. This painting now hangs in the headquarters of the Federation, Beaumont House. The APS Archives is fortunate to have a complete scrapbook compiled by Luckhardt of the 1939 annual meeting in Toronto dedicated to the use of the historian of 1989.

BIBLIOGRAPHY

1. ANONYMOUS. Arno Benedict Luckhardt, 1885–1957. *Physiologist* 13: 1–2, 1970.
2. HOWELL, W. H., AND C. W. GREENE. *History of the American Physiological Society Semicentennial, 1887–1937.* Baltimore, MD: Am. Physiol. Soc., 1938, p. 146–147.
3. MCLEAN, F. C. Arno B. Luckhardt, physiologist. *Science* 127: 509, 1958.

14 (1934–35)

Charles Wilson Greene
(1866–1947)

Charles Wilson Greene, elected a member of APS in 1900, was long active in the service of APS. Born in Indiana, he received his A.B. and A.M. degrees from Stanford University and his doctorate from Johns Hopkins in 1898. He returned to Stanford to teach physiology from 1893 to 1900 before commencing his long tenure as professor of physiology and pharmacology at the University of Missouri from 1900 to 1936. He reorganized the Physiological Department into a center for research and teaching and set up the first laboratory of experimental pharmacology in the Mississippi Valley. His edition of *Kirke's Handbook of Physiology* (1922) was widely used. Greene's research interests were exceedingly varied. He published studies on the physiology of Pacific salmon as special investigator for the Bureau of Fisheries; he carried on research in high-altitude physiology during and after World War I; and he wrote numerous papers in cardiovascular physiology—on the influence of inorganic salts on inorganic tissues, on pharmacological reactions of the mammalian heart, on changes in the human heart in hypoxic conditions, and on cardiac nerve control of the coronary blood vessels.

As secretary of APS from 1915 to 1923, during a period of rapid growth of the Society, Greene instituted many of the mechanisms that allowed the Society to function smoothly in the period before it acquired an executive officer. His duties were many: he sent out annual notices, organized the program, collected abstracts of the papers presented, channeled communication among the members of Council, acted as the interface between the Council and the managing editor of the journals, and maintained the minutes of the Society. For many years he was the Society's representative to the Union of American Biological Societies, which founded *Biological Abstracts*. He was a constant attendee at annual meetings; Meek recalled in 1938 that Greene had been on the program of all but four meetings since his election. One of his last services to the Society was as author of the second quarter century of the Society's history. Meek wrote of him, "Throughout his long connection with the Society Doctor Greene has been constantly relied upon by the Council and officers for advice and help in initiating all kinds of administrative policies."

BIBLIOGRAPHY

1. ANONYMOUS. Charles Wilson Greene, 1866–1947. *Physiologist* 14: 1–2, 1971.
2. HOWELL, W. H., AND C. W. GREENE. *History of the American Physiological Society Semicentennial, 1887–1937.* Baltimore, MD: Am. Physiol. Soc., 1938, p. 153–154. [Biographical sketch by W. J. Meek.]

15 (1935–37)

Frank Charles Mann
(1887–1962)

During Frank C. Mann's term as president, APS was preoccupied with preparations for its Semicentennial Celebration to take place at the annual Federation meeting to be held in Baltimore in 1938. Born on the family farm in Indiana, Mann received his B.A. (1911), M.D. (1913), and M.A. (1914) degrees from Indiana University. His early work in surgical shock while instructor of experimental surgery at the University of Indiana led to his appointment in 1914 as Director of Experimental Medicine and

Pathological Anatomy at the Mayo Clinic. In 1915, when the Mayo Foundation was created as part of the Graduate School of the University of Minnesota, Mann became assistant professor of experimental surgery. He became associate professor in 1918 and held the rank of professor from 1921 to his retirement in 1952. All together, several hundred graduate students, mostly fellows of the Mayo Foundation, worked in his laboratory.

A prolific investigator, Mann was noted for his exceptional surgical skill. He was a pioneer in the experimental removal of the liver. By using hepatectomized animals, he was able to establish the crucial role of the liver in supplying glucose to the body and in the formation of urea. He also contributed to the pathogenesis of diseases of the digestive system, including peptic ulcer, pancreatitis, and cholecystitis. During both World War I and World War II, he was associated with the National Research Council's efforts to study traumatic shock.

Before his election to the presidency in 1935, Mann served as secretary of the Society from 1932 to 1935. He was a member of the Board of Publication Trustees from 1946 to 1951 and also served as chairman of the board of *Annual Review of Physiology* for many years during the period that the Society exercised joint control of its publication. In his prefatory chapter for *Annual Review of Physiology* in 1955, Mann recalled:

> I received the honor of membership in the American Physiological Society in 1916, and gave my first paper before the society at the Minneapolis and Rochester meeting in 1917. I found so much inspiration, pleasure, and value in these meetings that I attended every one for 35 years.

BIBLIOGRAPHY

1. ANONYMOUS. Frank Charles Mann, 1887–1962. *Physiologist* 15: 1–3, 1972.
2. ESSEX, H. E. Dr. Frank C. Mann. *Physiologist* 6: 66–69, 1963.
3. HOWELL, W. H., AND C. W. GREENE. *History of the American Physiological Society Semicentennial, 1887–1937.* Baltimore, MD: Am. Physiol. Soc., 1938, p. 160–161.
4. MANN, F. C. To the physiologically inclined. *Annu. Rev. Physiol.* 17: 1–16, 1955.
5. VISSCHER, M. B. Frank Charles Mann, September 11, 1887–September 30, 1962. *Biogr. Mem. Natl. Acad. Sci.* 38: 161–204, 1965.

16　(1938–39)

Walter Eugene Garrey
(1874–1951)

Walter E. Garrey, president of APS during its Semicentennial Celebration in Baltimore in 1938, was born in Wisconsin and received his initial training in physiology with Jacques Loeb at the University of Chicago. He received his Ph.D. degree in 1900. In 1899 he assisted Loeb in organizing the first course in physiology at Woods Hole and remained an instructor in the course until 1925. Garrey continued his training in Berlin and Paris and received a medical degree at Rush in 1909. He was a member of the faculties of Cooper Medical College, later incorporated into Stanford University (1900–10), of Washington University, St. Louis (1910 12), and of Tulane (1912–25) before becoming professor of physiology at Vanderbilt University from 1925 until his retirement in 1944. He spent almost all his summers at Woods Hole and served as a trustee of the Marine Biological Laboratory from 1920 to 1944.

Garrey's research covered a wide range of subjects, including artificial parthenogenesis, salt balance and tropisms, mammalian cardiac fibrillation (he was said to be first to describe clearly the phenomenon), and the action of the vagi on the heart. He advanced a theory of inhibition of nerve cells, investigated the rhythm of the neurogenic heart of the *Limulus*, studied parathyroid tetany, and carried out a study of adaptation of salivary secretion to diet.

Elected a member of APS in 1910, Garrey served as a member of Council in 1915–16 and 1924–28 before his election to the presidency. He served on the Editorial Board of the *American Journal of Physiology* from 1936 to 1945 as expert in the fields of blood and general physiology. F. P. Knowlton described him as a man who stood "tall and erect." "With his head of white hair he was a noteworthy figure in any gathering. A somewhat gruff exterior covered a friendly and sympathetic personality."

BIBLIOGRAPHY

1. ANONYMOUS. Walter E. Garrey, 1874–1951. *Physiologist* 16: 1–2, 1973.
2. HOWELL, W. H., AND C. W. GREENE. *History of the American Physiological Society Semicentennial, 1887–1937.* Baltimore, MD: Am. Physiol. Soc., 1938, p. 173–174.
3. KNOWLTON, F. P. Walter Eugene Garrey. *Biol. Bull.* 103: 13–14, 1952.

17 (1939–41)

Andrew C. Ivy
(1893–1978)

Andrew Conway Ivy served as president of the Society for the two-year term just before the dislocations caused by World War II. He had previously been secretary for five years.

Born in Farmington, Missouri, he was educated primarily at the University of Chicago, where he received his Ph.D. degree in 1918 under A. J. Carlson. He received an M.D. degree from Rush Medical College in 1922 while associate professor of physiology at Loyola University School of Medicine (1919–23). Other academic positions were at the University of Chicago (associate professor, 1923–25), Northwestern University Medical School (head of the Division of Physiology and Pharmacology, 1926–45), the University of Illinois (vice-president, 1946–53; and distinguished professor of physiology, 1953–62), and Roosevelt University (research professor of biochemistry, 1962–66). He also served as scientific director of the Naval Medical Research Institute (1942–43), executive director of the National Advisory Cancer Council (1947–51), and director of the Ivy Cancer Research Foundation.

Author of approximately 2,000 scientific articles (over 1,500 by 1955), his contributions were primarily in gastrointestinal physiology and pharmacology but also included the physiology of reproduction, applied physiology (aviation medicine), and physiological resistance to cancer.

His interest in cancer increasingly dominated his career after 1946. This included work on a highly controversial drug, "krebiozen," which led to a temporary estrangement from his colleagues at APS. During the mid 1970s, however, he began attending Society meetings again and displayed the same vigor characteristic of him in former years.

After his term as president, Ivy continued to serve the Society on the Board of Publication Trustees (1945–48), and in this capacity he is credited with recruiting Milton O. Lee, who in 1947 became the first employed executive secretary-treasurer of the Society as well as managing editor of the publications.

In his obituary in 1978, his former colleague, Morton I. Grossman, said of him:

> Dr. Ivy was known to be a man of much determination and courage. Physiologists who worked with him closely had a warm friendship with him and knew him as a

man of high ideals and broad vision, with a wide knowledge of physiology and much wisdom and skill as an executive. In Chicago, he was particularly vigorous and effective in the defense of the use of animals for medical research. He worked long and faithfully for the Society and its publications and for the advancement of physiology.

BIBLIOGRAPHY

1. ANONYMOUS. Andrew C. Ivy. 1893–. *Physiologist* 17: 11–14, 1974.
2. DILL, D. B. A. C. Ivy—reminiscences. *Physiologist* 22(5): 21–22, 1979.
3. FENN, W. O. *History of the American Physiological Society: The Third Quarter Century, 1937–1962.* Washington, DC: Am. Physiol. Soc., 1963, p. 5–7.
4. GROSSMAN, M. I. Andrew Conway Ivy, 1893–1978. *Physiologist* 21(2): 11–12, 1978.
5. WARD, P. S. "Who will bell the cat?" Andrew C. Ivy and krebiozen. *Bull. Hist. Med.* 58: 28–52, 1984.

18 (1941–46)

Philip Bard
(1898–1977)

Elected to his second term at the 1942 APS Meeting, Philip Bard served as president for four additional years, because the austerity of World War II precluded meetings at which a successor could have been selected. At the last meeting (Boston, 1942) $5,000 from the Society's publication funds and from individual contributions were donated to the Royal Society for support of British physiological journals. Also the Society agreed to participate in the recruitment of "younger" physiologists to assist the U.S. armed forces. Both of these actions showed the Society's overriding concern for wartime contributions. No further meetings of the Society were possible until 1946, and whatever Society or Federation business was done was by mail. Under these conditions, it was agreed that the "present officers" continue to serve.

Born in Port Hueneme, California, Bard received his higher education at Princeton (A.B., 1923) and Harvard (A.M., 1925; Ph.D., 1927). His doctoral thesis was supervised by W. B. Cannon. It is of interest that when only sixteen years of age he had obtained and read completely the 1905 edition of the *Textbook of Physiology* by W. H. Howell, the founder of the chair in which he spent his entire career. Before entering college he had served as a volunteer member of the U.S. Army Ambulance

Corps in France during World War I (1917–19). After postdoctoral research activities at Princeton (1928–31) and Harvard (1931–33), in 1933 he became professor and director of the Department of Physiology at Johns Hopkins University, a post he held for thirty-one years. He was dean of the medical faculty from 1953 to 1957.

Bard's scientific work was devoted to the functions of the nervous system, notably to brain stem involvement in emotional excitement, the vestibular system, and motion sickness. He made significant contributions in localization of function and in the conception of the "neural center," which had a profound effect on later disclosures of neural function.

Elected to APS in 1929, Bard's service to the Society included the Editorial Board of *Physiological Reviews* (chairman, 1950–53), the Board of Publication Trustees (chairman, 1959–61), and first chairman of the new Publications Committee (1961–62). His colleague, Vernon Mountcastle, described him as

in his person tall and powerfully built, his features regularly formed in heavy granite, his eye a piercing, pale blue. He possessed great charity for the opinions of others, and avoided disputation; in counsel he was wise, modest, and persuasive. He radiated an ambient spirit of good humor, friendliness, and a fond concern for those about him.

BIBLIOGRAPHY

1. BARD, P. Limbic elements in the publication policies of the APS. *Physiologist* 6: 324–327, 1963.
2. BARD, P. The ontogenesis of one physiologist. *Annu. Rev. Physiol.* 35: 1–16, 1973.
3. FENN, W. O. *History of the American Physiological Society: The Third Quarter Century, 1937–1962.* Washington, DC: Am. Physiol. Soc., 1963, p. 8–9.
4. MOUNTCASTLE, V. B. Philip Bard, 1898–. *Physiologist* 18: 1–5, 1975.
5. MOUNTCASTLE, V. B. Philip Bard, 1898–1977. *Physiologist* 20(3): 1–2, 1977.

19 (1946–48)

Wallace O. Fenn
(1893–1971)

Wallace O. Fenn took office after election at the 1946 meeting in Atlantic City. He had previously served nine years on the Board of Publication Trustees (1933–42), four years as treasurer (1936–40), and four years as secretary of the Society (1942–46).

Born in Lanesboro, Massachusetts, he was educated at Harvard (A.B., 1914; M.S., 1916; Ph.D., 1919). After a tour of duty in the Sanitary Corps of the U.S. Army during World War I, he held the position of instructor of applied physiology at Harvard (1919–22). With the aid of a Traveling Fellowship from the Rockefeller Institute, he worked with A. V. Hill and H. H. Dale in England (1922–24). On his return, he became professor and chairman of physiology at the new School of Medicine and Dentistry of the University of Rochester. In 1924 he became a member of APS. On his retirement from the chairmanship in 1959, he was appointed Distinguished University Professor and in 1962 also director of the Space Science Center, positions he held until his death in 1971.

Although his graduate work was in plant physiology, Fenn early shifted to animal studies. He worked in four general areas: muscle, electrolytes, respiration, and the effects of high and low barometric pressure. After having made notable contributions to the respiratory metabolism of muscle, he was intrigued by the two new domains that opened up in the 1950s, exploration of the new environments of space and the oceans. He studied the role of oxygen from hypoxia to hyperoxic toxicity and the mechanism of high inert gas pressure effects on various living organisms from unicellular organisms to vertebrates. His last research paper (posthumous) was a theoretical work on partial pressures of gases under high pressure (*Science* 176: 1011, 1972).

During Fenn's term as president, the Society underwent fundamental changes. The term of president was constitutionally limited to one year, a permanent position of executive secretary-treasurer was established, and annual election of a president elect, who would serve successively as president elect, president, and past president was instituted. Fenn became the Society's first past president and gave the first past president's address. The Society's activities were increased by the establishment of the fall meeting and the *Journal of Applied Physiology*. Also during his tenure the

American Institute of Biological Sciences was founded with APS as its first member society. Fenn's active interest in the institute continued for many years; he was APS representative to the American Institute of Biological Sciences from 1947 to 1963 and served as president in 1957–58.

After his term as president of APS, Fenn served again on the Board of Publication Trustees as chairman (1949–55), was cochairman with Ralph Gerard of the Survey of Physiological Science, and was coeditor with Hermann Rahn of the section on respiration of the *Handbook of Physiology* (1964–65). He worked with later APS presidents toward the establishment of the International Union of Physiological Sciences (IUPS) and served actively in its governance as secretary general from 1959 to 1965, president of the XXIV IUPS Congress held in Washington, D.C., in 1968, and president of IUPS from 1968 to 1971. His terminal illness prevented his attendance at the XXV IUPS Congress in Munich, where he would have presided over the IUPS. Among Fenn's numerous valued contributions to APS were his excellent history of the third quarter century (1963) and his history of the international congresses (1968).

A sincerely modest man, Fenn avoided the spotlight and never dominated a meeting or conversation, but he was forceful when required and had a warm, outgoing nature with a delightful sense of humor. He was parsimonious in his use of words. One of my own (O. E. Reynolds) best recollections of this is his telegram to me in the first months of his study of the effects of hyperbaric nitrogen in frogs. The wire said in effect, "Positive effect of high pressure N_2 noted—frog nodded."

BIBLIOGRAPHY

1. Daggs, R. G. Wallace O. Fenn, 1893–1971. *Physiologist* 14: 301–303, 1971.
2. Fenn, W. O. Physiology on horseback. Past-president's address. *Am. J. Physiol.* 159: 551–555, 1949.
3. Fenn, W. O. Born fifty years too soon. *Annu. Rev. Physiol.* 24: 1–10, 1962.
4. Fenn, W. O. *History of the American Physiological Society: The Third Quarter Century, 1937–1962.* Washington, DC: Am. Physiol. Soc., 1963.
5. Fenn, W. O. (Editor). *History of the International Congresses of Physiological Sciences, 1889–1968.* Washington, DC: Am. Physiol. Soc., 1968.
6. Rahn, H. Wallace O. Fenn, president of the American Physiological Society, 1946–1948. *Physiologist* 19: 1–10, 1976.
7. Rahn, H. Wallace Osgood Fenn, August 27, 1893–September 20, 1971. *Biogr. Mem. Natl. Acad. Sci.* 50: 141–173, 1979.

20 (1948–49)

Maurice B. Visscher
(1901–83)

Maurice Visscher's presidency was the first to be limited to one year by provisions of the bylaws. The same changes in the bylaws also provided for the first president elect of the Society and the last elected secretary.

Visscher received Ph.D. and M.D. degrees from the University of Minnesota (1925 and 1931). Between these degrees, in 1925–26, on leave of absence, he worked with Professor Starling at University College, London, on the "law of the heart" and cardiac oxygen consumption. Other academic appointments included the University of Minnesota, the University of Chicago, the University of Tennessee, the University of Southern California, and the University of Illinois.

In 1936 he returned to the University of Minnesota as professor and chairman of the Department of Physiology, a position he held for forty one years; on retiring as chairman in 1968, he became Regents' Professor of the university. In addition to developing a prestigious department, Visscher was active in extra-university activities. He was president of the Board of Trustees of *Biological Abstracts*, president of the National Society for Medical Research, and secretary general of IUPS.

Visscher became a member of APS in 1927. He served on Council (1940–50) and as secretary (1947–48), as president (1948–49), and as a member (1954–59) and chairman (1956–59) of the Board of Publication Trustees. He was the initiator of the *Handbook of Physiology*. Visscher was concerned with societal issues in general, especially with areas involving civil liberties, and he had a well-deserved reputation as an effective and fearless activist.

BIBLIOGRAPHY

1. BROWN, E. B., AND E. H. WOOD. Maurice B. Visscher. *Physiologist* 20(1): 1–2, 1977.
2. FOX, I. J. Maurice B. Visscher (1901–1983), scientist and humanitarian. *Physiologist* 27: 1–3, 1984.
3. VISSCHER, M. B. Musings of a physiologist. *Am. J. Physiol.* 159: 556–560, 1949. [Past president's address.]
4. VISSCHER, M. B. A half century in science and society. *Annu. Rev. Physiol.* 31: 1–18, 1969.
5. VISSCHER, M. B. A half century as a scientist-citizen. *Physiologist* 22(3): 15–21, 1979.

21 (1949–50)

Carl J. Wiggers
(1883–1963)

Carl J. Wiggers became president of the Society in 1949, after what must be a record of earlier service to APS: four terms on the Council (1920–23, 1927–30, 1941–42, and 1948–51). In the interim he was elected to the offices of treasurer (1941) and secretary (1942), which resulted in the brief duration of his third Council term. He was also a member of the Board of Editors of the *American Journal of Physiology* from 1933 to 1941.

Wiggers was born in Davenport, Iowa, and received the M.D. degree from the University of Michigan in 1906. He became a member of APS in 1907. After five years as an instructor at Michigan, a year of which was spent in the Physiologisches Institüt at Munich, he joined Graham Lusk's department at Cornell Medical School in 1911. In 1917 he became professor and director of the Department of Physiology at Western Reserve University in Cleveland, Ohio. He retired from Western Reserve in 1953.

Wiggers' substantial contributions to cardiovascular physiology were almost overshadowed by his teaching and leadership qualities; he personally supervised the training of over thirty-eight future department heads and research directors in prestigious institutions. He authored several texts in circulatory physiology: *Circulation in Health and Disease* (1915), *Pressure Pulses in the Cardiovascular System* (1928), *Physiology of Shock* (1950), *Circulatory Dynamics* (1952), and *Circulation in Health and Disease* (first edition, 1934). In 1952 he founded *Circulation Research* of which he was editor for five years. Wiggers received many honors in recognition of his scientific achievements, including several honorary doctorates, the Gold Heart Award of the American Heart Association, and the Albert Lasker Award.

A listing of Carl Wiggers' contributions to APS would be incomplete without reference to his founding of the Circulation Group, which is now still active as the Cardiovascular Section. The highest honor of the section is the Carl Wiggers Award, an annual invited lecture. Many of Wiggers' former associates and students have become APS presidents, including Louis Katz, Robert Berne, Ewald Selkurt, and Walter Randall.

BIBLIOGRAPHY

1. FENN, W. O. *History of the American Physiological Society: The Third Quarter Century, 1937–1962.* Washington, DC: Am. Physiol. Soc., 1963, p. 18–20.

2. LANDIS, E. M. Carl John Wiggers, May 28, 1883–April 29, 1963. *Biogr. Mem. Natl. Acad. Sci.* 18: 363–397, 1976.

3. RANDALL, W. C. Carl J. Wiggers. *Physiologist* 21(3): 1–5, 1978.

4. WIGGERS, C. J. Prefatory chapter: physiology from 1900 to 1920: incidents, accidents and advances. *Annu. Rev. Physiol.* 13: 1–20, 1951.

5. WIGGERS, C. J. *Reminiscences and Adventures in Circulation Research.* New York: Grune & Stratton, 1958.

22 (1950)

Henry Cuthbert Bazett
(1885–1950)

Henry Cuthbert Bazett served as president of the Society for less than three months. After assuming office on 20 April 1950, he died tragically on 11 July 1950 aboard ship en route to the International Physiological Congress in Copenhagen.

Born in Gravesend, England, Bazett obtained his education at Oxford (B.A., 1908; M.B., 1911; B.Ch., 1911; M.S., 1913; and M.D., 1919). After service as a medical officer in the British Army during World War I, he accepted a professorship in physiology at the University of Pennsylvania in 1921. He held this position until his untimely death in 1950. Bazett's scientific work was largely concerned with temperature control, circulation, and blood volume. He contributed greatly to the study of circulation in humans by using catheterization, and he had the reputation of serving himself as the first subject on new and potentially hazardous experimental techniques.

Before assuming the presidency of APS, Bazett served as president elect for one year and as a member of Council for the previous two years. His service to APS also included eight years on the Board of Publication Trustees (1936–44). While on the Council and as president elect he contributed much to the formation of IUPS and was to be a delegate to the congress to which the proposed constitution was presented. As chairman of the APS Committee on Scientific Aid he was responsible for providing much-needed books and equipment to the war-depleted physiology laboratories of Europe.

BIBLIOGRAPHY

1. FENN, W. O. *History of the American Physiological Society: The Third Quarter Century, 1937–1962.* Washington, DC: Am. Physiol. Soc., 1963, p. 20–21.

2. PETERSON, L. H. Henry C. Bazett. *Physiologist* 22(1): 4–5, 1979.

23 (1950–51)

David B. Dill
(1891–1986)

David Bruce Dill succeeded to the office of president after the death of Henry Cuthbert Bazett and served the remainder of the term; he presided at the fall meeting in Columbus, Ohio, in 1950 and the spring meeting in Cleveland in 1951. He had previously served as treasurer (1947–48) and for two years as a member of Council before his election as president elect.

Dill was born in Kansas, but after the death of his parents he was raised from an early age by relatives, first in Iowa and then in Santa Ana, California. He received his B.S. degree from Occidental College in California in 1913 and an M.A. degree from Stanford University in 1914. After teaching chemistry in high schools in California for two years, he was employed by the Bureau of Chemistry, U.S. Department of Agriculture, from 1916 to 1923. He returned to Stanford in 1923 as a fellow in chemistry and received the Ph.D. degree in 1925. He then went to Harvard University as a National Research Council Fellow to work with L. J. Henderson.

In 1927 he became one of the founding faculty of the Harvard Fatigue Laboratory in the School of Public Health, with which he was associated until 1947. Dill was elected to membership in APS in 1941. During World War II he served first with the U.S. Army Air Corps and later with the Quartermaster Corps, which awarded him the Legion of Merit. In 1947, with the discontinuance of the Fatigue Laboratory, he accepted appointment as scientific director of the Medical Division of the Army Chemical Center in Edgewood, Maryland. After his retirement from this position in 1961 he held research professorships at the University of Indiana and the University of Nevada.

During his APS presidency he proposed the establishment of the president elect's tour, which became a Society tradition. Eugene Landis, his successor as president elect, began the practice by visiting a number of institutions, lecturing, and exchanging views with their physiologists and administration. Dill also completed the activities of the Committee on Scientific Aid, which had been begun by Bazett and carried on by Walter Root. One of his most lasting contributions to the Society is the Senior Physiologists Committee, which he initiated in 1951 and chaired until 1980.

Throughout his career, Dill made contributions in the fields of exercise and environmental physiology. In the process, he led expeditions to high-altitude, tropical, and desert environments to study the effects of environmental extremes under natural as well as laboratory conditions. He continued his research in the Nevada desert to the age of ninety-five. At the 1986 Spring Meeting of the Society in St. Louis, a report of his research on aerobic capacity and aging was presented, and at the business meeting he received from President Howard E. Morgan the Society's Ray G. Daggs Award.

BIBLIOGRAPHY

1. BEAN, E. Dr. D. B. Dill. *Physiologist* 17: 449–450, 1974.
2. FENN, W. O. *History of the American Physiological Society: The Third Quarter Century, 1937–1962.* Washington, DC: Am. Physiol. Soc., 1963, p. 21–23.
3. HORVATH, E. C., AND S. M. HORVATH. David Bruce Dill. *Physiologist* 22(2): 1–2, 1979.

24 (1951–52)

Ralph W. Gerard
(1900–74)

Ralph Waldo Gerard became president after the Cleveland meeting in 1951 and presided at the fall meeting in Salt Lake City (1951) and the spring meeting in New York City (1952).

Born in Harvey, Illinois, he received B.S. (1919) and Ph.D. (1921) degrees from the University of Chicago. He then received the M.D. degree from Rush Medical College in 1925. In the interim he was professor of physiology at the University of South Dakota (1921–22). After receiving the M.D. degree he worked with A. V. Hill in London and Otto Meyerhof in Kiel on a National Research Council Fellowship (1926–27). He returned in 1928 to the University of Chicago, where he remained in the Physiology Department until 1952. He then became for three years professor of neurophysiology and physiology in the College of Medicine, University of Illinois, and then professor of neurophysiology at the Mental Health Research Institute in Ann Arbor, Michigan (1955–63). From 1963 he helped organize the Irvine Campus of the University of California and served as dean of its Graduate Division until his retirement in 1970.

Gerard's scientific contributions ranged from the metabolism and heat production of nerve (with A. V. Hill) to the behavioral and social sciences. One of his most pervasive contributions, with Ling and Graham, was the introduction of the intracellular recording capillary microelectrode. In addition to many research and review publications, he was author of several books; *Unresting Cells* (1940) and *Food For Life* (1952) are among the best known.

He was elected to APS in 1927 and to the Council in 1949. As president he was responsible for establishing the first standing committees of the Society, for initiating an ambitious Survey of Physiological Sciences (he was the author of the survey report, *Mirror to Physiology*, published by APS in 1958), and for dealing with the sensitive issues of animal experimentation and "loyalty clearance" of scientists. He also established and chaired a committee to revise the Society's constitution and bylaws. He was active in several aspects of governmental relations to science and helped establish the system and procedures for peer review of proposals for government research grants and contracts.

BIBLIOGRAPHY

1. ANONYMOUS. Ralph W. Gerard (1900–1974). *Physiologist* 23(1): 3, 1980.
2. FENN, W. O. *History of the American Physiological Society: The Third Quarter Century, 1937–1962.* Washington, DC: Am. Physiol. Soc., 1963, p. 23–26.
3. GERARD, R. W. By-ways of the investigator: thoughts on becoming an elder statesman. Past president's address. *Am. J. Physiol.* 171: 695–703, 1952.
4. GERARD, R. W. Prefatory chapter: the organization of science. *Annu. Rev. Physiol.* 14: 1–12, 1952.
5. GERARD, R. W. International physiology. *Physiologist* 6: 332–334, 1963.
6. LIBET, B. R. W. Gerard, born October 7, 1900—died February 17, 1974. *J. Neurophysiol.* 37: 828–829, 1974.
7. LIBET, B., AND O. E. REYNOLDS. R. W. Gerard, born October 7, 1900—died February 17, 1974. *Physiologist* 17: 165–168, 1974.

25 (1952–53)

Eugene M. Landis
(1901–87)

Eugene Markley Landis took office in July 1952 and presided at the fall meeting in New Orleans in 1952 and at the spring meeting in Chicago in 1953. He served as a member of Council from 1947 to 1951.

Born in New Hope, Pennsylvania, Landis received his A.B. (1922), M.S. (1924), M.D. (1926), and Ph.D. (1927) degrees from the University of Pennsylvania. He was a National Research Council Fellow (1926–27) and a Guggenheim Fellow (1929–31), working with Thomas Lewis in London and August Krogh in Copenhagen, and then held various research and faculty positions at the University of Pennsylvania (1931–39). He was elected to APS in 1928. In 1939 he became professor of internal medicine at the University of Virginia, but in 1943 he left to succeed Walter B. Cannon as George Higginson Professor of Physiology at the Harvard Medical School, where he remained until his retirement in 1967. From 1967 to 1971 he was adjunct professor of biology at Lehigh University in Pennsylvania. Landis's career mirrors his broad interests from fundamental biology to clinical medicine. His scientific accomplishments were many, mostly dealing with the cardiovascular system. He was the first directly to measure capillary pressure and make observations on the flow of water through capillary walls.

As president of APS he was the first to adopt measures against racial segregation at Society meetings. He served as chairman of a Society Committee on Loyalty, Clearance and Academic Freedom and dealt in a conservative manner with the then-explosive issue of animal research. He was responsible for several initiatives that became traditions. During his presidency the first APS newsletter (later *The Physiologist*) was issued. At the spring meeting after his last Council meeting as president, he entertained Council members at dinner and presented to his successor a gavel made with his own hands. The gavel, still in use today at Society business meetings, was made from wood historically associated with Henry P. Bowditch and Walter B. Cannon. It was later encased in a box, designed by Louis N. Katz, from wood associated with A. J. Carlson and Carl J. Wiggers, to which Hallowell Davis subsequently attached metal discs associated with Joseph Erlanger and Herbert S. Gasser.

BIBLIOGRAPHY

1. FENN, W. O. *History of the American Physiological Society: The Third Quarter Century, 1937–1962.* Washington, DC. Am. Physiol. Soc., 1963, p. 26–29.
2. LANDIS, E. M. APS and youth. *Physiologist* 7: 3–5, 1964.
3. PAPPENHEIMER, J. R., AND A. C. BARGER. Eugene Markley Landis. *Physiologist* 24(1): 1–2, 1981.

26 (1953–54)

Edward F. Adolph
(1895–1986)

Edward Frederick Adolph succeeded to the presidency in the spring of 1953. There was no fall meeting in 1953 because of the International Congress of Physiological Sciences at Montreal in August, but APS Council did meet in Montreal on 20 August and again in Washington on 9 November 1953. Adolph presided at the spring meeting in Atlantic City in April 1954.

Edward Adolph was born in Philadelphia and received his A.B. (1916) and Ph.D. (1920) degrees from Harvard University, where he worked under L. J. Henderson. He pursued graduate studies at Yale University (1916–18) and served with the Army Medical Corps (1918–19). From 1920 to 1921 he held a fellowship at Oxford University, where he worked with J. S. Haldane. He returned to appointments as instructor in zoology at the University of Pittsburgh (1921–24), a National Research Council Fellow at Johns Hopkins, and then assistant professor of physiology at the University of Rochester (1925), where he became professor in 1948. For over sixty years he used the same office, from 1975 in emeritus status.

Adolph is best known for his research in environmental physiology, particularly in adaptation to hot and cold environments. His wartime work in the field led to a book, *Physiology of Man in the Desert* (1947), a well-known classic. However, these interests were rooted in a more general concern with physiological regulation and integration. Examples of his scope of interest can be found in his *Physiological Regulations* (1943) and his *Physiological Integrations in Action* (1982) published as a supplement to *The Physiologist.*

Elected a member of APS in 1921, Adolph was a member of the Society for sixty-five years. His complete set of APS programs from 1921 through 1941 (after which they were printed in *Federation Proceedings*) were recently donated to the APS Archives.

Adolph was much concerned both before and after his term as president with the problems of recruitment and training of future physiologists. It is not surprising therefore that in 1945 he initiated the Society's first survey of the status of physiology and that he later took responsibility for organizing a Committee on Education, which he served as first chairman from 1953 to 1958. Other activities in which he played a part as president were the exact timing of papers given at the meetings and the purchase by the Society of the Hawley estate as a permanent home for the Society and the Federation. On his retirement from the University of Rochester he was honored by the establishment in his name of an award made annually to a medical student showing superior accomplishment. Other honors he received included the U.S. Presidential Certificate of Merit (1948), the Alumni Gold Medal of the University of Rochester (1964), and the Ray G. Daggs Award of APS (1984) for contributions to physiology and to the Society.

Adolph, who wrote the introduction to this volume, was looking forward to the Centennial Celebration. He died in December 1986 at the age of ninety-one.

BIBLIOGRAPHY

1. ADOLPH, E. F. The physiological scholar. Past president's address. *Am. J. Physiol.* 179: 607–612, 1954.
2. ADOLPH, E. F. Educational activities in the Society. In: *History of the American Physiological Society: The Third Quarter Century, 1937–1962*, edited by W. O. Fenn. Washington, DC: Am. Physiol. Soc., 1963, p. 146–154.
3. ADOLPH, E. F. Prefatory chapter: research provides self-education. *Annu. Rev. Physiol.* 30: 1–14, 1968.
4. ADOLPH, E. F. Growing up in the American Physiological Society. *Physiologist* 22(5): 11–15, 1979.
5. ANONYMOUS. Ray G. Daggs Award, 1984. *Physiologist* 27: 150–151, 1984.
6. FENN, W. O. *History of the American Physiological Society: The Third Quarter Century, 1937–1962.* Washington, DC: Am. Physiol. Soc., 1963, p. 30–32.
7. FREGLY, M. J., AND M. S. FREGLY. Edward F. Adolph, 26th president. *Physiologist* 25: 1, 1982.

27 (1954–55)

Hiram E. Essex
(1893–1978)

Hiram Eli Essex became president of APS in July 1954. He presided at the 1954 fall meeting at the University of Wisconsin (Madison) and at the 1955 spring meeting in San Francisco.

He was born in Glasford, Illinois, and received his B.S. degree from Knox College (Illinois) in 1919. His first career was as a school teacher in Illinois (1911–14 and 1921–23). After he earned M.S. (1924) and Ph.D. (1927) degrees from the University of Illinois, he became instructor in zoology (1927–28) and in experimental biology (1928–32) at the University of Minnesota; later he headed the laboratory of physiology in the Institute of Experimental Medicine of the Mayo Foundation, where he became professor in 1944. He was cochairman (with Charles Code) of the Section on Physiology of the Mayo Foundation and Clinic from 1952 until his retirement in

1958. In 1959 and 1960 after retirement from Mayo, he was director of undergraduate research for a program in experimental biology at St. Mary's College in Winona, Minnesota.

Essex was elected to APS in 1932. He served on the APS Council from 1941 to 1946 and again in 1951 and 1952 before his election as president elect. He also served on many Society committees: the Board of Publication Trustees (1955–58); the Committee on Loyalty, Clearance and Academic Freedom; and the Committee on Animal Care and Experimentation (1952–56), which he served as first chairman. On this latter committee he was responsible for revising and extending the "Guiding Principles" originally developed in 1912 by Walter B. Cannon for animal research. These principles were adopted by the Federation, the National Society for Medical Research (as "Principles for Laboratory Animal Care"), the American Psychological Association, the American Medical Association, and the American Society for the Prevention of Cruelty to Animals. Essex later became president of the National Society for Medical Research. After serving as president of APS, Essex was a member of the Finance Committee (1963–64) and the Senior Physiologists Committee.

Essex's scientific work, though in diverse areas, largely dealt with cardiovascular function. Shock was one of his early interests. From this he moved to measurement of regional blood flow. Techniques developed here have permitted new understanding of function in a wide range of organ systems from the heart and lungs to the intestine, liver, and kidney.

In addition to his scientific career, Essex had two others, as a farmer and as a painter. Starting in 1940, he achieved considerable success as a Holstein breeder and served as president of the Minnesota Holstein Association in 1958. As a painter he developed a very individual style and techniques. An example of his artistry graces the main lobby of the Milton O. Lee Building on the Beaumont campus in Bethesda.

BIBLIOGRAPHY

1. DONALD, D. E. Hiram Eli Essex, 27th APS president. *Physiologist* 26: 1–3, 1983.
2. ESSEX, H. E. The philosophy of physiology. Past president's address. *Am. J. Physiol.* 183: 583–590, 1955.
3. FENN, W. O. *History of the American Physiological Society: The Third Quarter Century, 1937–1962.* Washington, DC: Am. Physiol. Soc., 1963, p. 32–34.

28 (1955–56)

William F. Hamilton
(1893–1964)

William F. Hamilton presided at the 1955 fall meeting at Tufts University in Medford and at the 1956 spring meeting in Atlantic City. He had previously served on Council (1942–49) and as a member of the Board of Publication Trustees (1951–54 and 1957–59). He had been elected to APS in 1924.

Hamilton was born in Tombstone, Arizona. He received the A.B. degree from Pomona College (1917) and Ph.D. degree from the University of California in 1921. He was an assistant in biology at Pomona College (1914–17), served in the Army Medical Corps (1917–19), and was a laboratory assistant at the University of California (1917–21). After serving as instructor in biology at the University of Texas (1920–21) and at Yale (1921–23), he joined the faculty of the University of Louisville, where he became professor in 1930. From 1932 to 1934 he was at George Washington University. Finally he became professor of physiology at the University of Georgia, a position he held until his retirement to emeritus status in 1960.

After an early interest in animal behavior, Hamilton's research turned first to sensory physiology and later to the circulation. Methods he and his collaborators developed were crucial in establishing pressure-flow and volume relationships in the circulation that led directly to many of our present clinical capabilities.

Among his many contributions to the Society were his codiscovery with Milton Lee (1953) of the Hawley estate in Bethesda as a potential "home" for APS, his inspiration to name the property Beaumont and the mansion Beaumont House, his establishment of the executive secretary of APS as a full-time appointment independent of FASEB, and his design of the APS seal. He was a member of the Committee on the Use and Care of Animals (1952–58) and the Senior Physiologists Committee (1957–60 and 1962–64) and was section editor of the volumes on circulation of the first APS Handbook series. Hamilton was a strong advocate of the concept that the Board of Publication Trustees should remain a body one step removed from immediate Society exigencies and politics and thought it unfortunate that a small minority of members could bring about its downfall in the 1961 business meeting.

BIBLIOGRAPHY

1. BAKER, C. E., ET AL. William F. Hamilton. *Physiologist* 27: 64–65, 1984.

2. Dow, P. William F. Hamilton. *Physiologist* 8: 95–96, 1965.

3. FENN, W. O. *History of the American Physiological Society: The Third Quarter Century, 1937–1962.* Washington, DC: Am. Physiol. Soc., 1963, p. 34–37.

4. HAMILTON, W. F. A day dream. Past president's address. *Am. J. Physiol.* 187: 579–581, 1956.

5. HAMILTON, W. F. Personal recollections as to the affairs of the Society from 1921—to date. *Physiologist* 7: 38–41, 1964.

29 (1956–57)

Alan C. Burton
(1904–79)

Alan Chadburn Burton, a member of Council since 1953, was elected president elect in 1955 at the meeting in San Francisco and began service as president in 1956. He was the second Canadian to hold that office. (J. J. R. Macleod was president in 1921–22.)

Born in London, England, he received his B.Sc. degree from University College (London) in physics in 1925. His M.A. and Ph.D. degrees, also in physics, were from the University of Toronto (1929 and 1932). He held a series of fellowships from the National Research Council of Canada (1928–32), the University of Rochester (1932–34), the Rockefeller Foundation (1934–37), and the Johnson Foundation (1937–40). He was elected to APS in 1937. It was during his postdoctoral fellowships that he was attracted to physiology through his interest in skin temperature and heat exchange. This led him to wartime work in aviation medical research under assignment by the Canadian National Research Council. After the war, he joined the Department of Medical Research at the University of Western Ontario, where he became professor of biophysics in 1948.

In addition to the presidency of APS, Burton served as president of the Federation (1957–58) (of which he initiated a reorganization) and of the Canadian Physiological Society (1959), the Canadian Federation of Biological Societies (1963), and the Biophysical Society (1966).

During his president-elect tour, Burton discussed changes in APS that he thought would prove necessary when the membership exceeded 3,000. (This number was reached in 1967.) Included in his proposals for change were the institution of several meetings per year similar to the APS fall meeting and a Society news publication. During his term, the decision was made to convert the "President's

News Letter" into *The Physiologist.* The first issue contained Burton's past president's address, "The human side of the physiologist, prejudice and poetry," which included three of his original poems. Also during his term as president, the associate member category was approved by Council for submission to the membership.

BIBLIOGRAPHY

1. BURTON, A. C. The human side of the physiologist, prejudice and poetry. *Physiologist* 1(1): 1–5, 1957.
2. BURTON, A. C. Variety—the spice of science as well as of life: the disadvantages of specialization. *Annu. Rev. Physiol.* 37: 1–12, 1975.
3. FENN, W. O. *History of the American Physiological Society: The Third Quarter Century, 1937–1962.* Washington, DC: Am. Physiol. Soc., 1963, p. 37–41.
4. GROOM, A. C. Alan Chadburn Burton (1904–1979). *Physiologist* 23(1): 17–18, 1980.
5. GROOM, A. C. Alan Chadburn Burton: biophysicist extraordinary. *Physiologist* 28: 66–68, 1985.

30 (1957–58)

Louis N. Katz
(1897–1973)

Louis Katz assumed office in July 1957 after serving on Council for six years, the last as president elect. Born in Poland in 1897, he arrived in the United States in 1900. His education was at Western Reserve University (A.B., 1918; M.D., 1921; and M.A. in medicine, 1923). He received a National Research Council Fellowship in 1924 to work with A. V. Hill in London. He then served in the physiology program at Western Reserve until 1930, when he was appointed director of the Cardiovascular Research Department (designated the Cardiovascular Institute in 1941) of the Michael Reese Hospital in Chicago. Concomitantly he held professorial appointments in physiology at the University of Chicago from 1930 to 1969.

Elected to APS in 1924, Katz served on the Membership Advisory Committee (1952–54; chairman, 1953–54) and was chairman of the Program Advisory Committee (1954–57). During his term as president of APS, Katz was responsible for establishing the Bowditch Lecture with sufficient financial support to continue it for several years. He was especially interested in the organizational and financial stability of the Society; he initiated the Operational Guide (a handbook of Society procedures supplementing the constitution and bylaws), clarified committee responsibilities, and worked to bolster the General Operating Fund of the Society. After his presidency, Katz served on the Finance Committee (1962–65), which he chaired for a year (1964–65).

Widely respected for his contributions to the physiology of the heart and clinical cardiology (over 500 papers in journals), Katz played leading roles in the American Heart Association and in a large number of national and international organizations devoted to the heart and circulation. He became director emeritus of the Cardiovascular Institute in 1967 and remained active in study and writing until his death in 1973.

BIBLIOGRAPHY

1. FENN, W. O. *History of the American Physiological Society: The Third Quarter Century, 1937–1962.* Washington, DC: Am. Physiol. Soc., 1963, p. 42–43.
2. FISHMAN, A. P. Louis N. Katz, M.D. (1897–1973): an appreciation. *Physiologist* 16: 691–696, 1973.
3. KATZ, L. N. Physiology and physiologists: a swan song. *Physiologist* 1(5): 18–25, 1958.

31   (1958–59)

Hallowell Davis
(b. 1896)

Hallowell Davis became president of APS in July 1958 after having served on Council for two years, the second as president elect. He had previously served on Council from 1942 to 1946.

Born in New York City, Davis was educated at Harvard (A.B., 1918; M.D., 1922). This was followed by a year with E. D. Adrian at Cambridge University. He held progressive academic appointments at Harvard Medical School from 1923 to 1946 (associate professor of physiology, 1927–46). In 1946 he joined the Central Institute for the Deaf in St. Louis, Missouri, as director, a position he retained throughout his subsequent career, and became emeritus director of research in 1965 until his retirement in 1985. He also held appointments at Washington University as research professor of otolaryngology and professor of physiology.

A pioneer in the development of electroencephalography, he attained recognition in this field, but his principal scientific activity was in the fundamental physiology of the sense of hearing and its practical application.

Davis became a member of APS in 1925. He served the Society as treasurer during World War II for four years (1942–46), as APS representative to the Division of

Medical Sciences of the National Research Council in the 1950s, as a member of the Board of Publication Trustees (1954–55), and as chairman of the Membership Advisory Committee (1956–57 and 1960–61). As a member of the Board of Publication Trustees, he was a strong advocate for the establishment of *The Physiologist* as the Society house organ. After his term as president, Davis was a member of the first Finance Committee (1961–63), which he chaired in his second year, and was a long-time member of the Senior Physiologists Committee (1966–81).

The author of this brief biography was handicapped by not having in hand a biography like those published in *The Physiologist* for most former presidents. Since he is still living, no obituary has appeared, nor has he been featured in a biographical article. (Alan Burton, the nineteenth president, was featured in 1985.) With his usual prescience, Hallowell Davis sent the centennial office an unsolicited "biographical sketch" (APS Archives, Bethesda, MD) just in time to round out his career in the years since 1962.

BIBLIOGRAPHY

1. FENN, W. O. *History of the American Physiological Society: The Third Quarter Century, 1937–1962.* Washington, DC: Am. Physiol. Soc., 1963, p. 44–46.

32 (1959–60)

Robert F. Pitts
(1908–77)

Robert Franklin Pitts became president in July 1959. Born in Indianapolis, he received a B.S. degree from Butler University. He earned his Ph.D. degree from Johns Hopkins under S. O. Mast in 1932, and from 1932 to 1938 he served on the staff of the Department of Physiology of New York University College of Medicine while earning his M.D. degree. He was elected to APS in 1934. From 1938 to 1942 he held research fellowships at the Neurological Institute of Northwestern University and the Johnson Foundation for Medical Physics at the University of Pennsylvania. He was assistant and associate professor of physiology at Cornell University Medical College from 1942 to 1946. In 1946 he became professor and departmental chairman at Syracuse University. He returned to Cornell as professor and chairman of physiology in 1950 and remained there until his retirement in 1974. After retirement he served as research professor at the University of Florida until his death in 1977.

Pitts was most noted for his contributions to our knowledge of kidney function. Application of his principles of acid-base, electrolyte, and water balance led to standard therapies in daily use in medical practice. It was his universal practice to serve as the first human subject in his own new research procedures.

Known also as a dedicated teacher, he was awarded the first Distinguished Teaching Award of the Association of Chairmen of Departments of Physiology in 1978. He authored two major texts: *The Physiological Basis of Diuretic Therapy* (1959) and *The Physiology of the Kidney and Body Fluids* (1974).

Pitts became president elect in 1958, after having served on the APS Council since 1955. He had previously been a member of the Board of Publication Trustees (1948–53) and chairman of the Membership Advisory Committee (1955–56), the Committee on the Use and Care of Animals (1957–59), and the Porter Fellowship Committee (1956–59). During his presidency, the membership category of "sustaining associate" was established for individuals and institutions making special contributions to the Society. It has proved to be a valuable asset both in terms of finances and in promoting communication between the Society and industry. Also during his presidency, a committee was established to study the relation of the Board of Publication Trustees to Council (see Comroe biography in this chapter).

In public manner, Pitts was reserved and austere, but closer association revealed him to be warm and thoughtful with a provocative sense of humor. He was much admired by his students and colleagues.

BIBLIOGRAPHY

1. ALEXANDER, R. S. Pitts and urine acidification. *Physiologist* 26: 364–366, 1983.
2. FENN, W. O. *History of the American Physiological Society: The Third Quarter Century, 1937–1962.* Washington, DC: Am. Physiol. Soc., 1963, p. 47–48.
3. PITTS, R. F. Past president's address: the teacher and the ferment in education. *Physiologist* 3(4): 20–29, 1960.
4. SELKURT, E. E., ET AL. Robert Franklin Pitts (1908–1977). *Physiologist* 20(5): 9–11, 1977.

33 (1960–61)

Julius H. Comroe, Jr.
(1911–84)

Julius Comroe became president of APS in July 1960 after having served on Council since 1956. Elected to membership in the Society in 1943, he participated in Edward Adolph's survey (1946) and then served on the Education Committee and the Central Committee of the Survey of Physiology (1952–56).

Born in York, Pennsylvania, Comroe entered the University of Pennsylvania, where he received his B.A. (1931) and M.D. (1934) degrees; he remained associated with that institution for thirty years. He was appointed instructor in pharmacology in 1936 and rose through the ranks to become professor and chairman of physiology and pharmacology of the Graduate School of Medicine in 1946. In 1957 he moved to the University of California, San Francisco, to become director of the new Cardiovascular Research Institute. He retired as director in 1973 but continued as professor of physiology; he was given emeritus status in 1978.

Noted for his scientific contributions in cardiovascular and respiratory function, Comroe was perhaps most interested in teaching and communications. He was an initial member of the APS Education Committee (1953–58). Not only was he a highly effective teacher personally (Association of Chairmen of Departments of Physiology Teaching Award in 1974), he also conceived and carried out a number of special programs and publications promoting better communication of physiological knowledge to a broader audience. Among these were the first APS "Refresher Course" for teachers (1954), an APS Postgraduate Course for Physicians (1960), *Physiology for Physicians* (1963, later the series "Physiology in Medicine"), and his popular book, *Retrospectoscope* (1977), which traced the scientific background of prominent "medical" discoveries.

As a member of Council, Comroe promoted changes in the bylaws to create the two new categories of membership, associate and sustaining associate. The most notable, certainly the most controversial, event of Comroe's presidency was a change of the bylaws, which he strongly advocated, abolishing the Board of Publication Trustees and creating a Publications Committee and a Finance Committee more directly responsible to the elected Council.

Comroe was presented the Ray G. Daggs Award in 1977 for contributions to physiology and to the Society.

BIBLIOGRAPHY

1. BERLINER, R. W. Julius Hiram Comroe, Jr. (1911–1984). *Physiologist* 28: 3–4, 1985.
2. FENN, W. O. *History of the American Physiological Society: The Third Quarter Century, 1937–1962.* Washington, DC: Am. Physiol. Soc., 1963, p. 48–49.

34 (1961–62)

Horace W. Davenport
(b. 1912)

Horace Willard Davenport became thirty-fourth president in July 1961 after service on Council for five years (1951–55 and 1959–60). Elected to membership in 1942, he had previously served on the Central Committee for the Survey of Physiology (1952–56), the Membership Advisory Committee (1951–53), the Committee on Use and Care of Animals (1952–59), the Porter Fellowship Committee (1952–56), the Education Committee (1958–59), and the Committee on Matters Related to Loyalty (1953–54). Subsequent to his presidency Davenport has been on the Editorial Board of the *American Journal of Physiology* and the *Journal of Applied Physiology* (1961–67), the Senior Physiologists Committee (1978–81), the Centennial Celebration Committee (1978–80), and the Honorary Membership Committee (1980–83). (For one who appears to find committee activities onerous, Davenport has certainly put up with a great deal of pain in service to the Society.)

Born in Philadelphia, he received his B.S. (1935) and Ph.D. (1939) degrees from the California Institute of Technology and B.A. (1937) and B.Sc. (1938) degrees from Oxford. After two years of fellowship at Rochester and Yale, he became instructor in physiology at the University of Pennsylvania (1941–42) and Harvard (1943–44). In 1945 he was appointed professor and chairman of physiology at the University of Utah and chairman of the Division of Biology in 1948. In 1956 he moved to the University of Michigan as professor and chairman of physiology. He became emeritus as William Beaumont Professor of Physiology in 1978.

Davenport's scientific work has concentrated on gastrointestinal physiology, especially the secretion of HCl by the stomach. In 1947 he authored *ABC of Acid-Base Chemistry*, a very popular text that went through six editions.

During Davenport's administration, the Society was faced with the complex and stressful job of reorganization brought on by the demise of its Board of Publication

Trustees. This required complete restructuring of the finances of the Society, new contracts with those involved in the publication operation, and implementation of the new committee structure (Publications and Finance Committees). Also during Davenport's presidency, the *Journal of Neurophysiology* was purchased, a move that had been considered before but without success and that required delicate negotiations.

In recent years, especially since becoming emeritus, Davenport has directed his activities toward the history of physiology and medicine. He is a painstaking and demanding student of history who has made numerous contributions of professional quality in the field. Recently a number of historical papers drawn from lectures given to students have appeared in *The Physiologist*. At the request of the Centennial Celebration Committee, he prepared an article on how to write a departmental history, and his supplement to *The Physiologist*, "Physiology, 1850–1923: the view from Michigan" (1982), has provided a model for other departmental historians.

Few members of APS have made contributions to the Society covering such a broad range of interests as has Horace Davenport.

BIBLIOGRAPHY

1. DAVENPORT, H. W. Human voices. Past president's address. *Physiologist* 5: 265–269, 1962.
2. DAVENPORT, H. W. A. N. Richards; or, why I don't have an M.D. *Physiologist* 21(6): 25–30, 1978.
3. DAVENPORT, H. W. Some notes on preparing a history of a department of physiology. *Physiologist* 22(1): 30–31, 1979.
4. DAVENPORT, H. W. Physiology, 1850–1923: the view from Michigan. *Physiologist Suppl.* 24(1), 1982.
5. DAVENPORT, H. W. The apology of a second-class man. *Annu. Rev. Physiol.* 47: 1–14, 1985.
6. FENN, W. O. *History of the American Physiological Society: The Third Quarter Century, 1937–1962.* Washington, DC: Am. Physiol. Soc., 1963, p. 51–53.

35 (1962–63)

Hymen S. Mayerson
(1900–85)

Hymen Samuel Mayerson became president in July 1962. Born in Providence, Rhode Island, he received his education at Brown University (A.B., 1922) and Yale (Ph.D., 1925). After a year as instructor in physiology, he went to the Department of Physiology at Tulane University in 1926. He remained there with increasing responsibilities and rank, serving as chairman from 1945 until his retirement in 1965. After retirement from the university, he served for ten years as Associate Director for Research and Education of Touro Infirmary, a large hospital in New Orleans.

Mayerson is best known for his research on lymphatic and capillary function, especially the factors influencing movement of macromolecules (protein) from plasma to lymph and the consequences of this process.

Mayerson was elected to APS in 1928. Before his election as president elect in 1961, he had served on Council since 1957. Also he served on the Editorial Board of *Physiological Reviews* (1958–61) and on the Education Committee (1959–61). For this latter committee, he developed and operated a Visiting Scientist Program in Mississippi and Texas that was judged a great success. During his presidency, the APS offices were moved from Beaumont House to the newly built office building on the Beaumont campus (later named the Milton O. Lee Building), *Physiology for Physicians* began publication, and plans were underway for publication of the Society's history and the celebration of the Society's seventy-fifth anniversary at the fall meeting in Coral Gables, Florida. At that meeting, Mayerson gave a humorous past president's address on physiologists in the gay nineties. Quoting E. C. Andrus, Mayerson introduced the since oft-repeated phrase, "There is nothing that gets paster faster than a past president." After serving as APS president, he took office as president of FASEB (1963–64). Other service to the Society included the International Physiology Committee (1961–66), the Finance Committee (1964–67; chairman, 1965–67), and the Senior Physiologists Committee (1971–80; chairman, 1977–80).

In 1979 Mayerson was honored by a symposium in his name on capillary permeability and mechanisms of transport at the APS Fall Meeting in New Orleans.

Mayerson, who was unable to attend because of illness, contributed a paper entitled "A Chance to Reminisce" which he began, "Fellow capillarians and lymphomaniacs—greetings!" He briefly reviewed the background of his own contributions to the subject with appropriate and charming reference to the work of his colleagues. Mayerson is remembered as an excellent teacher, with an unflappable demeanor and wonderful sense of humor, who was always ready to give of himself for others.

BIBLIOGRAPHY

1. FENN, W. O. *History of the American Physiological Society: The Third Quarter Century, 1937–1962.* Washington, DC: Am. Physiol. Soc., 1963, p. 53–54.
2. MAYERSON, H. S. Physiology and physiologists in the gay nineties. *Physiologist* 6: 328–348, 1963.
3. MAYERSON, H. S. A time to reminisce. *Physiologist* 23(1): 41–43, 1980.
4. TAYLOR, A. E. Dr. Mayerson, the scientist. *Physiologist* 23(1): 38–40, 1980.
5. WASSERMAN, K. Professor H. S. Mayerson, Ph.D.—the man. *Physiologist* 23(1): 35–37, 1980.
6. WASSERMAN, K. 35th President, Hymen Samuel Mayerson (1900–1985). *Physiologist* 29: 197, 1986.

CHAPTER 7

Presidents, 1963–1987

JOHN R. BROBECK

In earlier volumes on the history of the American Physiological Society (APS, the Society), the chronology of the Society's affairs was set forth mainly in brief biographies of the presidents in the order in which they held office. This practice is continued here for the most recent twenty-five years. Presidents or former presidents were asked to submit information about their research training, most important publications, and participation in APS business and programs. Hermann Rahn (1963–64) was the last of the presidents considered in Wallace Fenn's volume in a somewhat abbreviated write-up; he is consequently the first to be included in the current series.

36 (1963–64)

Hermann Rahn

(b. 1912)

When APS celebrated its seventy-fifth anniversary at the fall meeting in Coral Gables in August 1963, Hermann Rahn presided over the session. He first introduced his friend and colleague, Wallace O. Fenn, nineteenth president of APS (1946–48), who reviewed the most recent twenty-five years of the Society's history. Rahn then called on former presidents to speak briefly on topics of historical, contemporary, and prospective interest to members of the Society and their guests. He began as follows:

> Our Presidents do not wear uniforms and medals and ribbons, but they wear a halo which is not easily recognized by outsiders. It shines with a soft blue light which can be seen only by those "in the know."

The twelve talks that constituted this program were subsequently published in volumes 6 and 7 of *The Physiologist.*

Hermann Rahn was born in East Lansing, Michigan, but from his graduation from high school in Ithaca, New York, in 1929, his professional life has been in general identified with central New York State. He received his A.B. degree from Cornell University in 1933 and his Ph.D. degree from the University of Rochester in 1938; he joined the staff of the Department of Physiology at the University of Rochester School of Medicine and Dentistry in 1941 and eventually became vice-chairman of that department. In 1956 he moved west only a few miles to become professor and chairman of the Department of Physiology at the University of Buffalo (since 1962, the State University of New York at Buffalo). In 1973 he was honored by appointment as Distinguished Professor of Physiology.

Rahn has not always been found "at home," however. He has been a visiting professor at San Marcos University, Lima, Peru (1955); at Dartmouth Medical School (1962); at the Laboratoire de Physiologie Respiratoire, Centre Nationale Recherche Scientifique, Strasbourg, France (1971); and at Max-Planck Institüt für experimentalle Medizin, Göttingen, West Germany (1977). He has received the honorary degrees of Docteur (H.C.), University of Paris (1964); LL.D., Yonsei University, Seoul, Korea (1965); D.Sc. (Hon.), University of Rochester (1973); Titulo de Profesor Honorario, Universidad Peruana, Lima, Peru (1980); and Doctor Medicinae honoris causa, Universität Bern, Switzerland (1981). He was elected to honorary membership in the Harvey Society of New York in 1960, the American Academy of Arts and Sciences [AAAS (Boston)] in 1966, the National Academy of Sciences (NAS) in 1968, and the Institute of Medicine of the NAS in 1971. In 1976 he received a Senior U.S. Scientist Award of the Alexander von Humboldt Foundation.

Beginning with his first appointment to the Editorial Board of the Society's journals in 1953, Rahn has served APS in many different capacities. Among other appointments he has been a member of the Board of Publication Trustees (1959–61) and the Editorial Board for the *Handbook of Physiology* (1958–66). With Wallace Fenn, he edited the respiration section of the *Handbook of Physiology.* He was a member and/or chairman of the Education Committee (1958–61), the Perkins Memorial Fund Committee (1968–80), the Daggs Award Committee (1980–83), and the Honorary Membership Committee (1979–84). Elected to Council in 1960, he was chosen as president elect two years later and so continued on Council until he finished his term as past president in 1965. In 1978 he was recipient of the Society's Ray G. Daggs Award.

Another of Rahn's interests has been the International Union of Physiological Sciences (IUPS). He has been a member of the Council (1965–74) and has served as vice-president (1971–74). While serving on the U.S. National Committee (1966–74), he was a member of the Organizing Committee for the XXIV IUPS Congress (1968) in Washington, D.C. (1965–68). As a member of this committee he was given the responsibility of organizing the satellite symposia. This was the first time these symposia were officially recognized as part of the IUPS Congress. He served as chairman of the Satellite Symposium Committee for six years. There is no doubt that these symposia have contributed greatly to the viability of our IUPS Congresses and to the advancement of physiology as a science.

Rahn's graduate study and first three postdoctoral years are marked by a series of sixteen papers on the endocrinology of the avian pituitary gland and the biology of rattlesnakes—the latter a subject of interest during the two years he was an instructor at the University of Wyoming at Laramie. When Wallace Fenn invited him back to Rochester, endocrinology lost a promising young investigator. With Fenn and Arthur Otis he began the research on respiration and pulmonary ventilation for which these three men and their associates have become so well known. By 1955 Rahn and Fenn had arrived at their concept, *A Graphical Analysis of the Respiratory Gas Exchange: the O_2-CO_2 Diagram*, published by APS (3). Here the partial pressure of CO_2 is plotted against partial pressure of O_2. The graph can represent the composition of any gas mixture of physiological significance, as well as any combination of CO_2 and O_2 tensions in blood, plasma, lymph, or other body fluids.

For this centennial history, former presidents were asked to describe their training in science and their experience in their laboratories. Rahn responded to this invitation as follows:

> Throughout my career it has been my special privilege to be associated with mentors and peers of unusual talents and imagination in exploring uncharted areas. It started in 1941 with Wallace Fenn and Arthur Otis in Rochester where we described the first modern version of the pressure-volume diagram of the chest (1) in support of our fighter pilots equipped with positive-pressure breathing masks. These adventures in our primitive, home-made high-altitude chamber eventually led Fenn and me to publish *A Graphical Analysis of the Respiratory Gas Exchange: the O_2-CO_2 Diagram* (3). The O_2-CO_2 diagram was the centerpiece of this book, equivalent to a map on which few roads had been charted, and only one's imagination limited the future paths that would later be explored and charted on this diagram in the area of high-altitude physiology, deep and shallow diving (with Lanphier and Hong), gas bubble resorption (with H. Van Liew), space travel (with L. E. Farhi), and artificial and insect gills (with C. V. Paganelli).
>
> The O_2-CO_2 diagram also provided a map charting the alveolar gas concentration as determined by different ventilation-perfusion ($\dot{V}_A/\dot{Q}$) ratios (2) and, with L. E. Farhi, predicting the alveolar-arterial O_2 differences as a consequence of a logarithmic distribution of $\dot{V}_A/\dot{Q}$ ratios in the lung (4). During the late years in Rochester and early years in Buffalo I had the good fortune to work with P. Dejours, P. Sadoul, J. Knowles, T. Finley, J. Piiper, P. Haab, T. Velasquez, C. Lenfant, E. Agostoni, L. E. Farhi, J. West, and P. Cerretelli, while continuing to chart the various consequences of $\dot{V}_A/\dot{Q}$ distribution on the O_2-CO_2 diagram, and to work with a young medical student, F. Klocke, who demonstrated the existence of a normal arterial-alveolar N_2 difference in humans (5).
>
> When gas exchange limitations of water breathers were mapped for the first time on the O_2-CO_2 diagram (6), it showed why their CO_2 partial pressure (P_{CO_2}) would not exceed 5 Torr. This opened up new adventures in unraveling the evolutionary sequences of gas exchange organs from gills to lungs and, with Kylstra and Lanphier, led to the demonstration that dogs could successfully breathe water (saturated with O_2 at 5 atm) and fully recover. It led also, with B. Howell, to the curious observations

that at various body temperatures lower vertebrates do not regulate their acid-base balance to maintain a constant pH but rather a constant relative alkalinity or OH^-/H^+ ratio (7, 8). The explanation was later provided by R. B. Reeves who showed that all vertebrates, including humans, do not regulate acid-base balance to preserve a constant pH, but rather to regulate a constant protein net charge through the properties of a special protein buffer, imidazole groups of histidine.

With O. D. Wangensteen it was demonstrated that gas exchange of the hen's egg is limited to diffusive transport. In more recent years, with C. V. Paganelli and A. Ar, we have charted this gas exchange on the O_2-CO_2 diagram (9, 10) and gained new insights in the limitations of gas transfer operating under Fick's law and the importance and limitations set by diffusion coefficients. In the egg, O_2 and CO_2 concentrations are determined by diffusion-perfusion ratios instead of ventilation-perfusion ratios, and this may eventually serve as a model for the gas transfer processes at the alveolar gas-capillary interface of the lung.

Much of this work was carried out under unusual circumstances, which in themselves provided new adventures for me. It started out in the high-altitude chamber at Rochester and led to high-altitude expeditions in the Rocky Mountains with A. Otis, S. M. Tenney, R. S. Stroud, and H. Bahnson, and in Peru with T. Velasquez and A. Hurtado. At Buffalo these carried me into the high-pressure chambers with E. Lanphier and to the diving Ama of Korea with S. K. Hong, and in Japan with T. Yokoyama. With W. Garey I went to the Amazon to study acid-base balance in the electric eel, and during the last decade, with C. V. Paganelli and many others, I have explored strange islands in Alaska, Mexico, the Marshall and Midway Islands of the Pacific, and Spitsbergen in the high Arctic to study diffusive gas transport in eggs of various native birds.

The names of many of Rahn's associates and disciples can be found in his complete bibliography, together with titles of all his papers (on file in APS Archives). Something of the range and variety of his investigations, however, is evident from the account given above. Rahn's friends know that it is both high professional competence and unusual breadth of interest that intensify the "soft blue light" of his halo.

BIBLIOGRAPHY

1. Rahn, H., A. B. Otis, L. E. Chadwick, and W. O. Fenn. The pressure-volume diagram of the thorax and lung. *Am. J. Physiol.* 146: 161–178, 1946.
2. Rahn, H. A concept of mean alveolar air and the ventilation-blood flow relationships during pulmonary gas exchange. *Am. J. Physiol.* 158: 21–30, 1949.
3. Rahn, H., and W. O. Fenn. *A Graphical Analysis of the Respiratory Gas Exchange: the O_2-CO_2 Diagram.* Washington, DC: Am. Physiol. Soc., 1955.
4. Farhi, L. E., and H. Rahn. A theoretical analysis of the alveolar-arterial O_2 difference with special reference to the distribution effect. *J. Appl. Physiol.* 7: 699–703, 1955.
5. Klocke, F. J., and H. Rahn. The arterial-alveolar inert gas ("N_2") difference in normal and emphysematous subjects, as indicated by the analysis of urine. *J. Clin. Invest.* 40: 286–294, 1961.
6. Rahn, H. Aquatic gas exchange: theory. *Respir. Physiol.* 1: 1–12, 1966.
7. Rahn, H. Gas transport from the external environment to the cell. In: *Development of the Lung*, edited by A. V. S. de Reuck and R. Porter. London: Churchill, 1966, p. 3–23.
8. Howell, B. J., F. W. Baumgardner, K. Bondi, and H. Rahn. Acid-base balance in cold-blooded vertebrates as a function of body temperature. *Am. J. Physiol.* 218: 600–606, 1970.

9. Rahn, H., and O. Prakash (Editors). *Acid-Base Regulation and Body Temperature.* Dordrecht, Holland: Martinus Nijhoff, 1985.

10. Rahn, H., and C. V. Paganelli. Transport by gas-phase diffusion: lessons learned from the hen's egg. *Clin. Physiol. Oxf.* 5, *Suppl.* 3: 1–7, 1985.

37 (1964–65)

John Richard Pappenheimer
(b. 1915)

With the diamond anniversary of the Society properly celebrated the preceding year, as John Pappenheimer assumed the presidency in 1964 APS was already planning for the XXIV IUPS Congress to be held in Washington, D.C., in 1968. Pappenheimer later (1967–68) served as chairman of the Program Committee of the congress. His first contribution to the success of these meetings, however, was in persuading Council and then the members of APS to agree to a voluntary assessment that eventually provided some $50,000 for preliminary expenses of the IUPS Congress. At the Ninety-fourth Business Meeting of APS on 11 April 1965, members present "voted to assess themselves $10 each a year for three years to provide unrestricted funds for the International Congress." At that time the "regular" dues were fifteen dollars per year. Pappenheimer wrote of the proposal:

Your Council strongly supports this measure for the following reasons. . . .

1. The assessment would strengthen both the hand and the spirit of our National Committee in approaching Government and Industry for further support.

2. The assessment would allow each of us to feel that we are fulfilling our role as hosts to physiologists who will be visiting us—many for the first time—from all parts of the world. For many of us this would be an opportunity to return, in small measure, the hospitality we have enjoyed in other lands.

3. Finally, the assessment would enable us to live up to the precedent and faith of our forebears who contributed personally to their Congress in 1929, at a time when there was no government support whatsoever. A reaffirmation of our faith in the International Congress of Physiology is specially important today as outside sources of support falter under the barrage of requests from those who have come to rely wholly on subsidies.

Born in New York City, John Pappenheimer followed his older sister, Anne (Pappenheimer Forbes), and brother, Alwin Max, Jr., to college in Cambridge,

Massachusetts, where all three subsequently became professors—Anne in medicine at Massachusetts General Hospital and Alwin in biochemistry/biology at Harvard. John received his B.S. degree from Harvard College in 1936. In the summer following his junior year of college he took the physiology course at the Marine Biological Laboratory at Woods Hole, where his instructors included Laurence Irving, C. Ladd Prosser, and J. K. W. (Ken) Ferguson. Pappenheimer wrote:

> Alan Burton (APS president, 1956-57) was a fellow student, and we competed for the *Collecting Net* prize. Ferguson had just published his important work with Roughton on carbamino hemoglobin, and together we investigated carbamino reactions in fish blood. Later, I too went to Cambridge, England, to study with Barcroft, Roughton, and Winton. As a graduate student I worked on the thermodynamic efficiency of urine formation in isolated perfused dog kidneys [with Frank Winton and Grace Eggleton (2)]. As part of this work we developed spectrophotometric methods for oxygen saturation in flowing blood (with Glenn Millikan, then fellow of Trinity College). When the war came in 1939 it was natural for me to join forces with Millikan to work on the ear oximeter and on oxygen equipment for military aircraft under the aegis of Detlev Bronk.

Pappenheimer received his Ph.D. degree from Cambridge University in 1940, after having served for a year as demonstrator in pharmacology at University College, London. In his past president's address Pappenheimer wrote of this stage of his training:

> I was fortunate enough to take the Part II Honors course in physiology at Cambridge, England. This was full-time physiology for a year, and the course consisted of one lecture each day followed by reading, reading and more reading. There was relatively little laboratory work and what there was I have mostly forgotten. But we came to know the classical literature of three languages, leading up to the then frontiers in almost all sectors of physiology. I have been everlastingly grateful for this period of intensive study. It provided a framework for subsequent research and teaching and made possible the enjoyment of continued reading in fields outside one's own narrow research interests.

When he returned to this country Pappenheimer was for two years (1940-42) a research fellow and instructor in physiology in the department of Magnus K. Gregersen at the College of Physicians and Surgeons of Columbia University. He then devoted three years (1942-45) to research with Millikan in Bronk's laboratories in the Johnson Foundation at the University of Pennsylvania, as noted above. He moved to Harvard in 1946. In 1953 he was designated a Career Investigator of the American Heart Association (AHA) and thus became a visiting professor at Harvard. In 1969, however, he was appointed to the George Higginson Chair in Physiology while retaining the award of the AHA.

Most of the research for which Pappenheimer is known has been published from the Harvard laboratory. He described it as follows:

> After the war I was eager to resume prewar studies on edema formation in perfused muscle, and this led me to Landis's new department at Harvard. Ideas and techniques I had absorbed from Winton provided the basis for measurements of capillary pressure, pre- and postcapillary resistances, in vivo protein osmotic pressure, and transcapillary fluid movement in perfused muscle (3). The theory of restricted

diffusion, molecular sieving, and osmotic transients soon followed and with it the understanding that diffusion permeability differs from flow permeability and their ratio is directly related to the dimensions of aqueous channels through the membrane. By 1951 (4) we were able to characterize the passive permeability properties of muscle capillaries in terms of the molecular dimensions of permeant molecules, on the one hand, and the dimensions of aqueous channels through capillary walls, on the other. The general theory survives to this day (5).

This is the research by which Pappenheimer is represented in every modern textbook of physiology, although in many cases the name of the author is no longer stated. It has become "what every schoolboy knows."

More recently, having developed a technique for cerebral ventriculocisternal perfusion using unanesthetized goats, Pappenheimer and his associates studied chemical factors involved in control of breathing (6, 8) and induction of sleep (9, 11). They have not been able to isolate and identify the factor present in cerebrospinal fluid that induces slow-wave sleep, but they have discovered that a muramyl peptide present in brain or urine has a potent sleep-promoting action. Whether it is the inducer of natural sleep remains uncertain. This research led to Pappenheimer's current interest in hypoxic insomnia (10) and in the absorption of muramyl peptides from the small intestine. He hopes to report on one of these topics at the One Hundredth Anniversary Meeting of APS.

Pappenheimer became a member of APS in 1946. In 1956 he was selected to present the Society's first Bowditch Lecture in Rochester, New York. Elected to APS Council in 1961, Pappenheimer served there through his year as past president in 1966. Since 1968 he has served as chairman of the Perkins Memorial Fellowship Committee (see chapt. 21). He was a member of the Editorial Board of *Physiological Reviews* (1959–63), the *Handbook of Physiology* Editorial Committee (1961–67; chairman, 1972–78), and the Editorial Board of the *American Journal of Physiology* (1967–73). He also has been a member of the editorial boards of *Review of Scientific Instruments* (1949–51), *Proceedings of the Society for Experimental Biology and Medicine* (1955–61), and *Circulation Research* (1962–65). He has held office or served in an advisory capacity for the National Research Council (NRC) from 1956 to 1962, the National Heart Institute from 1961 to 1965, the American Institute of Biological Sciences (AIBS) from 1962 to 1966, *Annual Reviews* from 1971 to 1977, and IUPS from 1974 to 1983.

Elected a fellow of the AAAS (Boston) in 1954, Pappenheimer was made a member of NAS in 1965. He was a visiting professor at the Rockefeller University from 1960 to 1963. In 1971 he was given the Carl J. Wiggers Award by the Circulation Group of APS, and in 1979 he received the Ray G. Daggs Award of APS.

The association Pappenheimer began with British physiology when he was a graduate student at Cambridge has continued. He is an honorary member of the British Physiological Society and was an overseas fellow of Churchill College, Cambridge (1971–72), and also Eastman Professor at Oxford (1975–76). Whether his affinity for bicycling developed in the Cambridge of England or of Massachusetts is not known. It served, however, as both an introduction and framework for his past president's address (7). He described his daily "commute" along the Charles River to Boston and related the ride to those forces that seem to be troubling the discipline of physiology, by distributing it among subdisciplines to such a degree that little seems to be left of the parent science. He concluded by noting that his bicycle had gone 15,000 miles and probably would serve for another 15,000. In a more recent communication he reports, "I have already added another 30,000, and I'm projecting an additional 20,000 between now and the turn of the millenium."

One of Pappenheimer's most engaging qualities is the fun he has derived from a career in laboratories and classrooms—as well as from riding his bicycle.

SELECTED PUBLICATIONS

1. PAPPENHEIMER, J. R., M. P. LEPIE, AND J. WYMAN, JR. The surface tension of aqueous solutions of dipolar ions. *J. Am. Chem. Soc.* 58: 1851–1855, 1936.
2. EGGLETON, M. G., J. R. PAPPENHEIMER, AND F. R. WINTON. The influence of diuretics on the osmotic work done and on the efficiency of the isolated kidney of the dog. *J. Physiol. Lond.* 97: 363–382, 1940.
3. PAPPENHEIMER, J. R., AND A. SOTO-RIVERA. Effective osmotic pressure of the plasma proteins and other quantities associated with the capillary circulation in the hindlimbs of cats and dogs. *Am. J. Physiol.* 152: 471–491, 1948.
4. PAPPENHEIMER, J. R., E. M. RENKIN, AND L. M. BORRERO. Filtration, diffusion and molecular sieving through peripheral capillary membranes. *Am. J. Physiol.* 167: 13–46, 1951.
5. LANDIS, E. M., AND J. R. PAPPENHEIMER. Exchange of substances through the capillary walls. In: *Handbook of Physiology. Circulation*, edited by W. F. Hamilton. Washington, DC: Am. Physiol. Soc., 1963, sect. 2, vol. II, chapt. 29, p. 961–1034.
6. PAPPENHEIMER, J. R., V. FENCL, S. R. HEISEY, AND D. HELD. Role of cerebral fluids in control of respiration as studied in unanesthetized goats. *Am. J. Physiol.* 208: 436–450, 1965.
7. PAPPENHEIMER, J. R. Past-president's address. A bicycle in the age of jets. *Physiologist* 8: 341–347, 1965.
8. PAPPENHEIMER, J. R. The ionic composition of cerebral extracellular fluid and its relation to control of breathing. *Harvey Lect.* 61, 1966.
9. FENCL, V., G. KOSKI, AND J. R. PAPPENHEIMER. Factors in cerebrospinal fluid from goats that affect sleep and activity in rats. *J. Physiol. Lond.* 216: 565–589, 1971.
10. PAPPENHEIMER, J. R. Sleep and respiration of rats during hypoxia. *J. Physiol. Lond.* 266: 191–207, 1977.
11. PAPPENHEIMER, J. R. Induction of sleep by muramyl peptides. Bayliss-Starling Memorial Lecture. *J. Physiol. Lond.* 336: 1–11, 1983.

38 (1965–66)

John M. Brookhart

(b. 1913)

Born in Cleveland, Ohio, John Brookhart completed both his undergraduate and his graduate studies at the University of Michigan in Ann Arbor. He had intended to become a physician. In his senior year in college, however, he was introduced to experimental science by Alvalyn Woodward by way of an elective problem course. As a result, he enrolled in Robert Gesell's department as a graduate student in physiology and received his master's degree in 1936 and his Ph.D. degree in 1939. The following year he was a postdoctoral fellow in S. W. Ranson's Institute of Neurology at Northwestern University in Chicago. He then joined successively the faculties of Loyola University School of Medicine (1940–46), University of Illinois College of Medicine (1946–47), and the Department of Physiology at Northwestern University Medical School (1947–49). In 1949 he became an associate professor in the Department of Physiology of the University of Oregon Medical School in Portland, where he served only three years before being appointed professor and chairman of the department (1952–79). This was followed by four years (1979–83) as Acting Vice-President for Academic Affairs, Oregon Health Sciences University in Portland.

Brookhart has identified Robert Gesell and S. W. Ranson as important preceptors in his development as a scientist. He moved from Gesell's laboratory to Ranson's at a time when both of these senior physiologists were interested in neural control of pulmonary ventilation and had come to contrary conclusions on how the central control mechanisms are organized. Before he left Michigan, Brookhart had written a paper on the respiratory effects of faradic stimulation of the medulla oblongata (2). But when he reached Chicago, instead of going on with this research, he was asked to collaborate with Fred Day in the study of possible neural control of reproduction (3). During the war years he temporarily left neurophysiology and participated with others in studies of cardiac functions influenced by respiration (4). Later, in other settings, he returned to studies of the nervous system, where his interest was fostered by association with H. W. Magoun of Northwestern and G. Moruzzi, a neurophysiologist from Pisa, Italy, who collaborated with Magoun. Of his early laboratory experience, Brookhart wrote, "In truth, this business of research training never ceases and is really a continuous process for as long as one continues to be active." Then, after mentioning the names of Gesell, Ranson, Moruzzi, and Magoun, he added:

> Theodore Boyd [his chairman at Loyola] probably shaped me more importantly in a number of ways. We worked together during the war years in a poorly supported, poorly equipped, poorly housed small medical school. The teaching load was exceedingly heavy because of the increase in class size and the paucity of faculty members. Nevertheless I learned from him the value of patient persistence, the rewards of doing things for oneself, and ways of analyzing problems to select the best solution. . . . It was an influence on general attitudes and habits that certainly impacted my life as a scientist.

From the year of his first appointment to a National Institutes of Health (NIH) study section (physiology, 1951–55), Brookhart served on a succession of committees and councils, and as a consultant, to NIH, the National Institute of Neurological Diseases and Blindness, NSF, and the Office of Naval Research. For two terms (1959–62 and 1976–79) he was a member of the Physiology Test Committee of the National Board of Medical Examiners. In many of these appointments he represented neurophysiology or psychobiology, but he also provided a more generalized type of counsel (e.g., on the Advisory Council on Health Research Facilities for NIH, 1967–71). His participation in international scientific organizations is of long standing. In 1965 he was a delegate to the General Assembly of IUPS in Tokyo, Japan, as he was again at the IUPS Congress in 1968 in Washington, D.C., in 1971 in Munich, and in 1974 in New Delhi, India. From 1969 to 1975 he was a member of the U.S. National Committee of IUPS. Then for six years (1974–80) he was treasurer and a member of the Executive Committee of the parent organization, IUPS. After the International Brain Research Organization (IBRO) was formed in 1961, Brookhart was a member of the Central Council (1966–68 and 1974–77).

Among other honors he has received, Brookhart was elected a fellow of AAAS (Boston) in 1967. Ten years earlier he had been a Fulbright Research Scholar at the University of Pisa (1956–57) and had been elected to foreign membership in the Accademia delle Science dell' Instituto di Bologna. In 1974 he was the first recipient of the Ray G. Daggs Award of APS.

Brookhart's contributions to the Society have been in three somewhat different types of office. He was a member of Council (1960–64) and continued there as president elect, president, and past president until 1967. During this time (1961) the Society's constitution was revised to limit the independence of the Board of Publication Trustees and to establish the Council-dependent Finance and Publications Committees.[1] Secondly, Brookhart was a member of the Editorial Board of the *Journal of Neurophysiology* (1960–64) and thus was involved in negotiations for purchase of that journal from Yale University and Charles C Thomas in 1961.[2] For the following ten years (1964–74), a period when its eminence among scientific journals became firmly established, Brookhart was chief editor of the *Journal of Neurophysiology*. He was associated with the *Handbook of Physiology* as a member of the Editorial Board (1967–72). With Vernon B. Mountcastle, he organized the first revision of the *Handbook of Physiology* section on the nervous system. Finally, as chairman of the Finance Committee for APS for six years (1967–73), Brookhart assisted in a reorganization of the Society's business operations, which he described as follows:

> During several years as a member of Council, I learned a lot about the interrelations between the Federation [Federation of American Societies for Experimental Biology, FASEB] and the APS. At that time, all APS financial transactions, except those related to the Publications Contingency and Reserve Fund, were carried out through the FASEB Business Office on a fee-for-service basis. The income to the Society and publications operating accounts was received by the Federation, and disbursements were made by the Federation. Because of the periodicity of dues and subscription collections, sizable amounts would accumulate in these accounts for eventual payout over the year. The Federation Executive Director and the Comptroller looked upon this ... as money controlled by the Federation.... Annual reports were quite unsatisfactory. Any income from the short-term investment of these collections was credited to Federation accounts. Since we were dealing with annual turnovers approximating a million dollars, this income could have been a significant source of revenue for the Society....
>
> With vigorous and persistent effort on the part of Ray Daggs, Council was persuaded to plan during 1963–64 for the creation of its own business office and for discontinuing the use of the Federation's services. This was brought into being during my term as president.... The transition was smooth from the point of view of the APS, [and] I don't know of anyone who thinks now that we erred. The APS Business Office has provided the Society with greatly improved control and accounting, and the cost of the new operation ... was more than made up by income from prudent short-term investments. The important result has been that the APS has had the financial flexibility to engage in a number of innovative activities that would otherwise have been impossible.

As his bibliography indicates, from the time he left the influence of the Ranson laboratory and that of Boyd, Brookhart's interests have been focused on the central nervous system—in particular, on the control of skeletal muscle. He wrote:

> My research interests have centered primarily on ways in which the brain's functions are manifested in motor neuronal output.... The early studies of respiratory control, the venture into hypothalamic influences on mating behavior, the inquiry into some of the characteristics of the corticospinal system, the intracellular study of frog motor neurons, and finally, the last several years of work with postural control mechanisms all fit this basic formula. The various projects were all exciting and fun for different reasons. The one common characteristic is the residue of continuing friendship and affection that remains as a result of the professional collaboration and sharing of ideas and labor during the work.

Friends who served on Council or on major committees of APS while Brookhart was in any of his several offices know how well his devotion and good judgment served the Society and its journals. As a neurophysiologist he is recognized for the progress the *Journal of Neurophysiology* made under his guidance and respected for the care and the imagination he focused on the problem of control of skeletal muscle.

NOTES

[1] FENN, W. O. *History of the American Physiological Society: The Third Quarter Century, 1937–1962.* Washington, DC: Am. Physiol. Soc., 1963, chapt. 4, p. 78.
[2] FENN, p. 76.

SELECTED PUBLICATIONS

1. BROOKHART, J. M., E. H. STEFFENSEN, AND R. GESELL. Stellate ganglia and breathing. *Am. J. Physiol.* 115: 357–363, 1936.

2. BROOKHART, J. M. Respiratory effects of localized faradic stimulation of the medulla oblongata. *Am. J. Physiol.* 129: 709–723, 1940.

3. BROOKHART, J. M., F. L. DAY, AND S. W. RANSON. The abolition of mating behavior by hypothalamic lesions in guinea pigs. *Endocrinology* 28: 561–565, 1941.

4. BROOKHART, J. M., AND T. E. BOYD. Local differences in intrathoracic pressure and their relation to cardiac filling pressure in the dog. *Am. J. Physiol.* 148: 434–444, 1947.

5. BROOKHART, J. M., G. MORUZZI, AND R. S. SNIDER. Spike discharges of single units in the cerebellar cortex. *J. Neurophysiol.* 13: 465–486, 1950.

6. ZANCHETTI, A., AND J. M. BROOKHART. Measurement of electrical responsiveness of cortico-spinal efferents in cat and monkey. *J. Neurophysiol.* 18: 288–298, 1955.

7. BROOKHART, J. M., A. ARDUINI, M. MANCIA, AND G. MORUZZI. Thalamocortical relations as revealed by induced slow potential changes. *J. Neurophysiol.* 21: 499–525, 1958.

8. BROOKHART, J. M., AND E. FADIGA. Potential fields initiated during monosynaptic activation of frog motor neurons. *J. Physiol. Lond.* 150: 633–655, 1960.

9. KUBOTA, K., AND J. M. BROOKHART. Recurrent facilitation of frog motor neurons. *J. Neurophysiol.* 26: 877–893, 1963.

10. BROOKHART, J. M., S. MORI, AND P. J. REYNOLDS. Postural reactions to two directions of displacement in dogs. *Am. J. Physiol.* 218: 719–725, 1970.

11. BROOKHART, J. M., AND R. E. TALBOTT. The postural response of normal dogs to sinusoidal displacement. *J. Physiol. Lond.* 243: 287–307, 1974.

12. MIRKA, A., AND J. M. BROOKHART. Role of primary visual cortex in canine postural control. *J. Neurophysiol.* 46: 987–1003, 1981.

39 (1966–67)

Robert Elder Forster II
(b. 1919)

In planning for the XXIV IUPS Congress in Washington, D.C. (1968), Wallace Fenn chose Robert Forster to be chairman of the Finance Committee. The success of this committee has become legendary. With the help of what Forster called "the tremendous efforts of Wallace Fenn and of K. K. Chen," the committee raised enough money to leave the congress with a $65,000 surplus. (Friends have wondered whether Forster might have been a consultant to the U.S. Olympic Committee for the games in Los Angeles in 1984.) This surplus was turned over to APS as a trust fund, with income of the fund devoted to support of travel grants to future congresses.

Forster was born in St. David's, Pennsylvania, and has lived in that neighborhood for most of his life. He graduated from Radnor High School in Wayne in 1937 and with his family took up residence again in that area (Haverford) some twenty years ago. Much of his education and training, however, was completed in New England. From high school he enrolled in the Sheffield Scientific School of Yale University in biological sciences. At this time one could leave Yale in three years and get credit for the first year of medical school as the last year of college, so he entered the University of Pennsylvania School of Medicine in 1940. World War II accelerated medical school training, and he was able to graduate in December 1943 to take an internship in medicine at the Peter Bent Brigham Hospital in Boston. He had only an abbreviated period as a house officer, and in October 1944, having been discharged from the U.S. Navy as physically unfit, he entered the Army. After basic training at Carlisle, Pennsylvania, he was assigned to the Quartermaster Corps Climatic Research Laboratory in Lawrence, Massachusetts. This appointment was available because of the ill fortune of Clifford Barger, who had been at the laboratory but had contracted tuberculosis. At the Quartermaster Corps Laboratory, Forster did research on temperature regulation and heat exchange in the course of testing and helping design new field clothing and equipment for military personnel.

Forster had been interested in doing research as a medical student. Stimulated by his teachers, Julius Comroe and Carl Schmidt, he had measured changes in blood plasma viscosity with thiocyanate treatment for hypertension and had published his first article in the "yellow journal" of Lea and Febiger, the *American Journal of the*

Medical Sciences (1). During several summer terms he worked at the Alfred I. DuPont Institute in Wilmington, which was then affiliated with the University of Pennsylvania School of Medicine, in biochemical research. There he first learned to operate a Van Slyke manometric gas analyzer under the tutelage of Douglas Mac-Fadyen and Murray Angevine, who had been colleagues of D. D. Van Slyke at the Rockefeller Institute. Forster has described his subsequent training and tour of duty in the Quartermaster Corps Laboratory as follows:

> The laboratory was led by Colonel John H. Talbot, who had been on the high-altitude Andes expedition of 1935. Many distinguished individuals interested in the response of humans to extreme environmental conditions passed through the laboratory. Sir Hubert Wilkins, who had sailed in a submarine under the North Pole, acted as a consultant. He had the delightful trait of simply turning off his hearing aid when he thought the argument was going against him. Paul Siple, the Boy Scout who had been at Antarctica with Admiral Richard E. Byrd, also gave frequent advice. Cuthbert Bazett, chairman of the Department of Physiology at the University of Pennsylvania, came often for consultations.

> My time in the laboratory included a somewhat amusing experience. I had put together an electrical impedance plethysmograph, which measured the impedance of a segment of a finger at radio frequencies (and compared simultaneous changes in volume from a volume plethysmograph). The impedance plethysmograph was championed by Jan Nyboer, but unfortunately the instrument really does not measure changes in finger volume. I wrote to Alan Burton about the instrument and enclosed records of the impedance pulses. Burton replied that the impedance plethysmograph was of little account and that he could record better volume pulses from the projected shadow of a straw set in Plasticine on his toe. Bazett, when he was told of this put-down response, said that I should pay little attention; "Alan does not know much physiology and what he does know I taught him."

> Richard L. Day, a pediatrician, had been at the laboratory and had developed a plethysmocalorimeter, a closed box which could act as a venous occlusion plethysmograph and whose walls acted as gradient calorimeter. (We have a painting of it in our office.) This work resulted in a number of publications on the relation of body temperature to hand and foot blood flow, but the most important was the publication of Bazett, Love, Newton, Eisenberg, Day, and Forster, in which countercurrent cooling of the arterial blood to the hand was demonstrated. (I believe this was the first experimental demonstration of this mechanism.) This appeared as the first article of the first volume of the *Journal of Applied Physiology* (2).

Near the end of 1946 Forster was discharged from the Army and spent the next year as a graduate student in mathematics; he took courses in physics, mathematics, and physical chemistry in the graduate school at Harvard. When he returned in 1947 as a resident in internal medicine at the Peter Bent Brigham Hospital, he was permitted to measure the pulmonary arterial blood temperature with a wedge catheter in a patient of Lewis Dexter. He also borrowed an ear oximeter from Glenn Millikan for studies on the oxygenation of some of Samuel A. Levine's patients.

On a Life Insurance Medical Research Fund Fellowship, Forster then spent two years in the Department of Physiology at Harvard Medical School under Eugene M. Landis. He first worked on a thermostromuhr, a method for the measurement of intestinal blood flow. ("This really did not work very well.") He then moved on to measure changes in hypothalamic temperature in a chronic unanesthetized cat during changes in environmental temperature conditions. Because the hypothalamic temperature had to change nearly 0.5°C before the peripheral heat flow effectors were triggered, he found it to be a rather insensitive thermostat (3).

At the end of his posdoctoral fellowship Forster sought an academic post and visited several departments of medicine and physiology. (He was interviewed at Johns Hopkins Hospital under the impression he was applying for entrance to medical school.) He was offered, and he accepted, a position as assistant professor of physiology in anesthesiology under Robert Dripps and Julius Comroe in the Department of Physiology and Pharmacology of the University of Pennsylvania Graduate School of Medicine. Forster described his years in this department in these words:

> This was a very lucky choice, as I arrived at an exciting period in the short history of this department. Julius Comroe was rising to the peak of his reputation as a pulmonary physiologist and investigator of pulmonary function testing. The faculty included not only Comroe but Seymour Kety and later George Koelle, Ward S. Fowler, and Arthur DuBois. Kety and Comroe had already on order a respiratory mass spectrometer, to facilitate rapid measurements of inert respiratory gases, and an infrared meter, to measure low concentrations of carbon monoxide, which I was able to use to study the diffusing capacity of the lung. In the next few years a good deal of work was accomplished on diffusion exchanges in the lungs, including a description of the single-breath method for the measurement of the diffusing capacity of the lung, a widely used pulmonary function test (4). In collaboration with F. J. W. Roughton (5) I studied the theory and practice of measurements of pulmonary capillary blood volume (V_C) and the diffusing capacity of the pulmonary membrane (D_M). [During this period Forster was awarded a Lowell M. Palmer Fellowship (1954–56).]
>
> Roughton introduced me also to rapid-mixing techniques for observing kinetic processes (6), and over the next several decades I was able to develop a number of different types of rapid-mixing apparatus to measure physiologically important chemical reactions and rapid exchanges of red cells, at first CO, O_2, and CO_2 but later anions and water. These studies have formed the foundation of my research career. In 1977 Itada and I introduced a method for the measurement of carbonic anhydrase inside intact red cells with ^{18}O exchange (9), the only method so far reported for this purpose, because any other technique stops before significant measurements can be made because of accumulation of end products within the cell membrane. An offshoot of this interest in CO was work with Coburn and other colleagues that showed that a molecule of CO is produced for each heme group that is catabolized, at a rate of 0.4 ml/h in humans; this provides an index of the measurement of destruction of hemoglobin and other heme proteins (7).

During this period Julius Comroe organized several courses to teach the growing body of knowledge on respiratory physiology and pulmonary function testing to physiologists from other institutions. Based on this experience, Comroe, Forster, DuBois, Briscoe, and Carlsen produced *The Lung*, which is just entering its third edition and has been a medical best-seller.

In 1957 Julius Comroe left the Department of Physiology and Pharmacology in the Graduate School of Medicine of the University of Pennsylvania to set up the new

Cardiovascular Research Institute at the University of California School of Medicine at San Francisco. George Koelle became chairman of the Department of Physiology and Pharmacology but in several years took over the chairmanship of the Department of Pharmacology in the School of Medicine to succeed Carl Schmidt. At this point Forster became chairman of the physiology part of the Graduate School of Medicine, and there continued to be two departments of physiology, one in the Graduate School of Medicine and one in the School of Medicine, the latter under John Brobeck. In 1970 Brobeck resigned in favor of Forster, and the existing departments were fused.

Forster attended his first spring meeting of the APS in 1946. He served on the Publications Committee from 1963 to 1965, was elected to Council in 1963, and was president elect in 1965. He was on the Editorial Board of the *Handbook of Physiology* (1973–79) and was a member of the Perkins Memorial Fund Committee (1968–70), the Daggs Award Committee (1975–78), and the Finance Committee (1978–83; chairman, 1982–83). He was also on the Editorial Board of the *Journal of Clinical Investigation* (1962–67). He served on the Advisory Board of the Life Insurance Medical Research Fund from 1967 until 1970, when it was unfortunately dissolved. He served in several advisory capacities to NIH: the Cardiovascular Study Section (1960–64), the General Clinical Research Center Committee (1964–67), and the National Advisory Heart Council (1967–71). He was on the Editorial Board of *Annual Review of Physiology* (1984–), and he was a member of the U.S. National Committee of IUPS (1976–83).

Many colleagues and associates from abroad have visited in Forster's laboratory, and he has returned the visits on many occasions. Although he has never taken a sabbatical year's leave, he did spend a long vacation at Trinity College with F. J. W. Roughton in 1954. Forster has long been interested in the application of physiology to underwater biology and to high-altitude and space conditions. In 1966 he was chairman of the NAS-NRC Space Science Board Summer Study Group on Respiratory Physiology that concluded that an atmosphere of pure O_2 was dangerous and a fire hazard in spacecraft. The report of this group appeared almost simultaneously with the tragic fire on the launching pad at Cape Canaveral, where several astronauts lost their lives. Forster wrote that he consequently experienced a "transient notoriety."

One of the more notable events of Forster's association with APS was the fall meeting in 1976, the Bicentennial Year of the Declaration of Independence, held in Philadelphia and principally at the University of Pennsylvania. Brobeck happened to be the one who extended the invitation, but Forster was both officially and unofficially the host. The meeting was judged to be a grand success. Especially gratifying to Forster was the fact that it was "the first Fall Meeting in many years that did not require any subsidization by the American Physiological Society."

SELECTED PUBLICATIONS

1. FORSTER, R. E. The medical use of thiocyanates in the treatment of arterial hypertension. *Am. J. Med. Sci.* 206: 668–686, 1943.

2. BAZETT, H. C., L. LOVE, M. NEWTON, L. EISENBERG, R. DAY, AND R. E. FORSTER. Temperature changes in blood flowing in arteries and veins in man. *J. Appl. Physiol.* 1: 3–19, 1948.

3. FORSTER, R. E., AND T. B. FERGUSON. The relationship between hypothalamic temperature and thermoregulatory effectors in unanesthetized cat. *Am. J. Physiol.* 169: 255–269, 1952.

4. OGILVIE, C. M., R. E. FORSTER, W. S. BLAKEMORE, AND J. W. MORTON. A standardized breath holding technique for the clinical measurement of the diffusing capacity of the lung for carbon monoxide. *J. Clin. Invest.* 36: 1–17, 1957. (Abstr. *Federation Proc.* 14: 108, 1955 and *J. Clin. Invest.* 34: 917, 1955.)

5. ROUGHTON, F. J. W., AND R. E. FORSTER. Relative importance of diffusion and chemical reaction rates in determining rate of exchange of gases in the human lung, with special reference to true diffusing capacity of pulmonary membrane and volume of blood in the lung capillaries. *J. Appl. Physiol.* 11: 290–302, 1957. (Abstr. *Am. J. Physiol.* 183: 615–616, 1955.)

6. ROUGHTON, F. J. W., R. E. FORSTER, AND L. CANDER. Rate at which carbon monoxide replaces oxygen from combination with human hemoglobin in solution and in the red cell. *J. Appl. Physiol.* 11: 269–276, 1957.

7. COBURN, R. F., W. J. WILLIAMS, AND R. E. FORSTER. Effect of erythrocyte destruction of carbon monoxide production in man. *J. Clin. Invest.* 43: 1098–1103, 1964.

8. CONSTANTINE, H. P., M. R. CRAW, AND R. E. FORSTER. Rate of the reaction of carbon dioxide with human red blood cells *Am. J. Physiol.* 208: 801–811, 1965.

9. ITADA, N., AND R. E. FORSTER. Carbonic anhydrase activity in intact red blood cells measured with ^{18}O exchange. *J. Biol. Chem.* 252: 3881–3890, 1977.

10. DODGSON, S. J., R. E. FORSTER II, B. T. STOREY, AND L. MELA. Mitochondrial carbonic anhydrase. *Proc. Natl. Acad. Sci. USA* 77: 5562–5566, 1980.

11. DODGSON, S. J., R. E. FORSTER II, D. E. SCHWED, AND B. T. STOREY. Contribution of matrix carbonic anhydrase to citrulline synthesis in isolated guinea pig liver mitochondria. *J. Biol. Chem.* 258: 7696–7701, 1983.

40 (1967–68)

Robert W. Berliner
(b. 1915)

Berliner's term as president of APS was unusual—in a sense, unique. Most of that year's activities concerned preparation for the XXIV IUPS Congress to be held in Washington, D.C., 25–31 August 1968, with Wallace O. Fenn as president and chief organizer of the congress. Moreover there was no fall meeting following the congress. Consequently, Berliner was the only president of those serving since 1947 who was not called on to deliver a past president's address. He wrote that his year in office was uneventful, which can be interpreted to mean that many members of the Society, especially those located near the nation's capital, were preoccupied with planning for the congress and that the more mundane problems of APS were therefore given only routine consideration. Nevertheless it was while Berliner was president that the John Forbes Perkins, Jr., Memorial Fund in support of international exchanges in physiology was established, as described elsewhere in this volume.

From his birthplace in New York City, Berliner went to Yale for his undergraduate education (B.S., 1936). Three years later he was awarded his medical degree by the College of Physicians and Surgeons of Columbia University (M.D., 1939). After an internship of two years at Presbyterian Hospital in New York, he completed his residency/fellowship training on the Third Division Research Service of Goldwater Memorial Hospital, also in New York (1942–47). For most of this interval (1943–47) he held also a junior appointment on the faculty of the New York University College of Medicine. This was followed by three years (1947–50) as assistant professor of medicine at Columbia.

In 1950 Berliner began an association with NIH that lasted for twenty-three years. He was at first (1950–62) the Chief of the Laboratory of Kidney and Electrolyte Metabolism of the National Heart Institute, where he soon became also Director of Intramural Research (1954–68). He left the National Heart Institute to become for one year (1968–69) the Director of Laboratories and Clinics for NIH and then served for four years (1969–73) as Deputy Director for Science of NIH. During much of this time (1951–73) he was concurrently a member of the faculty of the School of Medicine of George Washington University, eventually with the rank of professorial lecturer (1964–73). He resigned from this position and from NIH in 1973 to become dean and professor of physiology and of medicine at the School of Medicine of Yale University. In 1984 he retired from the deanship but continues as a member of the faculties in physiology and in medicine.

Berliner was appointed to the Publications Committee of APS in 1961, where he served for five years (1961–66; chairman, 1963–66) and was reappointed for another term (1976–79). He joined the Council of APS in 1965 as president elect and continued there through his presidential years. He received the Ray G. Daggs Award in 1982. At various times he served as president of a variety of other scientific societies—the American Society for Clinical Investigation (1959), the American Society for Nephrology (1968), and the Society for Experimental Biology and Medicine (1978–81). In 1972 he was elected vice-president of the American Association for the Advancement of Science (AAAS). Other societies in which he holds membership include the Society of General Physiologists, the Harvey Society, the Philosophical Society of Washington, and the Washington Academies of Medicine and of Sciences.

In 1958 Berliner was chosen for membership in the Association of American Physicians, in 1961 in AAAS (Boston), in 1968 in NAS, and in 1971 in the Institute of Medicine. He was chairman of the Division of Medical Sciences of NRC (1976–78), and for three years (1978–81) he was a member of the Council of NAS.

Most of the institutions and societies with which Berliner has been affiliated have awarded him their highest honors. From the Department of Health, Education and Welfare (HEW) he received the Distinguished Service Award (1962). Three years later he was given the Homer W. Smith Award in Renal Physiology (1965). His medical school selected him for the Alumni Award for Distinguished Achievement in 1966, for the Bicentennial Medal for Achievements in Internal Medicine in 1967, and for the Joseph Mather Smith Prize in 1978. He had been honored by the Distinguished Achievement Award of *Modern Medicine* in 1969. This was followed by the AHA Research Achievement Award in 1970, the A. Ross McIntyre Award from the University of Nebraska Medical Center in 1974, the Service Award of the Association of Chairmen of Departments of Physiology (ACDP) in 1981, the David M. Hume Memorial Award of the National Kidney Foundation in 1983, and the George M. Kober Medal of the Association of American Physicians in 1984. He has received honorary degrees of doctor of science from the Medical College of Wisconsin in Milwaukee and from Yale University, both in 1973.

Berliner's early research during World War II was concerned with antimalarial drugs (2). His lifelong specialty, however, has been the kidney, fluid, and electrolyte exchange. This interest was stimulated mainly by Robert F. Loeb while Berliner was a medical student and intern in Presbyterian Hospital, and later by James A. Shannon at Goldwater Memorial Hospital. "Bob" Loeb, son of Jacques Loeb, the prominent general physiologist, was Lambert Professor of Medicine at Columbia and associate attending physician at Presbyterian Hospital until Berliner joined the staff as assistant professor (1947). At that time Loeb became Director of the Medical Service and Bard Professor of Medicine. He was one of the leading clinical investigators of his generation and one of the most influential medical educators.

"Jim" Shannon became famous for his directorship of NIH. From 1942 to 1946, however, he was director of the Research Service at Goldwater Memorial Hospital where Berliner completed his formal training (1942–47). Shannon left New York to become Director of Research for the Squibb pharmaceutical enterprise and subsequently became Associate Director in Charge of Research at the National Heart Institute (1949–52). He then moved into administration at NIH central offices, where he became director in 1955. He is given credit for establishing many of the principles and guidelines that have brought NIH to its preeminent position in biomedical research. Shannon's interest in the kidney began in association with Homer Smith at the New York University College of Medicine early in the 1930s. In Smith's 1951 monograph, *The Kidney* (Oxford Univ. Press, New York), twenty-six items in the bibliography begin with Shannon's name.

Berliner has written of his research experience in the following account:

> My first major, independent work was probably my most important discovery—potassium secretion by the renal tubules (4). The discovery was not accidental. We looked for indisputable evidence of secretion because of peculiarities of potassium excretion when the amount in the urine was far too low to indicate secretion by the usual criterion (i.e., excretion in excess of filtration). . . . In later years we demonstrated among other things that potassium is secreted in exchange for sodium. . . . A 1951 paper (5) pointed out the relationship between urine acidification and potassium excretion. Although the explanation was not entirely correct, it provided a useful working hypothesis. . . . The next paper (6) helped to define the relationship between urine pH and the excretion of weak acids and bases—so-called nonionic diffusion. . . . Two of the papers cited (7, 8) showed that vasopressin is not required

for production of hypertonic urine, if volume presented to the mechanism is small enough, and introduced a new hypothesis that although in error in not allowing for countercurrent multiplier character of the nephron loop, was first to point out that it is addition of salt to medullary interstitium that drives the urine concentration process and not water uptake by the loop. . . . On a somewhat different topic, Kety's inert gas method for measuring local blood flow was modified for continuous measurement (9). Application to the renal medulla showed high efficiency of the countercurrent exchange by the vasa recta (10). . . . The 1967 paper (11) was the first demonstration of the predicted difference in osmolality and salt concentration between thin limbs of the loop of Henle. . . . The last paper (12) was the first demonstration of the predicted difference between ascending and descending limbs of the loop of Henle in their permeability to water. Vasopressin acted only on collecting ducts.

Berliner's career as a physiologist has two remarkable characteristics. First, he was among the handful of investigators who came to correct conclusions about renal function before the discovery of the countercurrent concentrating mechanism and before the renaissance of micropuncture techniques. Yet when these developments occurred he utilized both to extend further the understanding of, especially, tubular mechanisms. Second, he continued productive laboratory research while simultaneously creating a distinguished record as an administrator of medical and educational institutions. He has demonstrated the efficiency and the efficacy of a relatively low-key, collegial approach to administrative responsibility.

SELECTED PUBLICATIONS

1. PERERA, G. A., AND R. W. BERLINER. Relation of postural hemodilution to paroxysmal dyspnea. *J. Clin. Invest.* 22: 25–28, 1943.

2. SHANNON, J. A., D. P. EARLE, JR., B. B. BRODIE, J. V. TAGGART, AND R. W. BERLINER. The pharmacological basis for the rational use of atabrine. *J. Pharmacol. Exp. Ther.* 81: 307–330, 1944.

3. EARLE, D. P., JR., AND R. W. BERLINER. A simplified clinical procedure for measurement of glomerular filtration rate and renal plasma flow. *Proc. Soc. Exp. Biol. Med.* 62: 262–264, 1946.

4. BERLINER, R. W., AND T. J. KENNEDY, JR. Renal tubular secretion of potassium in the normal dog. *Proc. Soc. Exp. Biol. Med.* 67: 542–545, 1948.

5. BERLINER, R. W., T. J. KENNEDY, JR., AND J. ORLOFF. The relationship between acidification of the urine and potassium metabolism: effect of carbonic anhydrase inhibition on potassium excretion. *Am. J. Med.* 11: 274–282, 1951.

6. ORLOFF, J., AND R. W. BERLINER. The mechanism of the excretion of ammonia in the dog. *J. Clin. Invest.* 35: 223–235, 1956.

7. BERLINER, R. W., AND D. G. DAVIDSON. Production of hypertonic urine in the absence of pituitary antidiuretic hormone. *J. Clin. Invest.* 36: 1416–1427, 1957.

8. BERLINER, R. W., N. G. LEVINSKY, D. G. DAVIDSON, AND M. EDEN. Dilution and concentration of the urine and the action of antidiuretic hormone. *Am. J. Med.* 24: 730–744, 1958.

9. AUKLAND, K., B. F. BOWER, AND R. W. BERLINER. Measurement of local blood flow with hydrogen gas. *Circ. Res.* 14: 164–187, 1964.

10. AUKLAND, K., AND R. W. BERLINER. Renal medullary counter-current system studied with hydrogen gas. *Circ. Res.* 15: 430–442, 1964.

11. JAMISON, R. L., C. M. BENNETT, AND R. W. BERLINER. Countercurrent multiplication by the thin loops of Henle. *Am. J. Physiol.* 212: 357–366, 1967.

12. MORGAN, T., AND R. W. BERLINER. Permeability of the long loop of Henle, vasa recta, and collecting duct to water, urea, and sodium. *Am. J. Physiol.* 215: 108–115, 1968.

13. BENNETT, C. M., B. M. BRENNER, AND R. W. BERLINER. Micropuncture study of nephron function in the rhesus monkey. *J. Clin. Invest.* 47: 203–216, 1968.

14. FALCHUK, K. H., B. M. BRENNER, M. TADOKORO, AND R. W. BERLINER. Oncotic and hydrostatic pressures in peritubular capillaries and fluid reabsorption by proximal tubule. *Am. J. Physiol.* 220: 1427–1433, 1971.

41 (1968–69)

Loren D. Carlson
(1915–72)

The APS Fall Meeting in Davis, California, in August 1969 was a delightful occasion, especially for participants from other parts of the country who envy colleagues able to live and work in California. Loren Carlson was unofficial host for all guests and provided his usual efficient management of all that took place. He also delivered the past president's address (10). Beginning with the observation that the souvenir volume for the IUPS Congress of the preceding year had been Cannon's *The Way of an Investigator* (1945), Carlson reviewed how science, and especially physiology, had changed in the past twenty-five years. In an editor's note, Ray Daggs explained that the address "was accompanied by an ever-changing series of background slides which served as a backdrop for points being made by the speaker . . . both color photographs and cartoons." The address concluded with this summary:

> The way of an investigator has changed.
>
> We have passed through an era of "blessedness" when support of research was unquestioned as providing a source of betterment and progress.
>
> We are faced with fragmentation of disciplines to subdisciplines.
>
> We are increasingly aware that we can no longer live in an ivory tower and insulate ourselves from political, economic and social pressures.
>
> We must accept the inevitability of change in university structure from an aristocratic one to a democratic form involving the student and faculty community in its decisions.
>
> Somehow, we must change the connotation of the conjunction between teaching and research to teaching with research so that the public and government recognize their inseparable nature in the university.

Because it came near the end of the student unrest that marked the latter half of the 1960s, Carlson's message was particularly appreciated.

Loren Carlson was born in Davenport, Iowa, and graduated from St. Ambrose

College there in 1937 with a degree in biology. Four years later (1941) he was awarded the Ph.D. degree in zoology by the University of Iowa in Iowa City, where he continued for a year as research associate in physiology. He then joined the professional staff at the Aeromedical Laboratory at Wright Field, Ohio, where he served from 1942 until 1946 as first lieutenant, captain, and then major in the U.S. Army Air Corps. With the end of the war he moved to the University of Washington, at first in zoology, and then from 1946 in the Department of Physiology of the School of Medicine. From 1955 to 1960 he held the rank of professor. During this interval he also served the university in various administrative posts, including Director of General Education (1949–51). In 1960 he began a six-year association with the College of Medicine of the University of Kentucky as the first professor and the founding chairman of the Department of Physiology and Biophysics. For a year (1965–66) he was simultaneously chairman of the Department of Zoology.

Carlson's career in science concluded at the University of California at Davis, where from 1966 until his death in 1972 he was chairman of the Division of Sciences Basic to Medicine. Once again he was also the first professor and the founding chairman of the department, this time of human physiology. Being recruited for this faculty early, he had major roles in development of the school and its curriculum. From 1966 to 1969 he served as assistant dean and then became Associate Dean for Research Development and Curricular Affairs. The appointments he held exhibit not only Carlson's commitment to the sciences of zoology and physiology but also his talent for administrative responsibility. Worthy of note is how, although he almost always held some administrative post, he continued virtually full time as a teacher and laboratory investigator.

Honors came to Carlson from many sources. The Alumni Association of St. Ambrose College gave him its Award of Merit in 1967, and in 1969 he received the degree doctor of philosophy honoris causa from the University of Oslo. In that same year he was elected to fellowship in AAAS (Boston). Although he was not a medical graduate, he was elected to charter membership in Alpha Omega Alpha, the national medical honor society, in 1972, when the Eta Chapter was established at Davis. Thirteen years later (1985) the university's biology and medical library was renamed the Loren D. Carlson Health Science Library.

Carlson's military service at Wright Field initiated a continuing interest and participation in national scientific affairs, especially in aeromedical problems and space flight. For his contributions he was listed in *Who's Who in Space* and in *Engineers of Distinction, Including Scientists in Related Fields.* The U.S. Army Air Corps (later the U.S. Air Force) awarded him the Legion of Merit (1946), the Exceptional Civilian Service Medal (1962), and the Outstanding Achievement Award (1970). He was an associate fellow of the American Institute of Aeronautics and Astronautics, a fellow of the Aerospace Medical Association, and a member of the International Academy of Astronautics. For five years (1957–62) he was chairman of the Aeromedical and Biosciences Panel of the Scientific Advisory Board of the U.S. Air Force, while for six years (1961–67) he was a member of two space technology subcommittees of the President's Scientific Advisory Committee. He also served from 1962 in a variety of positions, advisory or consultant, to the National Aeronautics and Space Administration (NASA), as well as on several boards and committees of NAS (from 1964). Other related appointments involved the Office of Naval Research (1965–68) and the Aerospace Medical Association (1967–72). If this listing of responsibilities seems to mark a career given almost exclusively to public service, such an impression is mistaken. Through all these obligations Carlson preserved his principal base of activity in his university positions and in his laboratory.

He was a member and/or fellow of many other scientific societies. For example, he was a charter member of the Biomedical Engineering Society and served on its Board of Directors (1970–72), a member of the Board of Trustees of *Biological Abstracts* (1971–72), and a fellow of AAAS. He was approved for membership in APS in 1945.

Appointed to the Membership Committee, Carlson began his official service for APS from 1962 to 1965, simultaneously with becoming section editor for environmental physiology of the *American Journal of Physiology* (1962–66). Elected to Council in 1964, three years later he became president elect, in time to serve as president of the Society during the XXIV IUPS Congress in Washington, D.C., and to serve for a year (1969–70) as president of FASEB. He joined the Editorial Committee of the *Handbook of Physiology* in 1967, and briefly in 1971 he represented the Society on the Council of Academic Societies of the Association of American Medical Colleges (AAMC).

Late in his career Carlson summed up his research interests:

a series of investigations directed toward a description and understanding of mechanisms involved in adaptation to temperature. Currently, these are directed toward measurements of changes in the response of the peripheral circulation following chronic cold exposure.

A second program is directed toward understanding of mechanisms involved in the change in the cardiovascular response to tilting or lower body negative pressure following the hypodynamic state, or weightlessness.

Between 1939 and 1941, Carlson published six papers in collaboration with Joseph Hall Bodine, professor of zoology and head of the department at the University of Iowa. The papers carried the running title "Enzymes in ontogenesis (Orthoptera)" and were concerned mainly with activators of protyrosinase (1). One additional study completed before the war analyzed the effect of hydrogen peroxide on frog skin (2). During his years at the Wright-Patterson Air Force Base, Carlson investigated problems of adequate oxygen supply for air crews, and thence for commercial aviation (3). From this he moved to problems of acclimatization (4) that began with a project on cold exposure funded by the Air Force during the winter of 1948–49. In 1954, with W. Cottle, he published the first of a series of papers describing adaptive changes in rats exposed to cold (5). This interest continued until the end of his laboratory career; in one of the last titles in his bibliography, on the calorigenic effect of norepinephrine in newborn rats (12), from

1957 to 1972 he collaborated with Emery and with A. C. L. Hsieh, with whom he had worked since 1957.

With a variety of coauthors, Carlson's complete bibliography lists well over a hundred titles. They include numerous review articles he was invited to prepare on, for example, equipment for providing oxygen, acclimatization to cold, respiratory exchanges, and human performance under stressful conditions. Many are summaries commissioned by the several federal agencies asking about human performance in cold climates. The list also includes papers on combined effects of ionizing radiation and heat or cold on white rats (6). In the early 1960s he published "Requirements for monitoring physiological function in space flight" (7) and "Necessity for biological experimentation in space" (8). Also published in 1963 was the paper that began Carlson's analysis of the relationship between skin temperature and blood flow in a rabbit ear (9). In a sense his scientific work was brought to a climax by publication with Hsieh of the small volume, *Control of Energy Exchange* (11). In 106 pages it provides both mathematical and block diagram models of energy balance, metabolism, energy expenditure, thermophysiology, and regulation of body temperature. Perhaps its most useful features, however, are appendices that serve as a handbook of reference for anyone interested in these subjects. The last ten pages of the appendix include models of energy exchange as they have been conceived by various investigators.

His past president's address, mentioned above, concluded as follows:

> I can add a happy and optimistic note in this discussion. It is the role of friends at home and abroad. Being a biologist has brought to me the benefits and pleasures of friendships in the United States and in many countries. At home, here in Davis, this has been particularly true for me. As a physiologist and as your president, I am grateful to those of you whom I know for your friendship and, to those of you who attend the business meetings, for your confidence in electing me your president. I accept with some sadness the edict that nothing gets paster faster than a past president.

Perhaps even more clearly than publications, achievements, positions, and honors, these lines disclose the character of Loren Carlson as his friends knew him.

SELECTED PUBLICATIONS

1. BODINE, J. H., AND L. D. CARLSON. Enzymes in ontogenesis (Orthoptera). XIV. The action of proteins on certain activators of protyrosinase. *J. Gen. Physiol.* 24: 423–432, 1941.

2. MARSH, G., AND L. D. CARLSON. The effect of hydrogen peroxide on the rate of oxygen consumption of frog skin. *J. Cell. Comp. Physiol.* 22: 99–114, 1943.

3. CARLSON, L. D., W. R. LOVELACE II, AND H. L. BURNS. Requirements for oxygen in commercial aviation. Some aspects of its use. *J. Aviat. Med.* 19: 399–413, 1948.

4. CARLSON, L. D., H. L. BURNS, T. H. HOLMES, AND P. P. WEBB. Adaptive changes during exposure to cold. *J. Appl. Physiol.* 5: 672–676, 1953.

5. COTTLE, W., AND L. D. CARLSON. Adaptive changes in rats exposed to cold. Caloric exchange. *Am. J. Physiol.* 178: 305–308, 1954.

6. CARLSON, L. D., AND B. H. JACKSON. The combined effects of ionizing radiation and high temperature on the longevity of the Sprague-Dawley rat. *Radiat. Res.* 11: 509–519, 1959.

7. CARLSON, L. D. Requirements for monitoring physiological function in space flight. *Astronautik* 2: 310–321, 1960.

8. CARLSON, L. D. The necessity for biological experimentation in space. *Adv. Astronaut. Sci.* 17: 1–20, 1963.

9. HONDA, N., L. D. CARLSON, AND W. V. JUDY. Skin temperature and blood flow in the rabbit ear. *Am. J. Physiol.* 204: 615–618, 1963.

10. CARLSON, L. D. The way of an investigator–reanalyzed. *Physiologist* 12: 425–432, 1969.

11. CARLSON, L. D., AND A. C. L. HSIEH. *Control of Energy Exchange.* London: Macmillan, 1970.

12. HSIEH, A. C. L., N. EMERY, AND L. D. CARLSON. Calorigenic effect of norepinephrine in newborn rats. *Am. J. Physiol.* 221: 1568–1571, 1971.

42 (1969–70)

C. Ladd Prosser
(b. 1907)

Writing of his service on Council and in presidential offices of APS, Prosser recalled many rewarding experiences—association with high-quality physiologists, Washington contacts, and successes of both the publication and the education programs. At the same time he reported that a prominent recollection is one of "frustration."

Before I was a member of Council, I was appointed to the first Education Committee, organized in 1953 under the chairmanship of Edward Adolph. The third member was William Amberson. At this time Milton Lee was turning the executive secretaryship over to Ray Daggs, who strongly supported educational activities. My own effort was to strengthen teaching of physiology to undergraduates. We obtained funds to organize summer workshops for college teachers. I directed the first, at the University of Connecticut at Storrs, where twenty teachers heard from several research physiologists and discussed laboratory experiments. We started to assemble a laboratory manual under the supervision of Louise Wilson of Wellesley College. Because the college teachers wanted to do research in which their undergraduate seniors could participate, we obtained a grant from NIH to APS for a program under which a college teacher might obtain up to $500 annually. Applications were evaluated by the Education Committee, supplemented by ad hoc reviewers. The program went well for several years, until bureaucrats at the NIH decided not to make grants which finally would be allocated by a non-NIH committee. The program died. . . . Frustration
. . . .

Another continuing interest of mine was the establishment of APS as a parent organization for all areas of physiology—not just for those of interest to medical physiologists. In the mid-1950s the biophysicists formed their own society, and the Society of General Physiologists was founded, largely by the Woods Hole group. At about this same time, for the American Society of Zoologists, I served on a committee that led to creation of some six semi-autonomous divisions of the society, each of which was permitted to expand according to membership need. One of the growing divisions was comparative physiology. Because its members wanted a publication, I tried to strengthen the section of comparative physiology in the *American Journal of Physiology*, which had become simply a catch-all for papers without obvious medical orientation. A proposal that this journal section be sponsored by the Division of Comparative Physiology of the American Society of Zoologists was rejected by Council. . . .

As president of APS, I pursued the constitutional changes needed for sectionalization of the Society along the lines of the American Society of Zoologists. Neurophysiologists were then threatening to form a society, and I did not want to see this large area of physiology become independent of APS. After extensive debate, Council decided to put the question of sectionalization to a vote at the Spring Business Meeting. When the vote went against my recommendation as president, I was genuinely disappointed.

Now, some fifteen years later, the Society has established sections that may help prevent further fragmentation of the APS. Meanwhile, we did manage to add representatives from general and from comparative physiology to the American Committee for IUPS and the international congresses. In hindsight, it probably was best that the biophysicists, the general, and then the comparative and the neurophysiologists did form separate societies. Each has become scientifically strong. I believe, however, that there remains a need for a parent organization that can represent all of physiology in Washington and to the public and also can strengthen teaching and recruitment at all levels. . . . Moreover the upgrading of education for the Ph.D. degree in physiology requires a continuous and united action by members of all these societies.

Ladd Prosser was born in Avon, New York, and earned his A.B. degree at the University of Rochester (1929), followed by the Ph.D. degree from Johns Hopkins University (1932). For two years he was a Parker Fellow, first at Harvard University Medical School and then at Cambridge University, England. He was a member of the faculty of Clark University in Worcester, Massachusetts (1934–37), until he joined the Department of Physiology of the University of Illinois in Urbana as an assistant professor (1937). By 1949 he was a professor there; from 1960 to 1969 he was head of the department, and in 1975 he became professor emeritus without perceptible change in his academic or scientific life.

Prosser's research interests encompass a variety of comparative fields. He wrote:

My interest in the nervous systems of invertebrates was initiated in 1927 by Herrick's book, *Neurological Basis of Behavior*. I learned general physiology from S. O. Mast at Johns Hopkins and electrophysiology at Harvard, principally from Hallowell Davis. In England, I learned much from E. D. Adrian and J. C. Eccles. My first job was at Clark University where Hudson Hoagland introduced me to academic politics. [Ed. note: Prosser is the only former president who identified where he received this type of training.] During World War II, I was active in radiation biology and biophysics as Associate Section Chief for Biology at the Metallurgical Laboratory of the Manhattan Project, with K. S. Cole as my section chief. Before the war and afterwards, contacts at Woods Hole in the Marine Biological Laboratory were important to me.

Topics that have been of greatest interest to Prosser include *1*) nervous systems of invertebrate animals; 2) comparative physiology of muscles; 3) theory of physiological adaptation; 4) temperature adaptation, both metabolic and neural, mostly in fishes; 5) the mechanism of rhythmicity in intestinal muscle; and 6) electrical properties of smooth muscle. Representative papers on these problems are included in the bibliography that follows. His complete bibliography includes summarizing articles on comparative physiology of nerve systems and sense organs (*Annu. Rev. Physiol.*, 1954), physiological variation in animals (*Biol. Rev.*, 1955), theoretical aspects of adaptation (*Handbook of Physiology*, 1968), smooth muscle (*Annu. Rev. Physiol.*, 1974), temperature compensation in poikilotherms (*Physiol. Rev.*, 1974), slow rhythmic activity in gastrointestinal muscles (9), and evolution and diversity of nonstriated muscle (13). Prosser noted:

> The series of papers on rhythmic activity in intestinal muscle has established the concept of rhythmic electrogenic sodium pump and has contributed to understanding of intercellular conduction in smooth muscle (5–11). . . . Recently a series of papers examines mechanisms of compensatory acclimation of fishes to heat and cold (cf. refs. 8 and 12). One aspect of these mechanisms deals with changes in activities of energy yielding enzymes, with explanations in terms of protein synthesis. Another aspect deals with changes in membrane phospholipids. Adaptations in behavior and the central nervous concomitants include synaptic failure as the most sensitive effect of temperature stress.

Prosser has enjoyed a more varied career than many of his contemporaries. He served at various times as visiting professor at the Universities of Washington, Stanford, Massachusetts, Arizona State, and Hawaii. In 1950, after many years of summer research at the Marine Biological Laboratory in Woods Hole, he was elected a trustee of that laboratory. For a year (1963–64) he was a Guggenheim Fellow at the University of Munich, and later (1971–72) he was International Exchange Fellow at Monash University in Australia. When he took office as president elect of APS (1968) he had already served as president of the Society of General Physiologists (1958–59) and the American Society of Zoologists (1961). In 1957 Prosser became a fellow of AAAS (Boston) and in 1974 was chosen for membership in NAS. He received the honorary degree of doctor of science from Clark University in 1975. He is a foreign associate of the Bavarian Academy of Sciences.

Elected to APS in 1935, Prosser's first committee assignment was as a member of the Education Committee (1953–59). At about the time of his election to Council in 1967, he became active in the service of APS publications, as a member of the Handbook Committee (1967–72), section coeditor for comparative and general physiology of the *American Journal of Physiology* (1968–71), and section editor for muscle physiology (1970–71). More recently he was a member of the Senior Physiologists Committee (1978–81). In 1983 he was the recipient of the Society's Ray G. Daggs Award.

As a champion of the importance of comparative physiology, Prosser has represented this discipline on numerous editorial boards and committees. These include the *American Journal of Physiology*, the *Journal of Comparative Physiology, Digestive Disease*, and *Science*. From 1975 he has been co-managing editor of *Physiolog-*

ical Zoology. Perhaps his most important work as author and editor, however, is the monograph and textbook, *Comparative Animal Physiology,* published in 1951 and revised in 1961 and 1973 (3). It has been translated into three foreign languages, including Russian, and has been a major factor in establishing comparative physiology as an independent discipline both in theory and practice.

In commenting on his association with APS, Prosser noted other significant events that came to pass. These included completion of the Lee Building, which permitted expanding the scope of the Society's publications. He also took part in debates over how strongly APS should support the AIBS and how generously NIH should support training programs. Finally, when Ray Daggs was about to retire from his office, Prosser proposed Orr Reynolds for the position and saw his nomination safely through confirmation by Council. Ladd concluded his recollections of his involvement with the Society with these words:

> The future of APS is again being debated. I urge my young colleagues not to lose sight of the breadth of the subject. The functional organization of living organisms continues to provide inspiration to all who appreciate the beauty of biological function and who recognize that we can better understand humans by studying physiology of all sorts of organisms at molecular, cellular, and organismic levels.

"Inspiration" in place of "frustration"!

SELECTED PUBLICATIONS

1. PROSSER, C. L. Action potentials in the nervous system of the crayfish. I. Spontaneous impulses. *J. Cell. Comp. Physiol.* 4: 185–209, 1934.
2. PROSSER, C. L. An analysis of the action of acetylcholine on hearts, particularly in arthropods. *Biol. Bull. Woods Hole* 83: 145–164, 1942.
3. PROSSER, C. L., F. A. BROWN, D. W. BISHOP, T. L. JAHN, AND V. J. WULFF. *Comparative Animal Physiology.* Philadelphia, PA: Saunders, 1951. [Prosser was general editor and author of 13 of the 23 chapters.]
4. PROSSER, C. L., C. E. SMITH, AND C. E. MELTON. Conduction of action potentials in the ureter of the rat. *Am. J. Physiol.* 181: 651–660, 1955.
5. KOBAYASHI, M., T. NAGAI, AND C. L. PROSSER. Electrical interaction between muscle layers of cat intestine. *Am. J. Physiol.* 211: 1281–1291, 1966.
6. PAPASOVA, M., T. NAGAI, AND C. L. PROSSER. Two-component slow waves in smooth muscle of cat stomach. *Am. J. Physiol.* 214: 695–702, 1968.
7. PROSSER, C. L., AND H. OHKAWA. Functions of neurons in enteric plexuses of cat intestine. *Am. J. Physiol.* 222: 1420–1426, 1972.
8. HAZEL, J. R. AND C. L. PROSSER. Molecular mechanisms of temperature compensation in poikilotherms. *Physiol. Rev.* 54: 620–677, 1974.
9. PROSSER, C. L., J. A. CONNOR, AND W. A. WEEMS. Types of slow rhythmic activity in gastrointestinal muscles. In: *Physiology of Smooth Muscle,* edited by E. Bulbring and M. F. Shuba. New York: Raven, 1976, p. 99–109.
10. PROSSER, C. L., J. A. CONNOR, AND D. L. KREULEN. Interaction between longitudinal and circular muscle in intestine of cat. *J. Physiol. Lond.* 273: 665–689, 1977.
11. MANGEL, A. W., D. O. NELSON, J. A. CONNOR, AND C. L. PROSSER. Contractions of cat small intestinal smooth muscle in calcium-free solution. *Nature Lond.* 281: 582–583, 1979.
12. NELSON, D. O., AND C. L. PROSSER. Temperature-sensitive neurons in the preoptic region of sunfish. *Am. J. Physiol.* 241 (*Regulatory Integrative Comp. Physiol.* 10): R259–R263, 1979.
13. PROSSER, C. L. Evolution and diversity of nonstriated muscles. In: *Handbook of Physiology. Vascular Smooth*

Muscle, edited by D. F. Bohr, A. P. Somlyo, and H. V. Sparks, Jr. Bethesda, MD: Am. Physiol. Soc., 1980, sect. 2, vol. II, chapt. 21, p. 635–670.

14. PROSSER, C. L., AND D. O. NELSON. The role of nervous system in temperature adaptation of poikilotherms. *Annu. Rev. Physiol.* 43: 281–300, 1981.

43 (1970–71)

A. Clifford Barger

(b. 1917)

During his tenure on Council and in presidential offices, Barger led APS into effective programs for improving professional opportunities for minority groups and for women. He has written:

> My most important contributions probably have been in the founding and funding of the Porter Physiology Development Program and the education of minority physiologists through the Porter Physiology Development Committee, as well as the first presidential tour of the predominantly black schools, the organization of a workshop for minorities in research, . . . and support of women in the affairs and offices of the Society.

Born in Greenfield, Massachusetts, as a young man Barger went east and found his professional career almost entirely within ten miles of the Boston Common. He graduated from Harvard University in 1939 and from Harvard's Medical School in 1943. In 1943 and again in 1945 he was a member of the house staff of Peter Bent Brigham Hospital, until he joined the research and then the teaching staff of the Department of Physiology in 1946 under the chairmanship of Eugene M. Landis. Moving steadily upward in rank, he became professor of physiology in 1961 and Robert Henry Pfeiffer Professor two years later. During much of his time in the department, Barger has held also appointments in clinical departments. For example, he was assistant in medicine at Peter Bent Brigham Hospital from 1946 to 1953, until he joined the associate staff of that hospital. In 1959 his title was changed to consultant, and from 1965 he has held a similar title at Children's Hospital Medical Center and at St. Vincent's Hospital. For two years (1974–76) Barger filled the position of chairman of the Department of Physiology.

Before he became identified with this department, Barger had begun fundamental research in three other settings.

I was fortunate as an undergraduate to be a student fellow in the Fatigue Laboratory when D. Bruce Dill (23rd president of APS) was director. I worked on the problem of anaerobic glycogenolysis in frog muscle with my Harvard College tutor, Robert E. Johnson (1). (Dr. Johnson later served for more than 20 years as head of the Department of Physiology at the University of Illinois in Urbana.) My next research experience, as an intern at the Peter Bent Brigham Hospital, was the study with George Thorn of the effect of testosterone on exercise endurance in patients with progressive muscular dystrophy. During World War II, I was assigned by the Army to the Climatic Research Laboratory (in Lawrence, MA) to do research on protection of soldiers exposed to cold climates.

After my release from the Army, as a research fellow with Dr. Landis I studied the response of the cutaneous blood vessels during treadmill exercise (2).

These experiments with Landis led naturally to thirty-five years of ongoing research on pathophysiology of congestive heart failure, renovascular hypertension, and coronary artery disease. Barger and his co-workers first attempted to explain why patients with congestive heart failure retain water and electrolytes and to this end developed a technique for inducing an analogous condition in dogs. They did this by combining operations to produce tricuspid valvular insufficiency with pulmonary arterial stenosis and so created right-sided congestive failure (3). When renal function of the animals was tested, a depressed response to intravenous salt loading was found before there was evidence of any decrease in glomerular filtration rate (4). The results suggested an increased tubular reabsorption of sodium. To test this hypothesis, methods for studying the output of the two kidneys separately were devised, with perfusion of one kidney with hypertonic sodium chloride solution while the other kidney served as a control. Increased tubular sodium reabsorption by the perfused kidney was thereby confirmed (5). Later experiments demonstrated for the first time that unilateral infusion of physiological doses of aldosterone in unanesthetized dogs produces a unilateral kaliuresis with no change in sodium excretion, although in adrenalectomized dogs the same infusion led to both kaliuresis and antinatriuresis (6).

While infusing solutions into renal vessels, Barger and his colleagues discovered that in dogs with congestive failure the cortical nephrons of the kidney had a reduced blood flow, from which they concluded that the outer, shorter nephrons might be "relative salt losers," whereas the inner nephrons with longer loops might be "relative salt retainers." To study differential blood flow more thoroughly, the laboratory developed an inert gas technique for measurement of regional blood flow in kidneys of unanesthetized dogs (7). Similarly unanesthetized animals were later utilized for experiments on the role of the renin-angiotensin-aldosterone system in compensation for congestive heart failure. Here the preliminary operations included permanent placement of catheters and implantation of inflatable cuffs on pulmonary artery and thoracic aorta. When the renin-angiotensin-aldosterone system was inactivated, these animals lost most of their capacity to compensate for the congestive failure (11).

The second of Barger's three primary research interests, renovascular hypertension, is represented in his bibliography by a series of papers, of which five are included here (8–12). The research was made possible by the prior development of methods for inducing chronic changes in blood flow of unanesthetized dogs and so studying sequentially changes in the renin-angiotensin system. The authors found that in salt-depleted dogs maintained on a low-salt diet to prevent water and electrolyte retention, renovascular hypertension is maintained primarily by the renin-angiotensin system (12, 13).

Current work of Barger's laboratory again concerns coronary arterial blood flow and coronary artery disease. New evidence has been provided for a significant role of the vasa vasorum in pathogenesis of coronary artery plaques (14), in confirmation of results published nearly fifty years ago by Winternitz and his colleagues in the Department of Pathology at Yale.

An unusual feature of research in Barger's laboratory has been the production of motion pictures of physiological responses. He and his associate, R. Beeuwkes, were able to photograph renal tubules as they were being injected with silicone rubber through micropipettes. These studies led to others on the vascular-tubular organization of canine and human kidneys and to photographs of neovascularization of atherosclerotic coronary arteries. Shown at the IUPS Meeting in Munich in 1971, at the International Congresses of Nephrology in Mexico in 1972 and in Florence in 1975, at the Annual Meeting of AHA in 1979, and at a FASEB Meeting in 1983, these films have been awarded altogether twelve prizes and medals. Included are three separate Golden Eagle Awards of the Council on International Nontheatrical Events (1973, 1975, and 1983) and a gold medal by the International Medical Film Festival held in Parma, Italy, in 1983 for the coronary film.

In his years of service to Harvard and to the Boston medical-scientific community, Barger has been appointed or elected to important offices. In 1957–58 he was president of the Boylston Medical Society and also president of the Massachusetts Society for Medical Research. In 1970 he became president of the Harvard Apparatus Foundation. From 1978 to 1982 he was first master of the Cannon Society of the Harvard Medical School, an academic society that seeks to perpetuate the ideals and legacies of Walter Bradford Cannon. An accomplished historian, Barger is coauthoring a two-volume life and letters of Cannon, the first volume of which is scheduled to appear in time for the centennial of APS.

Nationally, Barger served on committees of NRC (1955–57 and 1957–62) and was a member of the Physiology Study Section of NIH (1960–64) and the Board of Scientific Counselors of the National Cancer Institute (1969–72; chairman, 1972–73). During this period he also was on the Research Committee of AHA (1966–68). In 1971 he joined the Physiology Committee of the National Board of Medical Examiners (chairman, 1973–76). His editorial duties have included *Proceedings of the Society of Experimental Biology and Medicine* (1960–62), *Circulation Research* (1963–66), and "Physiology in Medicine" in the *New England Journal of Medicine* (1971–80).

Honors Barger has received include the Certificate of Merit of the National Society of Medical Research (1958), selection as the Goldblatt Memorial Lecturer (1978) and the Annual Sosman Lecturer of Peter Bent Brigham Hospital (1980), and the Carl J. Wiggers Award of APS (1982). He was elected a fellow of AAAS (Boston) in 1964 and a member of the Institute of Medicine of NAS in 1974. In 1977 the University of Cincinnati conferred on him the degree doctor of science.

Elected to membership in APS in 1949, by 1960–61 he was a member of the Editorial Board of the Society's two primary journals. He was appointed to the newly established Publications Committee (1961–63; chairman, 1962–63) to which he was later reappointed (1966–69). He was elected to Council in 1968, and only a year later he began his presidential terms as president elect. In 1985 he was chosen for the Society's Ray G. Daggs Award. The citation for the presentation read:

> Cliff has also been very deeply involved in the fight by our Society and its members to maintain access to appropriate animal models for our physiological research, working especially closely with the [National] Society for Medical Research and the IUPS.

Barger's designation as cochairman of the Porter Development Committee in 1966 marked the official beginning of his effort to apply resources available to the Society in the interests of minority scientists. Funds distributed by this committee originate from the Harvard Apparatus Foundation, successor to the original Harvard Apparatus Company founded by William Townsend Porter. In a graceful although brief biography of Porter, Barger wrote of how in 1929 Porter offered to give the Harvard Apparatus Company to APS, but Council declined to accept the gift. Porter's response was to set up a nonprofit foundation to run the company, with the net proceeds then turned over to APS to support the Porter Fellowships (16). The total thus made available now exceeds $750,000.

In his past president's address, Barger described the fascinating history of personal interactions among Bowditch, Porter, and Cannon and recounted how funds from the Harvard Apparatus Company were made available for the benefit of young minority-group physiologists (15). Barger conceived this idea and in 1965 persuaded Council to approve it. More detail is given elsewhere in this historical volume (see chapt. 20). Cliff Barger will certainly be remembered in the annals of APS, not alone for his contributions to physiology of the heart and kidneys, but perhaps even more for his dedication to the training and careers of scores of young investigators of minority-group backgrounds.

SELECTED PUBLICATIONS

1. BARGER, A. C., AND R. E. JOHNSON. Anaerobic glycogenolysis in the muscles of *Rana pipiens* living at low temperature. *J. Gen. Physiol.* 24: 669–677, 1941.
2. GREENWOOD, W. F., A. C. BARGER, J. R. DiPALMA, J. STOKES III, AND L. H. SMITH. Factors affecting the appearance and persistence of visible cutaneous reactive hyperemia in man. *J. Clin. Invest.* 27: 187–197, 1948.
3. BARGER, A. C., B. B. ROE, AND G. S. RICHARDSON. Relation of valvular lesions and of exercise to auricular pressure, work tolerance and to the development of chronic congestive failure in dogs. *Am. J. Physiol.* 169: 384–399, 1952.
4. BARGER, A. C., R. S. ROSS, AND H. L. PRICE. Reduced sodium excretion in dogs with mild valvular lesions of the heart, and in dogs with congestive failure. *Am. J. Physiol.* 180: 249–260, 1955.

5. RUDOLPH, A. M., S. N. ROKAW, AND A. C. BARGER. Chronic catheterization of the renal artery. Technic for studying direct effects of substance on kidney function. *Proc. Soc. Exp. Biol. Med.* 93: 323–326, 1956.

6. BARGER, A. C., R. D. BERLIN, AND J. F. TULENKO. Infusion of aldosterone, 9-α-fluorohydrocortisone and antidiuretic hormone into the renal artery of normal and adrenalectomized, unanesthetized dogs: effect on electrolyte and water excretion. *Endocrinology* 62: 804–815, 1958.

7. THORBURN, G. D., H. H. KOPALD, J. A. HERD, M. HOLLENBERG, C. C. C. O'MORCHOE, AND A. C. BARGER. Intrarenal distribution of nutrient blood flow determined with krypton[85] in the unanesthetized dog. *Circ. Res.* 13: 290, 1963.

8. GUTMANN, F. D., H. TAGAWA, E. HABER, AND A. C. BARGER. Renal arterial pressure, renin secretion, and blood pressure control in trained dogs. *Am. J. Physiol.* 224: 66–72, 1973.

9. TAGAWA, H., F. D. GUTMANN, E. HABER, E. D. MILLER, JR., A. I. SAMUELS, AND A. C. BARGER. Reversible renovascular hypertension and renal arterial pressure. *Proc. Soc. Exp. Biol.* 146: 975–982, 1974.

10. MILLER, E. D., JR., A. I. SAMUELS, E. HABER, AND A. C. BARGER. Inhibition of angiotensin conversion and prevention of renal hypertension. *Am. J. Physiol.* 228: 448–453, 1975.

11. WATKINS, L., JR., J. A. BURTON, E. HABER, J. R. CANT, F. W. SMITH, AND A. C. BARGER. The renin-angiotensin-aldosterone system in congestive failure in dogs. *J. Clin. Invest.* 57: 1606–1617, 1976.

12. ROCCHINI, A. P., AND A. C. BARGER. Renovascular hypertension in sodium-depleted dogs: role of renin and carotid sinus reflex. *Am. J. Physiol.* 236 (*Heart Circ. Physiol.* 5): H101–H107, 1979.

13. KOPELMAN, R. I., V. J. DZAU, S. SHIMABUKURO, AND A. C. BARGER. Compensatory response to hemorrhage in conscious dogs on normal and low salt intake. *Am. J. Physiol.* 244 (*Heart Circ. Physiol.* 13): H351–H356, 1983.

14. BARGER, A. C., R. BEEUWKES III, L. L. LAINEY, AND K. J. SILVERMAN. Hypothesis: vasa vasorum and neovascularization of human coronary arteries: a possible role in the pathophysiology of atherosclerosis. *N. Engl. J. Med.* 310: 175–177, 1984.

15. BARGER, A. C. Past-president's address. To assist young men and women in the study of physiology: the Porter Development Program. *Physiologist* 14: 277–285, 1971.

16. BARGER, A. C. The meteoric rise and fall of William Townsend Porter, one of Carl J. Wiggers' "Old Guard." *Physiologist* 25: 407–413, 1982.

44 (1971–72)

John R. Brobeck
(b. 1914)

In his past president's address (10) Brobeck identified cycles in medical education and in physiology since the time of Boerhaave in the early eighteenth century. In comparing physiologists with seekers after the mythical unicorn, he predicted that despite the uncertainties occasioned by recent changes in the medical curriculum, the cyclic curve representing the number of unicorn hunters was again on the rise and that physiology would have its renaissance. His closing paragraph is especially relevant to the mission of APS "to promote the increase of physiological knowledge and its utilization" as we look forward to our second century:

> For almost 300 years physiology has been a powerful science. Its strength is drawn in part from the inherent interest of biological mechanisms and processes, but also in part from the utility an understanding of these processes finds in medicine and the related professions. We must not be simplistic about our discipline. Intellectual curiosity does not need to be our only reason for existence. Neither is a practical application enough to insure the perpetuation of the science. The two go together— the history of even our most distinguished forefathers shows that they do. We can well be guided by their experience.

Brobeck's professional career has involved only three institutions, or four if Wheaton (Illinois) College is included. After graduating from college in 1936, he spent three years at the Institute of Neurology of Northwestern University in Chicago, where he received the Ph.D. degree in 1939. He was then able to continue his education at the School of Medicine at Yale University and was awarded an M.D. degree in March 1943. On the first day of April he began an association with John Fulton's Laboratory of Physiology at Yale that continued until 1952, when Brobeck moved to the Philadelphia area as professor and chairman of the Department of Physiology of the School of Medicine of the University of Pennsylvania. He was also chairman of the Graduate Group Committee in Physiology. At that time the university included also another department of physiology in the Graduate School of Medicine. Julius Comroe had made it one of the strongest departments in the country. In 1957, however, Comroe resigned from his positions at Pennsylvania to take up his new responsibilities at the University of California in San Francisco. Two years later, Robert Forster became chairman of this department. Brobeck meanwhile held office in the School of Medicine until 1970. He then resigned so that the two departments could be brought together under Forster's direction. From 1970 until his official retirement in 1982, Brobeck held the title of Herbert C. Rorer Professor in the Medical Sciences.

Brobeck wrote of his training in science and his scientific interests:

> Although my training took place in the laboratories of three world-class scientists, Stephen Walter Ranson at Northwestern University and John Farquhar Fulton and C. N. H. Long at Yale University, the predominant influence on my career as an investigator and teacher was not the heads of the laboratories so much as the younger persons they attracted to work with them. At Northwestern these included principally H. W. Magoun, our preceptor in stereotaxic surgery, and Frank Harrison, George Clark, and Albert W. Hetherington, fellow graduate students. At Yale, where for four years I was a medical student, Jay and Helen Murphy Tepperman, with other students and research fellows, continued my education in experimental science. It was Donald Henry Barron, however, then newly appointed as associate professor of physiology, who most largely contributed to my understanding of the academic life, the responsibilities and opportunities open to teachers of science, and the international community of physiologists. Two other names should be mentioned, although I never worked or published with either. E. F. Adolph of Rochester, through his monograph

on *Physiological Regulations* (1943), turned my interest in that direction. And Merkel H. Jacobs, senior member of this department when I arrived here in 1952, introduced me to membrane phenomena I previously had not considered. Finally, I remain grateful to my teachers at Wheaton (Illinois) College, where my formal training in science began in a Christian context that continues as an important part of my life.

In celebration of the bicentennial of the founding of what became this school of medicine, in 1965 William S. Yamamoto and I edited *Physiological Controls and Regulations*, with chapters written mainly by current or former members of our faculty. The introductory chapter, "Exchange, control, and regulation," expressed my own research interests (9). In particular, I have been studying control of energy exchange and energy balance. Having learned from my own observations and the work of other laboratories that stimulation or lesions of the hypothalamus may alter body temperature regulation, food intake, body weight, or motor output, I proposed some years ago that the hypothalamus might be the part of the brain responsible for integration of these several variables into patterns of energy exchange. The basis for the integration might be thermal signals. In adult animals this integration usually leads to a balance between intake and expenditure and consequently to a stable body weight. Publications offering evidence for this proposal began with the first on my bibliography (1) and continued with the Yale series of papers (2–4), with Anand (5, 6), and with the paper with Gladfelter (8). In 1960 and again in 1981, I was given the privilege of summarizing my views on this subject at, first, the Laurentian Hormone Conference (7) and, second, a symposium on *The Body Weight Regulatory System: Normal and Disturbed Mechanisms* in Italy (10).

Elected to membership in APS in 1943, Brobeck's first assignment was as chairman of the Education Committee in 1960. From 1963 to 1972 he served as chairman of the Editorial Board of *Physiological Reviews*. He was elected to Council in 1967 and became president elect in 1970. In 1980 he received the Ray G. Daggs Award. He wrote of his experiences as an office holder of APS:

It is embarrassing to confess that my first responsibility with the Society was a complete fiasco, and terribly frustrating. In 1960 when I was asked to serve as chairman of the Education Committee, I did not know what the committee was doing, what it should do, or what it might do. Consequently I presided rather vaguely over meetings, while Ray Daggs kept everything in order and managed the several projects the committee had earlier initiated. It was a relief to me, and probably to Ray, when I had to resign to go on sabbatical leave to Taiwan in 1962.

While I was president in 1971, the Council began to plan how to honor Dr. Daggs on his retirement in 1973 and formally invited Orr E. Reynolds (coeditor of this volume) to continue in his position of education officer and assistant executive secretary, in the expectation that two years later he would succeed Ray in the combined office of executive secretary-treasurer. Conditions of the appointment were summarized in a two-page letter to Orr. In his reply he wrote, simply, "I am very honored by the Society's offer and most pleased to accept the conditions as expressed in your letter." This decision was no doubt the most significant of the years I was associated with the Council and in my judgment one of the most important of the twenty-five years covered by this history of the Society.

Wheaton College has conferred three honors on Brobeck: the Distinguished Service Award of the Alumni Association (1953), a Centennial Award (1959), and the degree doctor of laws (1960). In 1959 he received a Centennial Merit Award from Northwestern University. He is a member of the American Society for Clinical Investigation, the Halsted Society, AAAS (Boston) (1969), and NAS (1975). In 1962–63 he and most of his family, with a grant from the China Medical Board of New York, were able to spend nine months at the National Defense Medical Center in Taipei, Taiwan. They visited also the major medical centers in Korea, Hong Kong, the Philippines, Bangkok, and New Delhi, India.

After he had summarized his training, research interests, and participation in affairs of the APS as noted above, Brobeck concluded by writing that what most of his friends seem to remember about him is that in spring, summer, and fall he rides a bicycle from Swarthmore to the university and that he was born and reared in Steamboat Springs, Colorado.

SELECTED PUBLICATIONS

1. MAGOUN, H. W., F. HARRISON, J. R. BROBECK, AND S. W. RANSON. Activation of heat loss mechanisms by local heating of the brain. *J. Neurophysiol.* 1: 101–114, 1938.

2. BROBECK, J. R., J. TEPPERMAN, AND C. N. H. LONG. Experimental hypothalamic hyperphagia in the albino rat. *Yale J. Biol. Med.* 15: 831–853, 1943.

3. TEPPERMAN, J., J. R. BROBECK, AND C. N. H. LONG. The effects of hypothalamic hyperphagia and of alterations in feeding habits on the metabolism of the albino rat. *Yale J. Biol. Med.* 15: 855–874, 1943.

4. BROBECK, J. R., J. TEPPERMAN, AND C. N. H. LONG. The effects of experimental obesity upon carbohydrate metabolism. *Yale J. Biol. Med.* 15: 893–904, 1943.

5. ANAND, B. K., AND J. R. BROBECK. Localization of a "feeding center" in the hypothalamus of the rat. *Proc. Soc. Exp. Biol. Med.* 77: 323–324, 1951.

6. ANAND, B. K., AND J. R. BROBECK. Hypothalamic control of food intake in rats and cats. *Yale J. Biol. Med.* 24: 123–140, 1951.

7. BROBECK, J. R. Food and temperature. *Recent Prog. Horm. Res.* 16: 439–459, 1960.

8. GLADFELTER, W. E., AND J. R. BROBECK. Decreased spontaneous locomotor activity in the rat induced by hypothalamic lesions. *Am. J. Physiol.* 203: 811–817, 1962.

9. BROBECK, J. R. Exchange, control, and regulation. In: *Physiological Controls and Regulations,* edited by W. S. Yamamoto and J. R. Brobeck. Philadelphia, PA: Saunders, 1965, p. 1–13.

10. BROBECK, J. R. A reconsideration of the "Biological Clock in the Unicorn." *Physiologist* 15: 327–337, 1972.

11. BROBECK, J. R. Models for analysing energy balance in body weight regulation. In: *The Body Weight Regulatory System: Normal and Disturbed Mechanisms,* edited by L. A. Cioffi et al. New York: Raven, 1981, chapt. 1, p. 1–9.

45 (1972–73)

Robert M. Berne
(b. 1918)

Beginning his past president's address with a three-stanza limerick, Berne referred to his two predecessors in a line that ran, "Barger, Brobeck, and Berne / As President they each had a term . . ." and concluded:

Barger, Brobeck and Berne
Can never ever return,
The rules on election
Make no exception
So now it is Tosteson's turn.

Of greater significance, however, Berne was first in a succession of three graduates of Harvard Medical School who served as presidents of APS—Berne, Tosteson, and Guyton. If not for the intervention of Brobeck, the string would have included four in a row, beginning with Barger.

Born in Yonkers, New York, Berne attended the University of North Carolina (A.B., 1939) in preparation for his professional education at Harvard. On graduation he interned at Mount Sinai Hospital in New York City, was an assistant resident for nine months, and then served for two years mostly as battalion surgeon with an infantry unit of the U.S. Army. On his discharge from military service he returned to Mount Sinai as resident in medicine for a little more than a year (1947–48) and then transferred to the Department of Physiology at Western Reserve University in Cleveland, Ohio, for training in cardiovascular research under Carl J. Wiggers (21st president of APS, 1949–50). After deciding to stay longer in physiology, Berne became progressively assistant professor (1952–55), associate professor (1955–61), and professor (1961–66) in Wiggers' department. From 1957 to 1966 he held also an appointment in the Department of Medicine. A sabbatical leave in 1959–60 took him to the laboratory of E. C. Slater in Amsterdam, Holland, while another in 1965–66 found him with G. V. R. Born at the Royal College of Surgeons in London, England.

In response to questions about how he decided to become a physiologist, Berne wrote:

After graduation from medical school, serving some time in the military, and receiving excellent clinical training in internal medicine, I decided to become a cardiologist. To prepare for such a career I felt it necessary to obtain some basic science training in cardiovascular physiology. In this regard I was fortunate in obtaining a postdoctoral fellowship in physiology with Carl J. Wiggers at Western Reserve Medical School. My original expectation was to spend one to two years with Dr. Wiggers doing basic research and then return to a clinical cardiology setting. As things developed, I became so engrossed in, and excited by, the research that I kept postponing my pursuit of cardiology. . . . Since I really enjoyed internal medicine, I occasionally now wonder what would have happened had I chosen the clinical route.

Following on this early interest in cardiology, Berne's research career has continued what he began with Wiggers. He has written:

My main research interests have been in cardiovascular physiology, in general, and in the local chemical regulation of tissue blood flow, in particular. For years our attention has been focused on adenosine as the primary mediator of metabolically induced increases in blood flow. . . . In the course of our studies we have utilized several biochemical tools, cell culture techniques, and electron microscopy, as well as more conventional physiological procedures in an attempt to define the role of adenosine and to understand how it elicits vasodilation. . . . The nucleoside has been shown to be of importance as a possible neurotransmitter and an inhibitor of pre- and postsynaptic impulse transmission. Also, there are membrane adenosine receptors in the brain and in other tissues, and their full physiological function remains to be elucidated.

Berne's first publication (1) in a major refereed journal appeared in 1950, in collaboration with M. N. Levy, on renal circulation in dogs with reduced cardiac output. He soon turned his attention to the hypothesis that coronary blood flow is controlled by local concentrations of adenosine (2). This work has led to a continuing series of papers (e.g., ref. 3), including reports that adenosine is involved in control of blood flow in skeletal muscle (4), brain (6), and kidney (7), at least with reduced oxygen supply. By 1980 Berne and his associates were able to demonstrate changes in myocardial adenosine concentration within a single cardiac cycle (8). In 1982 they showed that adenosine release from the hearts of unanesthetized dogs correlated well with coronary blood flow and myocardial oxygen consumption during physiological stimuli (10). Two years later, with the use of dipyridamole, a drug that blocks adenosine uptake, they observed that under a variety of interventions, cardiac adenosine release and coronary blood flow were closely related but independent of myocardial oxygen consumption, a finding in support of the adenosine hypothesis for the regulation of coronary blood flow (12). After studying the possible contribution of adenosine to disturbances of the conduction system of the heart (9), they found that adenosine could be used therapeutically for treatment of supraventricular tachycardia in human patients (11).

This lifelong interest in the cardiovascular system has been expressed outside the laboratory and classroom in the associations and societies in which Berne has participated, including APS (see later). He is a member of the Microcirculatory Society and served on its Council (1971–72). For AHA he has been a member of the Councils on Basic Science, on Circulation, and on Hypertension, as well as of the Committee on Medical Education (1963–65), the Executive Committee of the Council on Basic Science (1965–68), chairman of the Publications Committee

(1981–85), and a member of the Board of Directors (1979–80 and 1983–85). He served as a member of the Editorial Board of *Circulation Research* for five years (1962–67), as editor for another five years (1970–75), and then again on the Editorial Board (1975–). For NIH, the National Heart Institute, and later the National Heart, Lung and Blood Institute he has provided counsel on numerous evaluation committees. For *Annual Review of Physiology*, Berne was a member of the Editorial Committee (1976–81), an associate editor (1982), and an editor (1983–). He has also been, at various times, advisor to NAS (1963); the National Board of Medical Examiners (1969–71 and 1983–85); AAMC (1974, 1975–79, and 1977–80); the Ciba Foundation (1975–77); the Alfred I. DuPont Institute in Wilmington, Delaware (1978–82); AAAS (1980–); the Hazen (1984–) and the Pew (1984–) Award Committees; and other regional and local organizations.

As a member of ACDP, Berne served as president in 1970–71. Later (1978–79) he was chairman of the Council of Academic Societies of AAMC. He has received the honorary degree doctor of science from the Medical College of Ohio in Toledo (1973); the Carl J. Wiggers Award (1976); the Physiology Teaching Award of the ACDP; and the Research Achievement Award (1979), the Award of Merit, and the Gold Heart Award (1985) from AHA. In 1982 he was chosen for distinguished service membership in AAMC. Berne is a member of the American Society for Clinical Investigation and in 1979 was elected to membership in the Institute of Medicine of NAS.

When he was elected as councillor of APS in 1970 and as president elect the following year, Berne had already been a member of the APS Program Committee (1962–65), section editor for circulation for the *American Journal of Physiology* (1964–65), and a member of the Finance Committee (1967–70) and the Steering Committee of the Circulation Group (1969–72). Later he was reappointed to the Finance Committee (1975–76), served for four years on the Editorial Board of *American Journal of Physiology* (1976–80), on the Publications Committee (1976–82), on the Perkins Award Committee (1977–80), as chairman of the Long-Range Planning Task Force (1980–84), and as a member of the Long-Range Planning Committee (1984–). He was a member of the Steering Committee for revision of the cardiovascular section of the *Handbook of Physiology* and editor of volume I on the heart, published in 1979.

Procedures by which APS conducts its affairs probably changed more during Berne's presidential years than at any other time in recent history. Traditionally any item that required decision by the membership of the Society had been considered at the annual business meeting held during the spring assembly of FASEB. In consequence, election of president elect and councillors usually involved only the 150 to 300 members in attendance and voting at these sessions. Berne first undertook an informal mail poll of members as to their preference and then successfully sponsored amendments to the bylaws that initiated elections by mail. In the first such election (1975) 1,677 ballots were returned, representing about forty-four percent of the members eligible to vote. At the same time he began the publication of annual reports of officers and committees in *The Physiologist*, rather than verbally at the annual business meeting.

A second change concerned fall meetings. From their inception in 1948, these had been held in academic centers, at universities or medical schools, usually in a month convenient for family vacations of the membership. Another survey initiated by Berne showed a significant preference for meetings held later in the fall and in cities, but because the ballots were almost evenly split between these two options, it was decided to have on-campus and in-city fall meetings in alternate years.

Perhaps even more apparent, however, to officers and members alike was a third change. The end of the calendar year marked the end of Ray G. Daggs' years of service as executive secretary-treasurer and the beginning of the tenure of Orr E. Reynolds in that office. To recognize the innumerable and invaluable contributions Ray Daggs had made to the Society, Berne announced at the annual meeting held in Atlantic City on 17 April 1973 the establishment of the Ray G. Daggs Award "to be presented beginning next year to a physiologist who is judged to have provided distinguished service to the science of physiology and to the American Physiological Society."

In making this announcement, Berne quoted from a telegram received from Alan Burton (29th president, 1956–57) to indicate how each president came to evaluate Daggs' assistance: "Using his great sense of humor, integrity and reasonableness, Ray Daggs steered and successfully managed a succession of ignorant and opinionated Presidents . . ." (*Physiologist* 16: 111, 1973).

Among other departures from tradition that marked Berne's presidency was the appointment of a Task Force on Women in Physiology, with M. Elizabeth Tidball of George Washington University as chairperson. The task force made its first report at the Atlantic City Meeting held the following April (*Physiologist* 17: 135–137, 1974). By 1974 the Society had its first woman as president elect, Bodil M. Schmidt-Nielsen.

The concern Berne felt for the well-being of APS is clearly seen in his past president's address:

> My concern was essentially two-fold. Looking internally at the present membership, I wondered if the APS was meeting the needs of all its members, or at least of the majority. . . . Looking externally, I wondered about the relationship of the Society to the Federation and beyond that to the total pursuit of basic scientific knowledge in the world today.

More thoroughly than most similar speeches, Berne's address outlines the events that had brought the Society to 1973 and his thoughtful analysis of how he had tried to respond to needs and to opportunities. Reading what he wrote leaves one with at least a mild disappointment that a president "can never, ever return."

SELECTED PUBLICATIONS

1. BERNE, R. M., AND M. N. LEVY. Effect of acute reduction of cardiac output on renal circulation of the dog. *J. Clin. Invest.* 29: 444–454, 1950.
2. BERNE, R. M. Cardiac nucleotides in hypoxia: possible role in regulation of coronary blood flow. *Am. J. Physiol.* 204: 317–322, 1963.

3. KATORI, M., AND R. M. BERNE. Release of adenosine from anoxic hearts. *Circ. Res.* 19: 420–425, 1966.

4. DOBSON, J. G., JR., R. RUBIO, AND R. M. BERNE. Role of adenine nucleotides, adenosine, and inorganic phosphate in the regulation of skeletal muscle blood flow. *Circ. Res.* 29: 375–384, 1971.

5. BERNE, R. M. Past-president's address. The American Physiological Society—a piece of the continent, a part of the main. *Physiologist* 16: 511–519, 1973.

6. BERNE, R. M., R. RUBIO, AND R. R. CURNISH. Release of adenosine from ischemic brain: effect on cerebral vascular resistance and incorporation into cerebral adenine nucleotides. *Circ. Res.* 35: 262–271, 1974.

7. MILLER, W. L., R. A. THOMAS, R. M. BERNE, AND R. RUBIO. Adenosine production in the ischemic kidney. *Circ. Res.* 43: 390–397, 1978.

8. THOMPSON, C. I., R. RUBIO, AND R. M. BERNE. Changes in adenosine and glycogen phosphorylase activity during the cardiac cycle. *Am. J. Physiol.* 238 (*Heart Circ. Physiol.* 7): H389–H398, 1980.

9. BELARDINELLI, L., F. L. BELLONI, R. RUBIO, AND R. M. BERNE. Atrioventricular conduction disturbances during hypoxia: possible role of adenosine in rabbit and guinea pig heart. *Circ. Res.* 47: 684–691, 1980.

10. BACCHUS, A. N., S. W. ELY, R. M. KNABB, R. RUBIO, AND R. M. BERNE. Adenosine and coronary blood flow in conscious dogs during normal physiological stimuli. *Am. J. Physiol.* 243 (*Heart Circ. Physiol.* 12): H628–H633, 1982.

11. DIMARCO, J. P., T. D. SELLERS, R. M. BERNE, G. A. WEST, AND L. BELARDINELLI. Adenosine: electrophysiologic effects and therapeutic use for terminating paroxysmal supraventricular tachycardia. *Circulation* 68: 1254–1263, 1983.

12. KNABB, R. M., J. M. GIDDAY, S. W. ELY, R. RUBIO, AND R. M. BERNE. Effects of dipyridamole on myocardial adenosine and active hyperemia. *Am. J. Physiol.* 247 (*Heart Circ. Physiol.* 16): H804–H810, 1984.

46 (1973–74)

Daniel C. Tosteson
(b. 1925)

Born in Milwaukee, Wisconsin, Tosteson's professional career has been identified mainly with two eastern institutions, Harvard and Duke Universities. He graduated from Harvard College (1944) and from its medical school (1949) and then undertook the postdoctoral training described below. On his return from England (1958) he was appointed associate professor in the Department of Physiology of Washington University in St. Louis, but he remained there for only three years. In 1961 he became professor and chairman of the Department of Physiology and Pharmacology at Duke. Ten years later he was made also a James B. Duke Distinguished Professor and remained so until 1975 when he moved for two years to the University of Chicago. There he was simultaneously dean of the Division of Biological Sciences and of the Pritzker School of Medicine, vice-president for the Medical Center, and

Lowell T. Coggeshall Professor of Medical Sciences. In 1977 he assumed similar positions at Harvard as dean of the faculty of medicine, president of the Harvard Medical Center, and Caroline Shields Walker Professor of Physiology. Throughout this sequence of top-level administrative positions, Tosteson has retained his identity as a physiologist and his reputation as an investigator at the forefront of membrane phenomena.

In both his pre- and postdoctoral training, Tosteson encountered a remarkably distinguished succession of mentors. He wrote of these years:

> My research interests are in general physiology. As a medical student and resident, I was attracted by the thinking of C. Bernard, L. J. Henderson, J. Loeb, and others who created this discipline by connecting and integrating biology with chemistry and physics. More specifically, I am interested in cellular functions and molecular mechanisms of ion transport across biological membranes. My fascination with the roles of water and salt in living systems began when I studied inorganic chemistry in Harvard College. It was strengthened during my first years at Harvard Medical School when I had the opportunity to learn under the tutelage of Professors E. M. Landis, H. Davenport, A. B. Hastings, J. Gamble, and others. This developing interest prompted me to do research with Eugene Landis for a year between my third and fourth years in medical school. During that year, I encountered J. R. Pappenheimer who had just joined the Department of Physiology at Harvard Medical School and whose analytical mind and enthusiastic commitment to the search for truth made a lasting impression on me.

> It was during my two years as a medical resident at the Presbyterian Hospital in New York that I first began working with ion movements across red cell membranes. I feel a strong sense of respect and gratitude to Robert F. Loeb who was my chief mentor during those years and to Bert Mudge and Bob Darling who permitted me to work in their laboratories. My postdoctoral years at Brookhaven National Laboratory, when I had the opportunity to talk frequently with D. D. Van Slyke, increased my understanding of and interest in the physiology of red cells. I was further encouraged in conversations with E. Ponder, A. Parpart, and M. Jacobs, among others. It was during that time that I first met J. F. Hoffman with whom I have worked closely, though intermittently, ever since.

> From 1955 to 1957, I spent two wonderful years in Europe, first in Copenhagen with H. H. Ussing and later in Cambridge with A. L. Hodgkin. I continued to work with red cell membranes, even though they were not the main line of work either in Ussing's or Hodgkin's laboratories. I spent many educational hours brainstorming with Hans Ussing during the time when he was first developing a picture of the differences in transport properties of the inward- and outward-facing membranes of the epithelial cells in frog skin. At the laboratories on Downing Street in Cambridge, I not only had the privilege of encountering Alan Hodgkin, but also many other outstanding scientists such as A. F. Huxley, R. Keynes, W. Rushton, I. M. Glynn, B. Matthews, and the Adrians, father and son. The final part of my lengthy (7 years) postdoctoral research training was at NIH in the Laboratory of Kidney and Electrolyte Metabolism of the National Heart Institute, then under the direction of Bob Berliner. Through the efforts of Bob, Jim Shannon, and others, NIH was then, as now, a superb environment for the growth and maturation of young investigators. Many of the associations and acquaintances that I made then have persisted throughout my career.

Three papers (e.g., ref. 1) describe the first experiments Tosteson conceived and carried out independently—ion transport in red blood cells from patients with sickle cell anemia. The relation between physicochemical properties of hemoglobin and transport of ions across red cell membranes still occupies his attention, with current work on cation transport in red cells from individuals homozygous for hemoglobin C. Three later papers (e.g., ref. 2) helped open a fruitful line of investigation that eventually led to understanding how the co-transport of Na and K in avian red cells is hormone and volume regulated. Similar transport systems seem to be present in many types of mammalian cells.

In his paper published in 1959 (3), Tosteson wrote of his long-term interest in transport of Cl^- and other monovalent anions across red cell membranes. "I did most of the work during my happy visit to Copenhagen in 1955–56." Two papers (4, 7) were done with colleagues named Hoffman, the former with J. F. Hoffman when he and Tosteson were together at NIH, the latter with an M.D.-Ph.D. student at Duke, P. G. Hoffman. They represent a number of studies from Tosteson's laboratory on high-K^+ and low-K^+ sheep red cells. Tosteson said he began the study hoping to learn how membrane transport regulates the Na and K composition of cells. The theoretical basis of experiments he and J. F. Hoffman described (4) was done "in Cambridge on those cold, dark mornings in the winter and spring of 1957." The second of the papers (7) made two important points about cation transport in high-K^+ and low-K^+ sheep red cells: 1) the difference between the two genotypes is due not only to a greater membrane surface concentration of Na-K pumps in high K^+ cells, but also to a difference in kinetic properties of the pumps in the two types of cells, and 2) the inside-facing and outside-facing sides of the pump are kinetically isolated from one another. A paper published in 1975 (10) is the first report of the most important pathway regulating distribution of lithium between inside and outside of human red cells. The maximum rate of transport through this system varies from one person to another, and these interindividual differences can be correlated with syndromes such as mania and hypertension.

In 1967 Tosteson and Andreoli and their colleagues began a continuing effort to use lipid bilayers to analyze molecular events in transport of ions across biological membranes and to connect the primary, secondary, and tertiary structures of molecules with their capacity to promote transport of ions across membranes (5, 6). Most of this work has been done in collaboration with his wife, Magdalena T. Tosteson. Examples of molecules studied in this manner include valinomycin (6) and some of its analogues (e.g., ref. 12), cholera toxin (11), and mellittin (13). Finally, Tosteson wrote that another theme of his research on bilayers is work done with J. Gutknecht (8) on the role of unstirred layers in regulating transport across membranes.

Tosteson was elected to membership in the Institute of Medicine of NAS in 1975 and to fellowship in AAAS (Boston) in 1979 and also in the Danish Royal Society. He has held senior offices in the Society of General Physiologists (president, 1968–69), AAMC (1960–70; chairman, 1973–74), and the Biophysical Society (Council, 1970–73). He is a member of the Association of American Physicians. The honorary degree doctor of science has been awarded to him by the Universities of Copenhagen (1979) and Liège (1983), as well as by the Medical College of Wisconsin (1984).

Most of the major national and federal scientific organizations have called on Tosteson for counsel from time to time. These include several committees and boards of NAS and the NRC, NIH, the National Science Foundation (NSF), the National Board of Medical Examiners, the Universities of Texas and of California, and the National Kidney Disease Foundation. He is a founding member of the National Foundation for Depression. For APS, Tosteson was elected councillor in 1968 and president elect in 1972. He served also as a member of the Education Committee as a representative of the general physiologists (1961–67), of the Finance Committee (1977–79), and of the Editorial Board of *Physiological Reviews* (1976–78).

During his term as president, Tosteson was particularly concerned about the relationship of the Society to the rapidly expanding fields of molecular, cellular, and general physiology. He attempted to increase the representation of these specialties in annual and fall meetings and also in affairs of APS. One method for attaining this goal was to organize a three-day intersociety colloquium at the 1974 Spring Meeting of FASEB on the topic "Membranes, ions and impulses," in collaboration with representatives of the Biophysical Society and the Societies for Neuroscience and of General Physiology. In his past president's address (9), Tosteson summarized his conclusions about these problems and spoke to the changing concepts and definitions that affect physiology and physiologists. In a more recent appraisal he has written:

> At present, the epistemology of the biological sciences is arcane and confused. . . . One might ask whether the word, physiology, has served its purpose and should be gracefully retired to the archives. Aside from the obvious practical, sound, economic, and even political problems that such action would encounter, there is a deeper philosophical reason for sustaining and strengthening the discipline. . . . Because of new and exciting discoveries in recent decades, the conceptual and practical pathway toward complete characterization of every molecule that comprises a human being or any other living entity is now open. It is no longer a question of whether but of when such information will be available. But it is a serious, nontrivial question as to how we can best use this vast body of information. . . . It seems to me that this difficult theoretical and integrative work is at the center of our discipline. Physiology has long attracted individuals with a frame of mind to seek out such labors.

SELECTED PUBLICATIONS

1. TOSTESON, D. C., E. SHEA, AND E. C. DARLING. Potassium and sodium of red blood cells in sickle cell anemia. *J. Clin. Invest.* 31: 406, 1952.
2. TOSTESON, D. C., AND J. S. ROBERTSON. Potassium transport in duck red cells. *J. Cell. Comp. Physiol.* 47: 147, 1956.
3. TOSTESON, D. C. Halide transport in red blood cells. *Acta Physiol. Scand.* 46: 19, 1959.
4. TOSTESON, D. C., AND J. F. HOFFMAN. Regulation of cell volume by active cation transport in high and low potassium sheep red cells. *J. Gen. Physiol.* 44: 169, 1960.
5. ANDREOLI, T. E., J. A. BANGHAM, AND D. C. TOSTESON. The formation and properties of thin lipid membranes from HK and LK sheep red cell lipids. *J. Gen. Physiol.* 50: 1729, 1967.

6. ANDREOLI, T. E., M. TIEFFENBERG, AND D. C. TOSTESON. The effect of valinomycin on the ionic permeability of thin lipid membranes. *J. Gen. Physiol.* 50: 2527, 1967.

7. HOFFMAN, P. G., AND D. C. TOSTESON. Active sodium and potassium transport in high potassium and low potassium sheep red cells. *J. Gen. Physiol.* 58: 438, 1971.

8. GUTKNECHT, J., L. BRUNNER, AND D. C. TOSTESON. The permeability of thin lipid membranes to bromide and bromine. *J. Gen. Physiol.* 59: 486, 1972.

9. TOSTESON, D. C. Physiology and the future, past-president's address. *Physiologist* 17: 423–430, 1974.

10. HAAS, M., J. SCHOOLER, AND D. C. TOSTESON. Coupling of lithium to sodium transport in human red cells. *Nature Lond.* 258: 428, 1975.

11. TOSTESON, M. T., AND D. C. TOSTESON. Bilayers containing gangliosides develop channels when exposed to cholera toxin. *Nature Lond.* 275: 142–144, 1978.

12. LATORRE, R., J. J. DONOVAN, W. KOROSHETZ, D. C. TOSTESON, AND B. F. GISIN. Ion transport mediated by the valinomycin analog cyclo (L-Lac-L-Val-D-Pro-D-Val) (PV-Lac). *J. Gen. Physiol.* 77: 387–417, 1981.

13. TOSTESON, M. T., AND D. C. TOSTESON. The sting: mellittin forms channels in lipid bilayers. *Biophys. J.* 36: 109–116, 1981.

47 (1974–75)

Arthur C. Guyton

(b. 1919)

"Sectionalization" has a long history in APS; it extends back at least to the presidency of William F. Hamilton (1955–56). It was, and is, a sort of catchword for attempts to unify the Society by providing publications and membership sections desired by each of the subspecialties of the science. Earlier presidents left office frustrated by their inability to move the Society in this direction. Under the presidency of Guyton, however, both forms of sectionalization were approved and formally announced to the Society in the "President's report" (*Physiologist* 18: 79–82, 1975).

Guyton was born in Oxford, Mississippi, and received his early education there. He was an honor graduate of the University of Mississippi (1939), where he won prizes both for work in physics and in a short story contest. For his medical education he went to Harvard (M.D., 1943) and then became a surgical intern at Massachusetts General Hospital. By 1 January 1944, he was in the U.S. Navy and was assigned first to the National Naval Medical Center in Bethesda (4 months) and then to Camp Detrick in Maryland for research in bacterial warfare (22 months). In 1946 he was able to return to Massachusetts General Hospital to continue his surgical residency, but he contracted poliomyelitis and was obliged to give up these plans. A year later (1947) he joined the faculty of the School of Medicine at the University of Mississippi. The following year, at the age of twenty-nine, he was appointed professor and chairman of the Department of Physiology and Biophysics. He has held these appointments for nearly forty years.

Guyton's loyalty to his home state and to its region extends well outside the university. He is a member of the Southern Society for Clinical Investigation, was president in 1956–57, and received the Founder's Award in 1979. As a member of the Mississippi Heart Association, he was president in 1955–56 and received its Silver Medallion Award for Research in 1961. He was president of the Mississippi Academy of Science in 1967–68; from the academy he received major awards in 1976 and 1980. He is a charter member (1975) of the University of Mississippi Hall of Fame.

For AHA, Guyton served on its Policy Committee (1960–61), Board of Directors (1961–67), and Publications Board (1971–77), as well as on the Advisory Council of the Circulation Section and on the Council for High Blood Pressure Research. He is a member of the International Society for Hypertension. His lifelong interest in the circulation and major research contributions have brought him awards and honors from a remarkable variety of organizations: the Ida B. Gould Award of AAAS in 1959, the Wiggers Award of the Circulation Group of APS in 1972, and the Annual Research Achievement Award of AHA and the Dickinson Richards Award of the Pulmonary Council of AHA in 1975. He has been an honorary fellow of the American College of Cardiology from 1975. At the 400th Anniversary Symposium in honor of William Harvey in London, England, in 1978, Guyton delivered the Harvey Lecture and, in the following year, gave the First Annual Evan Jones Memorial Lecture at St. Thomas' Hospital in London. These honors were followed by the George Griffith Memorial Lectureship of the California Heart Association (1980), the Ciba Award for Research in Hypertension (1980), and the Jenssen Annual Lectureship of the Society of Cardiovascular Anesthesiologists (1982).

Early in his career, Guyton was selected by the U.S. Junior Chamber of Commerce as one of the Ten Outstanding Young Men of America for 1951. In 1956 he received the U.S. Presidential Citation for Development of Aids for Handicapped Persons. Other honors include recognition by the American Society of Anesthesiologists (1967), the Biomedical Engineering Society (1972), and the University of Nebraska (1972). He was given the Leonard and Lillian Ratner Award (1973) and awards by the American College of Chest Physicians (1973), ACDP (1975), the American Surgical Society (1975), and the Albany Medical College (1977). He was honored by the International Anesthesia Research Society (1977) and by the Medical College of Wisconsin (1977) where he gave the Walter E. MacPherson Lecture and received the D.Sc. degree. In 1979 he delivered the Einthoven Lecture and received the medal given biennially by the Einthoven Foundation of Leiden, Holland. The University of Western Ontario invited him to give the James A. F. Stevenson Memorial

Lecture in 1980. A year later he received the Mellon Award of the University of Pittsburgh, and in 1982 he received an honorary M.D. degree from the University of Pretoria in South Africa.

Guyton has long been an active member of the Biophysical Society, as well as of the Biomedical Engineering Society. For four years (1957–61; chairman, 1959–61) he was a member of the Physiology Test Committee of the National Board of Medical Examiners. He is a fellow of AAAS and served on the Council of the Society for Experimental Biology and Medicine for six years (1965–71). For the NIH he served on the Cardiovascular Study Section (1954–58), the Physiology Training Grant Committee (1958–64; chairman, 1961–64), and the National Heart and Lung Council (1971–75).

Elected to membership in APS in 1949, like many other former presidents, Guyton began his official service for the Society on its editorial boards (1958–63) and as chairman of the Education Committee (1967–70). He was elected councillor in 1972 and president elect the following year. (Because he was appointed chairman of the Finance Committee in 1977, he continued as an ex officio member of Council for another five years, 1977–82.) He is a member of three sections of APS: Neurophysiology, Circulation, and Respiration. In 1981 he received the Ray G. Daggs Award of the Society. As president of FASEB (1975–76), he steered the Federation Board and the member societies through a reorganization that resulted in a more flexible financial relationship.

Although his formal association with physiology came only after Guyton had all but completed his training as a surgeon, he had been carrying out experiments of one sort or another from the time he was a small boy. He described his research "training" as follows:

> Perhaps the most important aspect of my research training was the lack of any specific formal episode, but instead a continuing self-interest in research beginning rather early in life. Like many other young boys, I had my own shop, which later became more a laboratory especially in the field of electronics. I built devices such as an oscilloscope, code recorder, multiple amateur radio transmitters, multiple radio amateur receivers, an operational amplifier for multiplication in the days before analog computers began to be used, and so forth. This continued through college years. Then during medical school I had also a research fellowship working primarily on physical chemistry projects under Dr. A. Baird Hastings. However, Dr. Hastings was caught up in the war effort, so that most of my work there was likewise without direct supervision. My association with Hastings and with others in the department such as Oliver Lowry and Jack Buchanan provided a high degree of stimulus to continue research. Fortunately, during World War II, I was assigned to bacterial warfare research and immediately entered into a number of different research projects, which led to the first series of electronic and physiological research papers I published.
>
> During my stint as a surgical intern prior to World War II and as a surgical resident afterwards at the Massachusetts General Hospital, I worked on several projects of a surgical nature, especially projects related to intubation of patients, intermittent suction devices, and so forth. I was allowed to set up a laboratory and was about to

get a number of projects underway related to shock and other studies at the time I developed polio in 1946. Thus this was my background for research when I decided to go into academic and research life on a permanent basis beginning in 1947. However, aside from the above experience, probably the one single factor in my training that has been most important through the years was the several summers during my college days when I studied mathematical analysis of electronic circuits. This provided the basis for our subsequent analysis of circulatory, respiratory, renal, and body fluid mechanisms, all of which require the same type of mathematical analysis. . . . My principal interest over the years of research has been to provide a working analysis of total circulatory function, with special emphasis on all the important regulatory mechanisms.

In his first paper, published in 1946, Guyton described a method for measuring the size of particles in aerosols (1). By 1948, however, his bibliography begins the long and distinguished series of publications on the circulatory system. A typical paper is the one in 1951 by Guyton and J. W. Harris (2) on what had been regarded as "spontaneous" cycles of oscillation in arterial pressure. They attributed these waves to variations in vasomotion as a result of oscillation in the baroreceptor control system. In the next paper listed here (3), the control of cardiac output was approached via studies on venous return, and the factors that determine both venous return and cardiac output were analyzed here and eventually more fully in a monograph (6). Guyton's research on pulmonary capillary function is typified by the 1959 paper with A. W. Lindsey on pulmonary edema (4). They found that when left atrial pressure is raised, or when plasma protein concentration is lowered, pulmonary edema does not necessarily occur at once, because the system includes "safety factors" that defend the alveoli against fluid accumulation. One of Guyton's best-known discoveries, although he calls it "controversial," is the negative pressure in interstitial fluid surrounding capillaries and lymphatics (5).

Autoregulation of local blood flow is another major theme of Guyton's work. In 1964, with other authors (7), he described the linkage of decreases in oxygen supply with increases in local blood flow that has become a part of every textbook account. Autoregulation of renal glomerular blood flow by feedback control at the juxtaglomerular apparatus was proposed as a result of a theoretical computer analysis in 1964 (8). This has led to intensive investigation of the phenomenon and of the function of the juxtaglomerular apparatus in a number of different laboratories. More extensive but related analyses using computers were published in review articles (9–11) and in a monograph (13) describing regulation of the circulation as it involves body fluid volumes, cardiac output, local blood flow, arterial pressure, and other significant variables. Both the normal circulation and hypertension are included.

Of all the former presidents who have authored or edited textbooks of physiology, Guyton probably has had the broadest influence on students in medical school and college courses. Currently in print are three different textbooks of physiology, one of neurophysiology, and additional volumes he has written or edited. Especially in considering difficult topics, Guyton's style of writing can be paraphrased thus: "We do not know fully how this system operates. But one way it might function is as follows: . . ." Guyton's explanations, although hypothetical, are always reasonable and therefore easy for students to understand and remember. Their teachers appreciate the engaging honesty of his expositions.

SELECTED PUBLICATIONS

1. GUYTON, A. C. Electronic counting and size determination of particles in aerosols. *J. Ind. Hyg.* 28: 133, 1946.

2. GUYTON, A. C., AND J. W. HARRIS. Pressoreceptor-autonomic oscillation: a probable cause of vasomotor waves. *Am. J. Physiol.* 165: 158, 1951.

3. GUYTON, A. C. Determination of cardiac output by equating venous return curves with cardiac response curves. *Physiol. Rev.* 35: 123, 1955.

4. GUYTON, A. C., AND A. W. LINDSEY. Effect of elevated left atrial pressure and decreased plasma protein concentration on the development of pulmonary edema. *Circ. Res.* 7: 649, 1959.

5. GUYTON, A. C. A concept of negative interstitial pressure based on pressures in implanted perforated capsules. *Circ. Res.* 12: 399–414, 1963.

6. GUYTON, A. C. *Circulatory Physiology: Cardiac Output and Its Regulation.* Philadelphia, PA. Saunders, 1963.

7. GUYTON, A. C., J. M. ROSS, O. CARRIER, JR., AND J. R. WALKER. Evidence for tissue oxygen demand as the major factor causing autoregulation. *Circ. Res.* 14: 60–69, 1964.

8. GUYTON, A. C., J. B. LANGSTON, AND G. NAVAR. Theory for renal autoregulation by feedback at the juxtoglomerular apparatus. *Circ. Res.* 14: 187–197, 1964.

9. GUYTON, A. C., AND T. G. COLEMAN. Long-term regulation of the circulation: interrelationships with body fluid volumes. In: *Physical Bases of Circulatory Transport Regulation and Exchange.* Philadelphia, PA: Saunders, 1967.

10. GUYTON, A. C., T. G. COLEMAN, AND H. J. GRANGER. Circulation: overall regulation. *Annu. Rev. Physiol.* 34: 13–46, 1972.

11. GUYTON, A. C., T. G. COLEMAN, A. W. COWLEY, JR., R. D. MANNING, JR., R. A. NORMAN, JR., AND J. D. FERGUSON. A systems analysis approach to understanding long-range arterial blood pressure control and hypertension. *Circ. Res.* 35: 159–176, 1974.

12. GUYTON, A. C. Past-president's address. Physiology, a beauty and a philosophy. *Physiologist* 18: 495–501, 1975.

13. GUYTON, A. C. *Circulatory Physiology III: Arterial Pressure and Hypertension.* Philadelphia, PA: Saunders, 1980.

48 (1975–76)

Bodil M. Schmidt-Nielsen
(b. 1918)

Bodil Schmidt-Nielsen brought to the presidency of APS traditions different from those of her recent predecessors. Of the eleven presidents beginning with Pappenheimer (1964–65), seven of them received a major part of their education or training in laboratories at Harvard University. Schmidt-Nielsen's background, by contrast, was in a relatively small but unusually distinguished laboratory in Copenhagen, founded by her parents, August and Marie Krogh (11).

> I am the youngest of four children. My father and mother were both physiologists, and we children were daily exposed to conversations dealing with topics in physiology. Also, the many visitors and guests who came to the house were mostly scientists. During the first five years of schooling I was educated at home by a private teacher, together with my two-years older sister. This gave us the opportunity to have lunch daily with our parents as they came over to the house from the laboratory. . . . At the age of eleven I entered the Rysensteen Gymnasium from which I graduated in 1937 specializing in mathematics and natural sciences. I must acknowledge the superb teachers I had in the Danish gymnasium.

Although she had originally intended to study medicine, Schmidt-Nielsen decided rather to enter the School of Dentistry, where she quickly discovered that she was fascinated by the subject of physiology. She began to tutor fellow students in the subject and began also a research project on the exchange of calcium and phosphorus in human teeth (1). She was married to Knut Schmidt-Nielsen in 1939. (This marriage ended in divorce in 1965; in 1968 she married Roger G. Chagnon.) The first of her three children was born shortly after she received her D.D.S. degree in 1941. Her father then advised her to give up plans for continuing studies in medicine in favor of experimental work. She therefore continued research in the School of Dentistry, audited lectures in physiology at the university, and briefly practiced dentistry while she taught prosthetics and, later, dental surgery. In 1946 she became the first person to qualify in Denmark for the newly established degree doctor of odontology. She received the Dr. Phil. degree in 1955.

> In 1946 Laurence Irving and Per F. Scholander invited my husband and me to come to Swarthmore College as research associates. We sailed for the United States on 5 November 1946, in a rickety Liberty ship with our two children. We arrived in Swarthmore on 2 December. There I first worked with gas analysis using Scholander's

new micromethods. Then, in the spring of 1947 Irving suggested that we study water metabolism of kangaroo rats in southern Arizona. We worked in the desert with kangaroo rats, pocket mice, and the desert rat *Neotoma* during two summers (1947 and 1948). The experiments were continued at Swarthmore, and later at Stanford University, where I decided to learn about renal function. In my spare time I studied Homer Smith's book on the kidney (*Physiology of the Kidney*, Oxford, 1937). The next important event in my education came when, on Smith's invitation, I spent my first summer at Mount Desert Island Biological Laboratory working with Roy Forster on urea secretion by frog tubules. My background with my father's laboratory had already prepared me for becoming a comparative physiologist, but I did not know how much the comparative approach dominated my thinking until I started working in comparative renal physiology at Mount Desert Island Biological Laboratory. Further, the many discussions with other scientists there, including Homer Smith and E. K. Marshall, immensely stimulated my interest.

Schmidt-Nielsen was appointed a research associate at Swarthmore College (1946–48), at Stanford University (1948–49), at the Kettering Institute of the University of Cincinnati (1949–52), and at Duke University (1952–54; senior research associate, 1954–57). She then became an associate research professor at Duke, at first in zoology (1957–61) and then in zoology and physiology (1961–64). Her next position was as professor of biology at Case Western Reserve University in Cleveland (1964–71), where she was chairman of that department for one year (1970–71), until she accepted the rank of adjunct professor on her move to Mount Desert Island as a research scientist on a year-round basis (1971–). At that time she became also an adjunct professor at Brown University. She had already served for fourteen years as a trustee of the Mount Desert Island Biological Laboratory (1955–69) and was appointed to this board again in 1975. Since 1978 she has been a member of the laboratory's Executive Committee, was deputy director for three years (1979–82) and vice-president for a year (1980–81), and has served as president since 1981.

A fellow of AAAS (Boston) since 1973, Schmidt-Nielsen is also a fellow of the New York Academy of Science (1958) and of AAAS (1959). She is a member of both the American and the International Society of Nephrology and of the International Society of Lymphology. For twenty-six years (1950–76) she was a member of the Society for Experimental Biology and Medicine and served on its Council from 1967 to 1971. She is a member of the American Society of Zoologists. Her research has been honored by her appointment as a John Simon Guggenheim Memorial Fellow (1953–54) and as an Established Investigator of AHA (1954–62), as well as with a Career Award of NIH (1962–64). She was the Alvin F. Reick Memorial Lecturer at the Medical College of Wisconsin (1982) and received the honorary degree, D.Sc., from Bates College of Lewiston, Maine, in 1983.

Not all her honors have come from American sources. As early as 1945–46 she received an award from the King Christian X Fund of Denmark for work on the role of saliva in protection against caries. Twenty years later she was invited to lecture at five of the Swedish universities (1965), and in 1974 she delivered the Jacobus Lecture at her alma mater, the University of Copenhagen.

In the following paragraphs, Schmidt-Nielsen has summarized her research interests:

My early interests centered on calcium and phosphorus metabolism. This included the role of calcium oxalate and vitamin D in calcification of bones and maintenance of plasma calcium. It also included the solubility of tooth substance (hydroxyapatite) in saliva.

Since 1947 when I first worked on the water metabolism of desert rodents, my main interest has been the physiology of the kidney and the role of the kidney and other excretory organs in regulating the osmolality and volume of the extra- and intracellular compartments. My approach has been comparative and has involved structure as well as function. I have worked with amoebae, invertebrates, fish, amphibians, reptiles, birds, and a variety of mammals from extreme habitats. My interests have primarily centered on the ability of various animals to form dilute or concentrated urine, the anatomical structures associated with these functions, and their relationship to habitat and evolution.

A major interest has been the countercurrent system of the mammalian kidney and the role of urea in production of a concentrated urine. I have also been specifically interested in handling of urea by mammalian and lower vertebrate nephrons and how the mammalian kidney conserves urea when dietary protein intake is restricted.

Schmidt-Nielsen's studies of desert rodents include a complete accounting of water metabolism in kangaroo rats, where the authors calculated output of water and input from all sources for animals on a completely dry diet. They showed that the kangaroo rat is able to survive without access to water (2). The study was comparative in that it compared these data with the same parameters for laboratory rats. In her Bowditch Lecture (3), Schmidt-Nielsen extended the comparative significance of her results by describing adaptive mechanisms for water conservation and desert survival in camels. A camel cannot escape the heat of a desert, whereas a kangaroo rat can hide underground.

When mammals are fed diets low in protein content, their kidneys help compensate for the low nitrogen intake by conserving urea. Schmidt-Nielsen and her associates studied this adaptation in various animals and found it more pronounced in ruminants than in others. In an intensive study using sheep, the authors reported that over a wide range of urine flows the filtered urea is reabsorbed differently by renal tubules when the animals are fed a low-protein versus a normal-protein diet (4). Another study in the same series (not cited here) dealt with distribution of urea and salt in the renal medulla on the two kinds of diet. The experiments showed that specific renal regulatory mechanisms conserve nitrogen and maintain blood urea concentration when dietary protein intake is low.

Following early publications by other authors on the countercurrent mechanism for concentrating urine, Schmidt-Nielsen and O'Dell (5) reported a clear-cut relationship between the length of the inner medulla in mammalian kidneys and the ability to concentrate urine. Stimulated by a chance remark by Homer Smith to the effect that at that time nothing was known about kidney function in invertebrates, Schmidt-Nielsen turned her attention to freshwater invertebrates, and then to amoebae, where excretion involves nothing more complicated than the contractile vacuole (6). Fluid within the vacuole proved to be dilute, as expected, but how the organism maintains an osmotic gradient between the contents of the vacuole and the surrounding cytoplasm could not be explained. Crocodiles cannot make a dilute urine, but their renal tubules can synthesize and secrete ammonium and bicarbonate ions. These ions then can be exchanged for Na^+Cl', with the result that they can excrete water without undue loss of extracellular electrolyte (7). In the spiny dogfish, Schmidt-Nielsen and her associates found that over an enormous range of reabsorptive rates there is a strict stoichiometric relationship between reabsorption of Na^+ and of urea (8). This indicates that in the shark the reabsorption of urea is essentially passive but also Na^+ dependent, even though the final urea concentration in urine is substantially lower than that of plasma. A countercurrent system is believed to be responsible.

The old question, What is the function of smooth muscle in walls of the renal pelvis? was at least partially answered when Schmidt-Nielsen and Graves reported that the renal pelvic wall actually milks the renal papillae (9). Mechanical forces in the pelvic wall have a profound effect on all events in the papilla, including flow in collecting ducts, capillaries, vasa recta, loops of Henle, and interstitial "spaces." These pelvic contractions may aid in transfer of fluid from the lumen of a collecting duct through the cells and into the interstitium. Finally, in a return to adaptation to low-protein intake, her laboratory has begun new studies of urea handling in rats fed low-protein diets. Physiological changes (i.e., urea conservation) are accompanied by morphological changes in the tubules that indicate which tubular segments, in addition to the collecting ducts, are involved in urea reclamation.

In 1949 Schmidt-Nielsen was approved for membership in APS. Her Bowditch Lecture at the fall meeting in 1957 (3) was the second in this series. She was elected to Council in 1971 and became president elect in 1974. Among the achievements that marked her term of office was the initiation of activities that eventually made sections of the Society functional. From 1976 to 1981 she was associate editor of the *American Journal of Physiology: Regulatory, Integrative and Comparative Physiology*. In addition, she served as a member or as chairman of the Perkins Fellowship Committee (1972–74 and 1980–83), the Ray G. Daggs Award Committee (1984–87), and the Honorary Membership Committee (1983–86). Of her presidential years she has written:

> During my presidency, Council became committed to the strengthening of specialty groups within the APS, and we asked members to form task forces to formalize existing groups into sections. I was personally eager to establish a section in comparative physiology and fortunately had an ally in Orr Reynolds who was equally interested. The section thus established became a joint section with the American Society of Zoologists, and this resulted in several joint meetings between the two parent societies. The task force on a section for gastroenterology was also formed in 1976.
>
> To improve the programming of APS meetings, I suggested that the program committee be reorganized so that advisory members should be named from each section. This would encourage specialty groups to become more active in the

meetings. To organize the new program committee, a very productive meeting was held at Mount Desert Island in the summer of 1976. To me the sectionalization of both the journal and the Society were the most important events during my years on Council.

Election of Bodil Schmidt-Nielsen as president was a unique event in the history of the Society. Recognizing the significance of her office for the rapidly growing number of women in physiology, she was heard to say, "I think the best way I can represent women in physiology is to do my best possible job as president." Her election, and then the obvious success of her presidency, gratified physiologists, both inside and outside the Society who had been following the progress of her research and career, and simply delighted her many friends.

SELECTED PUBLICATIONS

1. PEDERSEN, P. O., AND B. SCHMIDT-NIELSEN. Exchange of phosphorus in human teeth. *Acta Odontol. Scand.* 4: 1–20, 1942.

2. SCHMIDT-NIELSEN, B., AND K. SCHMIDT-NIELSEN. A complete account of the water metabolism in kangaroo rats and an experimental verification. *J. Cell. Comp. Physiol.* 38: 165–182, 1951.

3. SCHMIDT-NIELSEN, B. The resourcefulness of nature in physiological adaptation to the environment. *Physiologist* 1(2): 4–20, 1958.

4. SCHMIDT-NIELSEN, B., H. OSAKI, H. V. MURDAUGH, JR., AND R. O'DELL. Renal regulation of urea excretion in sheep. *Am. J. Physiol.* 194: 221–228, 1958.

5. SCHMIDT-NIELSEN, B., AND R. O'DELL. Structure and concentrating mechanism in the mammalian kidney. *Am. J. Physiol.* 200: 1119–1124, 1961.

6. SCHMIDT-NIELSEN, B., AND C. R. SCHRAUGER. *Amoeba proteus*: studying the contractile vacuole by micropuncture. *Science Wash. DC* 139: 606–607, 1963.

7. SCHMIDT-NIELSEN, B., AND E. SKADHAUGE. Function of the excretory system of the crocodile (*Crocodylus acutus*). *Am. J. Physiol.* 212: 973–980, 1967.

8. SCHMIDT-NIELSEN, B., B. TRUNIGER, AND L. RABINOWITZ. Sodium linked urea transport by the renal tubule of the spiny dogfish, *Squalus acanthias. Comp. Biochem. Physiol. A Comp. Physiol.* 42: 13–25, 1972.

9. SCHMIDT-NIELSEN, B., AND B. GRAVES. Changes in fluid compartments in hamster renal papilla due to peristalsis in the pelvic wall. *Kidney Int.* 22: 613–625, 1982.

10. SCHMIDT-NIELSEN, B. Historical sketch, August and Marie Krogh and respiratory physiology. *J. Appl. Physiol.: Respirat. Environ. Exercise Physiol.* 57: 293–303, 1984.

11. SCHMIDT-NIELSEN, B., J. M. BARRETT, B. C. GRAVES, AND B. CROSSLEY. Physiological and morphological responses of the rat kidney to reduced dietary protein. *Am. J. Physiol.* 248 (*Renal Fluid Electrolyte Physiol.* 17): F31–F42, 1985.

49 (1976–77)

Ewald Erdman Selkurt
(b. 1914)

Selkurt's year as president of APS essentially began with the "last" of the fall meetings to be held in August on a college or university campus. The University of Pennsylvania was the site of the meeting; the date was the year of the nation's bicentennial celebration. All the medical institutions of Philadelphia were hosts and helped plan and carry out the several activities (*Physiologist* 19: 73–80, 1976). Recalling the occasion, Selkurt wrote:

> Announcement of termination of the summer (August) campus-type of meeting was met with dismay by some. This was tempered by the statement that going to the October city-resort format for 5–6 years could be viewed as a trial basis; return to the original format could then be considered. In fact, specialty groups might want to take up the August dates. As it turned out, Michigan State was planned for an August 1978 meeting of a specialty group.

It also turned out that in 1983 in Honolulu, and in 1984 in Lexington, Kentucky, the meetings were held again in August. Moreover there continues to be a close association with academic institutions in the cities chosen for the "city" meetings.

Born in Edmonton, Alberta, Selkurt received his formal higher education at the University of Wisconsin; he graduated in 1937 with a major in zoology and received the Ph.D. degree in 1941 in physiology. For the next three years (1941–44) he was an instructor in Homer Smith's department at New York University School of Medicine, until he moved to Carl Wiggers' department in what is now the Case Western Reserve University School of Medicine. By 1949 he was associate professor there and from 1953 to 1955 was coordinator of phase I of the revised medical curriculum. He then moved to Indianapolis (1958) as professor of physiology in the School of Medicine of Indiana University, where he served as chairman of the department from 1958 until 1980. In 1976 his contributions to the school and to the university were recognized by his appointment as Distinguished Professor of Physiology.

Selkurt has been active in a variety of professional societies. He became a member of the Harvey Society (New York) in 1942 and of the Society for Experimental Biology and Medicine in 1946; he served later (1978–81) as a member of its Council. A member of AHA from 1948, he was a delegate of the Assembly to the Basic Science

Council in 1966 and 1967 and a member of the Executive Committee of this Council from 1966 to 1970. He took part in two of the conference series organized by the Josiah Macy, Jr., Foundation: on kidney (1949–53) and on shock and circulatory homeostasis (1951–55). Since 1973 he has been a member of the Council of Academic Societies of AAMC. In 1971–72 he served as president of ACDP and was given its Service Award in 1979.

For five years (1953–58) Selkurt was a member of the Subcommittee on Shock of the Committee on Medical Sciences of NRC. A few years later he served on the Cardiovascular Study Section of the National Heart Institute (1963–68) and, in 1963, on the Panel for Evaluation of Science Faculty Fellowships of NSF. For a year (1964–65) he was NSF Fellow in Göttingen, West Germany. He has held visiting professorships at the University of Puerto Rico in San Juan (1968) and at the University of California at Irvine (1969 and 1970) and also was Centennial Visiting Professor at Ohio State University College of Medicine in 1970.

Before he left New York in 1944, Selkurt was elected to membership in APS. He served as a member of the Finance Committee for four years (1967–71), was chairman of the Steering Committee of the Society's Circulation Group in 1969–70, and was then elected to Council (1971–74). He was chosen as president elect in 1975.

Selkurt was a member of the editorial boards of the Society's journals from 1954 to 1973. He has held editorial positions also with the *Annual Review of Physiology* (1965–68), the *American Heart Journal* (1966–77), the *Proceedings of the Society for Experimental Biology and Medicine* (1967–), and *Circulatory Shock* (1973–85). Because of his interest in blood flow, circulatory shock, and the renal circulation, he has been invited to take part in conferences in Stockholm, in Switzerland, and in Mexico City, as well as in sessions of the IUPS Congresses and in numerous conferences in this country.

Asked about his training as an investigator and his research interests, Selkurt replied:

> My main research interests lay broadly in renal and cardiovascular physiology. More specifically, in renal physiology, I was interested in renal hemodynamics and its relation to electrolyte handling, mainly sodium. In the cardiovascular area, my greatest interest was in the hemodynamics of the splanchnic bed (intestine, liver). In particular, one important phase was the study of these organs (kidney, splanchnic) in hemorrhagic and ischemic shock (work done on dogs and monkeys). I utilized indirect (renal clearances) and direct (rotameter, electromagnetic flow probe) methods for blood flow measurements. In more recent years, my interest has been directed to the mechanisms of renin release (dogs) and the role of histamine. Currently we are studying the role of H-receptors on the renin mechanism.
>
> My introduction to physiology (and first research experience therein) was in the laboratory of Dr. Walter J. Meek at the University of Wisconsin, Madison, where I participated as a graduate assistant. Dr. Meek engendered a keen interest in physiology. A liaison between Dr. Meek and Dr. Homer W. Smith eventuated in my first position in physiology, in 1941, as instructor in Dr. Smith's department in New York University. This was an opportune time, because Dr. Smith's laboratory and associates

(R. F. Pitts, J. Shannon, H. Lauson, S. Bradley, and others) were working in the heyday of new indirect renal clearance methods that could be used for studies in unanesthetized dogs and humans: the clearance of inulin for measuring glomerular filtration rate and the clearance of diodrast and PAH (*p*-aminohippuric acid) for measurement of renal plasma flow. I learned these methods and continue to apply them during the rest of my career. The scholarly stimulation of Homer Smith opened up renal physiology as a fascinating field.

During the war years, I was concerned about making some sort of contribution, so I looked around to see which laboratories were doing relevant work. Standing out strongly was work being done at Western Reserve University (Cleveland) in the department of Carl J. Wiggers. I came there in 1944 and stayed beyond Wiggers' retirement as chairman (1954) until I took the chairmanship at Indiana University (1958).

Heading an active group studying many aspects of shock, primarily hemorrhagic, Wiggers convinced me to study renal function in hemorrhagic and in ischemic shock. Later, we decided to examine other regions of the splanchnic bed and studied hemodynamics of gut and liver, both normal and in shock. This was continued into the early 1960s. My interest then went strongly to kidney (e.g., the influence of hemodynamic factors on electrolyte handling). Here the training obtained under Homer Smith again took over. We studied the influence of changes in arterial, renal venous, and ureteral pressure.

But the study of shock continued through the years at Indiana University, with a shift to study of primates (monkeys). Finally, about 1975, our interest was directed to the prostaglandin, histamine, and renin interrelationship in the dog kidney. How histamine stimulates renin release is the focus of my current research.

Publications that typify these experiments include two papers in 1946. The first demonstrated that renal blood flow and renal clearances do not necessarily vary in parallel during hemorrhagic shock (1). The second, on the relation between renal blood flow and effective arterial pressure, contained the first evidence for the possibility of autoregulation of renal blood flow (2). Clearances, including sodium ion clearance, were further utilized for evaluation of renal function during graded decrements of arterial pressure (3). The authors found that as arterial pressure decreased, sodium gradually decreased in the urine until at low pressure levels the urine was sodium free. This observation laid the basis for a clinical test for unilateral kidney function. A converse result was discovered in another study of renal handling of electrolyte when elevations of mean arterial pressure were found to increase sodium excretion, with no change in renal blood flow or in glomerular filtration rate (4). At about this time Selkurt returned to study of hemorrhagic shock, by measuring blood flow and pressure in the splanchnic vasculature, together with oxygen uptake, in the presence of the failing circulation (5). A curious feature of creatinine excretion turned up in experiments published in 1969 (6). As an amphoteric substance, creatinine is secreted by tubules of the guinea pig kidney as either an organic base or an organic acid. As noted in Selkurt's own account above, recent work involves the possible role of prostaglandins in hemorrhagic shock (7) and the interrelation of histamine and prostaglandins in evoking renin release in canine kidneys (9).

Two of Selkurt's reviews on renal circulation and electrolyte excretion have been widely quoted by other authors: one in *Physiological Reviews* (1954) and the other in section 2 on the circulation in the *Handbook of Physiology* (1962). He is editor of two textbooks, *Physiology* and *Basic Physiology for the Health Sciences* (now in editions 5 and 2, respectively, Little, Brown, Boston).

Events of particular interest while Selkurt was president of APS include a proposal to change the classes of membership to create a student member, different from the older associate member classification. In continuation of the process begun while Schmidt-Nielsen was president, a Program Executive Committee of three persons and a Program Advisory Committee of ten were organized to include representatives of specialty groups. One consequence was a relatively large number of symposia at the 1977 FASEB Meeting in Chicago. An endowment fund was established for the Society in 1976–77, the initial gift of $100,000 coming to the Society as the bequest of Dr. Caroline tum Suden. It was during Selkurt's tenure of office that the future of FASEB was threatened by a possible withdrawal of the biochemists (see chaps. 10 and 16). As Bowditch Lecturer for the 1977 fall meeting, Selkurt chose Franklyn G. Knox, who in 1985 was to become president elect of APS.

Selkurt's past president's address was unusual in its scholarly treatment of what at first seemed to be a whimsical title, "Is the dodo bird *really* extinct?" (8). The addresses of five former presidents were referred to, and portions of two others were actually quoted, together with Robert Pitts' call for a renaissance of laboratory teaching on the occasion when he received the teaching award of ACDP. James Reston, Maurice Visscher, J. E. Dunphy (surgeon at University of California, San Francisco), Vernon W. Lippard (former Yale dean), Harvey Cushing, and Sir William Osler were among other distinguished authorities cited. Selkurt concluded by quoting from chapter 3 of Ecclesiastes, "There is a time for everything . . .", by adding, "There is a time to look backward, a time to look ahead." This is not, however, the last word of the address as it was published. The talk ended, as it began, with a cartoon. In this one a physician seated behind an enormous desk is telling an anxious patient, "Selkurt, what you need is a massive infusion of federal funds!"

SELECTED PUBLICATIONS

1. SELKURT, E. E. Renal blood flow and renal clearances during hemorrhagic shock. *Am. J. Physiol.* 145: 692, 1946.
2. SELKURT, E. E. The relation of renal blood flow to effective arterial pressure in the intact kidney of the dog. *Am. J. Physiol.* 147: 537, 1946.
3. SELKURT, E. E., P. N. HALL, AND M. P. SPENCER. Influence of graded arterial pressure decrement on renal clearance of creatinine, para-aminohippurate and sodium. *Am. J. Physiol.* 189: 369, 1949.
4. SELKURT, E. E. Effect of pulse pressure and mean arterial pressure modification on renal hemodynamics and electrolyte and water excretion. *Circulation* 4: 341, 1951.
5. SELKURT, E. E., AND G. A. BRECHER. Splanchnic hemodynamics and oxygen utilization during hemorrhagic shock in the dog. *Circ. Res.* 4: 693, 1956.
6. ARENDSHORST, W. J., AND E. E. SELKURT. Renal tubular mechanisms for creatinine secretion in the guinea pig. *Am. J. Physiol.* 218: 1661–1670, 1970.

7. JOHNSTON, P. A., AND E. E. SELKURT. Effect of hemorrhagic shock on renal release of prostaglandin E. *Am. J. Physiol.* 230: 831–838, 1976.

8. SELKURT, E. E. Past-president's address. Is the dodo bird *really* extinct? *Physiologist* 20(5): 1–8, 1977.

9. SELKURT, E. E., G. M. HOCKEL, AND M. H. WEINBERGER. Some evidence for interrelationship of histamine and prostaglandin on renal function. *Proc. Soc. Exp. Biol. Med.* 60: 328, 1979.

50 (1977–78)

William F. Ganong
(b. 1924)

Known to his friends usually as "Fran," or occasionally as "Bill," Ganong is a mammalian physiologist, an endocrinologist, and a neurobiologist. In fact, he belongs to a discipline that intersects all three of these parent fields, the specialty of neuroendocrinology. His career developed pari passu with the emergence and maturing of research in this area (13). Whereas only a few years ago he was regarded as a "pioneer," he is now recognized as one of the founding fathers of the science.

My first exposure to intensive research was with Peter Forsham in Thorn's laboratory at the Peter Bent Brigham Hospital when I was a senior medical student. This stimulated my interest in the neural control of pituitary secretion, and at that time, the only neuroendocrine research in the United States was being conducted by David Hume in the Surgical Research Laboratory at Harvard. Subsequently, I became his postdoctoral fellow and spent three years in his laboratory. My research interest remains the broad field of neuroendocrinology, including the production of hormones by neurons and endocrine-brain interactions.

I set as my initial research goal elucidation of the mechanisms regulating aldosterone secretion. With the demonstration that renin via angiotensin II was a major regulator of aldosterone, I became interested in the control of renin secretion, and then in the neural components of this process. In the last few years, I have begun to work on the extravascular renin-angiotensin systems in the brain and the pituitary gland.

My first paper (1) is a favorite, partly because it was my first but also because it introduced a treatment for the Guillain-Barré syndrome that proved to be useful. I had always had a secret ambition to be "immortalized" by having my name attached to a syndrome, and I am pleased that the Lown-Ganong-Levine syndrome grew out of my second paper (2). I also published several papers on Korean hemorrhagic fever before settling down to fundamental physiological research.

The years with Hume were active and stimulating. I was able to show that hypothalamic lesions block compensatory and stress-induced adrenocortical hypertrophy (3). Don Fredrickson and I showed that in the dog, as in the rat, the hypothalamus regulates TSH [thyroid-stimulating hormone] secretion (4).

In California, I focused on regulation of aldosterone secretion. Mulrow and I found that aldosterone takes 30–60 minutes to exert its effect on the kidney (5). This, and similar findings by Barger and associates, led Edelman to experiments that showed that aldosterone acts by way of DNA [deoxyribonucleic acid], RNA [ribonucleic acid], and synthesis of new protein. Mulrow and I also found that there is a humoral factor other than ACTH [adrenocorticotropin] involved in regulation of aldosterone secretion and that it comes from the kidney (6). It is renin acting via angiotensin II (8). This work was carried out independently of, and yet simultaneously with, experiments of J. O. Davis and his associates.

We have also studied factors that regulate sensitivity of the zona glomerulosa. In animals on a low-sodium diet the sensitivity of the zona glomerulosa to ACTH is markedly increased (10). Sensitivity to angiotensin II is also increased, and exogenous renin can duplicate the effect of sodium depletion in sodium-replete dogs. Thus, angiotensin II is the "ACTH of the zona glomerulosa." Later (11, 12) we began to study the regulation of renin secretion with particular emphasis on neural components (as did Winer and his associates in Kansas City).

We have recently become interested in other aspects of the interactions between the brain and the renin-angiotensin system, as well as the distribution of renin and of angiotensin II and of converting enzyme (14). We have found also that the brain-renin-angiotensin system plays an important role in regulation of anterior pituitary secretion and that inhibiting the action of angiotensin II in the hypothalamus prevents the LH [luteinizing hormone] surge and ovulation (16). Finally, we have studied the anterior pituitary-renin-angiotensin system and found that gonadotropes contain angiotensin II-like immunoreactivity (15). Recently we have shown that norepinephrine acts via postsynaptic α_2 receptors in the hypothalamus to inhibit ACTH and stimulate growth hormone secretion (17). I can add that I am one of the few investigators who has successfully hypophysectomized a deer (9) and, with Clegg and others, demonstrated that light penetrates into the brain of mammals (7). I do not view this last as one of my more weighty achievements, but it has certainly attracted attention and generated peculiar comments.

Ganong was born in Northampton, Massachusetts, and entered Harvard College (1941) just in time to be caught up in wartime disjointing of college and professional education. Drafted into the U.S. Army and assigned to the infantry, he finally was enrolled in the ASTP in Georgetown College in Washington, D.C., for a year of premedical study and then assigned for two years to the University of Virginia Medical School in Charlottesville (1945–47). Harvard meanwhile decided that they could award him an A.B. degree in 1946, and at long last (1947) he was able to transfer to the Harvard Medical School, from which he graduated in 1949. For the next two years he was a member of the house staff in medicine at Peter Bent Brigham Hospital and then was recalled to the Army as a medical officer assigned to duty in Japan and Korea. From 1952 to 1955 he was able to return to Harvard as a research fellow in medicine and surgery for training in the laboratories of George W. Thorn and David M. Hume, respectively. His clinical research preparation came to an end in 1955, however, when he accepted a position in the department of Leslie

L. Bennett at the University of California at Berkeley. Three years later the department was moved to San Francisco, where by 1964 Ganong held the rank of professor. In 1982 he became the Jack D. and DeLoris Lange Professor of Physiology. He served some years as vice-chairman of the Department of Physiology, and from 1970 he has been its chairman.

A member of APS from 1957, Ganong has served the Society in many different capacities. In 1969-70 he was chairman of the Neuroendocrine Discussion Group, one of the informal groups associated with APS. He was elected to APS Council in 1975 and as president elect the following year. He belongs to the category of former presidents who had originally intended to practice medicine, as well as to the group who are graduates of Harvard Medical School. As president, he initiated the Financial Development Committee (chairman, 1980-83), which seeks alternative sources of support for the Society to alleviate partially the dues burden on the membership. His most recent services to APS are as a member of the council of the Section on Endocrinology and Metabolism (1983-) and as a member of the Publications Committee (1984-).

During the year Ganong was president of APS, three movements that had been in progress earlier finally came to realization. One was broadening the bases of membership, so that the Society might become more representative of physiology as a whole, and inclusive rather than exclusive. Another was a more active participation of members in both state and federal governmental policies relating to biomedical research. The third was final settlement of the question of equity of the several societies in the assets of FASEB (see chapt. 16).

Ganong has held major office also in most of the other professional and scientific societies of which he is a member. He has been a member of Council (1970-73) and chairman of the Nominating Committee (1980-81) of the Endocrine Society, a member of Council (1972-76) and vice-president of the International Society for Neuroendocrinology (1976-80), and a member of Council (1974-75) and president (1976-77) of ACDP. For six years (1975-81) he was a member of the U.S. National Committee for IUPS (chairman, 1976-79). A member of the Council of Academic Societies of AAMC, he was on their Administrative Board from 1981 to 1987. The California Heart Association (1965-70), the Ciba Award Committee of AHA (1980-83), the Lawrence Berkeley Laboratory (1975-82), and the Smokeless Tobacco Research Council (1983-) are among the institutions he has served in consultation. For NIH he has been a member of numerous study sections and panels, including the Neurology A Study Section (1971-75) and the Task Force on Hypertension (1975-78). Besides these he has served on various committees of the Society for Experimental Biology and Medicine, the International Society of Endocrinology, FASEB, and the Society for Neuroscience. Two of his memberships are uncommon in APS: the Society of Brigham Surgical Alumni and the 38th Parallel (Korea) Medical Society. During his military service in Korea he received a combat zone promotion from first lieutenant to captain.

Editorial responsibilities Ganong has fulfilled include the journals of the Society (1960–66), *Endocrinology* (1961–73), *Proceedings of the Society for Experimental Biology and Medicine* (1962–69), *Medcom* (1968–78), *Neuroendocrinology* (1968–73), the *Journal of Pharmacology and Experimental Therapeutics* (1969–75), and those he currently holds for *Neuroscience*, the *Italian Journal of Physiological Sciences*, and *Excerpta Medica*; he is currently editor-in-chief of *Neuroendocrinology*.

Ganong earned distinction in each of the colleges where his peripatetic career as a student took him. With his first formal venture into neuroendocrinology he was awarded the Boylston Medical Society Prize in 1949 for his paper on "Control of ACTH secretion." Later honors include a corresponding membership in the Chilean Endocrine Society (1966) and receipt of the Golden Hippocrates Award presented by the Instituto Farmacoterapico Italiano (1970). He presented the Sherrington Society Lecture in London (1976) and the Starling Memorial Lecture in Jamaica (1978), received the teaching award of ACDP (1978), and delivered the Centennial Distinguished Lecture at the University of Arkansas (1979), the Nelson Lecture at Rutgers University (1980), the Macallum Lecture at the University of Toronto (1980), and the Jane Russell Wilhelmi Lecture at Emory University (1982). He became a fellow of AAAS in 1980 and an honorary member of the Argentine Society of Physiological Sciences in 1982.

Physiologists unfamiliar with the scope of Ganong's research will nevertheless know of his textbooks and other volumes he has written or edited: *Review of Medical Physiology* (Lange Medical Publications, Los Altos, CA), now in its twelfth edition; the two-volume work on *Neuroendocrinology* edited by Martini and Ganong (1966, Academic Press, New York); and nine volumes of the series *Frontiers in Neuroendocrinology* (first with Oxford Univ. Press, New York, and then with Raven Press, New York). In adding these titles to his list of favorite publications, Ganong wrote, "It is obvious that I have invested a fair amount of libido in *Review of Medical Physiology.*" Prospective textbook writers will be pleased to learn this recipe for success.

SELECTED PUBLICATIONS

1. STILLMAN, J. S., AND W. F. GANONG. The Guillain-Barré syndrome: report of a case treated with ACTH and cortisone. *N. Engl. J. Med.* 246: 293–296, 1952.
2. LOWN, B., W. F. GANONG, AND S. A. LEVINE. The syndrome of short P-R interval, normal QRS complex and paroxysmal rapid heart action. *Circulation* 5: 693–706, 1952.
3. GANONG, W. F., AND D. M. HUME. Absence of stress-induced and "compensatory" adrenal hypertrophy in dogs with hypothalamic lesions. *Endocrinology* 55: 474–483, 1954.
4. GANONG, W. F., D. S. FREDRICKSON, AND D. M. HUME. The effect of hypothalamic lesions on thyroid function in the dog. *Endocrinology* 57: 355–362, 1955.
5. GANONG, W. F., AND P. J. MULROW. Rate of change in sodium and potassium excretion after injection of aldosterone into the aorta and renal artery of the dog. *Am. J. Physiol.* 195: 337–342, 1958.
6. GANONG, W. F., AND P. J. MULROW. Evidence of secretion of an aldosterone-stimulating substance by the kidney. *Nature Lond.* 190: 1115–1116, 1961.

7. Ganong, W. F., M. D. Shepherd, J. R. Wall, E. E. Van Brunt, and M. T. Clegg. Penetration of light into the brain of mammals. *Endocrinology* 72: 962–963, 1963.

8. Lee, T. C., E. G. Biglieri, E. E. Van Brunt, and W. F. Ganong. Inhibition of aldosterone secretion by passive transfer of antirenin antibodies to dogs on a low sodium diet. *Proc. Soc. Exp. Biol. Med.* 119: 315–318, 1965.

9. Hall, T. C., W. F. Ganong, and E. B. Taft. Hypophysectomy in the Virginia deer: technique and physiologic consequences. *Growth* 30: 383–392, 1966.

10. Ganong, W. F., and A. T. Boryczka. The effect of a low sodium diet on the aldosterone-stimulating activity of angiotensin II in dogs. *Proc. Soc. Exp. Biol. Med.* 124: 1230–1231, 1967.

11. Loeffler, J. R., J. R. Stockigt, and W. F. Ganong. The effect of α- and β-adrenergic blocking agents on the increase in renin secretion produced by stimulation of the renal nerves. *Neuroendocrinology* 10: 129–138, 1972.

12. Nolly, H. L., I. A. Reid, and W. F. Ganong. The effect of theophylline and adrenergic blocking drugs on the renin response to norepinephrine in vitro. *Circ. Res.* 35: 575–579, 1974.

13. Ganong, W. F. The brain and the endocrine system: a memoir. In: *Pioneers in Neuroendocrinology II*, edited by J. Meites, B. T. Donovan, and S. M. McCann. New York: Plenum, 1978, p. 189–200.

14. Brownfield, M. S., I. A. Reid, D. Ganten, and W. F. Ganong. Differential distribution of immunoreactive angiotensin and converting enzyme in the brain. *Neuroscience* 7: 1759–1769, 1982.

15. Steele, M. K., M. S. Brownfield, and W. F. Ganong. Immunocytochemical localization of angiotensin immunoreactivity in gonadotrops and lactotrops of the rat anterior pituitary gland. *Neuroendocrinology* 35: 155–158, 1982.

16. Steele, M. K., R. V. Gallo, and W. F. Ganong. A possible role for the brain renin-angiotensin system in the regulation of LH secretion. *Am. J. Physiol.* 245 (*Regulatory Integrative Comp. Physiol.* 14): R805–R810, 1983.

17. Ganong, W. F., J. Challett, H. Jones, Jr., S. L. Kaplan, M. Karteszi, R. D. Stith, and L. D. Van de Kar. Further characterization of the putative α-adrenergic receptors in the brain that affect blood pressure and the secretion of ACTH, growth hormone and renin in dogs. *Endocrinol. Exp.* 16: 191–204, 1982.

51 (1978–79)

David Francis Bohr
(b. 1915)

As president elect of APS, in the fall of 1977 Bohr visited medical schools and especially departments of physiology in Cuba (11). He reported that under Communist management the national rate of illiteracy has been reduced from twenty-five percent to three percent; that the 3,000 physicians who left Cuba after the revolution have been replaced threefold, resulting in a current physician-to-population ratio higher than that in the United States; and that medical education is not only of good quality, but also is free, with students receiving in addition an allowance for living expenses. He found physiology to be an active profession in both teaching and research. The president of the Cuban Physiological Society, F. R. Dorticos, seemed interested in the possibility of joint meetings of his society and APS, possibly in cooperation with the Mexican Physiological Society or with the Latin American Society of Physiological Sciences. As a result of Bohr's initiative, the Latin American Society was invited to take part in the APS Fall Meeting in San Diego in October 1982. There the president of the Latin American society, Cesar Tim-Iaria of Sao Paulo, Brazil, met with Council to discuss ways to foster better communication between the Americas.

Although for more than fifty years Bohr's career has been identified with the University of Michigan, he was born in Zurich, Switzerland, lived for five years as a boy in Cuba, and received most of his primary school education in the southernmost part of California. In 1933 he entered the University of Michigan, matriculated in 1936 in its medical school, and graduated in 1942 after having spent two years as a teaching and research assistant in the Department of Physiology where Robert Gesell was chairman. Bohr interned at Henry Ford Hospital for a year before he was assigned by the U.S. Army to a Dutch hospital ship for three years of duty as laboratory officer and detachment commander (1943–46). He then spent two years (1946–48) as a research fellow at the University of California in San Francisco before returning permanently to the University of Michigan. In 1957 he was promoted to the rank of professor.

On two occasions Bohr has served as visiting professor, first in 1955–56 at the Department of Pharmacology at the University of California at San Francisco and again in 1961–62 at the Physiologische Institüt at Heidelberg. Michigan chose him for a Distinguished Faculty Achievement Award in 1973 and for the Distinguished Faculty Lectureship in Biological Research in 1983. In 1977 he gave the Wiggers Lecture for APS, and in 1984 he received the Ciba Award for Hypertension Research. He is a member of the International Society of Hypertension, the InterAmerican Hypertension Society, and the Council for High Blood Pressure Research of AHA (1968–; chairman, 1978–80). For five years (1969–74) he was a member of the Research Advisory Committee of AHA and also chairman of one of its cardiovascular study committees.

As a member of the Physiology Study Section, Bohr began association with NIH advisory groups in 1960–64. He served with the NIH Hypertension Task Force (1978–79) and on the Study Section B of the Heart, Lung and Blood Institute (1981–86). He has been a member of the Committee on Physiology of the National Board of Medical Examiners (1965–68) and also of the Cardiovascular Review Panel for the Space Science Board of NAS (1968–72). Most of his editorial responsibilities have similarly involved the circulatory system. He served the Society on the Editorial Board of its journals in 1966–69, in 1969–75 as coeditor of the circulation section of the *American Journal of Physiology*, and in 1983–86 as associate editor of the *American Journal of Physiology: Heart and Circulatory Physiology.* He also served as editor of the *Handbook of Physiology, Vascular Smooth Muscle*. He was a member of the editorial boards of *Circulation Research* (1960–65 and 1968–74), *Proceedings of the Society for Experimental Biology and Medicine* (1978–81), *Blood Vessels* (1974–), and *Hypertension* (1979–81).

Bohr joined APS in 1949. His earliest committee responsibility was with the Membership Committee (1966–69; chairman, 1967–69). For a year (1969–70) he was chairman of the Subcommittee on Undergraduate Education in Physiology and then joined the Education Committee (1970–73). In 1970–73 he was a member of the Steering Committee of the Circulation Group of the Society. From 1974 to 1977 he served with the Perkins Memorial Fund Committee, after his election to Council in 1973. He became president elect in 1977. Regarding his presidential years, Bohr has written:

> There are two things for which I feel some satisfaction during my tenure at the helm. One was the institution of the Standing Committee on Career Opportunities in Physiology. Walter Randall agreed to serve as first chairman of the committee, and at least half of the members are to be under the age of forty years. Anything we can do to help those who are getting launched in our field will be of value to our profession and will be much appreciated by those we are helping. The other memorable event was the Society's support of my visit to Cuba (11). It initiated a regrettably abortive relationship with physiologists in Cuba, which I certainly would like to see rekindled. It will not be easy, but I would be glad to help.

A third important event was a meeting Bohr encouraged between members of the APS Animal Care Committee and representatives of animal welfare groups. In February 1980, Helene Cecil, chairwoman of the APS committee, met with Leon Bernstein and Christine Stevens and others of the Animal Welfare Institute for discussion of the use of animals and alternatives to such use in research and teaching. Christine Stevens is the daughter of Bohr's mentor, Robert Gesell. The meeting was reported briefly in *The Physiologist* [23(3): 16, 1980].

In regard to his research interests, training, and publications, Bohr wrote:

> [At first] I worked with John Bean in the Department of Physiology on oxygen toxicity (1). It was obvious then and still is that you can get major rewards in physiology from problem solving by using physical and chemical tools plus common sense. . . . I then began to be interested primarily in the contractile machinery of vascular smooth muscle, which made me want to know what causes the pressure to go up in hypertension. I have been working at these two problems for the last thirty years.
>
> It is an interesting coincidence that both of my favorite publications appeared in *Science*. The first (2) demonstrated that calcium not only causes contraction of vascular smooth muscle but also in higher concentrations decreases excitability. The second, published two years later (3), quantified the calcium requirement for contractile activity of vascular smooth muscle and skeletal muscle and showed that the contractile apparatus of the two machines has identical calcium dependency. Recently, I have been trying to understand those changes in contractile machinery of vascular smooth muscle that are responsible for the increase in total peripheral resistance in hypertension. . . . Here are thumbnail sketches of later papers.
>
> Reference 4. Small coronary arteries from the dog have little or no α-adrenergic

activity but respond to catecholamines with relaxation resulting from activation of β-adrenergic receptors. Large coronary arteries have both α- and β-adrenergic activity. The β-receptors of coronary vessels appear to differ from those in vessels supplying skeletal muscle.

Reference 5. There is an as yet unidentified vasoactor in plasma that causes contraction of isolated vascular smooth muscle. This factor may play a role in the maintenance of normal vascular tone.

Reference 6. Sensitivity of vascular smooth muscle to constrictor agonists was found to be elevated in deoxycorticosterone acetate, renal, and spontaneously hypertensive rats. This increase in sensitivity reflects a lessening of calcium binding to the cell membrane in vascular smooth muscle from rats with these types of hypertension.

Reference 7. The increase in sensitivity of vascular smooth muscle from hypertensive rats was demonstrated to be primary. It occurred in the hindlimb vasculature that was protected from the hypertension by ligation of the iliac artery.

Reference 8. Vascular smooth muscle made to contract in a potassium-free medium undergoes a relaxation when potassium is added back to the muscle bath. This relaxation was demonstrated to be due to membrane hyperpolarization resulting from activation of the electrogenic sodium pump. This phenomenon was subsequently used extensively to evaluate sodium pump activity in isolated vascular smooth muscle.

Reference 9. Pressor responses to intravenous infusions of norepinephrine or of angiotensin were elevated as early as two days after the beginning of treatment of the pig with deoxycorticosterone acetate.

Reference 10. Blood pressure elevation in the pig began two days following deoxycorticosterone acetate treatment and reached a plateau three weeks later. The pressure elevation was caused in some pigs by an elevation in cardiac output and in others by an elevation in total peripheral resistance, but in most animals it was caused by an elevation in both of these determinants of arterial pressure.

Reference 13. Evidence is presented to support a hypothesis that the primary fault in the pathophysiology of hypertension is a defect in the calcium binding of the plasma membrane of the cells of a pressure-regulating center in the hypothalamus.

Reference 14. Basilar arteries from spontaneously hypertensive rats (but not from normotensive controls) were characterized by spontaneous contraction which resulted from a membrane leak of extracellular calcium.

Reference 15. Administration of deoxycorticosterone to the sheep results in hypertension, polydipsia, hypokalemia, and the development of a "salt appetite."

Reference 16. Treatment of isolated vascular smooth muscle with serotonin results in the following sequence of events that causes an attenuation of the response resulting from subsequent stimulation of the muscle with norepinephrine: 1) increased membrane permeability to sodium, 2) elevated intracellular sodium, 3) stimulation of the sodium efflux pump, 4) membrane hyperpolarization, and 5) depressed calcium influx via the norepinephrine-operated receptor.

One of the themes that occurs and recurs in past-presidential addresses is the responsibility of physiologists in the overall enterprise of medical education and perhaps in what is known as "delivery" of health care. Bohr (12) pointed out that problems in medical practice have two parts: 1) economy or finances and 2) attitude. After summarizing how medical costs have become so high, he spoke of the contrast between what has happened in the United States and what he observed in Cuba. There health care is available to all without charge. Yet even though Cuban education and health care have moved forward rapidly, any visitor can see that "the economy is clearly an unsurmounted hurdle. Housing is poor, manufactured items are in scarce supply, clothing is rationed, and very few people have their own automobiles" (11). Later he added, "and besides, there is virtually nothing to buy" (12).

Bohr ended his remarks with a few words about improving attitudes of physicians and how medical education might be modified to preserve throughout their training the sensitive, ethical dedication students bring to medical school. "Sensitivity and feeling add nothing to the cost of health care." As physiologists working within medical schools, because "we train future clinical doctors, nurses, and dentists, we hold a measure of responsibility for the caliber, the quality, and the integrity of health care in America that we cannot disown."

SELECTED PUBLICATIONS

1. BEAN, J. W., AND D. F. BOHR. High oxygen effects on isolated striated muscle. *Am. J. Physiol.* 126: 188–195, 1939.

2. BOHR, D. F. Vascular smooth muscle: dual effect of calcium. *Science Wash. DC* 139: 597–599, 1963.

3. FILO, R. S., D. F. BOHR, AND J. C. RUEGG. Glycerinated skeletal and smooth muscle: calcium and magnesium dependence. *Science Wash. DC* 147: 1581–1583, 1965.

4. BOHR, D. F. Adrenergic receptors in coronary arteries. *Ann. NY Acad. Sci.* 139: 799–807, 1967.

5. BOHR, D. F., AND J. SOBIESKI. A vasoactive factor in plasma. *Federation Proc.* 27: 1396–1398, 1968.

6. HOLLOWAY, E. T., AND D. F. BOHR. Reactivity of vascular smooth muscle in hypertensive rats. *Circ. Res.* 33: 678–685, 1973.

7. HANSEN, T. R., AND D. F. BOHR. Hypertension, transmural pressure, and vascular smooth muscle response in rats. *Circ. Res.* 36: 590–598, 1975.

8. BONACCORSI, A., K. HERMSMEYER, O. APRIGLIANO, C. B. SMITH, AND D. F. BOHR. Mechanism of potassium relaxation of arterial muscle. *Blood Vessels* 14: 261–276, 1977.

9. BERECEK, K. H., AND D. F. BOHR. Whole body vascular reactivity during the development of deoxycorticosterone acetate hypertension in the pig. *Circ. Res.* 42: 764–771, 1978.

10. MILLER, A. W., D. F. BOHR, A. M. SCHORK, AND J. M. TERRIS. Hemodynamic responses to DOCA in young pigs. *Hypertension Dallas* 1: 591–597, 1979.

11. BOHR, D. F. President-elect's tour. Changes in Cuba. *Physiologist* 22(1): 9–11, 1979.

12. BOHR, D. F. Past-president's address. The health market and physiologists. *Physiologist* 22(6): 15–17, 1979.

13. BOHR, D. F. What makes the pressure go up? A hypothesis. *Hypertension Dallas* 3, Suppl. II: 160–165, 1981.

14. WINQUIST, R. J., AND D. F. BOHR. Structural and functional changes in cerebral arteries from spontaneously hypertensive rats. *Hypertension Dallas* 5, Suppl. III: 292–297, 1983.

15. MITCHELL, J., W. D. LING, AND D. F. BOHR. Deoxycorticosterone acetate hypertension in sheep. *J. Hypertension* 2: 473–478, 1984.

16. MORELAND, R. S., C. VAN BREEMEN, AND D. F. BOHR. Mechanism by which serotonin attenuates contractile response of canine mesenteric arterial smooth muscle. *J. Pharmacol. Exp. Ther.* 232: 322–329, 1985.

52 (1979–80)

Ernst Knobil
(b. 1926)

Knobil was born in Berlin, Germany, as a boy attended the Lycée Claude Bernard in Paris, and then continued his education in science in the United States at Cornell University (B.S., 1948; Ph.D., 1951) after two years of service with the U.S. Army (1944–46). From 1951 to 1953 he was Milton Research Fellow in the laboratory of Roy O. Greep of the Harvard School of Dental Medicine. This led to appointment in the Department of Physiology of the medical school (1953–61), which he left to become Richard Beatty Mellon Professor and chairman of the Department of Physiology at the School of Medicine at the University of Pittsburgh (1961–81). From 1974 to 1981 he was also director of the Center for Research in Primate Reproduction at Pittsburgh. In 1981 he moved again to become dean of the School of Medicine at the University of Texas Health Science Center (1981–84), H. Wayne Hightower Professor of Physiology, and director of the Laboratory for Neuroendocrinology (1981–). He now continues as a full-time investigator in his laboratory.

Knobil's membership in APS dates from 1955. He was a member of the Editorial Board of the Society's journals from 1960 to 1966 and of the Editorial Board for the *American Journal of Physiology: Regulatory, Integrative and Comparative Physiology* from 1976 to 1978. From 1979 to 1982 he was editor of the *American Journal of Physiology: Endocrinology and Metabolism.* With Wilbur Sawyer he edited the volumes on the pituitary gland and the hypothalamus of the *Handbook of Physiology.* His first committee service was on the Committee for the Use and Care of Animals (1962–67). He served on Council from 1969 to 1972 and later served on the Ray G. Daggs Award Committee (1976–79). He became president elect in 1978. As president he was instrumental in increasing APS activities in public affairs with a focus on the issue of animal experimentation.

Early in his career Knobil was honored by designation as a Markle Scholar at Harvard (1956–61). He received the Ciba Award in 1961 and the Fred Conrad Koch Award in 1982, both from the Endocrine Society. In 1983 he was chosen for the Carl G. Hartman Award of the Society for Study of Reproduction, and in 1985 he was given the Axel Munthe Prize in Reproduction. He has the degree Dr. Hon. Causa from the University of Bordeaux (1980) and an honorary Sc.D. from the Medical

College of Wisconsin (1983). He is a fellow of AAAS (Boston) and an honorary fellow of the American Association of Obstetricians and Gynecologists, as well as of the American Gynecological Society.

Outside APS, Knobil has served on editorial boards for *Endocrinology* (1959–75), *Annual Review of Physiology* (1968–77; editor-in-chief, 1974–77), *Psychoneuroendocrinology* (1974–77), *Neuroendocrinology* (1976–80), and *Endocrine Reviews* (1980–84). A recent addition to the list is *Chronobiology International* (1984–). For AAMC, Knobil served as a member of the Executive Council for three years (1971–74), of the Administrative Board of its Council of Academic Societies for four years (1970–74), and more recently as a member of the Steering Committee on Information Sciences and Medical Education (1984–). He was a member of the Council of ACDP (1968–71; president, 1969–70); in 1983 he received its Distinguished Service Award. Of the Deutsche Gesellschaft für Endokrinologie he is an honorary member, as he is of the Japan Endocrine Society; he holds membership also in Great Britain's Society for Endocrinology. Other memberships include the American Society of Zoologists, the International Society for Research in Biology of Reproduction, the Society for Study of Reproduction, the International Society for Neuroendocrinology, the Society for Experimental Biology and Medicine, and the National Society for Medical Research, which he served as a member of the Board of Directors (1977–79). For three years (1968–71) he was a member of the Council of the Endocrine Society (president, 1976–77). He has held a variety of responsibilities for the International Society of Endocrinology, especially in reference to its international congresses; from 1976 to 1984 he was chairman of its Executive Committee and serves as president from 1984 to 1988. From 1980 to 1983 he was a member of the U.S. National Committee for IUPS and is currently chairman of its Commission on Endocrinology (1982–).

Many organizations and institutions have asked Knobil to serve as consultant. For NIH he served on the Human Growth and Development Study Section (1964–66), the Reproductive Biology Study Section (chairman, 1966–68), the Advisory Committee on Primate Research Centers (1969–73), the Contraceptive Development Branch (1969–71), the Medical Advisory Board of the National Hormone and Pituitary Program (1980–83), and the Planning Committee for Developmental Endocrinology and Physical Growth of the National Institute of Child Health and Human Development (1984–), an institute he had earlier served as a member of the Population Research Committee of the Center for Population Research (1974–77). For three years (1966–69) he was a member of the National Advisory Council of the Institute of Laboratory Animal Resources of NAS. He served on the Physiology Test Committee of the National Board of Medical Examiners (1970–74) and on the Liaison Committee on Medical Education of the American Medical Association (AMA) and AAMC (1971–74). In addition, he has been asked to serve as consultant to the University of Michigan School of Medicine (1973), Harvard Medical School (1973–74), the Ford Foundation (1974–75), the Uniformed Services University of

Health Sciences (1975), the Human Reproduction Unit of the World Health Organization (1976–), the University of Texas Health Science Center at Houston (1979–
80), and the Advisory Committee of the Searle Scholars Program (1981–82).

Knobil's research has generated widespread interest both in this country and
abroad. He has been invited to deliver the following lectures: Bowditch Lecture of
APS in 1965; Gregory Pincus Memorial Lecture of the Laurentian Hormone Conference in 1973; and in 1974 the Upjohn Lecture of the American Fertility Society, the
Kathleen M. Osborn Memorial Lecture at the University of Kansas School of Medicine, and the First Annual Hopkins-Maryland Lecture in Reproductive Endocrinology. He gave the Karl Paschkis Lecture of the Philadelphia Endocrine Society in
1975 and in 1978 was Sigma Xi Lecturer at the University of Florida and Second Alza
Lecturer in Palo Alto. In 1979 he gave two lectures: the R. D. O'Brien Lecture at
Cornell University and the First Transatlantic Lecture of the Society for Endocrinology of Great Britain. The following year (1980) he presented the Lawson Wilkins
Lecture of the Pediatric Endocrine Society and the Scientific Lecture for the American
Gynecological Society. In 1981 he was at Johns Hopkins for the Second Bard
Lecture, in San Francisco for the Sixth Herbert McLean Evans Memorial Lecture, and
at the University of Uppsala for the Fourth Carl Gemzell Lecture. Since then he has
given the Fourth Annual James H. Leathem Lecture at Rutgers Medical School and
the Geoffrey Harris Memorial Lecture of the International Society of Neuroendocrinology, both in 1982; the Ayerst Lecture of the American Fertility Society in 1983;
and the Distinguished Guest Lecture for the Society for Gynecological Investigation
and the Potter Lecture of Thomas Jefferson University, both in 1984.

Knobil's training in endocrinology involved only two laboratories.

> I began my career as a graduate student in zoology in the laboratory of Samuel L.
> Leonard, an endocrinologist who was himself a student of F. L. Hisaw and P. E. Smith,
> two of the founders of American endocrinology. My mentor was a participant in the
> early development of this discipline in the United States and imparted to me a very
> personal sense of belonging to this branch of physiology. His Ph.D. thesis recounted
> the discovery of the ovulatory hormone, now known as LH (*Am. J. Physiol.* 98: 406–
> 416, 1931). From Leonard, I acquired an abiding love for endocrinology, which has
> given me immense pleasure and happiness throughout my career. My formal training
> was completed under the preceptorship of Roy O. Greep, who had been a fellow
> student of Leonard in Hisaw's laboratory at the University of Wisconsin.
>
> At Harvard, Greep introduced me to the world of academic medicine, which I
> entered as an ingenuous postdoctoral research fellow. His unfailing and often
> undeserved support is directly responsible for the path my professional life has
> followed. He introduced me to E. M. Landis in the medical school, who offered me
> an instructorship in his department (1953) the year before he became president of
> the APS. My years as a member of his faculty had a profound influence, in both style
> and substance, on the development of the Department of Physiology at the University
> of Pittsburgh School of Medicine, which I served as founding chairman for twenty
> years.
>
> My scientific contributions worthy of any note have all focused on the physiology
> of the primate adenohypophysis (e.g., refs. 2 and 3). My laboratory established the
> zoological specificity of growth hormone with the finding that only growth hormone

of primate origin is physiologically active in primates (4). This discovery led to the development of a radioimmunoassay for human growth hormone (6) and permitted a host of investigations dealing with control of growth hormone secretion (5, 7).

A decade or so ago we began a systematic investigation of control of the menstrual cycle, with the rhesus monkey as an experimental animal (10). This was occasioned by recognition of major species differences in control of reproductive processes and of the fact that findings obtained in one species could not necessarily be extrapolated to another. In the course of this effort, we were the first to describe the time course of progesterone during the ovarian cycle of any species (8, 9) and to elucidate the endocrinologic events of early pregnancy. This provided new insight into the control of the corpus luteum and the role of progesterone in regulation of gonadotropin secretion (11).

More recently we have delineated the neuroendocrine control system that governs secretion of gonadotropic hormones throughout the menstrual cycle of the rhesus monkey. In sum, it comprises the arcuate nucleus of the hypothalamus, which mediates an hourly discharge of the hypothalamic hormone, GnRH, with relatively simple negative and positive feedback loops between estrogens and the pituitary gland. We now understand control of the menstrual cycle for the first time. Consequently the ovarian cycle of monkeys is understood better than that of any other species (12, 13).

For his past president's address (14), Knobil abandoned the societal introspection of his immediate predecessors to consider the classic volume by William Harvey, *On the Generation of Animals.* Because spermatazoa were unknown, being at that time invisible, Harvey remained puzzled as to just what the male fluid may contribute to development and differentiation of the ovum. Knobil noted that Harvey's treatise is "tedious, repetitive, impossible to summarize, encapsulate, or even to sample adequately." It was begun when he was a student in Padua (ca. 1600), under the teaching of Fabricius, and apparently concluded only shortly before he was persuaded, somewhat against his better judgment, to permit its publication (1651). Knobil remarked that it is five times longer than Harvey's *De Motu Cordis*—and, one judges, five times less conclusive. Nevertheless, Knobil found it to be "vaguely reminiscent" of Claude Bernard's writings.

Commenting on his contributions to neuroendocrinology, one of his friends recently stated that Knobil has qualities that remind one of Claude Bernard, especially in his ability to analyze the integrative and relational aspects of endocrine systems and their controls. Knobil has done this for endocrinology of primates more successfully than anyone of his generation.

SELECTED PUBLICATIONS

1. GREEP, R. O., E. KNOBIL, F. G. HOFMANN, AND T. L. JONES. Adrenal cortical insufficiency in the rhesus monkey. *Endocrinology* 50: 664–676, 1952.

2. KNOBIL, E., A. MORSE, F. G. HOFMANN, AND R. O. GREEP. A histologic and histochemical study of hypophyseal-adrenal cortical relationships in the rhesus monkey. *Acta Endocrinol.* 17: 229–238, 1954.

3. KNOBIL, E., R. C. WOLF, R. O. GREEP, AND A. E. WILHELMI. Effect of a primate pituitary growth hormone preparation on nitrogen metabolism in the hypophysectomized rhesus monkey. *Endocrinology* 60: 166–168, 1957.

4. KNOBIL, E., A. MORSE, R. C. WOLF, AND R. O. GREEP. The action of bovine, porcine and simian growth hormone preparations on the costochondral junction in the hypophysectomized rhesus monkey. *Endocrinology* 62: 348–354, 1958.

5. KNOBIL, E., AND R. O. GREEP. The physiology of growth hormone with particular reference to its action in the rhesus monkey and the species specificity problem. *Recent Prog. Horm. Res.* 15: 1–69, 1959.

6. KNOBIL, E., AND R. O. GREEP. The detection of growth hormone in plasma. In: *Hormones in Human Plasma*, edited by H. N. Antoniades. Boston, MA: Little, Brown, 1960.

7. KNOBIL, E. Tenth Bowditch Lecture. The pituitary growth hormone: an adventure in physiology. *Physiologist* 9: 25–44, 1966.

8. NEILL, J. D., E. D. B. JOHANSSON, AND E. KNOBIL. Levels of progesterone in peripheral plasma during the menstrual cycle of the rhesus monkey. *Endocrinology* 81: 1161–1164, 1967.

9. NEILL, J. D., E. D. B. JOHANSSON, AND E. KNOBIL. Patterns of circulating progesterone concentrations during the fertile menstrual cycle and the remainder of gestation in the rhesus monkey. *Endocrinology* 84: 45–48, 1969.

10. MONROE, S. E., L. E. ATKINSON, AND E. KNOBIL. Patterns of circulating luteinizing hormone and their relation to plasma progesterone levels during the menstrual cycle of the rhesus monkey. *Endocrinology* 87: 453–455, 1970.

11. KNOBIL, E. On the regulation of the primate corpus luteum. *Biol. Reprod.* 8: 246–258, 1973.

12. KNOBIL, E. On the control of gonadotropin secretion in the rhesus monkey. *Recent Prog. Horm. Res.* 30: 1–46, 1974.

13. KNOBIL, E. The neuroendocrine control of the menstrual cycle. *Recent Prog. Horm. Res.* 36: 53–88, 1980.

14. KNOBIL, E. William Harvey and the physiology of reproduction. *Physiologist* 24(1): 3–7, 1981.

53 (1980–81)

Earl Howard Wood
(b. 1912)

On Wednesday, 31 July 1985, the *New York Times*, in a major editorial on the use of animals in biological research and testing, commended APS by name for its policy supporting humane care and use of laboratory animals. This policy is not something new for the Society. The result of deliberations extending over many years and involving many of the Society's members, committees, and officers, it expressed a long-standing concern. For example, four years earlier (1981) Ernst Knobil, as a recent past president, and Earl H. Wood, as past president of APS and president of FASEB, offered testimony before the U.S. House of Representatives Subcommittee on Science, Research, and Technology regarding the necessity *1*) of continuing utilization of animals for biomedical research and *2*) of avoiding any suspicion of abuse or neglect. Wood mentioned this problem in his past president's address (10) by noting, "The very active promotion of legislative proposals by these types of well-organized and financed animal rights groups is a most serious threat to continued progress in the biomedical sciences." More recently he added, "The ongoing, progressively increasing antivivisection activities seriously threaten continued progress in the biomedical sciences." In 1983 the Society selected Wood to serve as chairman of a blue ribbon panel that reviewed the case of Edward Taub, a Silver Spring, Maryland, researcher whose laboratory was raided by animal rights zealots; 117 charges of animal cruelty were filed against him (see chapt. 15).

Born in Mankato, Minnesota, after graduation from Macalester College in St. Paul in 1934 Wood entered the School of Medicine of the University of Minnesota but gave up his medical studies temporarily for training in Maurice Visscher's department, where he received the M.S. degree in 1939. Two years later (1941) he was awarded both the M.D. and the Ph.D. degrees, the latter for research on water and electrolytes of cardiac muscle, especially under the influence of digitalis (1). In fact, the year 1940–41 was spent at the University of Pennsylvania as a NRC fellow in the Department of Pharmacology, and for the following year he was instructor in pharmacology at Harvard. In 1942 he returned to Minnesota, to the Aeromedical Unit of the Mayo Foundation Laboratories, where he progressed steadily in rank in the Mayo Graduate School and then in the Mayo Medical School to become professor of physiology and of medicine in 1951. He officially retired from these positions in 1982.

Wood has written of his training, "My most important preceptors in science were Prof. M. B. Visscher (Minnesota), Prof. A. N. Richards (Pennsylvania), and Prof. Otto Krayer (Harvard). I had the greatest respect for Otto Krayer; he was a really exceptional gentleman and scholar."

Wood's research interests fall into three main classifications. He began, as noted above, by studying the action of cardiac glycosides in heart-lung preparations, and with Gordon Moe he demonstrated that the positive inotropic effects of digitalis glycosides are associated with loss of intracellular myocardial potassium (1). This loss may be accompanied by an increase in intracellular calcium, which in turn causes an increase in efficiency of excitation-contraction coupling and the positive inotropic effect. Twenty-seven years later (7) Wood and his associates showed that dramatic positive or negative inotropic effects can be produced at constant cardiac fiber length by relatively small induced changes in the plateau phase of the action potential. Again these effects were postulated to be caused by changes in intracellular calcium concentration and in efficiency of excitation-contraction coupling. This report was the first to describe use of the sucrose gap method of potential clamping for cardiac muscle fibers. Another fundamental study of cardiac physiology showed that there are position-dependent regional differences in pericardial pressures that influence cardiac function (8).

The second theme of Wood's research is protection against positive gravitational effects during high-speed aerial combat and dive-bombing maneuvers. An active program of investigation carried out with the Mayo human centrifuge and summarized relatively soon after the war (2) included the discovery that protection against blackout requires procedures or devices that maintain blood flow to the head by actually producing hypertension at heart level. This led to development of a simplified single-pressure suit and its associated G-activated and G-compensated pneumatic valve. In 1963 the later work of the group was described (6), with a theoretical model of changes in regional lung volume, ventilation perfusion, and pleural pressures caused by changes in the direction and/or magnitude of gravitational-inertial forces. The papers these reviews were based on made the Mayo laboratory a world leader in study of gravitational stress.

Thirdly, Wood and his associates have made notable contributions to laboratory methods and techniques, including 1) a Statham unbounded strain gauge for measuring blood pressure; 2) an oximeter for on-line, real-time measurement of oxygen saturation while blood is being withdrawn for diagnostic purposes; and 3) an absolute-reading ear oximeter; all three are generally used throughout the world. Their development and their properties have revolutionized procedures for diagnostic cardiac catheterization, as summarized by Wood in 1950 (7), and eventually they led to on-line, real-time monitoring during cardiac surgery (5). Studies of dye-dilution techniques began with continuous recording of Evans blue dye concentrations, then of indocyanine green, and finally led to an article summarizing these methods and their use (4). The most recent interest of Wood and his colleagues is a high-speed, computer-based X-ray scanning system that gives accurate and three-dimensional views of moment-to-moment changes in, for example, heart, lungs, or circulation of intact animals or humans. Known as the "dynamic spatial reconstructor," the machine is a high-speed, synchronous, volumetric whole-body computer-assisted tomographic (CT) scanner. It was described in 1977 (9) and in some detail in Wood's past president's address that speaks to the multiple grant requests and many site visits required to secure funding for construction of the machine (10).

For his development of the anti-G suit, Wood was awarded the Presidential Certificate of Merit by Harry Truman in 1947. He has received from Macalester College an honorary degree of D.Sc. in 1950 and a Distinguished Citizen Award in 1974. In 1963 he was given awards by the Aerospace Medicine Association and by *Modern Medicine.* The American College of Chest Physicians (1974), the Mayo Foundation (1978 and 1984), and the Biomedical Engineering Society (1978) have all honored him with lectureships. He is an honorary member of the Royal Netherlands Academy of Arts and Sciences (1977) and of the American College of Cardiology (1978). In 1982 he received an honorary degree, doctor of medicine, from the University of Bern, Switzerland, and in the following year he was given both the Humboldt Prize for Senior U.S. Scientists by the government of West Germany and the John Phillips Memorial Award of the American College of Physicians. Wood has been a visiting professor at the University of Bern (1965–66), an honorary research fellow in the Department of Physiology at University College, London (1972–73), and a visiting professor at the University of Kiel, West Germany (1983).

When he spoke to the Society as past president, Wood recalled that the first such address was given by Wallace Fenn at the University of Minnesota in 1948, and Wood added that he had heard all but two of the ensuing thirty-three addresses. Few members of APS can claim more. Wood became a member in 1943. He was active at first mainly in the Circulation Group and served as a member of its Steering Committee (1962–64; chairman, 1963–64). He received its Carl J. Wiggers Award in 1968. He was elected to APS Council in 1977 and became president elect in 1979. From 1978 to 1980 he was chairman of the Centennial Celebration Committee, and from 1982 to 1985 he served on the Finance Committee. Responsibilities with FASEB ran very much in parallel with those in the Society; in addition to his year as president of FASEB (1981–82), he was a member of the Long-Range Planning and the Development Fund Committees (1982–85) and the Public Affairs Committee (1984–85).

While he was participating in APS, Wood was likewise active in AHA. From 1962 to 1977 he was a Career Investigator of AHA. He was a member of its Basic Science Council and its Council on the Circulation from 1963, and for three years (1967–70) he was on both the AHA Research Fellowship Review Panel and the Physiology and Pharmacology Research Study Committee. In 1973 he was given the association's Research Achievement Award. A member of the Biomedical Engineering Society from 1970, he was elected president for 1983–84. He is a fellow of the Aerospace Medical Association (1964–) and its Space Medicine Branch (1976–). For his pioneer research on problems encountered in flight, he has been a consultant and served on committees for many federal agencies and ad hoc groups, beginning with the U.S. Air Force Aeromedical Center in Heidelberg, Germany, in 1946. The list includes also the Bioastronautics Panel of the President's Science Advisory Committee (1962–66); the Advisory Panel on Medical Biological Sciences (1962–67); a working group on Gaseous Environment for Manned Space Craft of NAS (1963–64); the Office of the Director of Defense Research and Engineering (1964–65); NASA panels and study groups (1964–65, 1967–80, and 1978–81); the U.S. Air Force Manned Orbital Laboratory Medical Advisory Group (1969–74); the American Institute of Biological Sciences Medical Program Advisory Council (1963–); and an AIBS ad hoc panel on medical selection and maintenance of crew health, created for NASA with Wood as chairman (1978–81). For a year he was a consultant for the Aerospace Corporation (1964–65), and for six years he served on the NASA Man in Space Committee (1964–70). He has served NIH on the Physiology Fellowship Review Panel (1963–65), the Research Career Development Award Committee (1963–65), the Artificial Heart/Myocardial Infarction Program Advisory Committee (1967–70), the Biomedical Engineering Special Study Section (1970), and the Computer and Biomathematical Sciences Study Section (1974–77).

In addition to membership in the usual, more generalized scientific societies (e.g., AAAS) Wood is a member of the American Society for Clinical Investigation and the Central Society for Clinical Research. He belongs to the Cardiac Muscle Society, the Minnesota Heart Association, and the Minnesota Academy of Sciences.

In 1979 he went to China as a member of a delegation representing the American College of Physicians.

Wood grew up in a family with a sister and four brothers, all of whom became distinguished in their several fields. Louise, an executive of the American Red Cross during World War II, eventually became director of all overseas activities of that organization and then served for eleven years as executive director for the Girl Scouts of America. Their brother, Harland, was professor of biochemistry at Case Western Reserve University. Chester (Ph.D., Stanford University), who taught in various schools in Minnesota, was a member of the staff of the University of Minnesota in Duluth and then of the Anchorage Methodist University. Delbert graduated from the St. Paul College of Law, was a member of the staff of the Federal Bureau of Investigation for six years, and then for twenty years was chief special agent for the Illinois Central Railroad. Wilbur received his medical education at Minnesota and after the war established a medical clinic in Littleton, Colorado. From 1942 the family and a few friends have come together each year in the Minnesota woods, since 1944 in their own camp, for a reunion that includes a fall deer hunt and other family traditions. Earl is said to be the one in charge; he sends out rules of the camp, four pages long. He also keeps statistics of each year's hunt and issues a "productivity" report after each season.

When he retired from his position at the Mayo, and from office in the APS, Wood with his sons, Andy and Mark, his close friend, Homer Warner (of the University of Utah), Warner's son, Steve, and a friend, Ray Skrocke, sailed from Hawaii to Seattle. Wood wrote of the rough parts of the trip, "winds about 30–40 miles per hour from the north. Seas very rough. Tremendous swells and breakers—at least 20 feet high. . . . Boat often heeling over to 40–45°. Terrifying. . . ."

This voyage must have been a grand way to "wind down" from three years of presidential service to APS.

SELECTED PUBLICATIONS

1. WOOD, E. H., AND G. K. MOE. Blood electrolyte changes in the heart-lung preparation with special reference to the effects of cardiac glycosides. *Am. J. Physiol.* 137: 6–21, 1942.

2. WOOD, E. H., E. H. LAMBERT, E. J. BALDES, AND C. F. CODE. Effects of acceleration in relation to aviation. *Federation Proc.* 3: 327–344, 1946.

3. WOOD, E. H. Special instrumentation problems encountered in physiological research concerning the heart and circulation in man. *Science Wash. DC* 112: 705–715, 1950.

4. WOOD, E. H., H. L. C. SWAN, AND H. W. MARSHALL. Technic and diagnostic applications of dilution curves recorded simultaneously from the right side of the heart and from the arterial circulation. *Proc. Staff Meet. Mayo Clin.* 33: 536–553, 1958.

5. WOOD, E. H., W. F. SUTTERER, AND D. E. DONALD. The monitoring and recording of physiologic variables during closure of ventricular septal defects using extracorporeal circulation. *Adv. Cardiol.* 2: 61–74, 1959.

6. WOOD, E. H., A. C. NOLAN, D. E. DONALD, AND L. CRONIN. Influence of acceleration on pulmonary physiology. *Federation Proc.* 22: 1024–1034, 1963.

7. WOOD, E. H., R. L. HEPPNER, AND S. WEIDMANN. Inotropic effects of electric currents. I. Positive and negative effects of constant electric currents or current pulses applied during cardiac action potentials. II. Hypothesis: calcium movements, excitation-contraction coupling and inotropic effects. *Circ. Res.* 24: 409–445, 1969.

8. Avasthey, P., C. M. Coulam, and E. H. Wood. Position-dependent regional differences in pericardial pressures. *J. Appl. Physiol.* 28: 622–629, 1970.

9. Wood, E. H. New vistas for the study of structural and functional dynamics of the heart, lungs, and circulation by non-invasive numerical tomographic vivisection. *Circulation* 56: 506–520, 1977.

10. Wood, E. H. Past-president's address. Four decades of physiology, musing, and what now. *Physiologist* 25: 19–32, 1982.

54 (1981–82)

Francis J. Haddy

(b. 1922)

In 1974 Haddy first assumed an office in APS when he became chairman of the newly established Committee on Committees. He was elected to Council in 1976 and as president elect in 1980. He continues as a member of the Finance Committee (1983–; chairman, 1985–). In these dozen years many changes have occurred in the Society and also in FASEB. Settlement of governance and equity in FASEB led to a mechanism whereby new societies might join the Federation and also to launching of the current new building program. The Society was sectionalized and its journals reorganized. Participation of membership of APS in meeting programming was increased through appointment of the Program Advisory Committee with representation from sections of the Society. Career opportunities, financial development, liaison with industry, the role of women in physiology, centennial planning, changes in the fall meetings and in plenary sessions and business meetings, relations with IUPS, and public affairs all occupied the attention of Council and affected membership at large. Recently Haddy wrote:

It was in the area of public affairs that I made my major contribution during my presidency (12). We reorganized the Public Affairs Committee, formed a state network, appointed a governmental affairs consultant, and directed a major effort toward improvement of animal care legislation. . . . Subsequent events have underscored the importance of this effort.

From his birthplace in Walters, Minnesota, Haddy studied first at Luther College in Decorah, Iowa, and then at the University of Minnesota (B.S., 1943; M.B., 1946; M.D., 1947). He has described his subsequent experiences in the U.S. Army and then in Visscher's department at the University of Minnesota and at the Mayo Clinic and his ultimate discovery that unwittingly he had become a physiologist in these words:

> I came into a career of research and into physiology by accident. After receiving my M.D. in 1947, I was placed on active duty by the U.S. Army and fully expected to remain there for two years. I intended to apply for a residency in internal medicine about midway through this tour of duty. Demobilization accelerated, however, and unexpectedly I was discharged after eight months. I had not yet applied for a residency. Because I lived in Minneapolis, I explored opportunities at the University of Minnesota and at the Mayo Clinic—together with hundreds of other returning physician-servicemen. There were no openings for fifteen months. I learned, however, that a year in a basic science could be applied toward the three years of formal training required by the American Board of Internal Medicine. After trying unsuccessfully to work in the Department of Pathology under E. T. Bell, I thought of physiology. I had not excelled in it in medical school but had found it interesting. I talked to Maurice B. Visscher, head of the department, and he immediately took me into his program as a candidate for a master's degree.
>
> His department was unique in that it was closely allied with the Department of Surgery, headed by Owen Wangensteen, and also related to physiologists at the Mayo Clinic (Earl Wood, for example). Visscher teamed me with Gilbert Campbell, one of the surgical residents working in the Department of Physiology, and with Visscher's exceptional technician, Wayne Adams. Together we developed a method for left-heart catheterization in dogs that we used for study of the pathogenesis of pulmonary edema. I wrote a thesis on this subject for my master of science degree. Then, according to plan, I went to the Mayo Clinic as a fellow in internal medicine. I quickly found that I missed research and decided to try to return to research after completion of my fellowship. Visscher sponsored my application for a research fellowship from the American Heart Association. When it was approved I returned to his department.
>
> At that point I was not interested in another degree, but Visscher said, "You might as well also work on a Ph.D. while you are here because it will be useful to you in many ways." I agreed, reluctantly. My thesis research involved measurement of pressures in small vessels in the peripheral circulation, and I fulfilled requirements for the degree in two years (1953). This completed my formal training, and I took my first job at Northwestern University in Chicago as assistant professor of physiology (1953–55). Later I was able to become board certified in internal medicine and still later became one of the first Clinical Investigators of the Veterans Administration.
>
> Visscher was right. My dual training increased my options, opened doors for me, and influenced the type of research I did and the way I chaired departments. It also created a problem or two, viz, another tour of military duty (1955–57) and an offer of a chairmanship of a Department of Physiology which forced me to declare what I was. After much reflection, I finally decided that I was, in fact, predominantly a physiologist and accepted the chair at the University of Oklahoma (1961). I did not, however, give up my connection with internal medicine and continued to hold a secondary appointment in that discipline.

As his account indicates, Haddy at first regarded himself as both an internist and a physiologist. When he returned from his second tour of military service (at the Army Medical Research Laboratory at Fort Knox, Kentucky, where Ray G. Daggs was director of research) and became a Clinical Investigator of the VA Research Hospital in Chicago, he was assistant professor of both medicine and physiology at Northwestern (1957–61). He gave up the investigatorship when he was appointed Assistant Director of Professional Services for Research at the VA Hospital in 1959. From Chicago he moved to Oklahoma City where he served as chairman of the Department of Physiology for five years (1961–66). Simultaneously he was associate professor of medicine. From Oklahoma he moved to East Lansing, Michigan, as professor and chairman of physiology in the new College of Human Medicine at Michigan State University (1966–76). He then transferred to another new school, the Uniformed Services University of Health Sciences in Bethesda, Maryland (1976–).

Haddy was elected to membership in the American Society for Clinical Investigation in 1960 and to fellowship in the American College of Physicians the following year (1961). He has been active in regional clinical research societies, both for the central and southern regions of the United States. He belongs to the Microcirculatory Society (1959–) and was a member of its Executive Committee (1961–67). For AHA he served as a member of its Medical Advisory Board of the Council for High Blood Pressure Research (1963–) and its Executive Committee (1974–80), the Physiology and Pharmacology Study Committee (1968–71), and the Research Committee (1974–80) and was cochairman of the Cardiovascular A Research Study Committee (1974–80) and a member of the Ciba Award Committee (1977–80). He is a member of the Board of Trustees of the National Hypertension Association (1979–). From 1964 to 1969 he served on the Cardiovascular Study Section for NIH; he was a member of the Cardiovascular Training Committee of the National Heart and Lung Institute from 1970 to 1973 and has served on its Atherosclerosis and Hypertension Advisory Committee since 1983. In 1982 he organized and chaired a Gordon Research Conference on magnesium in biochemical processes and in medicine. The following year (1983) he was appointed to a Basic Biomedical Sciences Panel of the Institute of Medicine of NAS.

Haddy joined APS in 1953 and became a member of the Circulation Group eight years later (1961). In 1966 he was given the Carl J. Wiggers Award by this group and from 1971 to 1974 was a member of its Steering Committee. In 1963 he began editorial responsibilities with the Society's journals that lasted until 1969; then in 1980 he joined the Editorial Board of the *American Journal of Physiology: Heart and Circulatory Physiology*. Other editorial boards on which he has served include those for the *Proceedings of the Society for Experimental Biology and Medicine* (1969–72), *Circulation Research* (1975–81), *Microvascular Research* (1978–81), *Hypertension* (1978–81), and *Microcirculation* (1980–). Haddy represented APS on the U.S. National Committee of IUPS from 1976 to 1979 and then again from 1981 to 1984. In 1983 he was appointed to the Audit Committee of AAMC.

Haddy noted that his research has always had "applied" (i.e., clinical) goals, including understanding of edema, shock, myocardial ischemia, and hypertension. Many of his review articles have been clinically oriented, because he has found that his training was ideally suited for bridging any communication gap between basic scientists and clinicians. He is an advocate of joint training and joint appointments. At one time he had six M.D.-Ph.D. and three D.V.M.-Ph.D. faculty members in his department, all with joint appointments. He described his own research interests as follows:

> My research activity has centered on three areas in the circulation: *1)* fluid flux across capillary membranes, e.g., pulmonary edema evoked by increased capillary pressure and peripheral edema evoked by increased capillary permeability to plasma proteins (the action of histamine, bradykinin); *2)* local regulation of blood flow, particularly the metabolic hypothesis in exercise hyperemia, reactive hyperemia, and autoregulation; and *3)* the role of naturally occurring cations in regulating peripheral resistance, particularly in relation to low-renin hypertension. Some of my favorite publications are described below.
>
> In collaboration with Visscher, Campbell, and Adams (1–3) my first three papers demonstrated that several forms of pulmonary edema result from increased pulmonary capillary hydrostatic pressure. They dispelled the then-current notion that most pulmonary edemas arose from increased capillary permeability to plasma proteins. In developing a miniature catheter technique for measuring pressures in a small (0.5 mm) vein, we found that veins are not simply passive conduits for blood but can contract spontaneously and in response to various vasoactive agents and thus raise capillary hydrostatic pressure (4). A paper published with colleagues at the Army Medical Research Laboratory in Fort Knox demonstrated that autoregulation of blood flow in kidney results from active vasomotion, rather than from changes in blood viscosity, and can occur equally well during perfusion with a cell-free fluid (5). In studies of vasoactive cations and anions, we found that potassium produces vasodilation, whereas calcium produces vasoconstriction in an intact, perfused vascular bed (6, 7). Changes in extracellular sodium concentration are vasoactive only by virtue of changes in osmolality. Potassium vasodilation can be blocked with ouabain, a potent Na^+, K^+-ATPase inhibitor (8). We hypothesized therefore that the vasodilation results from stimulation of the Na^+-K^+ pump, electrogenic hyperpolarization, and hence a decreased calcium influx into smooth muscle cells of the vascular walls. When we were able to measure net fluid flux across the capillaries, we showed among other things that histamine produces edema in part by increasing permeability to plasma proteins (9). In 1976 we presented the first definitive evidence that the Na^+-K^+ pump is suppressed in vascular smooth muscle of animals with low-renin hypertension (10) and showed that this suppression results from action of a humoral ouabain-like factor (11) that might be a natriuretic hormone released from the brain and capable of suppressing the Na^+,K^+-ATPase and, hence, the pump.

Because he took part in the expansion of two established medical schools (Northwestern and Oklahoma) and helped found two new schools (at Michigan State and in Bethesda), Haddy wrote that an appropriate summary of his career might be the statement, "I go around the country opening new medical schools." Now that the medical educational establishment is contracting rather than expanding, a career of this type seems to have a somewhat limited future!

SELECTED PUBLICATIONS

1. HADDY, F. J., G. S. CAMPBELL, W. L. ADAMS, AND M. B. VISSCHER. A study of pulmonary venous and arterial pressures and other variables in the anesthetized dog by flexible catheter techniques. *Am. J. Physiol.* 158: 89, 1949.

2. CAMPBELL, G. S., F. J. HADDY, W. L. ADAMS, AND M. B. VISSCHER. Circulatory changes and pulmonary lesions

in dogs following increased intracranial pressure and the effect of atropine upon such changes. *Am. J. Physiol.* 158: 96, 1949.

3. HADDY, F. J., G. S. CAMPBELL, AND M. B. VISSCHER. Pulmonary vascular pressure in relation to edema production by airway resistance and plethora in dogs. *Am. J. Physiol.* 161: 336, 1950.

4. HADDY, F. J., A. G. RICHARDS, J. L. ALDEN, AND M. B. VISSCHER. Small vein and artery pressures in normal and edematous extremities of dogs under local and general anesthesia. *Am. J. Physiol.* 176: 355, 1954.

5. HADDY, F. J., J. SCOTT, M. FLEISHMAN, AND D. EMANUEL. Effect of change in flow rate upon renal vascular resistance. *Am. J. Physiol.* 195: 111, 1958.

6. EMANUEL, D. A., J. B. SCOTT, AND F. J. HADDY. Effect of potassium on small and large blood vessels of the dog forelimb. *Am. J. Physiol.* 197: 637–642, 1959.

7. HADDY, F. J. Local effects of sodium, calcium and magnesium upon small and large blood vessels of the dog foreleg. *Circ. Res.* 8: 57, 1960.

8. CHEN, W. T., R. A. BRACE, J. B. SCOTT, D. K. ANDERSON, AND F. J. HADDY. The mechanism of the vasodilator action of potassium. *Proc. Soc. Exp. Biol. Med.* 140: 820, 1972.

9. HADDY, F. J., J. B. SCOTT, AND G. J. GREGA. Effects of histamine on lymph protein concentration and flow in the dog forelimb. *Am. J. Physiol.* 223: 1172–1177, 1972.

10. OVERBECK, H. W., M. B. PAMNANI, T. AKERA, T. M. BRODY, AND F. J. HADDY. Depressed function of a ouabain-sensitive sodium-potassium pump in blood vessels from renal hypertensive dogs. *Circ. Res.* 38, *Suppl.* II: 48, 1976.

11. PAMNANI, M., S. HUOT, J. BUGGY, D. CLOUGH, AND F. HADDY. Demonstration of a humoral inhibitor of the Na⁺-K⁺ pump in some models of experimental hypertension. *Hypertension Dallas* 3, *Suppl.* 2: 96–101, 1981.

12. HADDY, F. J. Past-president's address. I think I would rather watch them make sausage. *Physiologist* 25: 466–468, 1982.

55 (1982–83)

Walter Clark Randall
(b. 1916)

From his birthplace and boyhood home in the farming community of Akeley, Pennsylvania, Randall studied at Taylor University in Upland, Indiana, able to go to school during the Depression only because of financial sacrifices on the part of his parents and sisters and by virtue of the honor scholarships he received. Graduating in 1938, he enrolled for further study at Purdue University, where he received his Ph.D. degree in physiology in 1942. The following year he was a postdoctoral fellow in the laboratory of Carl J. Wiggers at Western Reserve University; Wiggers then recommended him for appointment as instructor in Alrick Hertzman's Department of Physiology at St. Louis University, and by 1949 he was an associate professor. Five years later (1954) he moved to Chicago as professor and chairman of the Department of Physiology in the Stritch School of Medicine of Loyola University. He held these positions until 1975, when he relinquished the chairmanship to continue essentially full-time research.

In 1962 Randall was a visiting scientist at the National Spinal Nerve Injuries Center in Aylesbury, England, and in 1965 he held a similar position at the National Spinal Injuries Center at the VA Hospital in Long Beach, California. During the summer of 1970 he was visiting professor of physiology at the University of Washington in Seattle. Taylor University designated him Alumnus of the Year in 1963, and since 1968 he has served that university in various capacities, most recently as a member of its Board of Trustees (1971–). He received the Stritch Medal from Loyola in 1971, was elected an honorary fellow of the American College of Cardiology in 1977, and was given the Carl J. Wiggers Award by the Circulation Group of APS in 1979.

Randall has been active both in national and in local and regional societies and programs. The latter include particularly the Chicago Heart Association and the Illinois section of the Society for Experimental Biology and Medicine (president, 1959–60). He is a fellow of AAAS. For three years he was a member of the Board of Governors of AIBS (1976–79). For AHA he served on the Scientific Council, the Basic Sciences Council, and the Research Review Committee (1969–75). Elected to membership in APS in 1943, he has long been active in the Temperature Regulation Group (chairman, 1957) and the Circulation Group (chairman, 1963). He was chosen for Council in 1976–80 and as president elect in 1981. He served on the Education Committee from 1963 to 1970 and later became the first chairman of the Committee on Career Opportunities in Physiology (1979–82), concerned with evaluation of training programs and providing attractive job opportunities for young physiologists. Most recently (1984–) he has been chairman of the Society's Long-Range Planning Committee.

For seven years (1968–74) Randall was a coeditor of the circulation section of the Society's journals. He had earlier (1964–68) served in a similar capacity for *Circulation Research*. He served the National Board of Medical Examiners on the Physiology Test Committee (1964–68; chairman, 1968) and also was a member of the Advisory Committee for the Medical College Aptitude Test of AAMC. Beginning in 1963 he has been a member of various committees for NIH, including the Heart Program Project Committee (1963–67), a Special Project Committee of the Heart and Lung Institute (1968–72), an ad hoc committee on myocardial infarction (1969–72), an ad hoc committee on electrical excitability of the heart (1971–72), and the Physiological Sciences Advisory Committee of the National Institute of General Medical Sciences (1968).

Of his decision to become a physiologist, Randall wrote that it came about simply because the teaching assistantship he needed to take up graduate study happened to be in that discipline.

At Purdue, I studied under a comparative physiologist, William A. Hiestand, trained at the University of Wisconsin. I was one of his three graduate students. He and I became interested in temperature regulation of birds and constructed small climate chambers that would accommodate different sizes and shapes of the many different species available to us. I exposed hundreds of hens and chicks to hot environments, and recorded respiratory rate and depth, blood pressure and heart rates, deep and surface temperatures, and behavioral reactions and thus published the first scientific paper of its kind (1). Because of my comparative interests, similar experiments were carried out on many avian species, as well as on virtually all orders of reptiles. Plethysmographs were constructed from long pieces of pipes to accommodate garter snakes, rattlers, or Florida blue snakes, or of a metal coffee can to accommodate a turtle. Different gas mixtures were administered with crude localization of chemo-receptors in upper respiratory passages. My dissertation research led to a paper on factors that influence body temperature of birds (2). Commercial chicken farmers noticed our early publications, and I was especially thrilled when an Australian farmer wrote that my paper allowed him to save his flock from heat exhaustion.

Hiestand was an avid journal reader and inculcated the habit in all his graduate students. I treasure my journal library today and feel privileged to be able to pass it along to my son, David, also a physiologist. I graduated from four years in Hiestand's laboratory with nine publications, including three in the *American Journal of Physiology*.

With Ph.D. credentials in comparative physiology, I applied in 1942 for a postdoctoral fellowship with Carl J. Wiggers, and in his laboratory I began a lifetime interest in cardiovascular physiology. At Western Reserve I encountered a veritable beehive of research on hemorrhagic shock. His faculty consisted of A. Sidney Harris, Harold Green, Paul Quigley, and Harold Wiggers, all destined eventually to head departments of their own. Green tutored me in elementary physics, while Harris showed me how his magic little wick electrodes operated to pick up epicardial action potentials, and we collaborated in a project on mechanisms underlying electrocardiographic changes associated with anoxia in the canine heart (3). There was a steady stream of visitors and former fellows—Gordon Moe, Louis Katz, Rene Wegria, Jane Sands Robb, Donald Gregg, Samuel Middleton, and dozens of others—with an equally outstanding group of younger students and fellows, including Robert Alexander, David Opdyke, Matthew Levy, Robert Berne, and Ewald Selkurt. We have enjoyed an esprit de corps that defies both time and advancing technology.

I was surprised when Wiggers informed me that Alrick Hertzman at St. Louis University wished to recruit a young cardiovascular physiologist for his faculty. The job paid $2,500 a year and would mean that I could be married and begin my own research program. Within a few days I was involved in the intricacies of the photoelectric plethysmograph with which I studied control of cutaneous blood flow (4). A short time later I stumbled onto a method for counting active sweat glands in a specified skin surface area. Iodine had slopped onto my palms (from use in the Miner test for sweating), and while chatting with a colleague I leaned my hands on a sheet of bond paper lying on the desk. On removing my hand I noted a clear palm print consisting of discrete lines of tiny black spots outlining the ridges along my finger pads and palmar surfaces. Each little spot was an active sweat pore (5).

Asked about his favorite publications, Randall replied:

They are all favorites, and I recall anecdotes and incidents relating to each.

I collaborated with W. F. Alexander, an exceptional neuroanatomist and surgeon, with J. W. Cox, my first Ph.D. student (later, Surgeon General of the U.S. Navy), and with Hertzman to write a paper requested by Wiggers for publication in the first issue of the new journal *Circulation Research* (6). This led to studies of the distribution of autonomic nerves carrying sweating and cutaneous vasomotor fibers to the upper extremities in humans via electrical stimulation of sympathetic rami of the T_1-T_2 spinal nerves in surgical patients (7).

At Loyola my research on nervous control of the heart provided immediate and exciting results (8). The cardiac nerves had long been known as accelerator nerves, but no one had explained their strongly augmentor action on cardiac contractility, much less their dromotropic influence. Initially we set up optical methods for recording cardiovascular pressure pulses, but we soon adopted electronic transducer and bridge techniques for multiple and simultaneous cardiovascular tracings. Robert Rushmer at the University of Washington was also describing cardiac control experiments similar to mine at APS meetings, and soon Stanley Sarnoff joined the fun to make it a three-ring circus. The years 1956-75 were extremely productive of quality research (12).

During the period I organized, edited, and contributed chapters to two books that brought together leaders in research on cardiac innervation (11, 13). The first of these volumes suggested literally hundreds of questions that needed to be answered experimentally, many of which were, in fact, answered when the second book appeared in 1977. A third volume published in 1984 (15) again updated our understanding of this same control system.

Randall never completely relinquished his interest in control of body temperature and, particularly, of sweating. Of the many papers that came from this research, two summary articles may be cited: one on control of sweating (9) and the other on central and peripheral factors in "dynamic" thermoregulation (10).

Simultaneously with his research programs, Randall developed what he has called "aggressive" teaching, in collaboration with Clarence N. Peiss, who accompanied him from St. Louis to Loyola. At the time of their move (1954) many schools were seeking to improve their teaching. Randall wrote:

> Some were introducing teaching machines; some were dropping "wet" laboratories entirely; others were applying concepts of control systems and emphasizing computers rather than animal studies. *Problem solving* became the magic word. With technical assistance from Robert McCook, a graduate student, we purchased our first Grass model 5 recorder for the student laboratory. With it students obtained records often far better than those in textbooks. Gradually we accumulated Grass recorders for each table of four students and probably were the first department in the world to perform all student laboratory experiments with commercially available electronic recorders. Shortly after, the Hoff-Geddes physiographs were marketed.

As president of APS, Randall was obliged to give more than the usual attention to legislation threatening use of laboratory animals. His past president's address (14) outlined in some detail how operations for coronary artery bypass, as one example, were developed only after generations of investigators had laid a sound foundation in experiments on laboratory animals. He concluded with a judgment that perhaps ninety percent of what is known in modern medicine has developed as a result of fundamental research, most of it in laboratories utilizing animal experiments.

In "retirement," Randall continues his research at Loyola, although he also has an offer of a place at Taylor University when he decides finally to leave his medical school laboratory. He closed his account of his activities by writing, "I anxiously and enthusiastically look forward to each day in the laboratory. I think the Lord still has much for me to do, and I anticipate continuing to be about His business."

SELECTED PUBLICATIONS

1. RANDALL, W. C., AND W. A. HIESTAND. Panting and temperature regulation in the chicken. *Am. J. Physiol.* 127: 761–767, 1939.
2. RANDALL, W. C. Factors influencing the body temperature of birds. *Am. J. Physiol.* 139: 39, 1943.
3. HARRIS, A. S., AND W. C. RANDALL. Mechanisms underlying electrocardiographic changes observed in anoxia. *Am. J. Physiol.* 142: 452–461, 1944.
4. HERTZMAN, A. B., W. C. RANDALL, AND K. E. JOCHIM. The estimation of the cutaneous blood flow with the photoelectric plethysmograph. *Am. J. Physiol.* 145: 716–727, 1946.
5. RANDALL, W. C. Sweat gland activity and changing patterns of sweat secretion on the skin surface. *Am. J. Physiol.* 147: 391–399, 1946.
6. RANDALL, W. C., W. F. ALEXANDER, J. W. COX, AND A. B. HERTZMAN. Functional analysis of the vasomotor innervation in the dog's hind footpad. *Circ. Res.* 1: 16–26, 1953.
7. RANDALL, W. C., J. W. COX, W. F. ALEXANDER, K. B. COLDWATER, AND A. B. HERTZMAN. Direct examination of the sympathetic outflows in man. *J. Appl. Physiol.* 7: 688–698, 1955.
8. RANDALL, W. C., H. MCNALLY, J. COWAN, L. CALIGUIRI, AND W. H. ROHSE. Functional analysis of the cardioaugmentor and cardioaccelerator pathways in the dog. *Am. J. Physiol.* 191: 213–217, 1957.
9. RANDALL, W. C. Sweating and its neural control. In: *Temperature: Its Measurement and Control in Science and Industry*, edited by C. M. Herzfeld. New York: Reinhold, 1962.
10. RANDALL, W. C., R. O. RAWSON, R. D. MCCOOK, AND C. N. PEISS. Central and peripheral factors in dynamic thermoregulation. *J. Appl. Physiol.* 18: 61–64, 1963.
11. RANDALL, W. C. Past and present hypotheses of cardiac control. In: *Nervous Control of the Heart*, edited by W. C. Randall. Baltimore, MD: Williams & Wilkins, 1965.
12. RANDALL, W. C., J. A. ARMOUR, W. P. GEIS, AND D. B. LIPPINCOTT. Regional cardiac distribution of sympathetic nerves. *Federation Proc.* 31: 1199–1208, 1972.
13. RANDALL, W. C. Changing hypotheses of cardiac control. In: *Neural Regulation of the Heart*, edited by W. C. Randall. New York: Oxford Univ. Press, 1977, chapt. I.
14. RANDALL, W. C. Crises in physiological research. *Physiologist* 26: 351–356, 1983.
15. WEHRMACHER, W. H., AND W. C. RANDALL. Performance of the heart in health and disease. In: *Neural Regulation of the Circulation*, edited by W. C. Randall. New York: Oxford Univ. Press, 1984, p. 3–20.

56 (1983–84)

Alfred P. Fishman
(b. 1918)

In 1961 Alfred Fishman began a more than twenty-five year association with the journals and publications of APS when he joined the Editorial Board of *Physiological Reviews*. He came with experience. From 1958 to 1963 he had been on the board of the *Journal of Clinical Investigation*, and in 1960 he was appointed to the boards of both *Circulation* (1960–65, 1966–70, and 1971–75) and *Circulation Research* (1960–65, 1966–70, and 1971–). By the end of his six years with *Physiological Reviews* (1961–67) he was serving as editor of *Physiology for Physicians* (1966–69) and of the series "Physiology in Medicine" (1969–79) in the *New England Journal of Medicine*, both sponsored by APS. He next became chairman of the Editorial Board of the *Handbook of Physiology*, where he served for five years (1967–72) and from which he moved to the Publications Committee of the Society (1972–81). From 1975 to 1981 he was chairman of this committee, and in 1979 he became editor of the Handbook volumes on respiratory physiology. In 1981 he was appointed editor of the *Journal of Applied Physiology*. From this rather considerable range of activities, Fishman will be known and remembered mainly for having sponsored and guided the reorganization of the Society's journals. His contributions to APS, however, have not been limited to its publications. He was chairman of the Program Committee (1965–68) and of the Task Force on Programming (1976), and he served on the Task Force on Clinical Physiology (1974–75), the Long-Range Planning Task Force (1980–84), and the Centennial Celebration Committee, which he has chaired since 1985. He also served as chairman of the committee to find a successor to Orr Reynolds. Chosen president elect in 1982, he became the only president in recent times who had not served previously as a member of Council.

As president of APS, Fishman made a deliberate attempt to alter the way its international counterpart, IUPS, is managed, with particular reference to composition of the IUPS Council and its programs and commissions. The U.S. National Committee was also "challenged" to assume its proper role in international science. To this end, the APS Council offered to collaborate with IUPS to create an international physiological journal.

He was born in New York City and educated first at the University of Michigan (A.B., 1938; M.S., 1939) and then at the University of Louisville (M.D., 1943). For two years he was a captain in the U.S. Army, for a time assigned to the Tropical Disease Center in North Carolina. Then after the series of fellowships he has described below he joined the faculty of the College of Physicians and Surgeons of Columbia University in 1953, where he held the rank of associate professor of medicine from 1958 to 1966. On his move to the University of Chicago in 1966 he became professor of medicine as well as director of the Division of Cardiovascular Disease in the hospital and director of the Cardiovascular Institute. Three years later (1969) he went to the University of Philadelphia as professor of medicine and director of the Cardiovascular-Pulmonary Division of the Department of Medicine. For seven years (1969-76) he served also as associate dean of the School of Medicine. Since 1972 he has been the William Maul Measey Professor of Medicine.

In 1980 Fishman was elected to membership in the Institute of Medicine of NAS and since 1982 has served on its Advisory Committee on Health Science Policy. He is an honorary fellow of the American College of Cardiology (1971) and the American College of Chest Physicians (1972) and has received the Jacobi Medallion from the Mt. Sinai Medical Center (1979), the Distinguished Achievement Award of AHA (1980), and the Distinguished Alumnus Award from the University of Louisville (1984). He has been honored by some twenty named lectureships, including two in honor of Louis N. Katz (1973 and 1975). He has been a visiting professor at the University of Sheffield in England (1965), Harvard (1970), the University of Tennessee (1971), Oxford University as the First Resident Litchfield Lecturer (1972-73), the School of Medicine of Washington University in St. Louis (1973), Johns Hopkins University and the University of Maryland (1974), two universities in Brazil (1976), Mt. Sinai Hospital in New York (1978), the University of Zurich (1978), Yale and Boston Universities (1979), and in Hawaii (1980).

Shortly after he became a member of APS in 1950, Fishman was elected to membership in the American Society for Clinical Investigation (1953) and the American College of Physicians (1954; fellow, 1959). He is also a fellow of AAAS and has served as chairperson of its Section on Medical Sciences (1985-86). He has been a member of the Harvey Society (New York, 1954-78) and the American College of Cardiology (1971-80); he is currently a member of the Royal Society of Medicine (London, 1958), the Association of American Physicians (1962-), the American College of Chest Physicians (1969-), and the American Thoracic Society (1970-). He, Weir Mitchell, Landis, and Berliner are the only former presidents of APS who have qualified for membership in that delightful and exclusive anachronism, the Interurban Clinical Club.

In regional associations and the parent AHA, Fishman has long been active in a variety of capacities. He has served on the Executive Committees of the Council on Cerebrovascular Disease (1966-70), the Basic Science Council (1969-80), and the Research Council (1977-79). He has been a member also of the Council on Circulation (1968-80), the Central Committee (1970-), and two different Program Committees (1970-72 and 1970-73); he is currently chairman of the Council on Cardiopulmonary Disease (1972-).

Editorial responsibilities Fishman has assumed outside the scope of the APS include the boards of *Medicina Thoracalis* (Switzerland, 1962–70), *Medcom* (1972–), *Merck Manual* (1972–), *Annual Review of Physiology* (1977–82), and *International Journal of Cardiology* (1981–); he also serves on the International Editorial Board of the *Cambridge Encyclopedia of the Life Sciences* (1984–).

Fishman began service to federal scientific agencies at the top, as consultant to the Executive Office of the President of the United States (1961–69). He has served with a number of study sections, task forces, and workshops of the National Heart, Lung and Blood Institute, in many instances as chairman. From 1977 to 1979 he was special advisor to the director of the Heart Institute. He has provided consultation to, among others, the chancellor of the University of Missouri in Kansas City, the Governor of Pennsylvania, the VA Hospital at Dartmouth Medical School, the National Center for Health Statistics, and the trustees of Lankenau Medical Research Center. For two years (1968–70) he was chairman of the Board of Cardiovascular-Pulmonary Training and Research Grants of the U.S. Veterans Administration.

The training that prepared Fishman for his career in science he has described as follows:

> My research training included a fellowship in pathology at Mt. Sinai Hospital in New York with Paul Klemperer, who at that time was busy discovering the collagen vascular diseases (1946–47). The following year (1948–49) I spent with Louis N. Katz (president of the APS, 1957–58) at the Michael Reese Hospital in Chicago and then became one of the first Established Investigators of the American Heart Association. A special board supervised my training for the next five years through successive experience in the laboratories of André Cournand and D. W. Richards at Bellevue Hospital, Homer W. Smith at New York University, E. M. Landis and J. R. Pappenheimer at Harvard, and then back to the Cournand-Richards laboratory. A few years thereafter the Nobel Prize in Physiology and Medicine was awarded to Cournand, Richards, and Forssman.
>
> In 1955 I established a cardiorespiratory laboratory in the Department of Medicine at the Columbia-Presbyterian Medical Center, where research was predominantly on humans, but where lungfish, dogs, goats, sheep, and pigs were also the subjects of study. In 1949 I began to spend each summer at the Mount Desert Island Biological Laboratory where I learned how to work with marine animals. After Homer Smith's death (1962), I assumed leadership of research in the "kidney shed." This lasted until 1966 when I left Columbia University to succeed Katz as director of the Cardiovascular Institute at the Michael Reese Hospital of the University of Chicago.
>
> One additional important element in my research training was the year (1964–65) spent with Geoffrey S. Dawes at the Nuffield Institute, Oxford, England. The year was given to studies of fetal circulation and the placenta. It rounded out the experience I had accumulated in regulation of the pulmonary circulation and introduced me to the other end of the spectrum where regulatory mechanisms are much brisker and technically much more difficult to examine.

As this account indicates, Fishman's main research interests have involved the integrated responses made by intact organisms, ranging from lungfish to humans. He has been particularly studying those regulatory mechanisms that promote the interplay among respiration, circulation, and blood. These obviously are examples both of homeostasis and of adaptation, and analyzing them relies heavily on the approach of comparative biology. An abbreviated summary of Fishman's comments on his favorite papers follows.

The first paper (1) introduced the artificial kidney to the United States. Kolff had developed a bizarre apparatus for dialysis of blood, which I undertook to test on humans about to die of renal failure. To everyone's surprise, the apparatus proved effective. The second paper (2) settled a controversy over the use of the Fick principle to determine cardiac output and redressed a serious mistake made earlier by the Cournand-Richards laboratory. My experience studying fish kidneys in Maine is typified by the next paper (3). The study of patients with pulmonary emphysema (4) began my interest in control of breathing, which continues now with use of much more sophisticated techniques (13). In 1956 we were successful in recording for the first time in an operating room the events of the cardiac cycle previously documented only in laboratory animals (5). Two years later we were using the Fick principle to study collateral circulation in human subjects (6). A systematic approach to control of the human pulmonary circulation began with a study of the effects of hypoxia and exercise (7) but eventually included hypercapnia, mechanical factors, and disturbances of acid-base balance. The paper with Dawes and his associates (8) was among the first to describe similarities and differences between the lung and placenta. My first important published experience with the lungfish is included (9); I believe ours is the only laboratory in this country that receives a regular supply of these fish from Africa. In 1969 we began to use tracers for studying movement of macromolecules across the alveolar-capillary barrier (11). Two papers, different from the others, are included: the *Physiological Reviews* article with Heinemann proved to be prophetic of current attention to nonrespiratory functions of the lung (10), whereas the other, a source of great personal pleasure, marks the splitting into sections of the *American Journal of Physiology* (12). Fortunately it seems to have worked out well. The remaining three papers represent our current research programs (13–15).

Fishman called the future of APS a "troublesome matter." He noted:

Left to its own devices, I believe that physiology is destined to languish. . . . But the sectionalization of the Society and its journals, with a greater involvement in international physiology, and a recognition that the Society must provide for cellular physiology, integrative biology, and regulatory biology will improve its chances for prospering in the decade ahead. . . . It seems inevitable, too, that physiology will continue to serve as a mainspring of clinical research and medicine.

No one has given the affairs of the Society a more thoughtful concern, or a more dedicated service, than Al Fishman.

SELECTED PUBLICATIONS

1. FISHMAN, A. P., I. G. KROOP, H. E. LEITER, AND A. HYMAN. Experiences with the Kolff artificial kidney. *Am. J. Med.* 7: 15–34, 1949.
2. FISHMAN, A. P., J. McCLEMENT, A. HIMMELSTEIN, AND A. COURNAND. Effects of acute anoxia on the circulation and respiration in patients with chronic pulmonary disease studied during the "steady state." *J. Clin. Invest.* 31: 770–781, 1952.
3. PUCK, T. T., K. WASSERMAN, AND A. P. FISHMAN. Some effects of inorganic ions on active transport of phenol red by isolated kidney tubules of the flounder. *J. Cell. Comp. Physiol.* 40: 73–88, 1952.
4. FISHMAN, A. P., P. SAMET, AND A. COURNAND. Ventilatory drive in cardiac pulmonary emphysema. *Am. J. Med.* 19: 533–548, 1955.

5. BRAUNWALD, E., A. P. FISHMAN, AND A. COURNAND. Time relationship of dynamic events in the cardiac chambers, pulmonary artery and aorta in man. *Circ. Res.* 4: 100–107, 1956.

6. FISHMAN, A. P., G. M. TURINO, M. BRANDFONBRENER, AND A. HIMMELSTEIN. The "effective" pulmonary collateral blood flow in man. *J. Clin. Invest.* 37: 1071–1086, 1958.

7. FISHMAN, A. P., H. W. FRITTS, JR., AND A. COURNAND. Effects of acute hypoxia and exercise on the pulmonary circulation. *Circulation* 22: 204–215, 1960.

8. CAMPBELL, A. G. M., G. S. DAWES, A. P. FISHMAN, A. I. HYMAN, AND G. B. JAMES. Placental gas exchange and oxygen consumption. *J. Physiol. Lond.* 180: 15P–16P, 1965.

9. JESSE, M. J., C. SHUB, AND A. P. FISHMAN. Lung and gill ventilation of the African lungfish. *Respir. Physiol.* 3: 267–287, 1967.

10. HEINEMANN, H. O., AND A. P. FISHMAN. Nonrespiratory functions of mammalian lung. *Physiol. Res.* 49: 1–47, 1969.

11. PIETRA, G. G., J. P. SZIDON, M. M. LEVENTHAL, AND A. P. FISHMAN. Hemoglobin as a tracer in hemodynamic pulmonary edema. *Science Wash. DC* 166: 1643–1646, 1969.

12. FISHMAN, A. P. Journals of the American Physiological Society. *Am. J. Physiol.* 232 (*Cell Physiol.* 1): C1–C2, 1977.

13. PACK, A. I., R. G. DELANEY, AND A. P. FISHMAN. Augmentation of phrenic neural activity by increased rates of lung inflation. *J. Appl. Physiol.: Respirat. Environ. Exercise Physiol.* 50: 149–161, 1981.

14. WEBER, K. T., J. S. JANICKI, S. SHROFF, AND A. P. FISHMAN. Contractile mechanics and interaction of the left and right ventricles. *Am. J. Cardiol.* 47: 686–695, 1981.

15. FISHMAN, A. P. Endothelium: a distributed organ of diverse capabilities. *Ann. NY Acad. Sci.* 401: 1–8, 1982.

57 (1984–85)

John B. West
(b. 1928)

Like the other former presidents, West was asked to identify significant events in APS while he was in office. He wrote:

I like to think that the establishment of the Section on the History of Physiology has been a significant advance and that I have been one of the main driving forces. Of course, the current interest in history because of the Centennial Celebrations helped a great deal as did the recruitment of Toby Appel as archivist. However, I think that the section helps form a focus for the many people in the Society who have an interest in the history of physiology. The other significant events are well known. The animal rights problem is a continuing cause of great concern. The future direction of the Society worries many people. This is because of increasing fragmentation as different specialty areas form their own groups (e.g., neurophysiology).

Another factor is that the cutting edge of research is moving more and more to cellular physiology, whereas integrative and organ physiology are so important in the teaching of medical students. This results in curious anomalies such as departments of physiology choosing chairmen who are not members of the APS.

My own attitude is that the discipline of physiology is not threatened so long as we all recognize that it includes many areas, such as cell biology, bioengineering, and biophysics. It may be, however, that the APS cannot include all these disciplines, just as biochemistry, which was originally part of physiology, eventually found its home elsewhere.

Now a citizen of the United States, West was born in Adelaide, Australia, where he received his early medical training (M.B.B.S., 1952) and where he was awarded M.D. (1959) and D.Sc. (1980) degrees. In the meantime he had gone to Hammersmith Hospital in London for an internship and residency, which he completed in 1960. That same year he received a Ph.D. degree from London University and joined Sir Edmund Hillary's Himalayan scientific and mountaineering expedition as a physiologist. In 1961-62 he was a postdoctoral fellow with Hermann Rahn at Buffalo; he then returned to London as director of the Respiratory Research Group at Postgraduate Medical School (1962-67). Advanced to the rank of reader in this school in 1968, he took a sabbatical leave for research at the NASA Ames Research Center at Moffett Field, California. The following year (1969) he was invited to become a member of the faculty of the new School of Medicine at the University of California at San Diego as professor of medicine and physiology.

When asked to tell about his training in physiology, West replied:

> My research training has been extremely haphazard by present-day standards. After I obtained my medical degree at the University of Adelaide and did the necessary internship year, I moved to London mainly because I wanted to see the world. Fortunately I gravitated to the Postgraduate Medical School where I first did a residency and then became a member of the Respiratory Research Group under Philip Hugh-Jones. It was at that time that the first respiratory mass spectrometer began working, and the first cyclotron designed for medical research also came on line at the same institution. The research opportunities offered by these two techniques were tremendous. There was a very stimulating intellectual atmosphere with chemists, physicists, and engineers all working in the same unit, but in terms of formal research training I grew like Topsy. My Ph.D. research was the analysis of regional blood flow and ventilation using the cyclotron-produced isotopes.

West has been able to continue his association with clinical disciplines, both inside and outside the University of California. He is a member of the American Society for Clinical Investigation and of the prestigious Association of American Physicians. He also belongs to the American Thoracic Society, AAAS, and ACDP. In Great Britain he holds membership in the Physiological Society and the Harveian Society of London. With Fishman, he initiated plans for the joint meeting of APS and the Physiological Society, held in Cambridge, England, in 1985. One is not surprised to find that he belongs to the Explorers Club.

In 1974 West was designated a Josiah H. Macy, Jr., Foundation Scholar, and he was given the Ernst Jung Prize for Medicine in Hamburg, West Germany, in 1977, as well as the Presidential Citation of the American College of Chest Physicians. In 1980 he won the Kaiser Award for Excellence in Teaching. He has held nearly twenty endowed lectureships, including the Wiltshire Memorial Lectureship at King's College, London (1971); the Brailsford Robertson Memorial Lectureship at Adelaide University (1978); the Brompton Annual Lectureship at Brompton Hospital, London (1979); the Harveian Lectureship in London (1981); a Centenary Lectureship at Auckland, New Zealand (1983); and a Telford Memorial Lectureship at Manchester University in England (1983). Invitations from societies in this country include those from the Anesthesia Research Society in 1975; AHA in 1978 and 1980; the American College of Surgeons in 1982; and the Aerospace Medical Association, the American Thoracic Society, and the Undersea Medical Society all in 1984. He has lectured at Washington University in St. Louis (1978), Loma Linda University Medical School (1979 and 1983), the College of Physicians and Surgeons of Columbia University (1981), and the Medical University of South Carolina (1982).

Editorial responsibility assumed by West began with the *American Journal of Physiology* (1969–75) and the *Journal of Applied Physiology* (1969–75; associate editor, 1979–81). He was on the board of *Respiration Physiology* as a founder member (1966–71). Currently he serves with editorial boards for the *American Review of Respiratory Disease* (1975–), *Circulation Research* (1975–), *Journal of Nuclear Medicine and Allied Sciences* (Italy, 1975–), *Microvascular Research* (founder member, 1968–), and *Clinical Physiology* (Sweden, founder member, 1980).

Some twenty years after he joined Hillary's expedition (1960), West led the American Medical Research Expedition to Everest in 1981 (see below). For NASA he has been chairman of the Science Verification Committee for Spacelab 4 in 1983 and a member of their Advisory Committee on Scientific Uses of Space Station in 1984. Also in that year he served as a member of a NAS Committee on Space Biology. Earlier in his career he was a member of the NIH Cardiovascular and Pulmonary Study Section (1971–75; chairman, 1973–75), the Physiology Committee of the National Board of Medical Examiners (1973–76), and the Cardiopulmonary Council of AHA (1977–78). Elected to membership in APS in 1970 and to Council in 1981, two years later he became president elect (1983).

West described his main research interests under five headings:

1. Investigation of pulmonary function, particularly ventilation-perfusion relationships, by analysis of expired gas with a respiratory mass spectrometer. This was my first project and came about partly because I had just joined a research group at the Postgraduate Medical School, London, where the first respiratory mass spectrometer had been constructed by K. T. Fowler.
2. Measurement of inequality of ventilation and blood flow in the lung by using short-lived radioactive gases. This was a very exciting project. It came about because the medical research cyclotron at Hammersmith Hospital in London had just begun to produce short-lived isotopes, including oxygen-15 (half-life 2 min). By having subjects inhale this isotope in various forms (e.g., molecular oxygen and oxygen-15-labeled CO_2) we were able to show for the first time that blood

flow is much greater at the bottom of the lung than at the top. I subsequently spent several years sorting out the reasons for the inequality of blood flow and ventilation.

3. High-altitude physiology. Again, this was serendipitous. I happened to be sitting next to someone at a meeting of the Physiological Society in England who told me of plans for the Himalayan Scientific and Mountaineering Expedition, which was to take place in 1960 and 1961. At that time I had no special interest in high altitude but was selected by Sir Edmund Hillary to go as one of the physiologists, and the expedition was a great success. We lived for several months during the winter at an altitude of 5,800 m, and I helped make the first measurements of maximal oxygen uptake at an altitude of 7,440 m on Mount Makalu.

I was able to continue this interest in 1981 when I led the American Medical Research Expedition to Everest. We were lucky enough to obtain some data on the summit, and generally the expedition was very productive from the scientific point of view.

4. Analysis of pulmonary gas exchange, particularly ventilation-perfusion relationships. This has been a continuing interest, and during the last ten years, my colleague Peter Wagner and I have developed the multiple inert gas elimination technique for measuring distributions of ventilation-perfusion ratios. One of my interests has been developing computer models for analyzing this difficult problem.

5. Effects of gravity on lung mechanics. At one stage I devoted a lot of time to a finite element analysis of lung distortion caused by gravity. This work was done with Frank Matthews of the Department of Aeronautical Structures in the University of London.

When asked to comment on favorite publications, West chose his first paper (1) and his most popular textbook (11). Then he noted:

In a paper published in 1960 (2), topographical distribution of blood flow in the lung was first described. In the next three titles (3-5), we wrote of work done during the Himalayan Scientific and Mountaineering Expedition of 1960-61. For the first time, in 1967, we demonstrated directly regional differences of alveolar size in lungs, whereas the next paper (8) describes a computer model of pulmonary gas exchange, which has been very influential. In 1972 (9) we described a finite element analysis which is not very well known but actually was very innovative at the time. The 1974 paper (10) gives the method for determining distributions of ventilation-perfusion ratios, a goal that many respiratory physiologists have had for some time. In 1978 (12) we described the first measurements of the effects of weightlessness on the distribution of ventilation and blood flow in the lung, an experiment carried out on a Learjet aircraft flying a Keplerian profile. The next three articles (13-15) describe some of the results of our recent Everest expedition. I am particularly fond of the paper describing the first measurements on the summit (14).

West is well known to medical students and residents who regularly use his accounts of pulmonary function and pathology. For many of his friends in APS, however, his name evokes another type of association. Although using techniques far superior to what they had available, he has nevertheless recreated in our time a heroic age of physiology marked by names such as Joseph Barcroft, J. S. Haldane, L. J. Henderson, and others, perhaps even of G. H. L. Mallory—almost legendary figures of remarkable achievement. Everest without supplementary oxygen!

SELECTED PUBLICATIONS

1. WEST, J. B., K. T. FOWLER, P. HUGH-JONES, AND T. V. O'DONNELL. Measurement of the ventilation-perfusion ratio inequality in the lung by the analysis of a single expirate. *Clin. Sci. Lond.* 16: 529–547, 1957.

2. WEST, J. B., AND C. T. DOLLERY. Distribution of blood flow and ventilation-perfusion ratio in the lung, measured with radioactive CO_2. *J. Appl. Physiol.* 15: 405–410, 1960.

3. WEST, J. B. Diffusing capacity of the lung for carbon monoxide at high altitude. *J. Appl. Physiol.* 17: 421–426, 1962.

4. WEST, J. B., S. LAHIRI, M. B. GILL, J. S. MILLEDGE, L. G. C. E. PUGH, AND M. P. WARD. Arterial oxygen saturation during exercise at high altitude. *J. Appl. Physiol.* 17: 617–621, 1962.

5. PUGH, L. G. C. E., M. B. GILL, S. LAHIRI, J. S. MILLEDGE, M. P. WARD, AND J. B. WEST. Muscular exercise at great altitude. *J. Appl. Physiol.* 19: 431–440, 1964.

6. WEST, J. B., C. T. DOLLERY, AND B. E. HEARD. Increased pulmonary vascular resistance in the dependent zone of the isolated dog lung caused by perivascular edema. *Circ. Res.* 17: 191–206, 1965.

7. GLAZIER, J. B., J. M. B. HUGHES, J. E. MALONEY, AND J. B. WEST. Vertical gradient of alveolar size in lungs of dogs frozen intact. *J. Appl. Physiol.* 23: 694–705, 1967.

8. WEST, J. B. Ventilation-perfusion inequality and overall gas exchange in computer models of the lung. *Respir. Physiol.* 7: 88–110, 1969.

9. WEST, J. B., AND F. L. MATTHEWS. Stresses, strains, and surface pressures in the lung caused by its weight. *J. Appl. Physiol.* 31: 332–345, 1972.

10. WAGNER, P. D., H. A. SALTZMAN, AND J. B. WEST. Measurement of continuous distributions of ventilation-perfusion ratios: theory. *J. Appl. Physiol.* 36: 588–599, 1974.

11. WEST, J. B. *Respiratory Physiology—the Essentials.* Baltimore, MD: Williams & Wilkins, 1974. [French ed., 1975; Spanish ed., 1976; Iranian ed., 1976; Portugese ed., 1977; Italian ed., 1978; 2nd English ed., 1980; Japanese ed., 1981; Dutch ed., 1981; Bahasa Malaysia ed., 1984.]

12. MICHELS, D. B., AND J. B. WEST. Distribution of pulmonary ventilation and perfusion during short periods of weightlessness. *J. Appl. Physiol.* 45: 987–998, 1978.

13. WEST, J. B., S. LAHIRI, K. H. MARET, R. M. PETERS, JR., AND C. J. PIZZO. Barometric pressures at extreme altitudes on Mt. Everest: physiological significance. *J. Appl. Physiol.: Respirat. Environ. Exercise Physiol.* 54: 1188–1194, 1983.

14. WEST, J. B., P. H. HACKETT, K. H. MARET, J. S. MILLEDGE, R. M. PETERS, JR., C. J. PIZZO, AND R. M. WINSLOW. Pulmonary gas exchange on the summit of Mt. Everest. *J. Appl. Physiol.: Respirat. Environ. Exercise Physiol.* 55: 678–687, 1983.

15. WEST, J. B., S. J. BOYER, D. J. GRABER, P. H. HACKETT, K. H. MARET, J. S. MILLEDGE, R. M. PETERS, JR., C. J. PIZZO, M. SAMAJA, F. H. SARNQUIST, R. B. SCHOENE, AND R. M. WINSLOW. Maximal exercise at extreme altitudes on Mount Everest. *J. Appl. Physiol.: Respirat. Environ. Excercise Physiol.* 55: 688–698, 1983.

58 (1985–86)

Howard Edwin Morgan

(b. 1927)

In the life of APS, little of significance happens solely within one year, confined to the twelve months of a single presidency. Rather, in the normal sequence from councillor through past president, each president takes part in important deliberations and decisions over a period of several years. For example, during the three years he was in presidential offices, Morgan became closely involved in planning for the Centennial Celebration because the long process of making these plans began to come to a focus in 1984–87. He was instrumental in making the final agreement for a project many years in the making—the joint publication with IUPS of *News in Physiological Sciences*. He also took an active part in the lengthy consideration of how to ensure a broader representation of the several sections by modifying governance of the Society. Finally it was in the year when Morgan was president elect that Orr E. Reynolds retired from the position of executive secretary-treasurer of APS and Martin Frank was appointed to that office. Morgan became therefore the first president to hold office in collaboration with Frank, as Berne had been the first to serve with Reynolds in 1973. Morgan brought to the office extensive experience not only with the Society's journals (see later) but also in the deliberations of the Porter Physiology Development Committee (1968–80).

Morgan was born in Bloomington, Illinois, and began his college education there with one year at the Illinois Wesleyan University (1944–45). He then moved directly into medical school at Johns Hopkins University, where he received his M.D. degree in 1949. His original intention was to become an obstetrician-gynecologist, a career he began on the house staff of the hospital of Vanderbilt University (1949–53). The following year (1953–54) he was instructor in these disciplines. He then became for a year a fellow in medical research in the unit of the Howard Hughes Medical Institute established in the Department of Physiology at Vanderbilt (1954–55). But the following year he was back in obstetrics and gynecology as assistant chief of that service on active duty in the U.S. Army Station Hospital at Fort Campbell, Kentucky. He then returned to Vanderbilt, and for the next ten years (1957–67) he was an investigator in the Hughes Institute, with faculty rank that progressed from assistant professor (1959–62), to associate professor (1962–66), and professor (1966–67). Morgan then became the first professor and chairman of the Department of Physiology in the Milton S. Hershey Medical Center of the Pennsylvania State University in Hershey, Pennsylvania. From 1973 he has been also Associate Dean for Research, and in 1974 was honored by designation as the Evan Pugh Professor of Physiology. In 1982 he was further honored by appointment as a scholar of the Howard Hughes Medical Institute. Morgan wrote briefly of his training:

> Because I entered physiological research after eight years of clinical training, research, and practice in obstetrics and gynecology, my training was entirely as a postdoctoral fellow. Charles R. Park served as my preceptor and guided me into studies of the effects of insulin on glucose uptake and sugar transport. With a solid background obtained in Park's laboratory, I later was able to undertake the new areas of investigation that have characterized the remainder of my career.

Before he became a member of APS (1965), Morgan had been elected to the Biochemical Society (Great Britain, 1960) and the American Society of Biological Chemists (1962). He holds membership also in the Biophysical Society (1965), the Cardiac Muscle Society (1969; president, 1976–77), ACDP (president, 1975–76), and AHA (Board of Directors, 1984–). In the Basic Science Council of AHA, Morgan served on the Executive Committee from 1973 to 1979 and again from 1981 to the present (chairman, 1983–). He has been a member of the Executive Committee of the American Section of the International Society for Heart Research (1976–79; president, 1979–82). From this office he became president elect of the International Society (1980–83) and has served as president (1983–86). Morgan therefore has served as chairman or president of a major scientific organization in all but two of the past eleven years, and in the year 1985–86 he held three such offices simultaneously.

In addition to the honors noted above, including those from the Howard Hughes Medical Institute and from his own university, Morgan has received an Award of Merit from AHA (1979), the Carl J. Wiggers Award from the Cardiovascular Section of APS (1984), and an honorary fellowship in the American College of Cardiology (1985). He was elected to APS Council in 1983 and became president elect the following year.

In areas related to cardiology, Morgan has provided scientific counsel to the Research Committee of the Pennsylvania Heart Association (1967–71; chairman, Research Peer Review Committee, 1983–); the Research Council of the New York City Heart Association (1974); the US:USSR Exchange Program for Problem Area 3, Myocardial Metabolism (coordinator, 1974–83); and AHA. He served as a member of the Physiological Chemistry A Research Study Group of AHA (1973–75; chairman, 1976–79) and of the AHA Research Committee (1974–79 and 1980–81). In 1977–78 he was vice-president for research, chairman of the Research Committee, and a member of the Board of Directors of AHA. NIH has called on him for membership in the Metabolism Study Section (1967–71), on an ad hoc committee for the National Heart Center Program (1973), on a Cardiology Advisory Committee (1975–78), and on the Advisory Council of the National Heart, Lung and Blood Institute (1979–83). In 1982 Morgan was asked to be chairman of a special panel appointed by this latter institute "to review alleged misconduct at Brigham and Women's Hospital/Harvard Medical School." Finally, he now holds membership on the U.S. National Committee for IUPS (1984–87).

Another important feature of Morgan's career is his association with scientific journals. Beginning with the Editorial Board of the *American Journal of Physiology* (1967–73), he became editor of *Physiological Reviews* (1973–78), associate editor of the *American Journal of Physiology: Endocrinology and Metabolism* (1979–81), and editor of the *American Journal of Physiology: Cell Physiology* (1981–84). For much of this time he served on the Publications Committee (1979–85; chairman, 1981–85). Other journals for which he has provided editorial assistance include *Circulation Research* (1971–76 and 1982–), the *Journal of Biological Chemistry* (1973–78 and 1980–85), the *Journal of Cardiovascular Pharmacology* (1977–82), and the *Journal of Molecular and Cellular Cardiology* (1974–; associate editor, 1979–83). Of this listing, his influence was perhaps the greatest on *Physiological Reviews.* During his tenure as editor it grew significantly in international reputation and influence.

Morgan's research interest is the physiological regulation of intermediary metabolism. For many of his studies he has used the isolated and perfused rat heart. He has described his work as follows:

> Initial studies dealt with the mechanism of action of insulin on glucose uptake and the nature of glucose transport. Insulin was found to accelerate glucose transport, a stereospecific, saturable process in the cell membrane (1). A kinetic model of sugar transport was proposed, based on studies in rabbit erythrocytes (2). This model and its mathematical description have been used by many other investigators in characterizing transport phenomena. Experiments measuring the rate of glycogen utilization led to investigation of the allosteric control of phosphorylase *a* and *b* and to the discovery that phosphorylase *b* activity was increased by 5′-AMP [adenosine 5′-monophosphate] and inhibited by ATP [adenosine triphosphate] and G-6-P [glucose 6-phosphate] (3). This mechanism of allosteric control accounted for the differential effects of anoxia and glucagon and for acceleration of glycogen utilization in working hearts.

> My interest in the effects of heart work on cardiac metabolism led to development of the isolated perfused working rat heart (4) that has been used extensively both in our laboratory and elsewhere for study of the effects of mechanical performance on carbohydrate, fat, and protein metabolism. In this model, perfusion medium is introduced into the left atrium over a range of atrial filling pressures and is pumped against a variable outflow resistance. With this model, myocardial oxygen consumption was found to depend on the aortic pressure to which the heart was exposed; greater oxygen consumption was accompanied by faster utilization of oxidative substrates.

> During the next phase of my research career, my interest shifted to identification of factors that control growth of the heart and that can lead to cardiac hypertrophy. Initiation of peptide chains on myocardial ribosomes was found to become a rate-controlling step during in vitro perfusion and to be accelerated by insulin, fatty acids, and other noncarbohydrate substrates, leucine, increased cardiac work, and exposure to higher aortic pressure (5, 6). A rigorous method for estimation of rates of protein

synthesis was developed that depended on measurements of the specific activities of phenylalanyl-tRNA (7). Protein degradation also was identified as a site of control of protein turnover that is affected by insulin, diabetes, energy availability, noncarbohydrate substrates, leucine, cardiac work, and increased aortic pressure (6, 8). The factor that links cardiac work to faster rates of protein synthesis and slower proteolysis (9) appears to be stretch of the ventricular wall, because these effects could be observed in hearts arrested with tetrodotoxin and containing a ventricular drain. In these preparations, an increase in aortic pressure stretched the ventricular wall, accelerated protein synthesis, and inhibited proteolysis. These events appear to represent early changes in the hypertrophy process.

After longer periods of exposure to pressure overload or to thyrotoxicosis in vivo, we found that content of cardiac RNA increased and accounted for much of the increment in protein synthesis. Since ribosomal RNA constitutes about eighty-five percent of cardiac RNA, these changes indicated that net ribosome production was increased, either by acceleration of rRNA transcription or processing or by inhibition of rRNA degradation (10). These events are the focus of my current research.

During the past twenty-five years the presidents of APS have often expressed concern about the apparent fragmentation of the science and the development of diverse and presumably independent interests by members of the Society. A countertendency is beautifully illustrated in the lecture Morgan gave when he received the Carl J. Wiggers award in 1984. He described experiments that began with a problem in classic physiology, the response of the heart to increased load. However, he pursued the response, not only by use of traditional physiological measurements such as oxygen consumption, but on through analysis of pathways of protein, carbohydrate, and lipid metabolism, until he reached the measurement of rates of synthesis and degradation of the several forms of RNA. The experiments moved clearly and easily from the whole organ to the level of molecular biology. His lecture illustrates how what seem to be old-fashioned problems can be studied by using the most sophisticated of modern techniques to provide a clearer understanding of what really takes place in living organisms.

SELECTED PUBLICATIONS

1. Morgan, H. E., M. J. Henderson, D. M. Regen, and C. R. Park. Regulation of glucose uptake in muscle. I. The effects of insulin and anoxia on glucose transport and phosphorylation in the isolated, perfused heart of normal rats. *J. Biol. Chem.* 236: 253–261, 1961. (Citation classic.) [Dr. Morgan's first paper in a series on the control of glucose uptake.]

2. Regen, D. M., and H. E. Morgan. Studies of the glucose-transport system in the rabbit erythrocyte. *Biochim. Biophys. Acta* 79: 151–166, 1964. [Mathematical model of glucose transport.]

3. Morgan, H. E., and A. Parmeggiani. Regulation of glycogenolysis in muscle. III. Control of muscle phosphorylase activity. *J. Biol. Chem.* 239: 2440–2445, 1964. [Discovery of the allosteric control of phosphorylase.]

4. Neely, J. R., H. Liebermeister, E. Battersby, and H. E. Morgan. Effects of pressure development on oxygen consumption by the isolated rat heart. *Am. J. Physiol.* 212: 804–814, 1967. [Development of in vitro working rat heart.]

5. Morgan, H. E., D. C. Earl, A. Broadus, E. B. Wolpert, K. E. Giger, and L. S. Jefferson. Regulation of protein synthesis in heart muscle. I. Effect of amino acid levels on protein synthesis. *J. Biol. Chem.* 246: 2152–2162, 1971. [Initial paper on control of protein synthesis.]

6. RANNELS, D. E., R. L. KAO, AND H. E. MORGAN. Effect of insulin on protein turnover in heart muscle. *J. Biol. Chem.* 250: 1694–1701, 1975. [Initial description of the effect of insulin on protein degradation in heart.]

7. McKEE, E. E., J. Y. CHEUNG, D. E. RANNELS, AND H. E. MORGAN. Measurement of the rate of protein synthesis and compartmentation of heart phenylalanine. *J. Biol. Chem.* 253: 1030–1040, 1978. [Discovery of a rigorous approach to measurements of protein synthesis.]

8. MORGAN, H. E., B. H. L. CHUA, E. O. FULLER, AND D. SIEHL. Regulation of protein synthesis and degradation during in vitro cardiac work. *Am. J. Physiol.* 238 (*Endocrinol. Metab.* 1): E431–E442, 1980. [Discovery of effects of cardiac work on protein turnover.]

9. KIRA, Y., P. J. KOCHEL, E. E. GORDON, AND H. E. MORGAN. Aortic perfusion pressure as a determinant of cardiac protein synthesis. *Am. J. Physiol.* 246 (*Cell Physiol.* 15): C247–C258, 1984. [Discovery of stretch as the mechanical factor affecting protein turnover.]

10. SIEHL, D., B. H. L. CHUA, N. LAUTENSACK-BELSER, AND H. E. MORGAN. Faster protein and ribosome synthesis in thyroxine-induced hypertrophy of rat heart. *Am. J. Physiol.* 248 (*Cell Physiol.* 17): C309–C319, 1985. [Discovery of role of increased ribosome content in cardiac hypertrophy.]

11. MORGAN, H. E., E. E. GORDON, Y. KIRA, D. L. SIEHL, P. A. WATSON, AND B. H-L. CHUA. Biochemical correlates of myocardial hypertrophy. Wiggers Award Lecture. *Physiologist* 28: 18–27, 1985.

59 (1986–87)

Franklyn G. Knox

(b. 1937)

As the most recent twenty-five year history of APS began in 1963 with Hermann Rahn presiding at the anniversary meeting in Coral Gables, Florida, it will come to a climax in 1987 at the Centennial Celebration at the FASEB Meeting in Washington, D.C., with Franklyn G. Knox in the chair. In 1963 Knox was an M.D./Ph.D. student with Donald Rennie in Hermann Rahn's department at the State University of New York at Buffalo and journeyed to the Coral Gables meeting with his wife and two small children. A scientific "generation," therefore, is exactly twenty-five years—at least in this instance. In anticipation of the coming anniversary, Knox has written:

As president of the American Physiological Society, I will have the distinct honor of presiding over the Centennial Celebration recognizing the One Hundredth Anniversary of the Society. The 1987 Spring Meeting of FASEB will be held in Washington, D. C., and will have the theme "A Century of Progress in Physiology." An impressive opening ceremony with the Marine Band is planned for the Washington Convention Center. The program for the meeting will be built around progress in physiology with both retrospective and prospective analysis of the field. These program activities will be complemented by publication activities of important historical books. In the meantime, the Society looks forward to the next century of progress through strengthening the roles of the sections of the American Physiological Society so that the Society can be in the strongest position to respond to the changing future.

Knox was born in Rochester, New York, and completed all his professional education at the now State University of New York at Buffalo. He received the B.S. degree in 1959 and the M.D. and Ph.D. degrees in 1965. The last of these represented training in the Department of Physiology and led to a position as staff associate at the National Heart Institute (1965–68). From there he moved to the University of Missouri, where he was promoted to an associate professorship in the Department of Physiology in 1970. The following year he joined the faculty of the newly organized Mayo Medical School. He became professor of physiology and of medicine and also chairman of the Department of Physiology and Biophysics in 1974. For five years (1978–83) he was also the Associate Director of Graduate Education: Research Training and Degree Programs of the Mayo Graduate School of Medicine. In 1983 he moved a step higher to become dean of the Mayo Medical School and Director for Education for the Mayo Foundation. Knox has written of his education, training, and faculty positions:

My decision to become a physiologist was preceded by my decision to have a research career. As an undergraduate student working in Gerhard Levy's laboratory in Buffalo, I was impressed by the power of the scientific method to make contributions to society. Subsequently I applied to medical school with the objective of further training toward a research career. It wasn't until I did summer research as a medical student in Donald Rennie's laboratory in the Department of Physiology that I began to consider a career as a physiologist. Consideration of this career track led to the development of an M.D./Ph.D. program. The Department of Physiology at Buffalo was under the leadership of Hermann Rahn and was noted for its particular strengths in respiratory physiology. With Rennie's interest in the kidney, it was natural that my thesis should be in the area of the respiration of the kidney. Rennie served as a role model for the kind of career that I envisioned in research and teaching in physiology.

My postdoctoral research training in the Laboratory of Kidney and Electrolyte Metabolism at the National Institutes of Health with Robert W. Berliner was a particularly exciting training experience. In addition to learning the micropuncture technique, there was interaction with a large number of enthusiastic young investigators who have subsequently developed leadership positions in renal physiology and nephrology. Berliner allowed for the individual creativity of these individuals, but in the context of intimate critique of the data as they unfolded day by day. Thus we would anticipate a noon flight of the "Eagle" in which he would ask, "Do you have any numbers?" It was a vibrant and exciting place to train.

James O. Davis invited me to join the faculty at the University of Missouri, where I had the opportunity to press the long-term objective of independent research and teaching into action. Given my background from Buffalo, I taught the respiratory physiology section of the medical school course and developed the teaching around a case of emphysema. I remember one student's essay, "Don't kiss me, I'm trying to breathe" in which the mechanics of respiration and emphysema were discussed and the necessity for pursed lips to maintain resistance in the airway to prevent airway collapse was explained.

John Shepherd and Jim [James C.] Hunt recruited me to Mayo at the time of the development of Mayo Medical School. This was a particularly exciting time because of the opportunities for developing a medical curriculum in an institution noted for its excellence in medicine and with the opportunity to develop innovative approaches with a small class of students. Shepherd served as a role model for outstanding effectiveness in administrative activities, and I subsequently succeeded him as department chairman. After approximately a decade as department chairman, I again succeeded Shepherd as Director for Education for the Mayo Foundation and dean of the Mayo Medical School. The Medical School, Graduate School, and School of Health-Related Sciences have grown and flourished, and the most significant accomplishment is the establishment of Mayo as an independent degree-granting institution.

Knox has been elected a fellow of AAAS (1985) and of the Council on Circulation of AHA. He has served AHA on the Board of Directors (1982–) and Executive Committee (1983) and in varied capacities on the Council on Kidney in Cardiovascular Disease (1977–) and the Council for High Blood Pressure Research (1974–). He is a member of AMA, the American Society for Clinical Investigation, and the Association of American Physicians. He has been a member of the Nominating Committee of the American Society of Nephrology (1979) and the Council of Deans of AAMC; he is currently on the Selection Committee for the AAMC Award for Distinguished Research in the Biomedical Sciences (1985–). He was president of ACDP (1981–82) and a member of the Physiology Test Committee of the National Board of Medical Examiners (1980–81); he currently is a member of the U.S. National Committee for IUPS (1985–88). He has performed review and advisory functions for the National Kidney Foundation (1980–83), the National Institute of Arthritis, Diabetes and Digestive and Kidney Diseases (1979–83), and for the NIH Division of Research Grants (1983–87; chairman of General Medicine B Study Section, 1986–87).

Knox was elected to membership in APS in 1969. He was a member of the Program Committee for eight years (1973–76 and 1977–82; chairman, 1981–82), chairman of the Renal Section (1975–77), APS representative to the AAMC Council on Academic Societies (1979–83), and chairman of the Committee on Committees (1983–85). He is a member of the Long-Range Planning Committee (1984–) and of the Subcommittee on International Physiology (1985–). Elected to Council in 1982, he became president elect in 1985.

Having served the Society as a member of the Editorial Board of the *American Journal of Physiology: Renal, Fluid and Electrolyte Physiology* (1976–80), Knox was appointed to the Publications Committee in 1984. He has provided editorial service also as a member of editorial boards or as a consultant for *Circulation Research, Journal of Clinical Investigation, Kidney International, Mineral and Electrolyte Metabolism, Contemporary Nephrology, Federation Proceedings, Seminars in Nephrology, American Journal of Kidney Diseases,* and the *Journal of Laboratory and Clinical Medicine* (editor, 1979–80). Many of these responsibilities developed from his own research interests.

Knox began research while, as an undergraduate student at Buffalo, he worked in Gerhard Levy's laboratory in the School of Pharmacy. His first paper (1) involved the bioassay of thyroid preparations, utilizing the capacity of the thyroid preparation to prevent propylthiouracil-induced goiters. Later in his career he returned to studies of thyroid and parathyroid hormones on renal function (6, 7). For his Ph.D. dissertation he studied the effects of osmotic diuresis on sodium reabsorption and oxygen consumption by the kidney in the laboratory of Donald W. Rennie in the School of Medicine (2). This began his ongoing interest in control of renal sodium excretion. He found that osmotic diuresis markedly decreases the number of sodium ions transported per mole of oxygen utilized and concluded that the osmotic diuresis enhances the backflux of sodium in proximal tubules. Many of his later publications have followed this theme.

While he was in Berliner's laboratory at the National Heart Institute, Knox learned micropuncture techniques and analyzed the effect of changes in blood volume on proximal sodium reabsorption (3). His first publication as an independent investigator was a study of furosemide natriuresis, carried out after he had moved to the University of Missouri (4). The earliest of the "favorite publications" he selected from the Mayo laboratories established the relative contributions of various nephron segments in control of sodium excretion (5). Of a paper published a year later (6), Knox wrote:

> This paper is particularly important because the control experiments were surprising and led to an entirely new phase of research on regulation of phosphate transport by proximal tubules. We were using hyperoncotic albumin solution preferentially to expand plasma volume at the expense of interstitial fluid volume and to determine the effects on renal sodium handling. Ultimately, because of control experiments with dextran solutions, we discovered that the effects of hyperoncotic albumin were not due to plasma volume expansion, but rather to binding of plasma calcium by the albumin and subsequent release of parathyroid hormone. This hormone, in turn, had an unexpectedly large effect on sodium ions in the proximal tubule, as well as on phosphate reabsorption. This led to a significant body of work dealing with intrarenal phosphate metabolism. Further work pointed the way to involvement of segments other than the proximal tubules in control of phosphate excretion (7). Our contributions to the understanding of renal phosphate handling were summarized in the Twenty-second Bowditch Lecture of the APS (8).

A new era of renal physiology opened with the discovery that nephrons deep in the kidney have functions surprisingly different from those located near the surface (9). Previously it was generally assumed that superficial nephrons are representative of nephrons throughout the kidney. On the contrary, proper interpretation of micropuncture results must include the possibility of nephron heterogeneity. This work was extended to include sodium, as well as phosphate, handling and served to help resolve the long-standing issue of the mechanism for escape from salt-retaining effects of mineralocorticoids (10). Finally, Knox selected a recent paper with Erik Ritman "because of its importance for the future" (11). It utilized high-speed dynamic spatial reconstruction techniques with advanced computers to bring renal physiology "into the modern era of noninvasive measurement of organ function."

The halo shining with a "soft blue light" that Hermann Rahn first identified in 1963 continues to illuminate the brow of presidents of APS. It seems especially appropriate that at the 1987 Centennial Meeting it will be worn by one of Rahn's most illustrious disciples, Franklyn Knox.

SELECTED PUBLICATIONS

1. LEVY, G., AND F. G. KNOX. The biological activity of orally administered thyroid. *Am. J. Pharm.* 133: 255–266, 1961.

2. KNOX, F. G., J. S. FLEMING, AND D. W. RENNIE. Effects of osmotic diuresis on sodium reabsorption and oxygen consumption of the kidney. *Am. J. Physiol.* 210: 751–759, 1966.

3. KNOX, F. G., S. S. HOWARDS, F. S. WRIGHT, B. B. DAVIS, AND R. W. BERLINER. Effect of dilution and expansion of blood volume on proximal sodium reabsorption. *Am. J. Physiol.* 215: 1041–1048, 1968.

4. KNOX, F. G. Effect of increased proximal delivery on furosemide natriuresis. *Am. J. Physiol.* 218: 819–823, 1970.

5. KNOX, F. G., E. G. SCHNEIDER, L. R. WILLIS, J. W. STRANDHOY, AND C. E. OTT. Effect of volume expansion on Na excretion in the presence and absence of increased delivery from the proximal tubule. *J. Clin. Invest.* 52: 1642–1646, 1973.

6. KNOX, F. G., E. G. SCHNEIDER, L. R. WILLIS, J. W. STRANDHOY, C. E. OTT, J. L. CUCHE, R. S. GOLDSMITH, AND C. D. ARNAUD. Proximal tubule reabsorption following hyperoncotic albumin infusion: role of parathyroid hormone and dissociation from plasma volume. *J. Clin. Invest.* 53: 501–507, 1974.

7. KNOX, F. G., AND C. LECHENE. Distal site of action of parathyroid hormone on phosphate reabsorption. *Am. J. Physiol.* 229: 1556–1560, 1975.

8. KNOX, F. G. The intrarenal metabolism of phosphate. *Physiologist* 20(6): 25–31, 1977.

9. KNOX, F. G., J. A. HAAS, T. BERNDT, G. R. MARCHAND, AND S. P. YOUNGBERG. Phosphate transport in superficial and deep nephrons in phosphate loaded rats. *Am. J. Physiol.* 233 (*Renal Fluid Electrolyte Physiol.* 2): F150–F153, 1977.

10. KOHAN, D. E., AND F. G. KNOX. Localization of the nephron sites responsible for mineralocorticoid escape in rats. *Am. J. Physiol.* 239 (*Renal Fluid Electrolyte Physiol.* 8): F149–F153, 1980.

11. KNOX, F. G., AND E. L. RITMAN. The intrarenal distribution of blood flow: a new approach. *Kidney Int.* 25: 473–479, 1984.

CHAPTER 8

Executive Secretary-Treasurers

JOHN R. BROBECK

Since 1948 four men have served the American Physiological Society (APS, the Society) as executive secretary-treasurer. Milton O. Lee held the office from 1948 to 1956, Ray G. Daggs from 1956 to 1972, Orr E. Reynolds from 1973 to 1985, and Martin Frank from 1985. The most recent twenty-five years therefore include much of Daggs' tenure, all of Reynolds', and the beginning of Frank's. All three men came to APS from responsible positions in federal research programs.

Ray G. Daggs

(b. 1904)

Born in McKees Rocks, Pennsylvania, Daggs graduated with a B.S. degree from Bucknell University in 1926. (His devotion and that of his family to Bucknell were recognized in 1955 when he was awarded an honorary doctor of science degree.) He became interested in biology and living things early in life and in college took all possible courses in biology, including human anatomy and physiology. He then entered the University of Rochester School of Medicine and Dentistry as a medical student, but after two years he was awarded a fellowship of the National Research Council for research in nutrition in John R. Murlin's Department of Vital Economics at Rochester. Required by the conditions of the fellowship to work on meat or meat products, he discovered that the amino acid cystine was the factor in certain meats that stimulated lactation in rats, dogs, and women. With a Ph.D. degree received in 1930, he served as instructor in physiology at Rochester for three years and assistant professor for three more (1933–36). He then moved to the University of Vermont as an associate professor and was made head of the Department of Physiology in 1941. He was the only member of the faculty without a medical degree. He found practically no research in progress in the medical school but was able to establish an animal house and colonies, and finally, with cooperation of research groups from the Agriculture College, the undergraduate faculty of the college, and certain members of the medical faculty, he was able to initiate a local chapter of Sigma Xi.

For five years during World War II, Daggs was a colonel in the U.S. Army Sanitary Corps, assigned as nutrition officer to the Eighth Service Command, where he worked with the Quartermaster Corps on diets for the troops and prisoners of war. At the end of the war he was invited to become the civilian Director of Research at the Army Medical Research Laboratory at Fort Knox, Kentucky. In addition to his laboratory responsibilities, he gave lectures in physiology at the Medical School of the University of Louisville. At Fort Churchill, Canada, with the Canadian Army Medical Corps, he investigated the effects of cold on soldiers—a research program also under way at Fort Knox. He was at Fort Knox when Milton Lee invited him to become executive secretary-treasurer of APS in 1956.[1]

On his retirement from office in December 1972, Daggs was honored by his many friends in the Society by establishment of the Ray G. Daggs Award, "an annual award to be presented beginning next year to a physiologist who is judged to have provided distinguished service to the science of Physiology and to the American Physiological Society." In a meeting where members of Council were searching for a suitable recognition of Daggs' many contributions to the Society, it was Daniel Tosteson (president, 1973–74) who proposed the award as it was eventually authorized. When President Robert Berne informed the business meeting on 17 April 1973 that the award had been established, he said:

> Ray Daggs was a guiding force in the American Physiological Society from 1956 until his retirement this (academic) year. Under his direction the Society has grown in size, extended its activities, and achieved considerable financial stability. . . . Ray Daggs has always been on hand to lead the officers down the right path. He has always been devoted to the Society and has placed its interests foremost. He is a direct individual who has the knack of getting to the heart of the matter quickly and succinctly. . . . The Society will very much miss Ray.

Asked to write of his experience in office for this volume, Daggs responded as follows:

> In 1956 I was at the Fort Knox Medical Research Laboratory when I received a phone call from Milton Lee asking if I was interested in the position of executive secretary-treasurer of APS. Because it was a position in physiology, my first love, I accepted at once. When I arrived in Bethesda I found that my office was a small room on the second floor of Beaumont House, with borrowed furniture; even the rug was lent [or donated] by Wallace Fenn. I was given a secretary, Grace Hamilton, by generosity of the Board of Publication Trustees of the Society. One of my first duties was to publish the newly formed *The Physiologist* (1957–58), an outgrowth of the "President's News Letter." It was made small in size so it could be carried in a jacket pocket. The staff grew when I obtained a second room across the hall, hired a clerk-

typist (Kelly Byars) to handle the archives (they were stored in a shower on built-in shelves) and the details of *The Physiologist*, and also hired a fiscal clerk.

The Federation [Federation of American Societies for Experimental Biology, FASEB] was handling all APS financial and publishing activities until 1956. One of my first actions, however, was to set up our own financial management. But this included only the transactions of the General Fund. Publications were under the control of an independent group called the Board of Publication Trustees (BPT), an outgrowth of the early days when APS did not own its publications. The mechanics of the publishing were managed by FASEB. Consequently the APS Council had very little to say about publications or their financial affairs. A few years of debates in business meetings were required before Council was finally authorized (1961) to recommend a change in the bylaws, with creation of a Publications Committee in place of the BPT (1). The new committee was directly responsible to Council.

At last the activities of the Society were united, under its own control, and completely under the jurisdiction of Council. The duties of the Finance Committee were therefore redefined, and a business manager was hired to supervise the much enlarged scope of financial transactions. The Society, likewise, was growing in numbers and in 1962 moved its headquarters into the new Milton Lee Building adjacent to the Beaumont House. By this time the APS staff had increased from an executive secretary-treasurer with one secretary to comprise also the business manager and staff and a publications manager and executive editor, as well as a staff of copy editors and an illustrator.

As executive secretary-treasurer and editor of *The Physiologist*, I wrote many editorials on physiology and physiologists. I set up a card file system of APS members; prepared an operational guide for officers, Council, and committees; and monitored the establishment of the APS logo designed by W. F. Hamilton (president, 1955–56). Most of my time and attention, of course, were given to assisting Council and its several committees in their deliberations and in implementing their decisions. I found that I was especially interested and involved in the affairs of the Education Committee.

I thoroughly enjoyed my years as executive secretary-treasurer of the APS. I came to know many physiologists and greatly appreciate their friendship. Perhaps the greatest tribute they could have paid me was the establishment of the Ray G. Daggs Award, granted each year to a person for distinguished service to the Society and to the science of physiology.

Orr E. Reynolds
(b. 1920)

Orr Reynolds brought to the office of executive secretary-treasurer a career remarkably varied in training, experience, and responsibility. Born in Baltimore, son of an organic chemist employed for most of his professional life in U.S. government laboratories, Reynolds developed an interest in biology as a schoolboy. In 1937 he entered Jones Ormond Wilson Teachers College in Washington, D.C., with scholarship support for study of chemistry and mathematics. In his sophomore year, however, while classifying earthworms for Professor Henry Olson, he was encouraged by Olson to transfer to the University of Maryland to study biology. Olson also introduced him to the Chesapeake Biological Laboratory at Solomons Island, Maryland, where Reynolds spent four or five summers. Here he planned to become a marine biologist under supervision of Reginald V. T. Truitt, director of the Chesapeake Laboratory and then chairman of the Department of Zoology at the University of Maryland. When Norman Phillips unexpectedly replaced Truitt as chairman, however, Reynolds found himself assisting in courses in vertebrate physiology, and eventually, with Phillips' encouragement and the advent of World War II, he settled on a thesis topic related to high-altitude physiology. A. C. Ivy, then scientific director of the Naval Medical Research Institute in Bethesda, agreed to serve as his advisor for research actually carried out at the National Institutes of Health (NIH) Laboratory of Industrial Hygiene. To this end Reynolds was employed as an "assistant physiologist" by that laboratory and did his own research after the usual working hours.

By 1944 he had completed the research for his dissertation, although the Ph.D. degree was not formally awarded until 1946. He applied for a commission in the U.S. Navy Hospital Specialist Corps and was assigned (1944) to the Pensacola Naval Air Station for training as a low-pressure chamber operator. But because of his experience at the NIH laboratory he was very shortly promoted from trainee to faculty status. Subsequently he was transferred to the U.S. Marine Corps Air Station in Quantico, Virginia; to the Naval Air Test Center in Patuxent, Maryland; and to the Aviation Physiology Research Administration, Research Division, Bureau of Medicine and Surgery of the Navy, Washington, D.C., then under the command of J. Newell Stannard, who had been Reynolds' supervisor two years earlier at NIH.

On discharge from the U.S. Navy, Reynolds accepted what he expected to be a temporary position with the U.S. Navy's Office of Research and Invention, for which he reviewed proposals in physiology received by that office. From 1946 to 1948 he was head of the physiology branch of the newly constituted Office of Naval Research (ONR) and from 1948 to 1957 was director of the Biological Sciences Division of the ONR. (The pattern of research grant evaluation and support established by Reynolds at ONR fortunately became a model for creation of the NIH granting policies.) From ONR in 1957 Reynolds became director of the Office of Science in the Office of the Director of Defense Research and Engineering of the U.S. Department of Defense (DOD). He served there for five years, and then in 1962 he was appointed director of Bioscience Programs for the National Aeronautics and Space Administration, a position he held for eight years. This period represented the most rapid development of the nation's astronautical effort. Reynolds was responsible for programs in exobiology, environmental biology, physical biology, behavioral biology, and bioscience communications.

Reynolds has been a member of most of the national committees, and of many international bodies, concerned with space biology. He holds membership in the Aero Medical Association (1945), International Academy of Astronautics (Paris, 1966), and American Society for Gravitational and Space Biology (charter member, 1984; vice-president, 1985), as well as in the more conventional societies such as the American Association for the Advancement of Science, American Institute of Biological Sciences, and the New York and the Washington Academy of Sciences. In 1957 he received the Meritorious Service Medal from the U.S. Navy, and in 1970 and 1976 he was honored when the National Aeronautics and Space Administration awarded him its Exceptional Service Medal and its Group Achievement Award, respectively. In 1978 the Semmelweis Medical University of Budapest bestowed on him the Semmelweis Medal.

A member of APS from 1948, Reynolds was asked to serve as director of the Survey of Physiological Sciences initiated by Ralph Gerard in 1952. Reynolds joined the staff of APS in 1970 as education officer, and the following year became also an assistant to Ray Daggs. On Daggs' retirement in 1973, Reynolds followed him in office and served until 1985. He continues his association with affairs of the Society as a volunteer directing the Centennial Celebration Task Force and as coeditor of this volume.[2]

Reynolds' bibliography (available from the APS Archives) begins with a study of metabolic changes associated with pregnancy and the fetal state in guinea pigs from the laboratory of Walter L. Hard at the University of Maryland. Much of the bibliography is on high-altitude physiology, including an analysis of "aerodontalgia" originating in the maxillary sinus but referred to a previously injured tooth or extraction site. From 1970 many of his papers concern teaching of physiology or activities of the Society. In particular, he has helped bring to realization the new section of the Society devoted to history of the science. It is appropriate therefore that on his retirement the Council established the Orr Reynolds Award, to be given annually "for the best historical article submitted by any member of the Society" (2).

Of his experience in office Reynolds has written:

I became executive secretary-treasurer of the APS during the last half of the presidency of Robert M. Berne. Berne considered himself somewhat outside the physiological establishment, and perhaps for that reason he viewed his role as president as requiring the analysis of Society activities from the standpoint of whether they were meeting the needs of the broad membership. In his past president's address (3), he listed his concerns as follows:

1. Representation of broad membership in the election of officers
2. Representation of minorities
   a. Minority subdisciplines
   b. Racial minorities
   c. Women in physiology
3. Relationship to FASEB
4. Relationships with medicine and basic science

During the first half of his term as president, Berne had conducted surveys of membership desires in the areas of election of officers and venue for the fall meeting. The other areas of his concern were studied by other means, primarily by setting up other committees and task forces. I believe, however, that in the period since 1973 the Council of the APS has been largely occupied with searching for solutions to the concerns expressed by Berne. In modification of the election procedure, for example, several efforts were required by Council in finding a mechanism of holding elections by mail that could be incorporated in bylaws acceptable to the business meeting of the Society. After several false starts, a procedure was adopted that permitted the whole process to be handled by mail while retaining many of the elements contained in the traditional system of election at the business meeting. Although this system is still in use, it is a frequent target of criticism from the membership, which finds it too cumbersome. It has had the advantage of increasing the total vote by approximately an order of magnitude. The total referendum is now in the vicinity of 1,500, whereas the business meeting elections normally did not count more than 150 members.

The second concern to be addressed was that of the venue for the fall meeting. Originally, on the basis of a nearly fifty-fifty distribution of votes in the survey, the Council decided to hold the fall meeting in alternate years on a university campus in August and in an "attractive city" hotel in October. After a brief trial this alternating system proved to be managerially impractical, and consequently Council decided to move to city meetings in October for a five-year period and then to reevaluate the decision. When the five-year period was completed in 1982, campus meetings were scheduled in August for 1983–85. Implementation of the campus meetings, however, turned out to be impractical for those years, and as a result, all were held as hotel-based meetings (in August 1983 and 1984 and in October 1985). The results were such that the Council decided that August campus meetings are not feasible. For the present at least, the plans are to hold the fall meeting in October each year for the foreseeable future, a decision that precludes meeting on campus. The change from August campus meetings to October "city" meetings has had the effect of about doubling the attendance at the fall meeting (see chapter on meetings).

The concern with subsidiary disciplines in the Society was met by establishment of task forces to consider individual subspecialties. During the period of Tosteson's, Schmidt-Nielsen's, and Knobil's presidencies, this evolved into a system of established sections in the Society with the various subject matter-oriented sections gaining progressively more voice in Society decisions. On a somewhat parallel track, the Publications Committee had been considering the desirability of sectionalizing the *American Journal of Physiology (AJP)* into separately bound sections. During the period of Alfred Fishman's chairmanship of the Publications Committee (1975–81), the sectionalization of *AJP* was achieved, and Fishman was encouraged by Bodil Schmidt-Nielsen to play a central role in the establishment of sections of the Society membership by following a pattern similar to that of sectionalizing the journal. These organizational steps gave specialty groups within the Society a greater ability to make their publications available to their peers, to control largely the content of the scientific program at meetings of the Society, and to inform the Society Council of special problems or discontent that representatives of the discipline have with Society governance.

The Porter Development Committee, which had been charged in 1968 with improving the access to physiology careers for racial minorities, has progressively stabilized its financial capabilities and broadened its approach to minority careers under the continuing leadership of A. C. Barger and E. W. Hawthorne (see chapter on Porter Development Program).

The status of women physiologists was first subjected to study by a special task force under the chairmanship of Elizabeth Tidball. Subsequently a social group or caucus of women in physiology was established, and finally a standing Committee of Women in Physiology was formed. Because the rate of growth of female membership in the Society has been spectacular in recent years, women may be expected to play an increasing role in Society activities, including governance. Another result of the Task Force on Women in Physiology was the establishment of a standing Committee on Careers in Physiology, to be concerned with all factors that influence the development of suitable careers in physiology, including graduate training, nontraditional career choices, and other problems.

Access to membership in the Society has been addressed by broadening the definition of associate member and by adding two new membership categories, corresponding membership and student membership. When corresponding membership was first proposed, its intent was to offer membership in the Society to persons in scientific specialty groups outside the conventional bounds of physiology, as well as outside North America. This dual concept, however, was not acceptable to members voting at the business meeting of the Society. Therefore the category of corresponding membership was limited to foreign physiologists. A subsequent activity on the part of APS to increase communication with disciplines adjacent to physiology has been to consider formal affiliation with other societies. This process is now actively under way as affiliation is being considered for the Microcirculatory Society. Other possible candidates for affiliation would include the Biomedical Engineering Society, the Society for Mathematical Biology, and the Society for Experimental Biology and Medicine. Each of these societies consists of a fair proportion of members of APS, but each also has other members from other disciplines.

The relationship with FASEB has been the subject of study by several ad hoc committees established by Council. The principal outcome of study has been the proposal to FASEB, and its adoption, of holding interdisciplinary "theme-oriented" sessions at the FASEB meeting to allow highly concentrated attention on subjects of particular interest to members of APS and other FASEB societies. This theme orientation has been so successful that it has spawned two "descendants." The first of these, a series of summer conferences managed by FASEB and organized similarly to the

Gordon Conferences run by the American Association for the Advancement of Science, has been highly successful and is now in its fifth year. The second, a recent plan for improving the fall meeting, includes the organization of theme orientation subject matter for each fall meeting.

Problems of relationships with medicine and basic science were addressed first with the establishment in 1975 of a Committee on Clinical Physiology. Among its recommendations, this committee proposed a series of three one-half-day symposia at each spring meeting on a scientific topic with strong clinical implications, the results of which were to be used as the basis for a monograph. This particular recommendation was sufficiently successful that the committee has now become a subcommittee of the Publications Committee, and eight monographs have resulted from the programs. Clinical affiliations with basic science have been strengthened by the theme orientation of the meetings and may be further advanced by affiliations with other societies if this proposal is successful.

Also starting in 1974, APS introduced the concept of "guest societies" to the FASEB meeting and later to the APS fall meeting. The practice of inviting guest societies has been followed every year since 1976. The "guests" at FASEB meetings have included the Biomedical Engineering Society, the Society for Mathematical Biology, the Society for Experimental Biology and Medicine, and the Society of General Physiologists. The American Society of Zoologists, the Canadian Physiological Society, the Canadian Society of Zoologists, the Commission on Gravitational Physiology of the International Union of Physiological Sciences (IUPS), and the American Society for Gravitational and Space Biology have held joint meetings with APS at several fall meetings.

The result of all this joint and guest meeting activity has been greatly to improve the communications opportunities for APS members to a wider group of members of closely related scientific groups.

Two important innovations that have developed in APS were not contemplated by Berne in his list of concerns. One has been the development of a public affairs initiative. This program, developed primarily in the United States and England, derived its motivation from the "animal rights" movement. It presents a challenge to animal experimentation somewhat different from the traditional antivivisectionist and animal welfare attacks. The animal rights movement disapproves of subjecting animals to any treatment to which animals do not give "consent." Therefore any use of an animal, other than as a pet presumably, is morally and ethically wrong. No medical benefit or other value to human beings is considered to be justification for limiting an animal's free will.

Ernst Knobil was the first APS president to express concern about the new challenge presented by the animal rights movement and strongly pressed APS to become active in the field. He had been, just previous to his election as president of APS, a member of the Board of the National Society for Medical Research. Because that organization had been experiencing serious reverses in finance and administration, Knobil felt it necessary for APS to enter the field itself rather than placing reliance on another organization that he considered ineffective and possibly moribund. As a result of his efforts, well supported by his successors and Council, APS employed William Samuels as a part-time consultant in public affairs, reorganized the Public Affairs Committee to make it representative of every state, and became active on both national and state levels in combating the legislative efforts of the animal rights groups. The effectiveness of this activity has been reflected in the facts that legislative proposals now likely to be enacted are very much in line with APS policy decisions and that the larger scientific community has been gradually drawn into a supportive position from an originally hostile attitude.

A second recent innovation has been the development of historical activities of APS. The Society has always been circumspect about maintenance of archival material. The quarter-century volumes have provided good historical milestones, and since 1957, with initiation of *The Physiologist*, APS has published items of historical interest.

With the approach of the centennial year, this effort has been intensified. Members have been encouraged to contribute articles of historical interest (e.g., vignettes, departmental histories), and a Section on History of Physiology has been established. Historians have been elected as associate members of the Society. I am pleased that the Orr Reynolds Award, an annual award for the best article on history of physiology, was established to encourage physiologists to engage in historical activity.

Martin Frank
(b. 1947)

In 1985 Martin Frank came to office at APS after seven years of service as executive secretary of the NIH Physiology Study Section. There, in addition to general supervision of the granting process, he had initiated workshops and symposia in emerging areas of physiology and served as spokesman for NIH at universities and national conferences on the subject of peer review of grant applications. He was born in Chicago, Illinois, and completed both his undergraduate and graduate study at the University of Illinois in Urbana. He received his Ph.D. degree in 1973 for an analysis of calcium storage sites in guinea pig atrium. The work was carried out in the laboratory of W. W. Sleator, then head of the Department of Physiology and Biophysics. Frank next spent a year with Samuel B. Horowitz at the Michigan Cancer Foundation in Detroit, where he studied nucleocytoplasmic interactions in amphibian oocytes. The following year he was with Tai Akera and Theodore M. Brody in the Department of Pharmacology at Michigan State University in East Lansing, where he investigated the effect of various pharmacological interventions on excitation-contraction coupling in cardiac muscle.

From 1975 until 1978 he was assistant professor in the Department of Physiology at the George Washington University Medical Center in Washington, D.C. He then accepted the post at NIH but continued an association with George Washington, first as an assistant professorial lecturer (1978–80) and then as an associate professorial lecturer (1980–). In 1983 he was selected for the Department of Health and Human Services Senior Executive Service Development Program designed to prepare "middle managers" for future positions of leadership and major responsibility. As a part of his training he was assigned to the office of the Deputy Assistant Secretary for Health (Planning and Evaluation) where he was concerned especially with the Food and Drug Administration and the Center for Disease Control.

In general, Frank's laboratory research continued the line he began with Sleator; he used the guinea pig atrium to study the influence of various ions (particularly, potassium) and a variety of organic compounds on ultrastructure and electromechanical properties of muscle. He has been a member of the American Heart Association's Nation's Capital Affiliate and holds membership in both the Biophysical Society and the Society of General Physiologists. He was elected a member of APS in 1976. Of his move to APS, he wrote as follows:

For APS and for physiology as a discipline, 1987 will prove to be a benchmark year. The Centennial Celebration, led by Alfred Fishman and Orr Reynolds, will mark the culmination of many years of planning. Scheduled for Washington, D.C., from 29 March, it will help focus the nation's attention on physiology and its role in acquisition of biomedical knowledge. In addition, under the leadership of Franklyn Knox (president, 1986–87), APS at that same time will be reevaluating its role and function as a society and as a proponent of the science.

While Howard Morgan was president (1985–86), the Society began a self-assessment designed to deal with several issues that might affect it in its second century. The two primary issues are the governance of APS and the vitality of physiology as a discipline. The present governance structure developed when the Society was not extensively sectionalized. Whereas the election of four councillors and the three "presidents" has provided a workable Council, it cannot be characterized as representative of the membership's broad range of physiological endeavor. The current Council, like those of past years, tends to favor the cardiovascular/respiratory segments of the Society. In part this is a result of the preferential ballot procedure as it reflects the numerical dominance of these groups in the Society.

Throughout its history, APS has spawned numerous other societies and groups devoted to physiological and biomedical endeavors. The current sectionalization of the Society and the organization of a Section Advisory Committee in 1984 can be viewed as a first step in providing poorly represented segments of the community with a voice in APS activities. The next step is likely to occur within the next several years. The Long-Range Planning Committee (Walter Randall, chairman) and the Section Advisory Committee (Marion Siegman, chairman) have been charged by the Council to review the present governance structure and make recommendations for the Society's second century.

It seems likely that the Council may be expanded to include sectional representation, as well as the current at-large members. In addition, the nominating and election procedures are likely to be revised to enhance sectional representation. A nominating committee might be created, and election to office might be by plurality, rather than by the current preferential ballot.

A second issue relates to the vitality of physiology as a scientific discipline. In the eyes of many members of the Association of Chairmen of Departments of Physiology (ACDP), physiology has become identified more with its teaching responsibilities in medical schools than with the excitement of the research endeavor. To ACDP, APS has been too closely associated with systems and organ physiology rather than with the "cutting edge" of physiological research. In conjunction with ACDP, attempts will be made to redefine the image of physiology. A first step will be a joint effort between APS, ACDP, and the Society of General Physiologists to develop a new careers brochure to highlight the current excitement of physiology. In addition, ACDP will be asked to accept a greater responsibility for the current image of the science, both as a discipline and for the Society. Chairmen are increasingly hiring for their departments investigators, including immunologists, molecular biologists, and biochemists, who answer physiological questions by other than traditional techniques. But if these chairmen want cutting-edge physiological research represented at APS meetings, they must encourage these individuals to participate in activities of the Society.

Nevertheless more and more physiological investigations employing cellular and molecular approaches are, in fact, being presented at meetings and in journals of the Society. To better incorporate these areas in meeting programs, APS developed a thematic approach for the fall meeting beginning in 1986. This should provide a new and exciting format for demonstrating the excitement of physiological research. Each theme will try to integrate cellular and molecular, as well as systems and organ level, points of view. Furthermore APS will strive to publicize the meetings and enlist participation of both physiologists and nonphysiologists.

The success of meetings, whether spring or fall, depends on the creativity of the Program Committee and the success of symposium organizers in attracting outstanding scientists as participants. For the latter, the resources available from APS for support of expenses are a critical consideration. With the assistance of Norman Marshall, chairman of the Committee for Liaison with Industry, APS has established a Second Century Corporate Founders Program Endowment Fund to provide stable funding for scientific programs. A steering committee headed by Theodore Cooper, vice-president of the Upjohn Company, is committed to helping raise the first $250,000 of a proposed million-dollar program endowment. The result will be secure funding that will underwrite exciting and innovative scientific programs.

In 1986 APS started a new publication jointly with the IUPS entitled *News in Physiological Sciences (NIPS)*. By providing current short reviews of physiological research in a trends-type format, *NIPS* will help strengthen interactions with physiologists from the international community. The organization of joint meetings with societies from other countries has a similar goal. For example, in September 1985 APS was the guest of the Physiological Society at meetings held in Cambridge, England. The APS Fall Meeting at Niagara Falls in 1985 included members of the Canadian Physiological Society as well as the International Union for Gravitational Physiology. Invitations for the 1986 APS Spring Meeting in St. Louis were sent to members of the Russian Physiological Society. American physiology was well represented at the XXX IUPS Meeting in Vancouver, British Columbia, in July 1986. In the centennial year, 1987, an international flavor will be achieved by the presence of twenty-five scientists from other countries invited to be guests of APS at the birthday celebration in Washington, D.C.

Furthermore the Latin American Physiological Society will be a guest society at the 1987 APS Fall Meeting in San Diego, California. This should provide stronger ties with physiologists from Latin America. In addition, an APS bylaw change has been proposed that would broaden the regular member category to include physiologists from all the Americas. This change was presented at the April 1986 Business Meeting by Aubrey Taylor, chairman of the Membership Committee, and by Alfred Fishman, chairman of the International Physiology Committee. It is believed that amending the bylaw will improve communication with Latin American physiologists and increase their participation in affairs of the Society.

The vitality of APS, as with any organization, resides in its membership. After experiencing a number of years of slow growth in the regular member category, it became the policy of the executive office to solicit new members. From the registration list at the spring and fall meetings, nonmembers were identified, informed of benefits of membership, and invited to apply for membership. The result was a surge in numbers that swelled the ranks of APS to approximately 7,000 members by 1987. Also contributing to this increase was a promotion of the student membership category. With the support of ACDP, graduate students as well as postdoctoral fellows were encouraged to participate in the meetings and join the Society. Student participation is a key element in maintaining the vitality of the Society.

The most difficult category to address is the emeritus members. For APS the impending growth of emeritus membership causes problems similar to the much publicized increase in Social Security recipients. A rapidly growing segment of the Society must be supported by a stable contributing group. In the past, emeritus members have paid no dues yet have received the same benefits as regular members. To reduce the burden on the dues-paying membership, however, beginning in 1986 the emeritus members will be asked to contribute twenty dollars to the Society if they wish to receive *NIPS*. Because it is a new venture, the cost of *NIPS* can be regarded as a reimbursable benefit for anyone who is not paying regular dues. Although some retirees may object, most of the emeritus members probably will be glad to pay for *NIPS* and thereby contribute their support to the Society.

The future of APS looks bright. During the past one hundred years, physiologists have been Nobel laureates, and many have been elected to the National Academy of Sciences. Their research has benefited all mankind by contributing to improved health and increased longevity. Much remains to be done, however, if the total working of living organisms is to be defined. Physiologists will be faced with many new scientific challenges for investigation. As in the past, in the second century they will meet these challenges and help provide the bases for understanding and for treatment of disease. For members of APS, the forums for communication of this understanding will continue to be the meetings and journals of the Society. Indeed, 1987 will mark the start of a "Second Century of Progress in Physiology."

NOTES

[1] The latter part of Daggs' service to APS can be gleaned from biographies of Presidents Rahn (1963–64) through Brobeck (1971–72) in this volume.

[2] Further information about Reynolds' years in office may be found in *Physiologist* 27: 383–384, 1984; and 28: 132–133 and 135–138, 1985.

REFERENCES

1. FENN, W. O. *History of the American Physiological Society: The Third Quarter Century, 1937–1962.* Washington, DC: Am. Physiol. Soc., 1963, p. 77–79.

2. *Physiologist* 29: 15, 1986.

3. BERNE, R. M. Past-president's address. The American Physiological Society—a piece of the continent, a part of the main. *Physiologist* 16: 511–519, 1973.

Top: *Council, Fall Meeting, Coral Gables, FL, 1963. Left to right: Hymen S. Mayerson, Ray G. Daggs (executive secretary-treasurer), John R. Pappenheimer, Robert E. Forster, John M. Brookhart, James D. Hardy, Louis N. Katz, Hermann Rahn (president).* **Center:** *Council, Spring Meeting, Atlantic City, NJ, 1973. Left to right: Ewald E. Selkurt, Arthur C. Guyton, Peter F. Curran (chairman, Publications Committee), John R. Brobeck, Bodil M. Schmidt-Nielsen, Daniel C. Tosteson, Robert M. Berne (president), Orr E. Reynolds (executive secretary-treasurer), Walter A. Sonnenberg (business manager), Grace Hamilton (secretary to APS), Jere Mead.* **Bottom:** *Council, Spring Meeting, St. Louis, MO, 1986. Seated left to right: Martin Frank (executive secretary-treasurer), Howard E. Morgan (president), Franklyn G. Knox, John B. West. Standing left to right: Paul C. Johnson (chairman, Publications Committee), Carl V. Gisolfi (chairman, Program Executive Committee), Aubrey E. Taylor, Harvey V. Sparks, Jr., Norman C. Staub, Shu Chien, Marion Siegman (chairwoman, Section Advisory Committee).*

APS publications. **Top left:** *Sara F. Leslie, publications manager and executive editor (1966–74).* **Top right:** *Stephen R. Geiger, publications manager and executive editor (1974–87).* **Bottom:** *group of exercise and environmental physiologists meeting at APS headquarters, Bethesda, MD, 10 January 1984, to discuss the future of the* Journal of Applied Physiology. *Seated left to right: Howard E. Morgan, Leo C. Senay, Jr., Alfred P. Fishman, Charles M. Tipton, H. Lowell Stone. Standing left to right: Jere H. Mitchell, John T. Shepherd, Loring B. Rowell, Carl V. Gisolfi, Jerome A. Dempsey, Philip D. Gollnick.*

Honors and awards. **Top:** *James D. Hardy (right) receiving the Daggs Award from Bodil Schmidt-Nielsen (president), Spring Meeting, Anaheim, CA, 1976. Seated at left: Ewald E. Selkurt and Orr E. Reynolds.* **Center left:** *John F. Perkins, Jr., namesake of the Perkins Memorial Awards.* **Center middle:** *G. Edgar Folk, Jr., in whose name the Folk Senior Physiologist Fund was donated.* **Center right:** *Roger Guillemin delivering the Physiology in Perspective: Walter B. Cannon Lecture, Spring Meeting, Anaheim, CA, 1985.* **Bottom:** *A. Clifford Barger (left) and Edward W. Hawthorne (right), cochairmen (1968–86) of the Porter Physiology Development Program.*

Honors and awards. **Top left:** *Hermann Rahn with Eugene Renkin, Bowditch lecturer, Fall Meeting, Coral Gables, FL, 1963.* **Top right:** *Howard Morgan with Yale E. Goldman, Bowditch lecturer, Fall Meeting, New Orleans, LA, 1986.* **Center:** *Helen J. Cooke, chairwoman of the Committee on Women in Physiology (right), with recipients of the Caroline tum Suden Awards, Spring Meeting, St. Louis, MO, 1986.* **Bottom left:** *award recipients at the Respiration Dinner, Spring Meeting, Anaheim, CA, 1985. Left to right: Richard W. Stow (Nernst Award), Poul B. Astrup (Henderson-Hasselbalch Award), Leland C. Clark, Jr. (Heyrovsky Award), John W. Severinghaus (master of ceremonies). The dinner celebrated thirty years of blood gas electrodes. (Photo by R. H. Kellogg.)* **Bottom right:** *James H. Jones receiving the Scholander Prize from Donald C. Jackson at the Comparative Physiology Luncheon, Fall Meeting, New Orleans, LA, 1986. (Photo by R. H. Kellogg.)*

Top: *platform party (left) at the opening session of the XXIV International Congress of Physiological Sciences, Washington, D.C., 1968. Left to right: Philip Bard, Maurice B. Visscher, Detlev W. Bronk, Hermann Rahn, Frank C. MacIntosh, Genichi Kato, Ulf S. von Euler, Jan W. Duyff, John R. Pappenheimer, Loren D. Carlson, Lindor Brown, James A. Shannon.* **Center left:** *Wallace O. Fenn, president of the XXIV International Congress of Physiological Sciences, Washington, D.C., 1968.* **Center right:** *Knut Schmidt-Nielsen, editor of* News in Physiological Sciences (NIPS) *inaugurated in 1986 and published jointly by APS and the International Union of Physiological Sciences (IUPS).* **Bottom:** *APS members in the People's Republic of China, 1983.*

Top left: *Francis J. Haddy (APS president), Congressman Doug Walgren (D-PA), and Orr E. Reynolds (APS executive secretary-treasurer) discussing animal care legislation, 1982.* **Top right:** *M. Elizabeth Tidball, chairwoman of the Task Force on Women in Physiology (1973–79).* **Center left:** *M. C. Shelesnyak, chairman of the Centennial Task Force (1979–84).* **Center middle:** *Ralph H. Kellogg, editor of historical vignettes series in* The Physiologist. **Center right:** *Peter A. Chevalier, chairman of the Centennial Committee (1980–84).* **Bottom left:** *Arthur B. Otis, editor of the departmental history series published in* The Physiologist. **Bottom right:** *Michael Jackson, chairman of the APS Program Executive Committee, at the first joint meeting of APS and the Physiological Society, Cambridge, United Kingdom, September 1985.*

Top: *gathering of past, present, and future presidents at the Presidents' Reception in honor of the retirement of Orr E. Reynolds, Spring Meeting, Anaheim, CA, 1985. Seated left to right: David F. Bohr, Earl H. Wood, A. Clifford Barger, Hermann Rahn, David Bruce Dill, Robert E. Forster, Orr E. Reynolds, Robert M. Berne, Ewald E. Selkurt. Standing left to right: Franklyn G. Knox, Harvey V. Sparks, Jr., Francis J. Haddy, Arthur C. Guyton, Alfred P. Fishman, Howard E. Morgan, John B. West, Ernst Knobil, Walter C. Randall, William F. Ganong.* **Bottom:** *past presidents at the Presidents' Reception, Spring Meeting, St. Louis, MO, 1986. Seated left to right: Franklyn G. Knox, Earl H. Wood, A. Clifford Barger, Bodil M. Schmidt-Nielsen, C. Ladd Prosser, Hermann Rahn. Standing left to right: Francis J. Haddy, Arthur C. Guyton, Howard E. Morgan, Alfred P. Fishman, John B. West, Walter C. Randall.*

Top: *artist's drawing of the enlarged Lee Building. Official opening is scheduled for the Spring Meeting, Washington, D.C., 1987, celebrating the centennial of APS and the diamond jubilee of FASEB.* **Bottom left:** *APS centennial medallion.* **Bottom right:** *FASEB diamond jubilee logo.*

CHAPTER 9

Membership

JOSEPH F. SAUNDERS AND AUBREY E. TAYLOR

In a reflective account of the third quarter century (1937–62) of the American Physiological Society (APS, the Society), Wallace Fenn (president, 1946–48) stated:

> No scientific society can professionally be better than the members of which it is composed. Compared to the maintenance of this standard of excellence, all administrative disputes within the Society pale into insignificance. The important effort must be to make sure that the American Physiological Society is not outranked in high scientific quality by any other similar group (10).

Some twenty-three years later this commitment to excellence, which goes back to the very founding of the Society, has not changed. Nevertheless the membership policy of the Society has continued to evolve, in allowing not only for continued growth, but also for a broader-based, more diversified, and therefore a more dynamic Society. The Society has gradually recognized its responsibility to develop excellence among those who are not yet regular members. The roots of this outreach were planted in the third quarter century, but considerable progress has been made in the fourth quarter century.

Membership Categories

Over the past twenty-five years, total membership in the Society has increased nearly threefold, from 2,454 in 1962 to almost 6,500 in 1985. Table 1 shows the total annual membership of the Society from 1887 to 1986 (18).

In the early years of the Society, there were only two categories of membership: ordinary (later changed to regular) and honorary. After 1939 it became possible for retired members, on request of Council, to be relieved of the payment of annual dues (10). After considerable debate, in 1958 the category of associate member was adopted, originally for

Table 1
ANNUAL MEMBERSHIP IN APS

Year	No. of Members	Year	No. of Members	Year	No. of Members	Year	No. of Members
1887	28	1912	215	1937	649	1962	2,454
1888	34	1913	230	1938	661	1963	2,498
1889	38	1914	245	1939	673	1964	2,650
1890	42	1915	265	1940	689	1965	2,819
1891	44	1916	270	1941	710	1966	2,966
1892	46	1917	282	1942	731	1967	3,273
1893	51	1918	295	1943	781	1968	3,545
1894	55	1919	305	1944	819	1969	3,758
1895	63	1920	323	1945	868	1970	3,792
1896	65	1921	327	1946	911	1971	3,986
1897	65	1922	352	1947	948	1972	4,166
1898	76	1923	372	1948	1,034	1973	4,342
1899	76	1924	393	1949	1,102	1974	4,500
1900	91	1925	410	1950	1,150	1975	4,759
1901	102	1926	431	1951	1,232	1976	4,899
1902	110	1927	453	1952	1,306	1977	5,032
1903	118	1928	476	1953	1,362	1978	5,267
1904	127	1929	487	1954	1,376	1979	5,492
1905	141	1930	522	1955	1,424	1980	5,708
1906	151	1931	558	1956	1,535	1981	5,927
1907	166	1932	587	1957	1,586	1982	6,077
1908	176	1933	591	1958	1,759	1983	6,144
1909	176	1934	614	1959	1,923	1984	6,195
1910	183	1935	634	1960	2,139	1985	6,248
1911	204	1936	637	1961	2,265	1986	6,303*

* As of 1 March 1986.

advanced graduate students in physiology at a pre-doctoral level, teachers of physiology at a pre-doctoral level, teachers of physiology, and investigators who have not yet had the opportunity or time to satisfy the requirements for full membership in the Society (1, 10).

Associate members were not members of the Federation of American Societies for Experimental Biology (FASEB, the Federation) and therefore could not present papers at the spring meeting without a sponsor (5). Their names did not appear in the annual Federation directory. In 1960, largely at the initiative of Julius H. Comroe, the Society accepted yet another new category, sustaining associate member, for individuals and organizations who had an interest in the advancement of biological investigation and who were invited to become members by the president with the approval of Council (10). As of 1986 there were thirty-four sustaining associates. (These institutional supporters are not reflected in the membership data in Table 1.)

Over the course of the fourth quarter century there have been a number of changes and additions relating to membership. The qualifications for associate membership have been greatly modified, and two more categories of membership have been instituted: student member and corresponding member.

Regular Membership and the APS Membership Committee

Fenn wrote:

The American Physiological Society has always been a rather exclusive organization. . . . Standards are high and all applications for membership are diligently screened by a sizeable committee, by the Council, and eventually by the whole membership . . . the Society has always taken membership problems very seriously (10).

Since 1951 the task of maintaining the quality of APS membership has fallen mainly on the Membership Committee, which reviews nominations of regular, associate, student, and corresponding members and makes recommendations to Council. This committee, composed of a chairman and five members, was first officially recognized in the bylaws of the Society in 1957 by the simple statement, "There shall be a Committee on Membership appointed by and advisory to Council." When the bylaws were revised in 1966 (5), the composition and duties of the Membership Committee were specified. There has been no change since 1966.

Article V. Standing Committees. *Section 3. Membership Committee.* A Membership Committee, composed of six or more regular members of the Society appointed by the Council, shall receive and review processed applications for membership and make recommendations for nomination to the Council. The term of each member of the Membership Committee shall be three years; a member shall not be eligible for immediate reappointment. The Chairman of the Committee shall be designated by the Council.

In April 1982 Council adopted the requirement that "the Chairman of the Membership Committee at the time of initial appointment will have been a past member of Council, having served within the previous three years." The purpose of this change was to avoid misunderstandings in principle between Council and the Membership Committee. The members of this committee from 1962 to 1986 are listed in Table 2.

Although the bylaws specify the general criteria for regular membership, it is left to the Membership Committee and Council to apply them. The standards have thus varied over the years, depending on the composition of the Membership Committee and Council. The criteria for regular membership as stated in the bylaws have been changed little over the past quarter century. Thus Article I. *Section 2* in 1962 stated:

Any person who has conducted and published meritorious original research in physiology and/or biophysics and who is a resident of North America shall be eligible for membership in the Society.

Early in 1966, occasioned by the impending retirement of Milton Lee, a proposed revision of the bylaws was published and then approved at the April 1966 business meeting. (With this revision the section on membership became Article III. The new Articles I and II concerned legal matters: the principal office and the corporate seal.) The new bylaws changed the member designation to "regular member" and slightly modified the definition. The words "and/or biophysics," which had been inserted in an unsuccessful effort to prevent the formation of the Biophysical Society in 1957, were eliminated in the 1966 revision. The phrase "who is presently engaged

Table 2
MEMBERSHIP ADVISORY COMMITTEE, 1962–86

	Year			Year	
	Member	Chairman		Member	Chairman
W. S. Root	1962–63	1962–63	C. M. Armstrong	1973–75	
J. D. Hardy	1962–64		B. R. Duling	1973–75	1974–75
R. B. Tschirgi	1962–64		P. C. Johnson	1974–76	
D. S. Farner	1962–65		B. P. Bishop	1975–78	1975–78
R. L. Riley	1962–65		E. O. Feigl	1975–77	
L. D. Carlson	1962–65	1963–65	E. M. Stephenson	1975–78	
C. P. Lyman	1962–65		G. A. Castro	1976–79	
H. D. Green	1963–66		S. Cassin	1976–77	
H. D. Patton	1964–67	1965–67	J. B. Scott	1977–79	
W. F. Ganong	1964–67		M. F. Dallman	1977–78	
D. E. Goldman	1965–67		J. S. Cook	1977–80	1978–80
W. B. Kinter	1965–68		I. J. Fox	1978–81	
A. B. Otis	1965–68		J. C. Fray	1978–81	
D. F. Bohr	1966–69	1967–69	R. E. Hyatt	1978–81	1980–81
C. Eyzaguirre	1967–70		A. E. V. Haschemeyer	1979–82	
G. F. Cahill	1967–70		M. C. Neville	1979–82	1981–82
I. S. Edelman	1967–70		S. M. Cain	1980–83	
L. E. Farhi	1968–71		E. L. Bockman	1981–84	
J. B. Preston	1968–71	1969–71	W. H. Dantzler	1981–84	
E. Knobil	1969–70		M. E. Freeman	1981–84	
W. C. Bowie	1970–73		S. McD. McCann	1982–85	1982–85
J. A. Herd	1970–73	1971–73	C. H. Baker	1982–85	
G. Hoyle	1970–72		J. A. Schafer	1983–86	
J. Metcalfe	1970–73		N. M. Buckley	1984–87	
J. P. Reuben	1970–73		D. N. Granger	1984–85	
S. Solomon	1971–74	1973–74	C. Levinson	1984–87	
E. R. Ramey	1973–75		A. E. Taylor	1985–88	1985–87
N. C. Staub	1973–76		S. Lahiri	1985–88	
B. D. Lindley	1973–76		L. C. Weaver	1985–88	

in physiological research" was added to reflect long-time policy in admitting new members. The term regular member was introduced to distinguish active members from retired members. "Retired members" then appeared for the first time as a separate category.

Article III of the new bylaws (adopted in 1966) reads as follows:

Section 1. The Society shall consist of regular members, honorary members, associate members, retired members, and sustaining associates.

Section 2. Regular Members. Any person who has conducted and published meritorious original research in physiology, who is presently engaged in physiological work, and who is a resident of North America shall be eligible for proposal for regular membership in the Society.

Section 3. Honorary Members. Distinguished scientists of any country who have contributed to the advance of physiology shall be eligible for proposal as honorary members of the Society.

Section 4. Associate Members. Advanced graduate students in physiology at a predoctoral level, teachers of physiology, and investigators who have not yet had the opportunity or time to satisfy the requirements for regular membership in the Society provided they are residents of North America. Associate members may later be proposed for regular membership.

Section 5. Retired Members. A regular or associate member who has reached the age of 65 years and/or is retired from regular employment may, upon application to Council, be granted retired member status.

Section 6. Sustaining Associates. Individuals and organizations who have an interest in the advancement of biological investigation may be invited by the President, with approval of Council, to become sustaining associates.

Interpretation of these criteria has been left to the Membership Committee. In general the tendency in the last quarter century has been toward greater inclusiveness and leniency. One question that especially concerned the committee in the 1970s under the chairmanship of Beverly Bishop and John S. Cook was the problem of what is meant by "original research." In the recent era, graduate students and even postdoctoral fellows generally do their work as part of a team rather than on their own. If original implied independent, then a physiologist would need an academic appointment and his/her own research grant to qualify. The committee upheld the bylaws requirement of "*meritorious original* research" but noted that "the word *independent* is not included." The committee concluded that "very few people publish solo papers, and the definition of independence becomes hard to define." Further, such a rigid requirement could lead to the "disqualification" of many second and third year postdoctoral fellows before they could assume full-time positions (12). Although most of the new APS members do have regular appointments, those in training positions, if they have published sufficiently, may still qualify.

Corresponding Members

The 1962 bylaws of the Society limited membership to North Americans. Despite this, there were, and still are, on the roster of active regular members a number of foreign scientists who were admitted to the Society because they were working in American laboratories or institutions and had expressed their intention to remain permanently in the United States. Once elected, they remained regular members, even though they later returned to their native countries.

The issue of whether and how to modify the bylaws to allow foreign physiologists to affiliate with the Society has been debated repeatedly. The first matter concerning membership policy to come before the Society in its fourth quarter century, it was discussed by Council in April 1962 (6). Council acted by instructing the Membership Advisory Committee to prepare a proposal for presentation to Council in advance of the Society's 1962 fall business meeting. Although Council discussed the deletion from the bylaws of the phrase "permanent resident of North America" and also considered a foreign member category, action was delayed pending additional review and consideration at the 1963 spring meeting. Council then referred the matter again to the Membership Advisory Committee for further study and preparation of a proposal to be presented to Council and the membership at the 1963 fall meeting (7). At that meeting a motion before Council to delete the phrase "resident of North America" from the bylaws was defeated. Another motion "to establish some provision for the association of foreigners" was, after considerable discussion, also defeated.

In 1973–74, foreign membership again appeared on the Council agenda in a different form. A new membership category, corresponding member, was proposed to include foreign physiologists as well as Americans in fields closely related to physiology, such as biomedical engineering. A rather detailed justification in support of this proposal, as well as appropriate amending of the Society's bylaws, was published for the membership before its presentation at the 1975 spring meeting of the Society (14). A response was solicited from the membership by mail. Of 1,066 respondents, 878 (82%) were in favor of including foreign scientists in the Society as corresponding members (16). Despite this enthusiastic support, the proposal was defeated by a two-thirds majority of members at the 1975 business meeting (17).

In October 1975 Council passed another resolution to consider incorporating into the Society's bylaws a "new membership category for persons who are not residents of North America." It differed from the earlier defeated proposal for corresponding member by dealing only with foreign physiologists. The "primary objective . . . was to provide a constitutional vehicle for the participation in Society science and education activities by scientists who are currently excluded because of their country of residence." In April 1976 this Council proposal for corresponding membership was accepted by an overwhelming vote at the business meeting (19). The Society's bylaws were amended to read:

> Article III. *Section 3. Corresponding Members.* Any person who has conducted and published meritorious research in physiology, who is presently engaged in physiological work and who resides outside of North America shall be eligible for proposal for corresponding membership in the Society.

Corresponding members are elected according to the same procedures as regular members, except "a corresponding or honorary member of the Society may substitute for one of the regular members in proposing a person for corresponding membership."

Student Membership and Modification of Associate Membership

In April 1976 the Membership Advisory Committee recommended to Council still another new membership category, student member. The committee reasoned that

> there are many individuals who should be associated with the Society but who do not meet the qualifications for Regular Membership . . . are in closely related fields or clinicians who may not be as deeply involved in basic research . . . it is these people that the Committee would like to include as Associate Members . . . however, the Committee does not feel that it can put an established professional into a category of membership that includes *graduate students.*

Therefore the committee suggested to Council that it establish a special category of membership for students. After several modifications, Council approved a bylaw amendment to establish a student membership category to be presented at the 1977 spring meeting of the Society.

At the 1976 fall meeting, further discussion during Council's deliberations on the associate and student member categories resulted in more definitive criteria, which were approved and adopted by Council. The proposals were published for review by the membership before the meeting (20) and approved at the April 1977 business meeting. Thus the bylaws were amended as follows:

Article III. *Section 5. Associate Members.* Persons who are engaged in research in physiology or related fields and/or teaching physiology shall be eligible for proposal for Associate membership in the Society provided they are residents of North America. Associate members may later be proposed for Regular membership.

Article III. *Section 7. Student Members.* Graduate students in physiology who have completed their preliminary examinations for the doctoral degree provided they are residents of North America. No individual may remain in this category for more than five years.

With these changes in the bylaws, associate members have become members of the Federation with the privileges of members, and their names appear in the annual directory.

Furthermore Council, almost at once, began a vigorous campaign to attract student members into the Society. A poster soliciting student members was widely distributed and also published on the back cover of the April 1978 issue of *The Physiologist.*

With the student member category in place, there was a general conviction that all matters related to associate and student membership had been resolved. That was not the case. In its report to Council during the 1977 fall meeting the Membership Committee expressed the conviction that "Student Membership was too restrictive" (4). As a result, Council proposed an amendment to Article III. *Section 7* of the bylaws, which was unanimously approved by the membership during the 1979 spring business meeting (11). The definition of student membership was changed to:

Article III. *Section 7. Student Members.* Any student who is actively engaged in physiological work as attested to by two Regular members of the Society and who is a resident of North America. No individual may remain in this category for more than five years, without reapplying.

During that same session, the designation of retired members was changed to emeritus members, and the definition was slightly modified.

Article III. *Section 6. Emeritus Members.* A regular or associate member may apply to Council to transfer to emeritus membership if that person (1) has reached the age of 65 and is retired from regular employment or (2) has been forced to retire from regular employment because of illness or disability. An emeritus member may be restored to regular membership status on request to Council.

Honorary Membership

The nomination of honorary members has been until recently entirely a function of Council. From 1962 to 1980 only three honorary members were elected, and by 1979 the number of honorary members had declined through death to a low of eight. In that year, an Honorary Membership Committee composed of three past presidents of APS was established to consider honorary membership requirements and to make recommendations to Council (12). Each year a new member has been rotated onto the committee and has become chairman in his/her third year. From 1981 to 1986, thanks to the work of this committee, ten distinguished honorary members have been elected. A complete list of the forty-six honorary members elected from 1904 to 1986 appears in Table 3.

Most Recent Changes

In December 1985 Council was requested to consider an amendment modifying the corresponding membership category by restricting it to individuals who reside outside the Americas. The proposed change was designed to allow applicants from South and Central America to become regular members. The amendment, which was approved unanimously in principle by Council, was submitted to the membership at the October business meeting in New Orleans in 1986 and was passed unanimously. For the categories of regular, corresponding, associate, and student, the words "North America" were replaced by "the Americas."

Thus the membership categories are well defined for residents of the Americas. In addition the establishment of corresponding membership allowed APS to accommodate members who reside outside the Americas, and the sustaining associate category allowed corporate organizations, individuals, and the military to support the Society.

Qualifications and Criteria for Membership

As the Membership Committee developed in purpose and grew in stature, a scoring system evolved to evaluate six different aspects of the candidate's training and performance as a professional physiologist (9): professional training, occupational history, interest in and commitment to the teaching of physiology, interest in the Society, contributions to the literature of physiology, and any other special considerations of professional activity. Criteria for each of these were established, along with a point-scoring system for each category. In August 1976 Council approved the committee's procedures, which continue in use as a principal means to evaluate a candidate's qualification for regular, corresponding, or associate membership. Student applications are graded simply acceptable or unacceptable.

The professional attributes and factors used in these evaluations are as follows:

1. Educational history. Academic degree and postdoctoral training are evaluated and assessed with regard to how closely the applicant's training has been related to physiology.

Table 3
HONORARY MEMBERS, 1887–1986

	Residence	Year		Residence	Year
A. Dastre (1844–1917)	Paris	1904	C. Monge (1884–1973)	Lima	1952
T. Englemann (1843–1909)	Berlin	1904	K. von Frisch (1886–1982)	Munich	1952
F. Hofmeister (1850–1922)	Strasburg	1904	F. J. W. Roughton	Cambridge	1957
J. N. Langley (1852–1925)	Cambridge	1904	(1899–1972)		
I. P. Pavlov (1849–1936)	Petrograd	1904	E. Braun-Menendez	Buenos Aires	1959
C. Sherrington (1857–1952)	Oxford	1904	(1903–59)		
O. Hammarsten (1841–1932)	Uppsala	1907	A. Hurtado (1901–84)	Lima	1959
E. Pflüger (1829–1910)	Bonn	1907	Y. Kuno (1882–1977)	Tokyo	1959
K. von Voit (1831–1908)	Munich	1907	G. Moruzzi	Pisa	1959
E. Sharpey-Shafer	Edinburgh	1912	H. H. Ussing	Copenhagen	1959
(1850–1935)			H. H. Weber (1896–1974)	Heidelberg	1959
B. A. Houssay (1887–1971)	Buenos Aires	1942	R. Granit	Stockholm	1963
E. D. Adrian (1889–1977)	Cambridge	1946	G. Kato (1890–1979)	Tokyo	1966
J. I. Barcroft (1872–1947)	Cambridge	1946	E. Gutmann (1910–77)	Prague	1971
A. V. Hill (1886–1977)	London	1946,	P. Dejours	Strasbourg	1981
		1950	R. A. Gregory	Liverpool	1981
A. Krogh (1874–1949)	Copenhagen	1946	A. F. Huxley	Cambridge	1981
L. Lapique (1866–1952)	Paris	1946	H. E. Huxley	Cambridge	1981
L. Orbeli (1883–1958)	Leningrad	1946	B. Folkow	Göteborg	1982
W. T. Porter (1862–1949)	Dover, MA	1948	T. P. Feng	Shanghai	1983
F. Bremer (1892–1982)	Brussels	1950	B. Katz	London	1985
W. R. Hess (1881–1973)	Zurich	1950	E. T. A. Teorell	Uppsala	1985
G. Liljestrand (1886–1968)	Stockholm	1950	K. J. Ullrich	Frankfurt	1985
J. C. Eccles	Canberra	1952	M. Ito	Tokyo	1986
A. L. Hodgkin	Cambridge	1952	J. Vane	London	1986

2. Occupation. Emphasis is given to those applicants who have a full-time position in a department of physiology or a closely allied field. Relatively high ratings are given to individuals with research positions in clinical departments and to those functioning as independent investigators in commercial or government laboratories.

3. Interest in and commitment to teaching physiology. This evaluation is based on *1*) the fraction of the applicant's time devoted to teaching; *2*) publications related to activities as a teacher, including production of educational materials; and *3*) special awards or other recognition received for outstanding teaching effectiveness.

4. Interest in the Society. Evaluation of this category is based on *1*) the meetings at which he/she reads papers and *2*) the Society meetings attended.

5. Contributions to the physiological literature. This category is of major importance. The applicant's bibliography is evaluated on the basis of publications in major refereed journals that are concerned with problems judged to be primarily physiological in nature. Emphasis is given to papers published as the result of original research. Publications on which the applicant is sole author or first author are accepted as clear evidence of an applicant's independence.

6. Special considerations. This category permits the Membership Committee to acknowledge any unique accomplishments of an applicant. These might be excellence in a specific area or unusual contributions to physiology resulting from talents, interest, or a background substantially different from the average.

All complete applications are sent to each committee member by the APS

Membership Services Department for independent evaluation and scoring on forms provided. These are returned to the APS Central Office where the scores given to each applicant are averaged by a computer program, and then a final list of all applicants is printed in rank order of highest score. Not only is the average total score for each candidate given on the printout, but the average score achieved in each of the six categories is also shown. These summary sheets are sent to the members of the Membership Committee for use in evaluating all candidates at the committee's biannual meeting. High and low scores are not used as absolute cut-off levels for membership. This scoring, however, provides an excellent means to identify those cases requiring special consideration in the evaluation process.

The chairman presents to Council all applications along with the committee's recommendations. Council either accepts the recommendations or overrides the committee's decisions. Reversal of the committee's recommendation by Council has occurred only in a very few instances.

Nomination and Election of Members

Over the past quarter century the nomination and election procedure for new members has become gradually more streamlined. In the first constitution of APS in 1887, candidates for ordinary membership were to be proposed in writing by two ordinary members of the Society who were not members of Council. The names of candidates approved by Council, together with the names of their proposers and a statement of their qualifications, were to be posted at the first session of the annual meeting. The candidates would be voted on by the membership at the last session, and "one black-ball in eight shall exclude." The "black-ball" procedure endured until 1958 when it was replaced by a two-thirds majority. The election process was at that time lengthy and cumbersome. Nominations of regular and associate members were presented by Council to the membership at a spring or fall meeting but not voted on until the following fall or spring meeting. The bylaws provided that

> the names of the candidates nominated by Council for membership and statements of their qualifications signed by their proposers shall be made available for inspection by members during the Society meetings at which their election is considered.

When the new set of bylaws was adopted in 1966 (2), a new and shortened procedure relating to nominations and elections of members was established under Article III. Membership. *Sections 7–9.*

> *Section 7. Nominations for Membership.* Two regular members of the Society must join in proposing a person for regular membership, honorary membership or associate membership, in writing and on forms provided by the Executive Secretary-Treasurer. The Membership Committee shall investigate their qualifications and recommend nominations to Council. Council shall nominate members for election at the Spring and Fall meetings of the Society. A list of nominees shall be sent to each regular member at least one month before the Spring and Fall meetings.

> *Section 8. Election of Members.* Election of regular members, honorary members and associate members shall be by secret ballot at Spring and Fall business meetings of the Society. A two-thirds majority vote of the members present and voting shall be necessary for election.

Section 9. Voting. Only regular members shall be voting members. Honorary, retired and associate members shall have the privilege of attending business meetings of the Society but shall have no vote.

In practice, this change in membership election (i.e., *Section 7*) required publication of Council's nominations for membership in the issue of *The Physiologist* that preceded the next meeting (i.e., the February issue or the August abstract issue). Thus a delay of several months still existed between nominations by Council and membership approval at the next meeting of the Society.

This impractical procedure was changed through an amendment to the bylaws in April 1975 (15). Membership responses had been solicited on the matter of expediting the election of members. Of 1,066 respondents, 901 (85%) favored a speedier election process, and Article III. Membership. *Section 7* was changed to allow members to be elected at the meeting at which they had been evaluated by the Membership Committee. According to the new rules, "A list of nominees shall be posted for consideration by the members attending the meeting two days prior to the Business Meeting at which the election occurs."

On election to membership in the Society, the applicants are informed by means of a congratulatory letter from the president elect of the Society. Included in the latter is a dues payment notice for the current year for the particular membership category. Accompanying the invitation and the dues notice are *1)* a membership records questionnaire, *2)* information related to membership in disciplinary/specialty section(s) of the Society, *3)* information on the APS journals and subscriptions procedures, and *4)* the current APS "Guiding Principles in the Care and Use of Animals."

Growth of the Society

Looking back at the membership growth over almost a century, we find that our current concept of broadening the membership structure of the Society is not a new one. The Society's objective has always been to increase the number of qualified members. Real momentum for this, however, did not occur until the waning of World War II (18). Sixty years of societal activity passed before the 1,000-member mark was achieved in 1948 (see Table 1). From a historical point of view, a year-by-year account of membership growth is useful. However, by looking at ten-year intervals one can see the possible effect of historical events on growth patterns become more apparent (Table 4).

The 1950–59 growth resulted from the tremendous influx of students into academic institutions related to GI Bill financing of education and to the postwar growth of degree programs in physiology. During the 1960–79 interval, a large increase in the college/university student population occurred and with it an increase in federal training and research funds. Continued increments in the support of biomedical research by federal agencies, such as the National Institutes of Health, Office of Naval Research, and National Science Foundation, and an expansion in the capacity of the universities to accommodate the student population explosion occurred during the 1970s. The figures in the 1980s reflect the leveling off of

Table 4
AVERAGE ANNUAL GROWTH IN APS
MEMBERSHIP, 1887–1985

Years	Historical Event	Members, no./yr
1887–99		4
1900–09		10
1910–19	World War I	13
1920–29		18
1930–39	Depression, Pre World War II	19
1940–45	World War II	65
1946–49	Post World War II	59
1950–59		82
1960–69	Space Age	178
1970–79		173
1980–85		141

Table 5
MEMBERSHIP STATUS OF APS, 1962–86

	No. of Members						
	Regular	Emeritus	Honorary	Associate	Corresponding	Student	Total
1962	2,141	133	16	164			2,454
1963	2,189	126	17	166			2,498
1964	2,320	133	17	180			2,650
1965	2,487	132	17	183			2,819
1966	2,615	137	18	196			2,966
1967	2,869	168	18	218			3,273
1968	3,122	162	17	254			3,545
1969	3,259	186	17	296			3,758
1970	3,282	182	17	311			3,792
1971	3,422	209	17	338			3,986
1972	3,542	220	15	389			4,166
1973	3,638	255	15	434			4,342
1974	3,737	288	14	461			4,500
1975	3,891	328	13	527			4,759
1976	3,978	349	13	559			4,899
1977	4,056	382	10	572	5		5,032
1978	4,115	441	9	619	24	59	5,267
1979	4,198	458	8	631	40	157	5,492
1980	4,286	490	8	648	60	214	5,706
1981	4,399	525	8	683	79	233	5,927
1982	4,456	553	11	703	94	260	6,077
1983	4,494	556	10	718	101	265	6,144
1984	4,515	593	11	750	117	209	6,195
1985	4,558	607	14	763	134	172	6,248
1986	4,632	634	16	773	159	186	6,400

funding. If these numbers are indicative of a trend, the membership growth rate is declining, though membership in the Society continues to increase each year.

An analysis of the annual membership growth by category of membership is presented in Table 5. In the period 1971–85 the average annual growth for the categories of total APS membership, regular, associate, and emeritus members, respectively, was 170, 89, 31, and 29. Expressed in terms of average percent annual increase, the figures indicate the smallest growth in regular members—2.18% per year compared with 5.28% for associate, 6.19% for emeritus, and 3.25% for total.

In general, there is evidence of continued and healthy growth in APS membership. Many members feel, however, that this growth is slow compared with the growth of more recently established physiological specialty societies, such as the Biophysical Society, the Society for Neuroscience, and the American Society for Cell Biology. Many agree with the belief of Orr Reynolds (executive secretary-treasurer, 1973–85) that active recruitment of new members, with concern for those who represent the utilization of physiological knowledge as well as its creation, is appropriate for the future. Reynolds stated in 1985:

> I believe the Society should consider some outright recruiting campaigns for membership. It would probably require some restructuring of present membership categories and probably some bylaw changes, but I feel that the Society will be in a much better position to deal with the problems it faces in the future if it has a broader membership base, including those whose physiological contribution is not in fundamental research but, rather, in teaching, applied research, and development and testing (21).

Agreeing with Reynolds' views, in the spring of 1985 APS President John West sent each chairman or director of a department of physiology a letter encouraging the nomination of faculty and students for APS membership. Reynolds' successor, Martin Frank, continues to solicit new members from a broad spectrum of scientists. Moreover during the summer of 1985 approximately 2,000 letters of invitation were mailed to nonmember attendees of the FASEB 1985 Spring Meeting and to nonmember contributors to the APS 1985 Fall Meeting in Niagara Falls. The result was a marked increase in the number of applications reviewed by the Membership Committee at the 1986 spring meeting. The purpose of such recruitment letters is to inform nonmembers of the benefits and activities of the Society. Acceptance for membership still requires that the applicant meet the criteria set forth by Council and the Membership Committee.

Council is currently evaluating the means whereby new members are screened, and the process has evolved as follows. The Membership Committee will review only regular and corresponding membership applications, while the APS Membership Office and the executive secretary-treasurer will review the student and associate memberships in consultation with the chairman of the Membership Committee. This process will be tested with applications received during the first half of 1987. This change in review procedures will be submitted to the full membership for their approval at the 1987 fall meeting.

REFERENCES

1. ANONYMOUS. Constitution and bylaws of the APS (amended April 1958). *Physiologist* 1(5): 3–6, 1958.
2. ANONYMOUS. Constitution (adopted 1953) and bylaws (adopted 1966) of the APS. *Physiologist* 9: 346–352, 1966.
3. BISHOP, B. Instructions for applying for APS membership. *Physiologist* 20(3): 11, 1977.

4. BOHR, D. E. American Physiological Society, 118th Business Meeting. V. Committee reports. A. Membership. *Physiologist* 20(6): 8, 1977.

5. BROOKHART, J. M. Homework for members—the President's message. *Physiologist* 9: 4–21, 1966.

6. DAGGS, R. G. Report on APS business meetings, 15 and 17 April 1962. *Physiologist* 5: 44–51, 1962.

7. DAGGS, R. G. Actions taken at spring meeting, 16–20 April 1963. *Physiologist* 6: 72, 1963.

8. DAGGS, R. G. Actions taken at spring meeting, 11–16 April 1966. *Physiologist* 9: 47, 1966.

9. DULING, B. R. Criteria for membership in the American Physiological Society. *Physiologist* 17: 417–418, 1974.

10. FENN, W. O. *History of the American Physiological Society: The Third Quarter Century, 1937–1962.* Washington, DC: Am. Physiol. Soc., 1963.

11. KNOBIL, E. American Physiological Society: 121st Business Meeting. VI. Amendments to the bylaws. *Physiologist* 22(3): 1, 1979.

12. KNOBIL, E. APS 121st Business Meeting. VII. Reports. D. Membership Advisory Committee. *Physiologist* 22(3): 2, 1979.

13. LEE, M. O. Associate membership approved. *Physiologist* 1(3): 3, 1958.

14. REYNOLDS, O. E. Proposal for corresponding membership and nominations for membership. *Physiologist* 17: 441–446, 1974.

15. REYNOLDS, O. E. Nominations for membership. *Physiologist* 17: 415–416, 1974.

16. REYNOLDS, O. E. Proposals for changes in bylaws regarding membership. *Physiologist* 18: 13, 1975.

17. REYNOLDS, O. E. Reports of actions taken at spring meeting, 14–18 April 1975. *Physiologist* 18: 63, 1975.

18. REYNOLDS, O. E. Annual membership in the APS, 1887–1975. *Physiologist* 19: 32, 1976.

19. REYNOLDS, O. E. Report of actions taken at spring business meeting, Anaheim, California, 13 April 1976. *Physiologist* 19: 53, 1976.

20. REYNOLDS, O. E. Proposed bylaw amendment to be voted on at spring business meeting. *Physiologist* 20(1): 3, 1977.

21. REYNOLDS, O. E. Reflections upon the past to provide a vision of the future of the American Physiological Society. *Physiologist* 28: 138, 1985.

CHAPTER 10

Spring and Fall Scientific Meetings

MICHAEL J. JACKSON AND JOSEPH F. SAUNDERS

A report of Council's activities published early in 1962 (1) stated that "Programs are beset with problems and are in a period of rapid evolution." In this chapter we summarize the attempts made in the quarter century following that statement to address the "problems" of the scientific programs at the meetings of the American Physiological Society (APS, the Society) and describe the changes that have contributed to the evolution of the program. Here we begin with the excellent survey of fall meetings prepared by Fenn (21) and review spring 1962 through fall 1985. This chapter is based on information derived from two main sources: *1*) reports of Council and of the standing and special committees concerned with program matters and published in *The Physiologist* and *2*) quantitative data from the meeting programs kept in the archives of the Society or of the Federation of American Societies for Experimental Biology (FASEB, the Federation).

Table 1 lists the annual meetings of the Society held during the period of review and includes data on attendance at those meetings. With the exception of 1968, when the fall meeting was omitted because of the International Union of Physiological Sciences (IUPS) Congress, the Society has held two regular scientific meetings each year. Traditionally the APS spring meeting has been presented under the organizational umbrella of the FASEB annual meeting. In contrast, the fall meeting usually has been organized independently, although from time to time other societies or groups have joined with APS in contributing to the program. For example, since 1972 the Society has been joined biennially for its fall meetings by the American Society of Zoologists' Division of Comparative Physiology and Biochemistry (ASZ/DCPB). In 1979 an APS tradition was introduced for triennial conjoint meetings with the IUPS Commission on Gravitational Physiology (IUPS/ CGP). Among other societies that occasionally have met with APS during its fall

Table 1
ATTENDANCE AT APS SCIENTIFIC MEETINGS, 1962–85

Spring*			Year	Fall			
Location	No. members registered	% Total membership		Location	Date	No. members registered	% Total membership
Atlantic City	882	38.84	1962	Univ. of Buffalo	28–31 Aug	285	11.61
Chicago	886	36.31	1963	Univ. of Miami†	27–30 Aug	NA‡	
Atlantic City	954	36.64	1964	Brown Univ.	7–11 Sep	NA‡	
Atlantic City	1,019	37.19	1965	Univ. of California, Los Angeles	23–27 Aug	NA‡	
Atlantic City	1,048	35.89	1966	Baylor Univ.	29 Aug to 2 Sep	312	10.52
Chicago	1,098	35.81	1967	Howard Univ.	21–25 Aug	388	11.85
Atlantic City	1,045	32.10	1968§				
Atlantic City	1,151	32.63	1969	Univ. of California, Davis	25–29 Aug	470	12.51
Atlantic City	1,238	33.10	1970	Indiana Univ.	27 Aug to 3 Sep	349	9.20
Chicago	1,063	27.05	1971	Univ. of Kansas	16–19 Aug	219	5.49
Atlantic City	1,133	27.63	1972	Penn State Univ.	27–31 Aug	NA‡	
Atlantic City	1,168	27.27	1973	Univ. of Rochester	20–24 Aug	419	9.65
Atlantic City	1,048	24.14	1974	State Univ. of New York, Albany	12–16 Aug	NA‡	
Atlantic City	1,099	23.83	1975	San Francisco	6–10 Oct	649	13.64
Anaheim	987	20.32	1976	Univ. of Pennsylvania	15–20 Aug	379	7.74
Chicago	1,080	21.67	1977	Hollywood, FL	9–14 Oct	432	8.59
Atlantic City	1,131	21.91	1978	St. Louis	22–27 Oct	563	10.69
Dallas	1,083	20.00	1979	New Orleans	15–19 Oct	633	11.53
Anaheim	939	16.74	1980	Toronto	12–17 Oct	740	12.97
Atlanta	1,077	18.59	1981	Cincinnati	11–16 Oct	532	8.98
New Orleans	1,225	20.35	1982	San Diego	10–15 Oct	706	11.62
Chicago	1,204	19.66	1983	Honolulu	20–24 Aug	496	8.07
St. Louis	1,458	23.57	1984	Lexington, KY	26–30 Aug	346	5.59
Anaheim	1,153	18.50	1985	Niagara Falls, NY	14–18 Oct		

* Usually held in the first half of April.   † Coral Gables, FL.   ‡ Not available from attendance record maintained by host university.   § No APS fall meeting because of XXIV IUPS Congress, Washington, DC.

meetings are the Canadian Physiological Society, the Biomedical Engineering Society, and the Latin American Association of Physiologic Sciences.

Organization and Structure of Meetings

Program Committees

Before and during the early part of the period of review, responsibility for organizing the Society's meetings was divided as follows. Presentation of symposia and invited speakers at the spring meeting were assigned to a Program Committee, which sometimes consisted of only one identified member. Program organization at fall meetings, including symposia, a refresher course, and specialty workshops, was carried out by a local committee of the host university, which changed from year to year with the venue of the meeting. Abstracts of volunteered papers were assigned to appropriate sessions by the executive secretary-treasurer and the APS central office staff.

An expansion of the responsibilities of the Program Committee was initiated in 1970 with a request from Council (8) that the committee (by then increased to five members) attempt long-range planning of symposia to provide continuity and balance over periods of several years. Subsequently, in 1973, with the proposal that alternate fall meetings be held at noncampus locations, the Program Committee was charged with organizing the programs for those meetings and was given representation on the local committee should planning call for a campus-based meeting (10).

Probably the most significant stimulus for change in Program Committee structure and responsibility has been the development of organized subspecialty sections within APS. Although sections were initially viewed with some reluctance (27), eventually they were recognized as an opportunity to accommodate the needs of members who might seek representation outside the Society (23). From its earliest consideration of this issue, Council appreciated the significant impact of sectional organization on program activities. Thus in 1976 an ad hoc committee of the organized APS sections was formed to advise the Program Committee (14), and members of the Program Committee were asked to open channels of communication with special interest groups within the Society, as well as with related societies, to ensure that the needs of these constituencies were represented adequately in the scientific programs (26).

In 1976 Council established a Task Force on Scientific Programs under the chairmanship of Alfred P. Fishman; it was charged with examining the scientific programs of APS with respect to the goals of the annual meetings and their accomplishment. The task force met at the home of the president of APS, Bodil Schmidt-Nielsen, in Bar Harbor, Maine, in July 1976 (APS Council minutes, 14–15 August 1976, p. 14); the participants spent two days considering optimal organization for specialty groups within the Society in the generation of programs of maximal acceptability to those special interests. The main recommendation of the task force was the establishment of a two-tier structure for program development: *1)* a Program Advisory Committee (Table 2), consisting of representatives of the organized sections of APS and other special interest groups, was charged with developing recommendations on program content and format; and *2)* a Program Executive Committee (Table 3) was appointed by Council to develop scientific programs based on the recommendations of the Program Advisory Committee. This structure was initiated in 1978 and has been maintained since then (15). The chairman of the Program Executive Committee served also as chairman of the Program Advisory Committee. Although the task force had suggested that ten members of the Program Advisory Committee might adequately represent the interests of specialty groups within APS, it has been found necessary to expand its size. Currently fifteen groups are represented on the committee (Fig. 1). Relationships between the Program Executive Committee and other organizational structures of the Society are also shown in Figure 1.

In addition to responsibility for organizing symposia as recommended by the task force, the Program Advisory Committee has assumed another role. Because the members of this committee are usually familiar with the directions of growth in their areas of interest, they have been assigned responsibility for organizing slide and poster sessions from abstracts of volunteered papers submitted for the spring and fall meetings.

The need for and benefits of a Program Advisory Committee are illustrated in

Table 2
PROGRAM ADVISORY COMMITTEE, 1962–77

	Years			Years	
	Member	Chairman		Member	Chairman
L. L. Langley	1962–65	1962–65	D. Kennedy	1970–73	
W. R. Anslow, Jr.	1962–64		M. F. Wilson	1970–73	
R. M. Berne	1962–65		S. Y. Botelho	1971–75	
W. L. Nastuk	1964–67		C. Lenfant	1973–76	
A. P. Fishman	1965–68	1965–68	G. G. Somjen	1973–76	
S. B. Barker	1965–69		F. G. Knox	1973–76	
N. B. Schwartz	1967–70		S. D. Gray	1973–77	
H. D. Lauson	1968–71	1968–71	A. P. Somlyo	1975–77	
F. F. Jobsis	1969–75	1971–75	P. H. Bogner	1975–76	
H. M. Goodman	1970–77	1975–77	L. R. Johnson	1976–77	

Table 3
PROGRAM EXECUTIVE COMMITTEE, 1977–86

	Years			Years	
	Member	Chairman		Member	Chairman
H. M. Goodman	1977–81	1977–81	M. J. Jackson	1982–85	1982–85
F. G. Knox	1977–82	1981–82	E. J. Masoro	1982–84	
M. J. Fregly	1977–80		B. L. Umminger	1983–86	
B. R. Duling	1980–83		P. D. Harris	1984–87	
A. P. Fishman	1981–82		C. V. Gisolfi	1985–88	1985–86

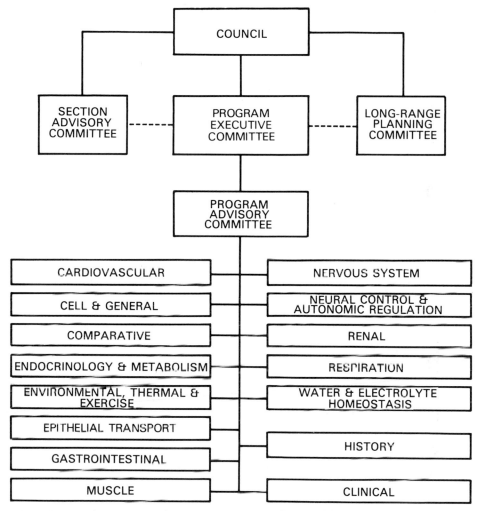

Fig. 1. Specialty groups and committee structure for scientific program development

Figures 2 and 3, which represent the distribution, by specialty, of volunteered papers submitted for the spring and fall meetings, respectively. The data represent the average number of volunteered papers for five-year intervals (1962–85) per specialty area corresponding to fifteen APS specialty sections/groups. History and clinical physiology are excluded. The former was not approved by Council until 1984; abstracts for historical papers were accepted as of 1985, but as yet there have been very few. Clinical physiology is now represented on the Program Advisory Committee by the chairman of the Subcommittee on Clinical Sciences of the Publications Committee. This subcommittee is responsible for the development of an annual state-of-the-art symposium as the basis for an APS publication of interest to physiologists in the basic and clinical sciences but has no responsibility for the organization of contributed paper sessions.

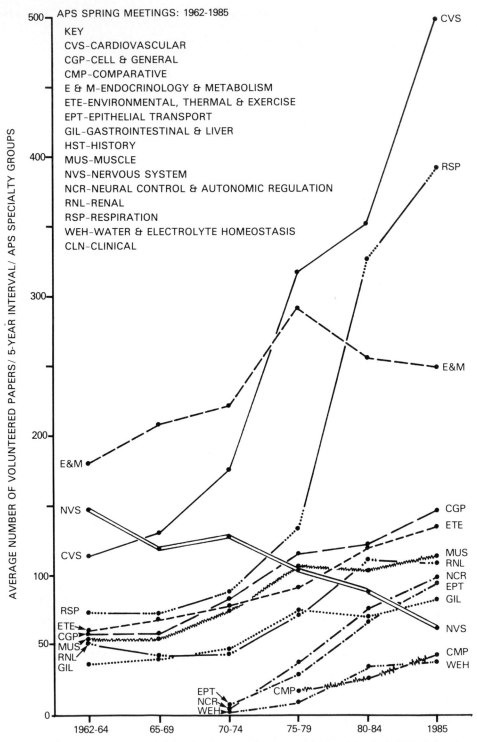

Fig. 2. Distribution, by specialty, of volunteered papers submitted for APS (FASEB) spring meetings, 1962–85

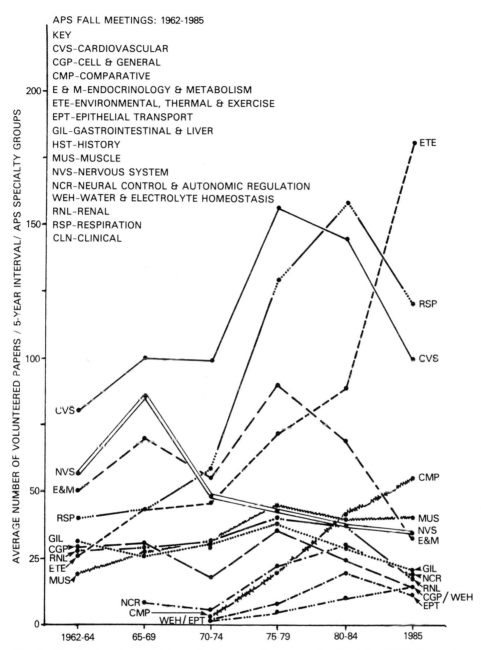

Fig. 3. Distribution, by specialty, of volunteered papers submitted for APS fall meetings, 1962–85

With respect to spring meetings (Fig. 2), the programs from 1962 to 1974 were dominated by volunteered papers in endocrinology and metabolism, cardiovascular physiology, and nervous system specialties. Since 1974 scientific reports in cardio-vascular physiology have doubled and the number of respiratory physiology papers has surged, whereas the number of endocrinology and metabolism papers has decreased. The decline in endocrinology and metabolism papers probably would have been more dramatic had it not been for the triennial contributions of members of the American Society of Biological Chemists (ASBC) and the annual contributions from members of the American Institute of Nutrition. The number of papers on the nervous system has been decreasing progressively since 1974, perhaps because of the attraction of more physiologists each year to the more interdisciplinary annual meetings of the Society for Neuroscience. Since 1974, on the other hand, contributions from the subdisciplines of cell and general physiology, environmental, thermal, and exercise physiology, neural control and autonomic regulation, and epithelial transport have steadily increased. A strong growth was evident for a time in the renal area; however, that specialty, along with muscle, appears to have plateaued.

In the distribution of volunteered papers for the fall meetings (Fig. 3), the areas of endocrinology and metabolism and cardiovascular and respiratory physiology show a growth similar to that for the spring meetings. The nervous system specialty appears to have lost its zest since 1974. With the exception of environmental, exercise, and thermal physiology, comparative physiology, and water and electrolyte homeostasis, all other specialty areas show a continuing decline since 1979. Perhaps the explanation for the steady increase in environmental, thermal, exercise, and comparative physiology papers relates to the triennial and biennial joint meetings with, respectively, IUPS/CGP and ASZ/DCPB.

Spring Meetings

Although consideration has been given from time to time to the suggestion that the APS spring meeting should be held independently (19), all the meetings held during the period of review have been presented as a component of the annual FASEB meeting. In 1980 a survey of APS members indicated overwhelming support for continuing this association (17). Thus the evolution of the APS spring meeting has been linked inextricably with that of FASEB, and many of the changes that have occurred in the APS meeting reflect changes in the structure of the umbrella organization.

A particular concern early in the period of review was the large size of the FASEB meeting and the increasing number of volunteered papers (20) (Fig. 2). This was not a new concern; the need to reduce the number of papers presented at spring meetings was one of the reasons that the annual fall meetings were instituted. It was felt that the demand for scheduling simultaneous sessions to accommodate the large number of volunteered papers at spring meetings, coupled with the increasing popularity of symposia and other sessions with invited speakers, inevitably would result in scheduling conflicts that would impact on the quality of the meeting. Among the more draconian measures proposed to address this problem was the total elimination of ten-minute papers (3). However, this proposal does not appear to have progressed beyond the consideration stage; more attention was given to other means of limiting the number of volunteered papers, as discussed below.

Data on the submission of volunteered papers, meeting locations, and meeting times for the spring and fall meetings of the Society from 1962 to 1985 are shown in Figure 4, which reflects stringent restriction on acceptance of volunteered papers from 1962 to 1972, despite the accommodation resource of Atlantic City. Although for the next three years in Atlantic City volunteered papers were accepted more liberally, the dramatic rise in submission and acceptance did not occur until 1976, the year after the initiation of the poster mode of delivery. This rise also coincided with the availability of facilities in cities other than Atlantic City for accommodating the enormous consortium of scientists for the FASEB spring meeting.

One approach to limiting the number of contributed papers that engendered considerable discussion at Society business meetings during the early part of the period of review was a limitation on authorship. At the 1962 spring business meeting, members were asked to approve a change in the bylaws to require a member to be listed among the authors of an abstract. This proposal was rejected in favor of one restricting a member to authorship, coauthorship, or sponsorship of only one abstract (2). Although apparently explicit, the full meaning of this restriction was not appreciated by the membership, and a number of abstracts submitted for the 1963 spring meeting had to be rewritten to ensure compliance (4). It was soon realized that this restriction did not suffice. Although the number of abstracts for the 1963 spring meeting decreased to 703 (from 837 in the previous year), the number resumed growth in subsequent years (Fig. 4).

The 1964 spring meeting presented a special problem in that accommodations in Chicago were more limited than those at previous meetings. A more stringent restriction was necessary to limit acceptable abstracts to those that included an APS member among the authors (5). Thus the number of volunteered papers was held at approximately 800, but the effect was transient; the trend toward increase was resumed by the 1965 meeting (22).

In a further effort to limit the number of volunteered papers, at the 1967 spring business meeting the Society approved a recommendation instructing Council to devise a nonselective procedure to limit the number of papers to a predetermined level. With this charge, Council decided to use a lottery system to limit the number of papers scheduled for presentation in 1968 and subsequent years to approximately the same number (i.e., 850) presented in 1967 (6). This approach provided the required level of control and was used, with some modifications, over the next several years. In 1970 members were provided the opportunity to vote for continuation of the lottery system for abstract selection or to choose a subjective committee evaluation procedure as an alternative. A clear majority of members voted to retain the lottery system (7).

Some relaxation of the numeric restriction on volunteered paper presentations was possible at the 1971 meeting in Chicago, because ASBC had elected to meet separately from FASEB. This allowed more space to be allotted to APS sessions, and the number of abstracts programmed for presentation was increased to nearly 950. However, the following year (1972), the limit was decreased to slightly over 800 to accommodate more symposia and invited speakers (9).

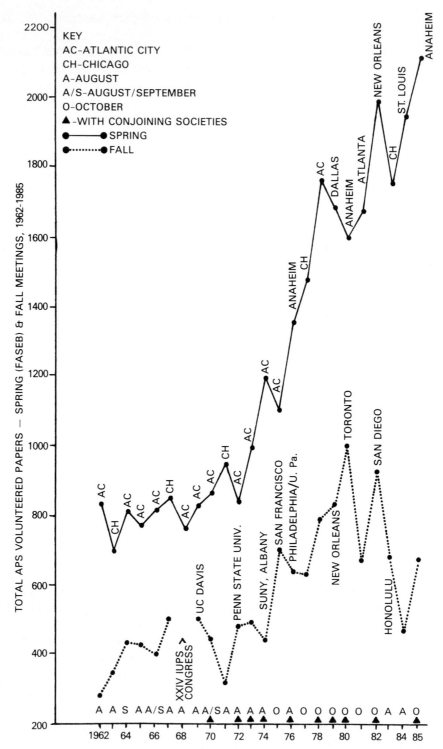

Fig. 4. Number of volunteered papers, meeting locations, and meeting times for spring and fall meetings, 1962–85

Experience with programming the 1971 and 1972 spring meetings demonstrated that selection of volunteered papers on the basis of space available for presentation was preferable to the arbitrary numeric limit used previously, and consequently, although no formal announcement of the change appears to have been made, this criterion was adopted for the selection of volunteered papers of later meetings (O. E. Reynolds, personal communication). Subsequently several factors contributed to a substantial improvement in the available space and allowed significant expansion of the number of volunteered papers and other elements of the program of the spring meeting.

Because their separate meeting experiment of 1971 was considered a success by ASBC members, a plan for regular independent meetings was adopted. Initially the plan called for ASBC to meet separately from FASEB in alternate years, but this was modified later so that ASBC would meet with the other FASEB societies only every third year. The impact of this change was partially offset in 1980 when APS invited members of ASBC to submit papers to the APS program. More than 300 papers, or approximately sixteen percent of the papers submitted to APS, were received in response to this invitation (16), and the practice was continued in subsequent years when ASBC did not meet with the other FASEB societies.

A second factor that has had a profound effect on the character of the spring meeting has been the change in its venue. For all but one of the years 1962–78, the FASEB societies met in Atlantic City or occasionally in Chicago. Other cities were unable to accommodate the large number of attendees and sessions. The transformation of Atlantic City from resort to gaming center resulted in a loss of many of the facilities, especially meeting rooms, that are essential to the FASEB meeting. An experimental format, the stretch meeting of 1979, in which the usual five-day period was extended to allow all the societies to meet under the FASEB umbrella, albeit not at the same time, was generally judged not to be successful, but by other mechanisms the problem has been well addressed and has not compromised the continued growth of the spring meeting.

The introduction of the poster session in 1975, as an alternative format for the presentation of volunteered papers, markedly relieved pressure on program scheduling and facilitated unrestricted growth in the last decade (Fig. 5). Posters do not compete for prime space, as do the slide and symposium sessions. From the beginning the poster session was greeted with unqualified acceptance (12), and the proportion of volunteered papers presented in this format increased from twenty-three percent of a total of 1,095 in 1975 to sixty-two percent of a total of 2,117 in 1985 (Fig. 6). However, the number of volunteered papers that can be presented in poster sessions is not without limit. While this review was in preparation, the FASEB Meetings Committee was beginning to confront problems associated with space and facilities for poster presentation. It appears that the growth that has characterized the volunteered paper segment of the program in the last ten years may be approaching its end and that the number of papers offered for presentation at the spring meeting may once again become an issue of concern.

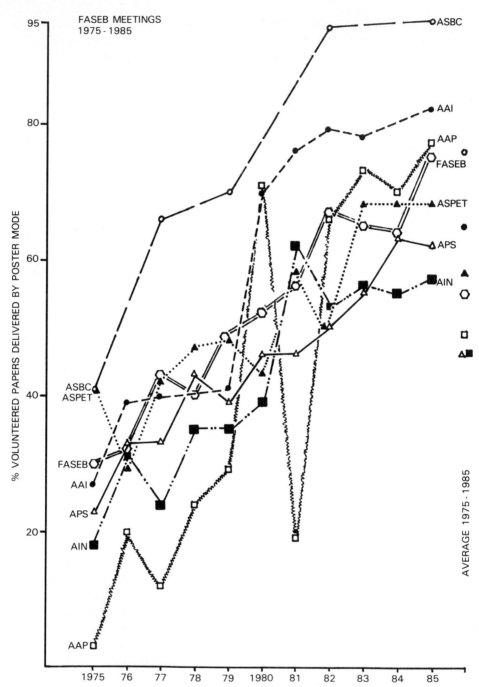

Fig. 5. Distribution, by FASEB member society, of volunteered papers presented by poster mode at FASEB meetings, 1975–85; ASBC, American Society of Biological Chemists; AAI, American Association of Immunologists; AAP, American Association of Pathologists; ASPET, American Society of Pharmacology and Experimental Therapeutics; AIN, American Institute of Nutrition

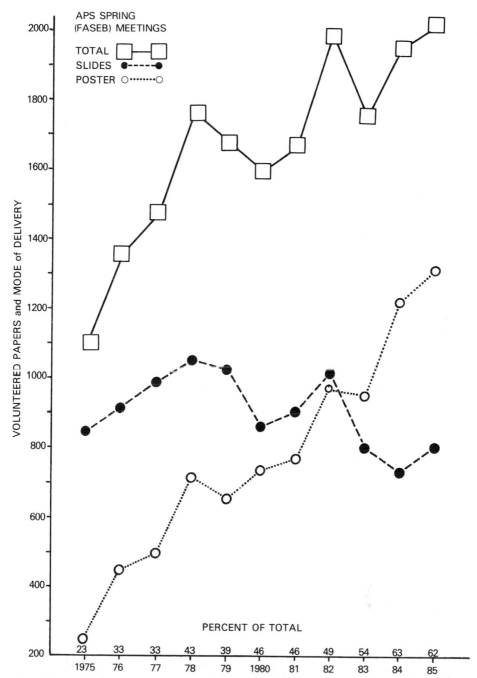

Fig. 6. Distribution, by mode of delivery, of volunteered papers presented at APS (FASEB) spring meetings, 1975–85

In 1973 the FASEB Meetings Committee made an innovative proposal for the organization of the 1974 meeting—that the six member societies of FASEB collaborate in the organization of a subject-oriented meeting (11). This proposal was rejected by the FASEB Executive Committee, and the matter was referred back to the Meetings Committee for further discussion. Despite this setback, the APS Council recognized the unique opportunity for examination of multidisciplinary programs of common interest to be presented at the FASEB meeting and authorized the organization of a three-day intersociety colloquium on "Membranes, Ions and Impulses" at the 1974 spring meeting (24). Subsequently an ad hoc committee of APS developed a proposal for the organization of future FASEB meetings on defined theme topics. This proposal was accepted in modified form by FASEB, and since 1981 the spring meeting has included multidisciplinary, intersocietal themes in addition to the contributions of individual societies.

Fall Meetings

In contrast to the vigorous debate and marked changes that have characterized the spring meeting in the last twenty-five years, the fall meetings of the Society have, until recently, continued a relatively uneventful development. The Society's records thus contain few references to questions of format, size, or content for these meetings.

From 1962 to 1974, the annual fall meetings were held on university campuses; this continued the tradition initiated in 1948. The first fall meetings were held in mid September, but with the shift in college schedules so that the first semester ended before Christmas, campus facilities were no longer generally available after mid to late August. In 1973 President Robert Berne surveyed the membership concerning their preference of a fall meeting at a campus location in August or a hotel-based meeting in an "attractive city" later in the fall. The vote was split nearly fifty/fifty between the two alternatives. On review of the survey, Council decided that a more attractive meeting might result if it were offered later in the year (i.e., October) than was possible with a campus location; in addition, city locations could offer amenities not found on or near college campuses (11). Even though local committees were still to be involved in the development of program elements, Council viewed such meetings as national undertakings and not regional in emphasis.

Accordingly the 1975 fall meeting was presented in October in a San Francisco hotel. Because Council intended to alternate campus and noncampus locations in the ensuing years, the 1976 fall meeting was held on a campus (13)—the University of Pennsylvania, an appropriate site in a year of bicentennial celebrations. Campus facilities in Philadelphia, however, were available only in August; it became clear that many members preferred an October date for the fall meeting. Though it was not intended to be so, the 1976 fall meeting was the last meeting held on a college campus.

Because of the managerial problem involved in alternating each year between a campus meeting and a nationally organized hotel meeting, Council decided in 1976 to schedule hotel-based meetings in October for a period of five consecutive years starting in 1977 and ending in 1982 (25, 26). In 1983 the XXIX IUPS Congress was to meet in Sydney, Australia. In part because of the timing of the congress, Council decided to schedule the next three meetings as campus meetings in August. The University of Hawaii was selected as the site for the 1983 meeting; Honolulu was considered an ideal location, because it could serve as a stopover for those who were going on to attend the congress the following week. The 1984 meeting was to be held at the University of Kentucky in Lexington, Kentucky, and the 1985 meeting at the University of Buffalo. However, as administrative arrangements were being developed with the respective universities, it became apparent that even in August, university campuses were no longer in a position to offer facilities sufficient to accommodate the APS fall meeting. One after another, the meetings were moved to off-campus facilities, albeit in proximity to the university. Thus the 1983 and 1984 meetings were held in hotels and convention centers in Honolulu and Lexington in August, and the 1985 meeting was held in Niagara Falls in October. After this experience, Council decided to retain the October meeting time because it attracted a larger number of attendees.

Another change during this period was the introduction of poster presentations in 1975. The rapid increase in poster presentations at fall meetings (Fig. 7), where there were no limitations on the number of volunteered papers, demonstrates the acceptance of the poster mode of report presentation.

During the past few years, however, another problem has arisen. The fall meeting appears to have become a "specialty meeting" in the sense that only some of the APS sections have participated, whereas other sections have offered little beyond the scheduling of a few volunteered papers. In essence the "invited speaker" component of the fall meetings has been dominated by only a few sections of the Society. Thus the APS fall meeting format has recently become a subject for considerable discussion by Council and committees.

In acknowledging the view that the fall meeting "is regarded by some physiologists as important as the Spring meeting," the Council in 1983 charged the Society's Long-Range Planning Committee to prepare an outline of a new fall meeting format for its consideration during the 1984 spring meeting. Thereafter the Long-Range Planning and Section Advisory Committees received the proposal for study of its feasibility and implementation. During the 1984 fall meeting, Council was presented with abundant information to support the view that "a city meeting, in a central location with desirable surroundings, is well attended as compared to an August campus meeting in an isolated area."

At the 1984 fall meeting in Lexington, Council charged the Program Executive Committee, under the chairmanship of Michael Jackson, in consultation with the Long-Range Planning Committee, to consider modifications in the format of the fall meeting that would "enhance the perception of physiology, improve the quality of scientific interaction and, thus, be more attractive to members and guests in meeting

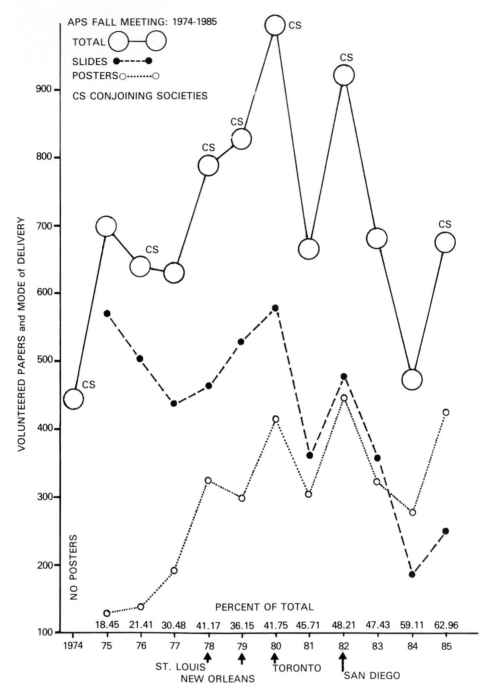

Fig. 7. Distribution, by mode of delivery, of volunteered papers presented at APS fall meetings, 1974–85

the needs of more significant numbers." The Program Executive Committee developed a proposal for the design of a new fall meeting format. It encompassed *1)* the selection of a place and time to ensure minimal conflict with competing activities, particularly the meetings of other societies and specialty groups, and to be attractive to members; *2)* inclusion of a meeting component specifically directed to the needs of trainees; and *3)* formulation of other components of the meeting based on a readily identifiable theme.

At the 1985 spring meeting, Council approved in principle the above recommendations of the Program Executive Committee "with details and date of adoption to be worked out." The thematic approach, as it has been proposed, offers an opportunity for all the APS sections to increase and strengthen their roles in the organization and presentation of scientific programs. Another benefit to the sections is the opportunity for "mixed activity" at the intersectional level, as well as interaction with other specialty interest (societal) groups. In essence, utilization of the sectional resource(s) in the organization of interesting and exciting fall meetings could be invigorating to physiologists and their science and enhance interdisciplinary activity in the face of the current "splintering phenomenon." The thematic approach was implemented for the first time at the 1986 fall meeting to be held in New Orleans in October.

Over the twenty-four years of this review, the Society has continued to try to meet, as closely as possible, all the needs of its members. Although there appear to be tendencies in certain areas of physiology toward a decline in the submission of volunteered papers for our meetings, the scientific interest in most areas of physiology continues to be intense.

It would be difficult to express the "goals of the annual meetings" more eloquently than did the Task Force on Scientific Programs in its 14 August 1976 report to Council.

> Chief among these is the transmission of scientific information. This occurs at several levels: at the ten-minute papers in the lecture hall, at symposia, at poster sessions and in the corridors and hospitality suites. These encounters entail the principles of peer review as well as scientific indoctrination. . . . While established scientists are being refreshed and refurbished, students and neophytes in science are often encouraged, by the proceedings and their personal involvement in the meetings, to probe further into the prospects of a career in physiology.

Editor's Note: Program Endowment Fund

The increase in the number of symposia and invited lectures at spring and fall meetings and the active participation of the sections in program planning have underscored the necessity for a more stable financial basis for the Society's programs. Solicitation for funding by individual members has sometimes led to conflicts when more than one member approached the same prospective donor for symposia support. For the 1985 spring meeting, more than $70,000 were received in support of symposia: $23,610 from the Society, $7,500 from FASEB, and $40,500 from industry. The recommendation to set up an endowment fund was first made by Michael Jackson to Council at the 1983 spring meeting. With the assistance of the Committee for Liaison with Industry, the Council has since appointed a Program Endowment Fund Committee. The goal of this committee, chaired by Theodore Cooper, is to raise, over a two-year period, a $250,000 endowment for APS program activities. Ultimately an endowment of $1,000,000 is needed to provide at least $80,000 a year in program support (18).

REFERENCES

1. ANONYMOUS. Program Advisory Committee. *Physiologist* 5: 31, 1962.
2. ANONYMOUS. APS business meetings. *Physiologist* 5: 44, 1962.
3. ANONYMOUS. Report of the Program Advisory Committee. *Physiologist* 5: 54, 1962.
4. ANONYMOUS. Editor's page. *Physiologist* 6: 3, 1963.
5. ANONYMOUS. Change in rules for presentation of papers at 1964 Spring Meeting. *Physiologist* 6: 318, 1963.
6. ANONYMOUS. Actions taken at the Spring Meeting, April 17–21, 1967. *Physiologist* 10: 49, 1967.
7. ANONYMOUS. Results of mail vote on selection of abstracts. *Physiologist* 13: 355, 1970.
8. ANONYMOUS. Call for APS program suggestions. *Physiologist* 13: 356, 1970.
9. ANONYMOUS. Abstracts for the Federation meeting. *Physiologist* 15: 3, 1972.
10. ANONYMOUS. Program Committee. *Physiologist* 16: 521, 1973.
11. ANONYMOUS. Report of Council activities—1972–1973. *Physiologist* 16: 521, 1973.
12. ANONYMOUS. Reaction to poster session at FASEB meetings. *Physiologist* 18: 65, 1975.
13. ANONYMOUS. APS Fall Meeting—Philadelphia. *Physiologist* 19: 73, 1976.
14. ANONYMOUS. Program Committee report to the Society Business Meeting, April 13, 1976. *Physiologist* 19: 93, 1976.
15. ANONYMOUS. Subspecialty groups to be represented on new program committees. *Physiologist* 20(1): 6, 1977.
16. ANONYMOUS. Program Committee's report. *Physiologist* 23(3): 12, 1980.
17. ANONYMOUS. Report of the Task Force on Future Meetings of the APS. *Physiologist* 24(3): 12, 1981.
18. ANONYMOUS. 134th APS Business Meeting. *Physiologist* 28: 465–467, 1985.
19. DAVENPORT, H. W. President's message. *Physiologist* 4(4): 27, 1961.
20. DAVENPORT, H. W. President's message. *Physiologist* 5: 29, 1962.
21. FENN, W. O. *History of the American Physiological Society: The Third Quarter Century, 1937–1962.* Washington, DC: Am. Physiol. Soc., 1963, chapt. 5, p. 85–91.
22. FORSTER, R. E. President's message. *Physiologist* 10: 13, 1967.
23. GUYTON, A. C. President's report. *Physiologist* 18: 79, 1975.
24. KNOBIL, E. President's message. *Physiologist* 23(3): 1, 1980.
25. SCHMIDT-NIELSEN, B. President's message. *Physiologist* 18: 503, 1975.
26. SELKURT, E. E. President's message. *Physiologist* 19: 471, 1976.
27. TOSTESON, D. C. Report of the president. *Physiologist* 17: 118, 1974.

CHAPTER 11

Publications

ALFRED P. FISHMAN AND STEPHEN R. GEIGER†

The American Physiological Society (APS, the Society), a major nonprofit pub-lisher of journals and books on physiology, engages in this activity as part of its mandate as a learned society. With the mantle of a scholarly publisher, the Society has assumed responsibility for setting standards for publications in the field. The publications program is now sizable, and the financial involvement is large. For a program of this magnitude to survive and prosper, academic excellence must be maintained; sound business practices must be followed; technological changes in publishing and printing must be exploited; and copyright, postal, and other govern-mental regulations must be continuously monitored.

The publications program of the Society operates under the aegis of a Publications Committee responsible to a Finance Committee and Council, with the help of a professional staff. The publications program functions without utilizing money from members' dues. The story of this program in the last twenty-five years, as outlined in the following pages, is one of unselfish, voluntary contribution by the members of the physiological community who devote much time as Publications Committee members, editors, editorial board members, and reviewers. It is also the story of a professional staff that shares the ambition of publishing the best in physiology.

Reorganization of Publications Management

On 1 July 1961 the Board of Publication Trustees was dissolved and replaced by a Publications Committee and a Finance Committee responsible to the Council of the Society. This was not an easy transition, and the implications were large: autonomy over publications was relinquished to Council, and both policy and finances became subject to control by the elected leaders of the Society rather than by a virtually independent Board appointed and charged to ensure outstanding publications. Milton O. Lee remained as managing editor and Sara F. Leslie became executive editor. From May 1962 until December 1965, publications were supervised

† Stephen R. Geiger died on 31 May 1987.

by the Publications Service Center of the Federation of American Societies for Experimental Biology (FASEB, the Federation). In December 1965 Lee began terminal retirement leave from his position as managing editor, which he had held since 1947.

In 1966 the APS editorial and business offices were reorganized. Leslie became publications manager and executive editor. Ray G. Daggs, executive secretary-treasurer since 1956, and Walter A. Sonnenberg, business manager since 1965, oversaw the reorganization of the Society's finances. This period of transition in the management of the Society's publications lasted from 1961 to 1968. Philip Bard was chairman of the Publications Committee in the first year of its existence (see Table 1). He was succeeded by A. Clifford Barger and Robert W. Berliner (1962–69).

Stephen R. Geiger joined the staff as executive editor of the *Handbook of Physiology* and assistant publications manager in July 1969. He was appointed acting publications manager in 1974, so Leslie could spend the last year before her retirement organizing the publications archives. (The fruits of that effort have been invaluable in preparing this chapter.) Geiger was appointed publications manager and executive editor in 1975, with Mary A. Phillips as managerial assistant from September 1974 until May 1975. Brenda B. Rauner succeeded her as managerial assistant in September 1975 and became production manager in 1978.

From the outset in 1961 until 1985, the Publications Committee consisted of three members appointed by Council, one of whom served as chairman. The president of the Society and the executive secretary-treasurer served as ex officio members. In 1967 three others were added to the committee as ex officio members: the chief editor of the *Journal of Neurophysiology*, the chairman of the Editorial Board of *Physiological Reviews*, and the chairman of the Handbook Editorial Committee. In 1976 the journals were reorganized, each under a separate editor, and the ex officio positions of the three editors on the Publications Committee were discontinued. In 1985, the number of members on the Publications Committee was increased from three to five; Jean McE. Marshall then became the first woman to be appointed as a member of the committee.

Journal Reorganization

With the reorganization of the management of the journals in 1962, the handling of manuscripts for the *American Journal of Physiology (AJP)* and the *Journal of Applied Physiology (JAP)* was also restructured. Section editors (Table 2) took over this responsibility from Lee. (The Editorial Boards of the *AJP* and *JAP* had been combined in July 1952.)

Throughout the 1960s and early 1970s, the Publications Committee struggled with the problem of how to handle the increasing number of new manuscripts being received and the increasing specialization in physiology. In 1967 *JAP*, which had been published bimonthly, became a monthly publication. New sections were added to *AJP* and *JAP*: Hematology in 1965, Other Aspects of Physiology from January 1968 through July 1969, and Muscle in 1970. The names of some sections were modified: Neurophysiology became Neurobiology in 1970 and Environmental Physiology

Table 1
PUBLICATIONS COMMITTEE

		Chairman	Members
Apr 1961–	1962	P. Bard	A. C. Barger
			R. W. Berliner
1962–Aug	1962		A. C. Barger
			R. W. Berliner
Feb 1963–Jun	1963	A. C. Barger	R. W. Berliner
			C. N. Woolsey
Jul 1963–Jun	1965	R. W. Berliner	R. E. Forster
			C. N. Woolsey
Jul 1965–Jun	1966	R. W. Berliner	D. S. Fredrickson
			J. Mead
Jul 1966–Jun	1969	A. C. Barger	D. S. Fredrickson
			J. Mead
Jul 1969–Jun	1971	D. S. Fredrickson	P. F. Curran
			J. Mead
Jul 1971–Jun	1972	P. F. Curran	D. S. Fredrickson
			P. Horowicz[a]
Jul 1972–Oct	1974	P. F. Curran[b]	A. P. Fishman
			P. Horowicz
Nov 1974–Jun	1975	A. P. Fishman	P. Horowicz
			F. E. Yates
Jul 1975–Jun	1978	A. P. Fishman	R. W. Berliner
			R. M. Berne
Jul 1978–Jun	1981	A. P. Fishman	R. M. Berne
			H. E. Morgan
Jul 1981–Jun	1982	H. E. Morgan	R. M. Berne
			L. E. Farhi
Jul 1982–Jun	1984	H. E. Morgan	L. E. Farhi
			E. E. Windhager
Jul 1984–Jun	1985	H. E. Morgan	W. F. Ganong
			L. R. Johnson[c]
			F. G. Knox[c]
			E. E. Windhager
Jul 1985–		P. C. Johnson	J. S. Cook[d]
			W. F. Ganong
			L. R. Johnson
			F. G. Knox[e]
			J. McE. Marshall

[a] Resigned. [b] Died 16 Oct 1974. [c] Appointed Sep 1984. [d] Appointed Jul 1986. [e] Term expired Jun 1986.

became Environmental and Exercise Physiology in 1971. Section coeditors were appointed. Several different ways were tried to list articles by sections in the table of contents. It was clear, however, that neither the increase in the number of sections nor the attempts at sorting were coping with the increasing number of papers and the evident trend to greater and greater specialization.

Under the chairmanship of Donald S. Frederickson and Peter F. Curran (1969–74), the Publications Committee devoted much effort to resolving the prevailing discontent over the organization of the journals. Early in 1974 the membership was presented with a ballot in which they were asked whether they favored the formation

Table 2
SECTION EDITORS—*AMERICAN JOURNAL OF PHYSIOLOGY* AND
JOURNAL OF APPLIED PHYSIOLOGY

Section	Editor	Term
Circulation	M. B. Visscher	Jan 1962–Dec 1962
	D. E. Gregg	Jan 1963–Dec 1963
	R. M. Berne	Jan 1964–Jul 1965
	B. F. Hoffman	Jan 1964–Dec 1966
	P. Dow	Jul 1965–Jun 1968
	T. Cooper	Jan 1967–Jun 1971
	W. C. Randall	Jul 1968–Jun 1974
	D. F. Bohr	Jul 1969–Jun 1975
	F. J. Klocke	Jun 1971–Feb 1975
	W. W. Parmley	Jul 1974–Dec 1976
	M. N. Levy*	Mar 1975–Dec 1976
	P. C. Johnson	Jul 1975–Dec 1976
Respiration	H. Rahn	Jan 1962–Sep 1962
	A. B. Otis	Jul 1962–Dec 1963
	J. Mead	Jan 1964–Dec 1966
	J. B. Severinghaus	Jul 1966–Jun 1969
	A. B. DuBois	Jan 1967–Jun 1969
	L. E. Farhi	Jul 1969–Dec 1973
	S. Permutt	Jul 1969–Jun 1972
	T. C. Lloyd, Jr.	Jul 1972–Dec 1976
	F. G. Hoppin, Jr.	Jan 1974–Dec 1976
Renal and Electrolyte Physiology	W. D. Lotspeich	Jan 1962–Dec 1963
	C. W. Gottschalk	Jul 1962–Dec 1962†
	J. Orloff	Jan 1964–Dec 1967
	R. W. Berliner	Jul 1964–Jul 1965†
	G. H. Giebisch	Jan 1968–Mar 1969
	E. E. Windhager	Jan 1969–Jun 1974
	W. B. Kinter	Apr 1969–Dec 1975
	J. H. Dirks	Jan 1973–Dec 1975
	W. E. Lassiter	Jul 1974–Jun 1976
	J. J. Grantham	Jan 1975–Jun 1976
Gastrointestinal Physiology	H. W. Davenport	Jan 1962–Jun 1963
	F. Hollander	Jul 1963–Mar 1966
	E. Grim	Apr 1966–Jun 1968
	P. F. Curran	Jul 1968–Jun 1971
	S. G. Schultz*	Sep 1971–Jun 1975
	J. G. Forte*	Jul 1975–Jun 1976
Endocrinology and Metabolism	J. A. Russell	Jan 1962–Mar 1967
	R. K. Meyer	Jan 1964–Dec 1966
	B. R. Landau	Jan 1967–Dec 1970
	R. O. Scow	Jan 1967–Jun 1968
	F. E. Yates	Jul 1968–Jun 1974
	N. S. Halmi	Jan 1971–Jun 1976
	D. S. Gann	Jul 1974–Jun 1976
Environmental Physiology and Exercise	L. D. Carlson	Jan 1962–Jun 1966

Table 2—*Continued*

Section	Editor	Term		
	J. D. Hardy	Jul	1962–Jun	1972
	H. S. Belding	Jul	1972–Aug	1973
	A. P. Gagge	Jul	1972–Jun	1976
	E. R. Buskirk	Jan	1974–Jun	1976
Comparative and General Physiology	K. Schmidt-Nielsen	Jan	1962–Dec	1964
	A. W. Martin	Jan	1965–Dec	1967
	C. L. Prosser	Jan	1968–Jun	1971
	L. B. Kirschner	Jun	1971–Jun	1976
Neurobiology‡	E. Henneman	Jan	1962–May	1963
	E. V. Evarts	Jun	1963–Jun	1966
	D. P. Purpura	Jul	1966–Jun	1970
	V. B. Brooks	Jul	1970–Aug	1971
	J. DeC. Downer	Jul	1971–Jun	1972
	O. A. Smith	Jul	1972–Jun	1976
Hematology	O. D. Ratnoff	Nov	1965–Dec	1972
	H. M. Ranney	Jan	1973–Dec	1976
Muscle Physiology	F. N. Briggs	Jul	1970–Jan	1975
	C. L. Prosser	Jul	1970–Jan	1971
	F. Fuchs		1975–Jun	1976
Biomedical Engineering	T. G. Coleman		1975–Jun	1976

* Acting editor.   † Approximate date.   ‡ Neurophysiology until Jul 1970.

of separate journals according to special fields of physiology. More than fifty percent of those who replied voted for the formation of separate journals. The remaining ballots were about equally divided between *1)* merging *AJP* and *JAP* and arranging articles by sections and *2)* no change. Curran reported the outcome to the Society in the May 1974 issue of *The Physiologist.* At its November meeting, the Publications Committee, with Alfred P. Fishman as chairman, in an attempt to ease gradually into sectionalization, proposed that *AJP* and *JAP* continue to be published as separate journals but that articles on circulation be published separately each month, as well as in the existing journals. Paul Horowicz, a member of the committee who had been unable to attend the meeting, objected to this plan as not accommodating the desires of those who favored full sectionalization. Because of the conviction that sectionalization was not an issue to be settled by a two-to-one vote, Fishman asked that the issue be placed on the agenda again for discussion at the first meeting in 1975.

In 1975 a new plan was developed by the Publications Committee: the two journals would continue to be published separately, but new publications would be generated to deal with special areas of physiology that were not reflected in the parent journals. The specialty journals would be directed primarily at members of APS; they would be sold to the members at cost, and subscription to at least one specialty journal (or a parent journal) would be mandatory for all members. The plan drew an unfavorable response at the Society's business meeting in April because of the requirement of mandatory subscription. Clearly some form of sectionalization according to specialties was in the offing, but the exact form that would be acceptable to the membership, financially sound, and in keeping with the scholarly aspirations of the Society remained to be settled.

During the summer and fall of 1975, the Publications Committee evolved the plan that was finally adopted. It included three elements: *1*) continued publication of *AJP* and *JAP, 2*) separate journals in specialized areas of physiology directed by separate editors and editorial boards responsible to the Publications Committee, and *3*) a consolidated *AJP* that would consist of the material in all the separate journals (i.e., a composite). In essence, it was a plan designed to satisfy both the "generalists" and the "specialists" in physiology. Noteworthy was the deliberate attempt to anticipate future directions in physiology by creating two new specialty journals, one in cellular physiology and the other in integrative physiology.

By the end of 1975, six editors had been appointed to lead the six specialty journals: Thomas E. Andreoli, Leon E. Farhi, Paul Horowicz, Rachmiel Levine, Matthew N. Levy, and F. Eugene Yates (Table 3). On 6 December they attended the

Table 3
EDITORS AND ASSOCIATE EDITORS

Journal	Term	Editors	Associate Editors
American Journal of Physiology			
Cell Physiology	Jan 1976–Jun 1981	P. Horowicz	P. DeWeer
			A. L. Finn
			H. A. Fozzard
			F. J. Julian
			C. F. Stevens
			J. S. Willis
			S. Winegrad
	Jul 1981–Jun 1984	H. E. Morgan	R. D. Berlin
			J. S. Cook[a]
			R. E. Fellows
			J. S. Handler
			P. A. Knauf
			M. J. Kushmerick
			M. Lieberman[a]
			A. E. Pegg
	Jul 1984–	P. A. Knauf	J. S. Bond
			J. S. Cook
			P. B. Dunham
			R. A. Frizzell
			M. Lieberman
			L. J. Mandel
			R. A. Murphy
			W. J. Pledger
Endocrinology, Metabolism and Gastrointestinal Physiology	Jan 1976–Dec 1980	R. Levine	R. N. Bergman
			L. R. Johnson
			D. Porte, Jr.
			R. P. Robertson
			D. Rodbard

Table 3—*Continued*

Journal	Term	Editors	Associate Editors
Endocrinology and Metabolism	Jul 1979–Jun 1982	E. Knobil	G. Sachs N. M. Weisbrodt R. N. Bergman M. F. Dallman[a] L. S. Jefferson[a] H. E. Morgan J. D. Neill J. M. Olefsky W. Tong[a]
	Jul 1982–May 1984	J. Neill	C. Desjardins[a] J. E. Gerich[a] M. Vranic[a]
	Jun 1984–	L. S. Jefferson	C. Desjardins J. H. Exton J. H. Gerich I. Reid M. Vranic
Gastrointestinal and Liver Physiology	Jul 1979–Jun 1985	L. R. Johnson	E. L. Forker M. J. Jackson G. Sachs J. H. Szurszewski
	Jul 1985	J. A. Williams	D. N. Granger W. G. Hardison T. E. Machen G. M. Makhlouf N. W. Weisbrodt
Heart and Circulatory Physiology	Jan 1976–Dec 1980	M. N. Levy	L. L. Hefner P. C. Johnson P. Martin R. A. Olsson W. W. Parmley K. Sagawa B. E. Sobel M. Vassalle
	Jan 1981–Dec 1986	E. Page	N. R. Alpert D. F. Bohr B. R. Duling E. O. Feigl H. A. Fozzard W. R. Gibbons J. R. Neely R. C. Webb
	Jan 1987–	V. S. Bishop	A. M. Brown H. J. Granger H. A. Kontos E. Morkin A. P. Shepherd, Jr. P. M. Vanhoutte
Regulatory, Integrative and Comparative Physiology	Jan 1976–Dec 1983	F. E. Yates	F. P. Conte W. H. Dantzler[a] J. J. DiStefano III D. S. Gann

Table 3—*Continued*

Journal	Term	Editors	Associate Editors
	Jan 1984–	D. J. Ramsay	P. R. McHugh[a] M. C. Moore-Ede[a] C. S. Pittendrigh B. M. Schmidt-Nielsen D. O. Walter E. J. Braun W. H. Dantzler P. R. McHugh M. C. Moore-Ede T. N. Thrasher
Renal, Fluid and Electrolyte Physiology	Jan 1976–Jun 1983	T. E. Andreoli	J. J. Grantham F. S. Wright
	Jul 1983–	J. A. Schafer	L. G. Navar D. G. Warnock
Journal of Applied Physiology	Jan 1976–Dec 1981	L. E. Farhi	N. R. Anthonisen E. R. Buskirk A. P. Gagge R. A. Klocke[a] C. Lenfant J. Milic-Emili S. M. Tenney J. B. West
	Jan 1982–	A. P. Fishman	V. S. Bishop[b] N. S. Cherniack R. P. Daniele J. O. Holloszy R. A. Klocke L. B. Rowell J. T. Sharp J. T. Stitt H. L. Stone[c] A. E. Taylor J. Widdicombe
Journal of Neurophysiology	Jan 1962–Jun 1964 Jul 1964–Jun 1974 Jul 1974–Dec 1977 Jan 1978–Jun 1983 Jul 1983–	V. B. Mountcastle J. M. Brookhart E. V. Evarts W. D. Willis L. M. Mendell	 P. A. Getting[d] P. L. Strick[d]
Physiological Reviews	1961–Oct 1963 Sep 1963–Dec 1972 Jan 1973–Dec 1978 Jan 1979–Dec 1984 Jan 1985–	C. F. Code[e] J. R. Brobeck[e] H. E. Morgan S. G. Schultz G. H. Giebisch	R. G. Daggs R. G. Daggs R. C. Rose R. A. Frizzell W. F. Boron

[a] Completed term under subsequent editor.   [b] Term ended 31 Dec 1986.   [c] Died 16 Nov 1984.   [d] Appointed Apr 1986.   [e] Designated chairman of Editorial Board through Dec 1972.

Publications Committee meeting for orientation. On 1 April 1976 the Society began to receive manuscripts for the separate new journals, each with its own editor, associate editors, and editorial board. In January 1977 *AJP* began to be published as separate specialty journals and as the consolidated *AJP*. At the same time, the scope of *JAP* was circumscribed, whereas the *Journal of Neurophysiology* and *Physiological Reviews* were left unchanged.

' In the first issue of each journal of 1977, the reorganization was described as consisting of the following essential elements:

1. The *American Journal of Physiology* and the *Journal of Applied Physiology* will continue to exist and to be available as separate publications of the American Physiological Society. This continuing arrangement will satisfy both the generalist in physiology and the librarian who will be able to preserve unbroken series of these two distinguished journals.

2. The *Journal of Applied Physiology*, under its own editorial board, will restrict itself to papers on respiration, exercise, and environmental physiology.

3. The *American Journal of Physiology* will henceforth be an umbrella for separate journals, each with its own editorial board and each devoted to specialized subject matter. Each component will be issued separately as well as part of the *American Journal of Physiology*. This arrangement will satisfy the specialist who is interested in having on hand outstanding papers in his own particular field of interest.

It was anticipated that because of this reorganization, the number of new manuscripts submitted to *AJP* would experience some increase: certain areas formerly covered in *JAP* would be incorporated into *AJP*, and two new journals would be formed. Moreover, since *JAP: Respiratory, Environmental and Exercise Physiology* was to become more specialized than *JAP* had been, it was anticipated that the number of new manuscripts submitted would decrease. Initially these predictions proved correct. But a subsequent phenomenal increase in the number of manuscripts received and those published (Figs. 1 and 2) was not anticipated. One happy result of the increased number of manuscripts was the split in 1980 of *AJP: Endocrinology, Metabolism and Gastrointestinal Physiology* into *AJP: Endocrinology and Metabolism* edited by Ernst Knobil and *AJP: Gastrointestinal and Liver Physiology* edited

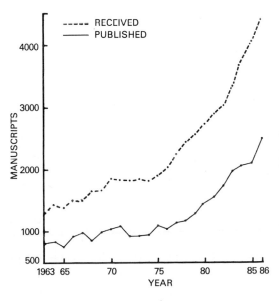

Fig. 1. Manuscripts received and published in journals of the American Physiological Society from 1963 to 1986

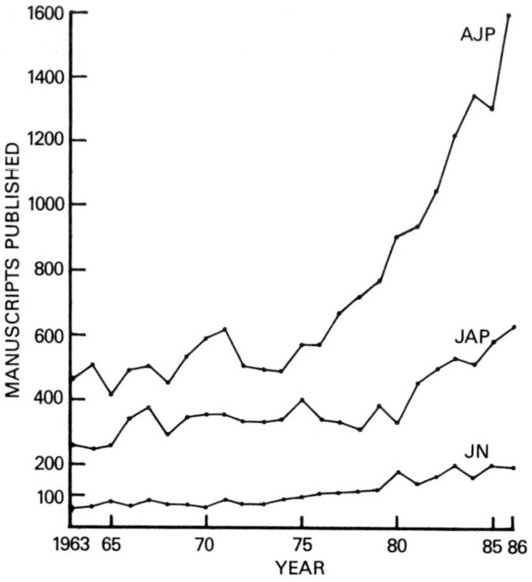

Fig. 2. Manuscripts published in American Journal of Physiology (AJP), Journal of Applied Physiology (JAP), *and* Journal of Neurophysiology (JN) *from 1963 to 1986*

by Leonard R. Johnson. Another was an increase in the frequency of publication from bimonthly to monthly for *AJP: Regulatory, Integrative and Comparative Physiology* in 1983 and for *AJP: Cell Physiology* in 1986.

Not only has the amount published increased, but new kinds of articles now appear in the Society's journals (e.g., theoretical articles, brief reviews, opinion papers, rapid communications, and a Modeling Methodology Forum). With separate specialty journals, each the responsibility of an editor, the journals have improved in many ways and problems are dealt with much more rapidly than before. Some concern has been expressed over the competition among Society journals for articles in particular areas (e.g., research on membrane transport appears in several of the specialty journals), but overall this appears to reflect an increase in options available to physiologists for publication in peer-reviewed journals of high quality.

About twenty percent of all papers published in *JAP* come from abroad. Whether greater editorial involvement of European scientists in *JAP: Respiratory, Environmental and Exercise Physiology* would be desirable was considered by an invited group during Fishman's president-elect trip to London in May 1985. This group of distinguished physiologists from different European countries was against establishing a separate editorial office in Europe but did advocate greater European involvement on the Editorial Board of the journal. Based on the responses to a questionnaire and a meeting of exercise and environmental physiologists at the Society's headquarters, the editor of the journal recommended to the Publications Committee that the subtitle of *JAP* be dropped. Thus, in January 1985, the title reverted to the *Journal of Applied Physiology* and the original purpose of the journal had been reaffirmed.

Prompted in part by the planning and then the success of dividing the journals according to specialties, interest was aroused in sectionalizing the Society. During a retreat held at Bar Harbor, Maine, in July 1976 at the invitation of President Bodil Schmidt-Nielsen, a committee chaired by Fishman recommended that steps be taken to initiate the Society's sectionalization. This move enabled the Publications Committee, under the chairmanship of Howard E. Morgan, to rely more heavily on advice from the newly formed sections of the Society before new editors were appointed. A "search committee" approach utilizing meetings with advisors and interviews with prospective candidates was developed.

By 1984 it was clear that the reorganization of the journals had added new vitality to the Society's periodicals. The number of pages published each year had grown from 8,121 in 1976 to 22,746 in 1986 and continues to increase. With growth, additional editorial board members and referees were necessary to ensure and improve quality. As of 1984, the 9 journals published by the Society depended on 9 editors, 40 associate editors, and 366 editorial board members, and here too the number continues to increase (Table 3).

New Societies and New Journals

During the past twenty-five years, new physiological societies have been formed as scientific research became more specialized and the needs of scientists for publication underwent change. New journals have evolved in parallel. The influence of new journals and of these changing needs may be illustrated by changes in the *Journal of Neurophysiology* and in the other Society publications that deal with the neurosciences.

The *Journal of Neurophysiology* was purchased after it was already established as a scholarly scientific publication (1). Under the auspices of the Society as owner and publisher, the journal has continued to grow in stature and in quantity published.

In 1973 the chief editor, John R. Brookhart, and the Publications Committee developed a plan to broaden the scope of the journal and for possible joint sponsorship with the newly formed Society for Neuroscience. With the approval of Council, Brookhart presented an informal proposal to the Communications Committee of the Society for Neuroscience. Items for consideration were a joint editorial board, enlarged scope of the journal, changes in frequency of publication (then bimonthly), change in the title of the journal, and joint financial responsibility. At the APS fall meeting, the Publications Committee discussed the matter with K. Krnjević, president of the Society for Neuroscience, and then Curran, chairman of the APS Publications Committee, outlined the committee's proposal in a letter to Krnjević for consideration by the Society for Neuroscience at their November meeting. The Society for Neuroscience failed to take action on the proposal.

A special APS Task Force on Neurophysiology met in March 1974. It recommended that the plan proposed by the Publications Committee be implemented and that discussions of joint sponsorship be continued with the Society for Neuroscience. When it became clear that the Society for Neuroscience would not reach a decision quickly, the Publications Committee decided that APS should go ahead with its own plan to expand the scope of the *Journal of Neurophysiology*. Edward V. Evarts was appointed chief editor in July and, with the journal's Editorial Board, developed a new stated aim of the journal.

The reorganization of *AJP* and *JAP* in 1976–77 did not affect the *Journal of Neurophysiology*. The Publications Committee, however, perceived a need to bolster this prestigious journal and to broaden its scope. Accordingly, William D. Willis, appointed chief editor in 1978, was encouraged to expand the Editorial Board and the coverage of the journal. This attempt was met with some success and, despite the appearance of numerous new journals in the neurosciences, the *Journal of Neurophysiology* became a monthly publication in 1980. Lorne M. Mendell, who succeeded Willis as chief editor in July 1983, has steadily moved this journal forward, and it continues to be a prestigious, highly cited publication.

Those involved in editing, reviewing, and publishing in the *Journal of Neurophysiology*, as well as other neurophysiologists within the Society, have generously given time and energy to the Society's other publications in neurophysiology: the *Handbook of Physiology* section on the nervous system, *Physiological Reviews*, and several of the specialty journals of *AJP*.

It is noteworthy that the names of certain members of the Editorial Board of the *Journal of Neurophysiology* also appear on the masthead of other neuroscience journals. Presumably these dual roles reflect the burgeoning of neuroscience and the need for alternatives to accommodate the explosive burst in this field. Also of interest is the continuing role of the Society's publications in fields that APS no longer dominates. In this capacity the publications clearly benefit science as a whole, as well as the particular interests of the Society.

International Cooperation

The journals and books of the Society are truly an international venture. Close to half of the subscribers to the journals and purchasers of the books are from outside the United States. Manuscripts come from all areas of the world where research is under way. Book editors, journal associate editors, members of the editorial boards, and reviewers from many countries share in the publishing activities. Perhaps nowhere else is this international cooperation more evident than in *Physiological Reviews*.

When H. S. Mayerson was chairman of the Editorial Board of *Physiological Reviews*, a European Editorial Committee was formally organized (1956). It was intended to extend the influence of the Editorial Board, which at that time was made up of appointees from APS and other United States societies. The European committee (Table 4) was supplemented in 1971 when corresponding members began to be appointed from other parts of the world. Mollie E. Holman, representing Australia, was the first corresponding member. Under the leadership of Howard E. Morgan and then Stanley G. Schultz (who, starting in January 1973, were designated as editor rather than chairman), the international nature was broadened still further. By the 1980s the journal had an American Editorial Board, representatives from other United States societies (e.g., American Society of Biological Chemists, Society of General Physiologists, American Institute of Nutrition, Biophysical Society, Society for Neuroscience, American Society for Cell Biology), a European Editorial Committee, and corresponding members from Australia and New Zealand, Japan, and Latin America. Each member of the board invites scientists to prepare manuscripts for the journal, sees to their review, and advises the author on necessary revisions before recommending to the editor that a manuscript be accepted for publication. This unique editorial approach has allowed *Physiological Reviews* to become a truly international journal.

Another aspect of international collaborative effort was the creation of a new journal, *News in Physiological Sciences* (see section Internal Communication).

Cumulative Indexes

Before 1976, cumulative author and subject indexes to *AJP* and *Physiological Reviews* were published irregularly. When they did appear, sales were small and availability was limited. In 1975 the Publications Committee decided that readily available indexes would enhance the use and value of the journals. At that time the decision was made to prepare indexes at appropriate intervals for all the Society's journals and to distribute them to the subscribers of the journal. The cost of the index would be added to the general expenses of producing the journal. Under this plan indexes were produced and distributed (Table 5).

Table 4
CHAIRMEN OF EUROPEAN EDITORIAL
COMMITTEE

Chairman	Term
E. Neil	1961–Dec 1969
D. R. Wilkie	Jan 1970–Dec 1973
D. S. Parsons	Jan 1974–Dec 1979
J. R. Bronk	Jan 1980–Dec 1985
B. R. Jewell	Jan 1986–

Table 5
CUMULATIVE INDEXES

Journal	Vol. No.	Years	Date of Publication
American Journal of Physiology	168–229	1952–75	Nov 1976
	230–240	1976–82	Nov 1983
Journal of Applied Physiology	1–39	1948–75	Nov 1976
	40–53	1976–82	Jul 1983
Journal of Neurophysiology	1–40	1938–77	Jan 1979
Physiological Reviews	47–61	1967–81	Oct 1982

Standardized Symbols, Abbreviations, and Glossaries

For clarity of communication the symbols, abbreviations, and glossaries commonly used in a scientific field should be standardized. Both international and national scientific societies have generally assumed the lead in this respect. APS, however, has played a minor role in this aspect of scientific communication, except for occasional publication of lists of symbols and abbreviations and glossaries in *JAP, AJP*, and the *Handbook of Physiology*. As a rule, these have been developed by the International Union of Physiological Sciences (IUPS) under the leadership of members of APS. On occasion, as in the case of respiration symbols, the initiative in standardization was taken almost entirely by members of the Society with *JAP* as the publication vehicle.

Internal Communication

A society needs some method for informing members of society affairs. For APS, *The Physiologist* was begun in 1958 as a newsletter to fulfill this purpose. Various departments were added to this publication, and supplements appeared periodically. Through 1985, it not only contained information about the Society, but it also served as a vehicle for dissemination of items on history, education, public affairs, and science. In 1986, with the initiation of *News in Physiological Sciences* (see below), *The Physiologist* became a sixteen-page newsletter designed to keep members informed of happenings in the Society and of events related to their academic pursuits.

The first editor of *The Physiologist* was Ray G. Daggs, the executive secretary-treasurer. The cost was borne by the Publications General Fund until 1972, when it was transferred to the Society Operating Fund. Daggs retired as editor at the end of 1972. Orr E. Reynolds became editor in 1973. The content was expanded, particularly with respect to the teaching and the history of physiology, and the format changed.

In 1979 the Publications Committee was prompted by Council to review *The Physiologist* and to make recommendations for changes. The stimulus for the initiative was the increasing scientific content of *The Physiologist*, its increasing cost, and its uncertain place in the total span of the Society's publications. The committee reaffirmed the need for a house organ but suggested greater circumscription in content. It favored inclusion of the Bowditch Lecture, the historical series, and the abstracts of the fall meeting.

Discussions between the Publications Committee and Council on the future of *The Physiologist* continued in 1980 and 1981. One result of these discussions was the charge by Council to the Publications Committee of providing help to Orr Reynolds in the processing of manuscripts for *The Physiologist*. In 1981 the Publications Office began logging manuscripts and by the end of the year was supplying copyediting help. Overall the coverage of *The Physiologist* remained unchanged. By the end of 1983, however, strong advocacy developed in the Publications Committee for transforming *The Physiologist* into a general review journal that would include not only news of the Society but would resemble in content the "trends" journals, which have become popular. Coincidentally IUPS was also weighing the prospects for publishing a journal that seemed to be similar in intent and content. Therefore representatives met to discuss joining forces for such a publication. Knut Schmidt-Nielsen, Klaus Thurau, and Heinz Valtin represented IUPS; Howard Morgan, Alfred Fishman, and Orr Reynolds were spokesmen for APS.

Negotiations were complicated by the fact that the two organizations had been considering different approaches to accomplishing similar goals. For example, a contract that IUPS was considering relied on APS member subscriptions, yet made no provision for the future of *The Physiologist*. IUPS urged APS not to convert *The Physiologist* into a competing journal and proposed that APS buy and distribute the IUPS publication. At the August 1983 Society Meeting in Hawaii, Council rejected the proposal that *The Physiologist* continue unchanged. Instead it requested that the Publications Committee

> submit recommendations of proposed changes in the format and content of *The Physiologist* including the organization of its editorial management with specific reference to including a Trends Section for consideration of Council at its retreat in November.

Subsequently the proposal was modified in anticipation of meetings scheduled with IUPS in Australia, but the inclination of Council to revamp *The Physiologist* was clear.

In November 1983, the Publications Committee presented to Council a draft agreement between IUPS and APS for a trends-type publication owned by IUPS and published by APS. Included with the draft agreement were financial projections for the joint project. An alternate proposal for changing *The Physiologist* into a trends-type publication with a truncated section of APS material, all owned and published by APS, was also presented. Council turned down the proposed IUPS-APS draft agreement. It instructed the Publications Committee to transform *The Physiologist* into a publication that would focus on reviews and Society affairs, to appoint an editor, and to assume responsibility for the new journal as it does for the other journals published by the Society. January 1986 was set as the date for the first issue of the new journal. At that meeting, Alfred Fishman was asked to organize a committee on international physiology and to continue discussions on publications with IUPS through that committee.

Discussions continued in 1984 between IUPS and APS in an attempt to develop a collaborative effort that would be of mutual benefit. To find an editor for the new journal, the Publications Committee formed a Search Committee, which would be useful whether the publication was to be produced by APS alone or jointly with IUPS. The Search Committee met in St. Louis in April and drew up an international roster of candidates. Meanwhile discussions continued between representatives from IUPS and APS in the attempt to draw up a contract that would satisfy both sides. In August the Publications Committee reversed itself and recommended that the Society not publish a trends-type journal at that time, either by itself or with IUPS.

Instead it offered to encourage and help IUPS start its own journal. It also recommended that *The Physiologist* continue in its usual form through 1987, while the Publications Committee continued to assess the needs of the Society with respect to the future of *The Physiologist*. The Council rejected these proposals and requested instead that Howard Morgan, chairman of the Publications Committee, undertake to mold the proposed agreement with IUPS into a mutually acceptable form.

After many discussions and considerations that focused sharply on financial projections, a working plan was set into motion. A Joint Managing Board with four members was formed to oversee the project. Its first meeting was convened in Washington, D.C., in December 1984. Representing APS were its president (John B. West) and the chairman of the Publications Committee, who incidentally was president elect (Howard E. Morgan); representing IUPS were the outgoing treasurer (Klaus Thurau) and his successor (Heinz Valtin). (It was decided that henceforth one of the APS representatives should be either the chairman of the Publications Committee or someone designated by the chairman.) Based on the recommendations of the Search Committee, chaired by Robert M. Berne, Knut Schmidt-Nielsen was appointed the first editor of the publication by the Joint Managing Board. At his suggestion, *News in Physiological Sciences (NIPS)* was selected as the name for the new journal.

A formal agreement was signed by the presidents of IUPS and APS in January 1985. It was agreed that *NIPS* would be a forty-eight-page publication and that *The Physiologist*, under the supervision of the Publications Committee, would become a sixteen-page newsletter. Each would appear bimonthly. Under the terms of this agreement, IUPS and APS are the joint owners and copyright holders of *NIPS*, and APS acts as the managing publisher. IUPS will provide $37,000 during the first two years for start-up and production of the journal. APS will cover all other expenses. The cost to APS members of *NIPS* and *The Physiologist*, including the fall meeting abstracts, will be no more than the cost of *The Physiologist* before *NIPS*. Additional costs will be recovered from publication funds.

Martin Frank was appointed by the Publications Committee as the editor of *The Physiologist* in July 1985.

Clinical Physiology

The Society ended its first one hundred years with two active projects to bridge clinical physiology and basic science: "Physiology in Medicine," appearing in *Hospital Practice*, and the Clinical Physiology Series of books (see section Evolution of Book Program).

"Physiology in Medicine" evolved from *Physiology for Physicians*, a succession of single articles published, sold, and distributed by the Society beginning in 1963. Julius H. Comroe, Jr., was the first chairman of its Editorial Board; Eugene A. Stead, Jr., succeeded him in 1964. Then, in 1966, APS and the American Society for Pharmacology and Experimental Therapeutics combined their efforts and merged *Physiology for Physicians* and *Pharmacology for Physicians* into *Physiology and*

Pharmacology for Physicians. Comroe was the cochairman for physiology in the first half of the year; then Irwin H. Page took over. Joint publication ceased in December 1966, and in 1967 *Physiology for Physicians* began to appear as a bimonthly section of the *New England Journal of Medicine,* with Page serving as editor.

In 1970 the editorship, orientation, and title of the series again changed—this time to "Physiology in Medicine" with the new editor, Alfred P. Fishman, directing the series more toward research-oriented physicians. In 1972 the series became a monthly feature of the journal. A change of editors of the *New England Journal of Medicine* in 1977 put the series in jeopardy; the new editor wished to change the nature and style toward the *Annual Reviews* format and broaden the coverage from physiology to all basic science. The series ended in May 1978, except for one additional article that appeared under the title in 1979, even though it was not invited by Fishman. Seventy-two articles were published between September 1970 and May 1978, and the series achieved great popularity and distinction.

For several years a new home was sought for the series. Episodic discussions with David W. Fisher, editorial director of *Hospital Practice,* and Howard E. Morgan, chairman of the Publications Committee, culminated in an agreement in 1982. Thomas E. Andreoli was appointed editor of "Physiology in Medicine," and the series was envisaged as emphasizing common ailments with a major focus on interests of general physicians. The series began to appear in *Hospital Practice* in 1984 and continues in an attractive format in this widely distributed journal.

Teaching Physiology

The Society's interest in the teaching of physiology has been expressed through the publication of *The Physiology Teacher,* begun to encourage laboratory instruction in physiology. It first appeared as a separate four-page publication in April 1971 under the editorship of Nancy S. Milburn. The publication was the direct responsibility of the Education Committee, to whom it had been delegated by Council.

In January 1973 Orr E. Reynolds became editor and William DeHart executive editor. So that *The Physiology Teacher* could be brought more to the attention of Society members, it was included in *The Physiologist* beginning in January 1978, but it could still be purchased separately. Unfortunately, because many of the readers were members of the Society, the number of separate subscribers declined precipitously. Because of financial problems and the difficulty of securing adequate material for publication, in April 1978 *The Physiology Teacher* was terminated as a separate publication and became a section of *The Physiologist.*

Mary Dittbrenner, who had served as executive editor of *The Physiology Teacher* for several years, was succeeded by M. C. Shelesnyak. He and Orr Reynolds continued to publish abstracts of review articles and educational materials in physiology (1976–80), reviews of books useful to those involved in teaching physiology, tutorial lectures, and, increasingly, information on the use of computers in teaching physiology. A separate publication, *Frontiers in the Teaching of Physiology: Com-*

puter Literacy and Simulation, edited by M. C. Shelesnyak and Charles S. Tidball (1981), was based on articles that appeared in *The Physiology Teacher*.

The April 1984 issue of *The Physiologist* was the last in which the names of the editor, Editorial Board members, or executive editor were listed. In February 1986, with the appearance of *NIPS* and the conversion of *The Physiologist* to a newsletter, the Society no longer had a vehicle for publishing material on the teaching of physiology.

At the same time, some of its members were pressuring the Society to become more involved in teaching, as well as research, in physiology. They were asking that a section of the Society be formed for those interested in the teaching of physiology, and the Association of Chairmen of Departments of Physiology was criticizing the Society for not being more active in teaching. Whether these constituencies will stimulate the rebirth of *The Physiology Teacher* or a related publication remains to be seen.

Evolution of Book Program

The Society entered its fourth quarter century with a blossoming *Handbook of Physiology* series under the direction of Maurice B. Visscher (1). He retained the chairmanship of the Handbook Editorial Committee through 1966, by which time Section 6, *The Alimentary Canal*, was nearing completion (Table 6). Alfred P. Fishman succeeded Visscher in 1967. Fishman revitalized the project, which was joined by Stephen Geiger as Handbook executive editor, initiated additional new sections, and began preparing revisions of the original sections. John R. Pappenheimer became chairman in July 1972 and held that position until his second term ended in June 1978. He oversaw the completion of additional books in the series and paid particular attention to the revision of Section 2, *Circulation*, retitled *The Cardiovascular System*. When his term ended, the Publications Committee decided to work directly with each editor or steering committee rather than to appoint a new chairman. Under this more direct supervision, new sections and revisions of existing sections continue. By the late 1970s, it had become clear that the total number of handbooks being sold, as was true for similar books being published, was not as large as it had been in the 1960s. Factors influencing this change are the more specialized nature of the revised volumes, increased price of the books, less money available for library acquisitions and purchase on grants, and more competition. Despite this more restricted distribution, the books remain invaluable, and many members of APS believe that their termination would be a great loss.

Based on the early success of the handbooks, and on previous experience by the Society in book publishing, the Publications Committee wished to expand its book-publishing program. Several projects were considered in the early 1970s. It was not until 1977, however, that the first book in what was to become the Clinical Physiology Series appeared (Table 7), when the Publications Committee recommended, and the Finance Committee and Council approved, the publication of *Disturbances in Body Fluid Osmolality*. These three groups and the Task Force on Clinical Sciences

Table 6
HANDBOOK OF PHYSIOLOGY

Section No.	Section Title	Editors Section	Editors Volume	Editors Executive	Vol. No.	Year of Publication
		First editions				
1.	*Neurophysiology*	J. Field[a]		V. E. Hall	I	1959
		H. W. Magoun			II	1960
					III	1960
2.	*Circulation*	W. F. Hamilton		P. Dow	I	1962
					II	1963
					III	1965
3.	*Respiration*	W. O. Fenn			I	1964
		H. Rahn			II	1965
4.	*Adaptation to the Environment*	D. B. Dill		C. G. Wilber		1964
		E. F. Adolph[b]				
5.	*Adipose Tissue*	A. E. Renold				1965
		G. F. Cahill, Jr.				
6.	*Alimentary Canal[c]*	C. F. Code		W. Heidel		
	Control of Food and Water Intake				I	1967
	Secretion				II	1967
	Intestinal Absorption				III	1968
	Motility				IV	1968
	Bile; Digestion; Ruminal Physiology				V	1968
7.	*Endocrinology*	R. O. Greep		S. R. Geiger		
		E. B. Astwood				
	Endocrine Pancreas		D. F. Steiner		I	1972
			N. Freinkel			
	Female Reproductive System		R. O. Greep		II	1973
	Thyroid		M. A. Greer		III	1974
			D. H. Solomon			
	The Pituitary Gland and Its Neuroendocrine Control		E. Knobil		IV	1974
			W. H. Sawyer			
	Male Reproductive System		D. W. Hamilton		V	1975
			R. O. Greep			
	Adrenal Gland		H. Blaschko		VI	1975
			G. Sayers			
			A. D. Smith			
	Parathyroid Gland		G. D. Aurbach		VII	1976
8.	*Renal Physiology*	J. Orloff		S. R. Geiger		1973
		R. W. Berliner				
9.	*Reactions to Environmental Agents*	D. H. K. Lee		S. R. Geiger		1977
		H. L. Falk[b]				
		S. D. Murphy[b]				
10.	*Skeletal Muscle*	L. D. Peachey		S. R. Geiger		1983
		R. H. Adrian[b]				
11.	*Cell and General Physiology*	J. F. Hoffman		S. R. Geiger		
		J. S. Cook				

Table 6—*Continued*

Section No.	Section Title	Editors Section	Editors Volume	Editors Executive	Vol. No.	Year of Publication
		Revisions				
1.	*The Nervous System*	J. M. Brookhart[d] V. B. Mountcastle		S. R. Geiger		
	Cellular Biology of Neurons		E. R. Kandel		I	1977
	Motor Control		V. B. Brooks		II	1981
	Sensory Processes		I. Darian-Smith		III	1984
	Intrinsic Regulatory Systems of the Brain		F. E. Bloom		IV	1986
	Higher Functions of the Brain		F. Plum		V	1987
	Development of the Nervous System		M. W. Cowan		VI	
2.	*The Cardiovascular System*	A. C. Barger[e] R. M. Berne[e] D. F. Bohr[e] A. C. Guyton[e] J. T. Shepherd[e]		S. R. Geiger		
	The Heart		R. M. Berne N. Sperelakis[b]		I	1979
	Vascular Smooth Muscle		D. F. Bohr A. P. Somlyo H. V. Sparks, Jr.		II	1980
	Peripheral Circulation and Organ Blood Flow		J. T. Shepherd F. M. Abboud		III	1984
	Microcirculation		E. M. Renkin C. C. Michel		IV	1984
3.	*The Respiration System*	A. P. Fishman		S. R. Geiger		
	Circulation and Nonrespiratory Functions		A. P. Fishman A. B. Fisher		I	1985
	Control of Breathing		N. S. Cherniack J. C. Widdicombe		II	1986
	Mechanics of Breathing		P. T. Macklem J. Mead		III	1986
	Gas Exchange		L. E. Farhi S. M. Tenney		IV	1987
6.	*The Gastrointestinal System*	S. G. Schultz		S. R. Geiger		
	Endocrinology		G. H. Makhlouf		I	
	Motility and Circulation		J. D. Wood		II	
	Salivary, Gastric, Pancreatic, and Hepatobiliary Secretion		J. G. Forte		III	
	Intestinal Transport		M. Field R. A Frizzell		IV	
8.	*Renal Physiology*	E. E. Windhager		S. R. Geiger		

[a] Editor-in-chief. [b] Associate editor. [c] Editorial Committee: J. R. Brobeck, R. K. Crane, H. W. Davenport, M. I. Grossman, H. D. Janowitz, C. L. Prosser, and T. H. Wilson. [d] Through vol. III. [e] Steering Committee member.

(later the Section on Physiology in Clinical Science) saw this type of book as a bridge between clinical medicine and basic science. Also, rapidly changing clinical areas could be dealt with more expeditiously than in the *Handbook of Physiology*.

Most of the early books in the Clinical Physiology Series were derived from

Table 7
CLINICAL PHYSIOLOGY SERIES

Title	Editor(s)	Year of Publication
Disturbances in Body Fluid Osmolality	T. E. Andreoli J. J. Grantham F. C. Rector, Jr.	1977
Disturbances in Lipid and Lipoprotein Metabolism	J. M. Dietschy A. M. Gotto, Jr. J. A. Ontko	1978
Pulmonary Edema	A. P. Fishman E. M. Renkin	1979
Secretory Diarrhea	M. Field J. S. Fordtran S. G. Schultz	1980
Disturbances in Neurogenic Control of the Circulation	F. M. Abboud H. A. Fozzard J. P. Gilmore D. J. Reis	1981
New Perspectives on Calcium Antagonists	G. B. Weiss	1981
High Altitude and Man	J. B. West S. Lahiri	1984
Interaction of Platelets With the Vessel Wall	J. A. Oates J. Hawiger R. Ross	1985
Effects of Anesthesia	B. G. Covino H. A. Fozzard K. Rehder G. Strichartz	1985
Physiology of Oxygen Radicals	A. E. Taylor S. Matalon P. A. Ward	1986
Atrial Hormone and Other Natriuretic Factors	P. J. Mulrow R. W. Schrier	1987

symposia held at the spring meeting under the auspices of the Section on Physiology in Clinical Science. Thomas E. Andreoli served as the first chairman of the section; he was succeeded by Francois M. Abboud. In 1981 Council reorganized this group as a subcommittee of the Publications Committee, because its main function has been to organize symposia with book publication as a primary goal. Abboud remained as chairman through June 1983, when George F. Cahill, Jr., assumed the leadership. Julien F. Biebuyck succeeded him in July 1986.

Table 8
SPECIAL PUBLICATIONS

	Editor(s)	Year of Publication
Animal Welfare		
Animal Pain: Perception and Alleviation	R. L. Kitchell H. H. Erickson E. Carstens* L. E. Davis*	1983
Animal Stress	G. P. Moberg	1985
People and Ideas		
Circulation of the Blood: Men and Ideas†	A. P. Fishman D. W. Richards	1982
Endocrinology: People and Ideas	S. M. McCann	1987
Renal Physiology: People and Ideas	C. W. Gottschalk G. H. Giebisch R. W. Berliner	1987
Membrane Transport in Physiology: People and Ideas	D. C. Tosteson	1987
Centennial Publications		
History of the American Physiological Society: The First Century, 1887–1987	J. R. Brobeck O. E. Reynolds T. A. Appel	1987
Physiology in the American Context, 1850–1940	G. L. Geison	1987
Others		
Excitation and Neural Control of the Heart	M. N. Levy M. Vassalle	1982
Voltage and Patch Clamping With Microelectrodes	T. G. Smith, Jr. H. Lecar S. J. Redman P. W. Gage	1985

* Associate editor.   † Reprint edition.

Further diversification of the book program occurred in 1982 with publication of *Excitation and Neural Control of the Heart* (Table 8). This volume contains updated versions of review articles that appeared in *AJP: Heart and Circulatory Physiology* between 1977 and 1981. This was followed by the reprinting of *Circulation of the Blood: Men and Ideas,* by Alfred P. Fishman and Dickinson W. Richards. Originally published by Oxford University Press in 1964, with the endorsement of the APS Centennial Celebration Committee, the Publications Committee oversaw the reprinting of this elegant, out-of-print book. A new preface emphasizes both the pivotal role of Harvey in the beginning of modern physiology and also the celebration of the Society's centennial.

The momentum resulting from the preparation and appearance of *Circulation of the Blood: Men and Ideas* and the historical interest of the Society engendered by the approaching centennial led to plans for other books of this type (i.e., a People and Ideas Series). Three books were commissioned: on endocrinology, on renal physiology, and on membrane transport.

Two additional books directly related to the centennial were authorized. One, of which this chapter is part, is a history of the first one hundred years of APS. The second book, *Physiology in the American Context, 1850–1940*, was derived from a workshop sponsored by the Society at the National Library of Medicine in January 1986.

As part of the Society's continuing interst in animal welfare, two books have been published. The first was the outgrowth of a symposium at the 1982 spring meeting on pain perception and alleviation. It was jointly sponsored by APS, the American Veterinary Medical Association, and the American Society for Pharmacology and Experimental Therapeutics. The book, *Animal Pain: Perception and Alleviation*, was published in 1983. The second book, *Animal Stress*, based on a symposium held in Davis, California, in 1983, was published in 1985.

Voltage and Patch Clamping With Microelectrodes, published in 1985, represents a further departure from the kind of book heretofore published by the Society. Rather than concentrating on physiological results, it is designed to help researchers select the best technique to use in studying the electrical properties of cells. It contains information on theory and practice, the advantages of each type of clamp, and "how-to" directions, as well as considerations of the limitations and artifacts of each system

In 1986 the Publications Committee undertook a major evaluation of the Society's book-publishing program and began to plan for the future.

Ethics

Ethical concerns haunt editors, publishers, and users of published material. Detection of misuse of animals, fabrication of data, plagiarism, selective reporting, and manipulation of data are distressing to reviewers, editors, and publishers and may be disastrous for the perpetrator. Also troublesome in recent years has been the uncovering of instances of undeserved authorship, multiple publication of the same material, abuse of confidentiality, and misuse of statistics.

Editors and reviewers have a responsibility to report unethical practices when they are detected. Although primary responsibility for prevention and detection resides with the scientists and the institutions in which they work, the editors and reviewers are charged with reporting to the Publications Committee any infringe-ments in ethical practice. The Publications Committee, in turn, handles such accusations by following policies and practices developed by the Society and similar scholarly publishers.

Peer Review

Manuscripts received by editors of Society publications normally are sent to two reviewers. If the editor receives divergent views from them, one or more additional reviewers are consulted to arrive at a decision. Reviewers serve as advisors to the editor; final decision on the scientific quality of a manuscript is the responsibility of the editor. Occasionally the Publications Committee has interceded to ensure dispassionate review on behalf of an author who believes his or her work has been wrongly judged. Each manuscript is treated as a confidential document, and every effort is made to reach a prompt decision on its acceptability or need for revision.

Fraud

Publications of the Society have been faced with several instances of fraud in recent years. These have ranged from outright plagiarism or tampering with data, on the one hand, to the fragmentation of publications and unjustified authorship, on the other. Each instance has been handled individually. But, as a general rule, the institution where the work was done has been asked to ascertain the validity of data in question and to certify that fraud was, or was not, involved. With this certification in hand, retractions or corrections for several articles published in APS journals have been published. Major abstracting and information retrieval services have been informed of the retractions and have added cross-references from the original abstract to the retraction in their computer data bases. Cross-references between the original article and retraction are included in the cumulative index. Howard E. Morgan's experience as chairman of a major National Institutes of Health panel investigating the fabrication of research findings was invaluable in helping the Publications Committee develop procedures for handling accusations of fraud.

Several editorials on the use and abuse of statistics have been published in the Society's journals, and general awareness has been heightened by reviews of relevant books. Specific issues have been highlighted by a series of statistics reports that began appearing in *AJP* in 1983.

Animal and Human Welfare

The history of the Society's concern about animal and human welfare is treated in detail elsewhere in this volume. As policies have been formulated by the Society, they have been distributed to departments of physiology and printed in the journals. Each issue of the research journals contains the "Recommendations from the Declaration of Helsinki" and the "Guiding Principles in the Care and Use of Animals." Reviewers are asked to examine each manuscript for violations of these guidelines through a specific question that appears on the sheets they use to comment on a manuscript under review. Editors are expected to refuse to publish manuscripts in which adherence to these principles is not apparent.

Technological Changes

In the 1970s and 1980s, great changes occurred in the printing industry that influenced how the Society publishes and prints its publications. In the early 1970s, the journals and books were hot-metal composed (Monotype and Linotype) and letterpress printed. By the mid 1970s photocomposition and offset printing were used to produce all the journals. Electronic composition then replaced photocomposition. Computers not only drive the composition equipment but are used for hyphenation and justification and for automated makeup of tables and pages. In 1983–84 the Society for the first time accepted manuscripts on disks for the book *Voltage and Patch Clamping With Microelectrodes.* By 1985 some journal articles were being set from disks without being rekeyed.

Computers were introduced to handle journal subscription fulfillment and to produce mailing lists in 1977. They began to be used to keep track of manuscripts and related functions toward the end of 1982 and for reprint ordering and for recording charges to authors in 1983.

A license agreement for the full-text storage and on-line retrieval of *JAP* and of *The Respiratory System* section of the *Handbook of Physiology* was negotiated by Howard E. Morgan, on behalf of the Society, with BRS/Saunders. The agreement was signed toward the end of 1985 and then amended to include *AJP* in 1986. The Society signed an agreement with University Microfilms International in 1986 for the research journals to become part of a similar program, based on optical disk storage.

The Society as Publisher

The Society is justifiably proud of its publications. During the past twenty-five years, particularly the last ten, the number of books and journals and their diversity have increased greatly. Quality has been maintained at the highest level. The membership, authors, editors, reviewers, and Publications Committee members have all featured prominently in this success, along with the dedicated commitment of the publications staff, printers, and others in the Bethesda office.

However, it should be noted that the journals and books of APS prospered during a period of upheaval in publishing, uncertainty about future directions of libraries, and dwindling university resources. Scientists and librarians have changed the ways in which they use and purchase journals and books. Inflation has placed severe strain on publishers.

Another dramatic change has been in the use of photocopying. The current indulgences were unimaginable twenty-five years ago. Duplicate subscriptions to journals have virtually disappeared as library budgets for journal and book acquisitions failed to keep pace with increases in the amount published and with inflation. Librarians have formed networks and consult with colleagues to avoid duplication of purchases by neighboring libraries. As a result, institutional departments and individuals rely less on their own journals and books, whereas scientists and librarians rely more heavily each year on information retrieval services. The current scene is characterized by libraries investing large amounts of money in the purchase or use of electronic equipment and in services that speed and increase the ability of scholars to retrieve information from the published literature and from various data bases. Simultaneously they are absorbing the cost of instructing patrons in their use.

Despite these new approaches to information storage and retrieval, the need persists for scholarly and timely publications. The finances of the Society's publications reflect this continuing need and remain strong. Over the years, interest income from the Publications Contingency and Reserve Fund (see chapter on finances) has enabled the Society to take on many projects that might not otherwise have been attempted because of the financial risk. In addition, the operating fund for the journals is now large enough to provide the needed working capital and leave some money for new opportunities and emergencies. In the past, publications funds have emerged as a recurrent temptation for Council to use for new ventures unrelated to publications. But the changing times are also a warning that circumspection will be required in years ahead if the Society is to maintain its lead in new approaches to information transfer. These funds must be used judiciously. It is the intention of the Publications Committee to review repeatedly the elements of its publishing program—books as well as journals—to keep the program trim and commensurate with the needs of physiologists. As the Society looks ahead, it must recognize that the role of publishers as distributors of scientific information is decreasing. Those who now provide access to this information, including distributors previously inexperienced in publishing, have changed scholarly communication. Because of these changes, new understanding and relationships with librarians and information retrieval groups are being explored. Changes are anticipated without loss of awareness by the Society of its responsibility as producer and guardian of what is published in physiology. Innovation and experimentation in publishing will be required in the years ahead if the Society is to maintain publications as one of its major reasons for being.

REFERENCE

1. FENN, W. O. *History of the American Physiological Society: The Third Quarter Century, 1937–1962.* Washington, DC: Am. Physiol. Soc., 1963, p. 76ff.

CHAPTER 12

Educational Programs

ARTHUR J. VANDER AND ORR E. REYNOLDS

The years 1963 to the present have been marked by a continuation of a variety of educational activities initiated earlier by the American Physiological Society (APS, the Society) (1), as well as by important ventures into entirely new educational realms. Throughout this period, the Education Committee has been the major focus of these activities, although two of the very important educational programs have been administered by the Publications Committee—the series known originally as *Physiology for Physicians* and the *Handbook of Physiology* series. Both are described elsewhere in this volume, as are the Porter Fellowship and Development Programs, with which the Education Committee was involved.

Education Committee and Education Office

The Society's Committee on Education, established in 1953, reflected the growing view by members that the Society should promote a variety of educational activities at all academic levels, both at the meetings of the Society and outside the context of those meetings. Membership of the committee since 1963 is shown in Table 1.

Beginning in 1956 with his appointment to the office of executive secretary-treasurer of the Society, Ray G. Daggs provided the Education Committee with extensive staff support. He acted also as executive director of the educational programs. The close involvement of Daggs with the Education Committee was a natural sequel of his early academic experience. After serving as a teaching assistant during graduate school, Daggs was called to the University of Vermont to assist in reorganizing and strengthening the Department of Physiology. Service in the U.S. Army sharpened his considerable administrative abilities. When he moved to Bethesda to the newly created full-time position at the APS secretariat, he immediately established the precedent of a close liaison between that position and the education activities of the Society.

In 1969 Council undertook an intensive long-range planning for the Society. A major conclusion was that one of the best contributions the Society could make to

Table 1
EDUCATION COMMITTEE, 1962–86

	Year		Organization Represented
	Member	Chairman	
J. R. Brobeck	1960–62	1960–62	
R. B. Tschirgi	1959–67	1963–67	
C. A. M. Hogben	1959–63		
R. R. Ronkin	1959–67		SGP
P. R. Morrison	1961–64		
L. H. Marshall	1961–69		
D. C. Tosteson	1961–67		SGP
J. R. Brookhart	1962–63	1962–63	
R. P. Forster	1962–63		ASZ
G. Richards	1962–63		ASZ
C. G. Wilber	1963–67		
W. C. Randall	1963–70		
D. M. Maynard, Jr.	1963–65		ASZ
D. K. Detweiler	1964–68		
I. J. Deyrup-Olsen	1966–74		ASZ
	1967–70		SGP
C. G. Stephens	1966–72		ASZ
A. C. Guyton	1967–70	1967–70	
J. A. F. Stevenson	1967–70		
E. W. Hawthorne	1967–69		
J. W. Green	1967–70		SGP
L. E. McDonald	1968–71		
A. R. Dawe	1968–71		
J. L. Kostyo	1970–76	1970–76	
L. Share	1970–76		
D. F. Bohr	1970–73		
B. A. Curtis	1970–75		SGP
P. B. Dunham	1970–73		SGP
	1973–75		
A. Ames III	1971–74		
F. N. Briggs	1971–77		
P. Farnsworth	1972–75		ASZ
	1975–78		
L. S. Jefferson	1973–79	1976–79	
B. Kaminer	1973–76		SGP
R. B. Gunn	1974–81		
R. H. Kellogg	1975–78		
B. M. Twarog	1975–81		
E. M. Renkin	1976–79		
F. N. White	1976–79		
S. G. Schultz	1977–79		
S. Ochs	1977–80		
J. H. Szurszewski	1978–81		
A. J. Vander	1978–82	1979–82	
J. B. Bassingthwaighte	1979–82		
E. O. Feigl	1979–82		
P. S. Timiras	1979–82		
J. B. West	1979–82		
J. C. Houk	1980–83		
J. A. Michael	1981–84		

Table 1—*Continued*

	Year		Organization Represented
	Member	Chairman	
J. E. Randall	1981–84		
J. A. Spitzer	1981–84	1982–84	
B. A. Horwitz	1982–85		
P. M. Hogan	1982–85		
A. H. Mines	1982–85		
M. Siegman	1983–88		SGP
M. Anderson-Olivo	1983–88		SGP
H. I. Modell	1983–86		
H. G. Hempling	1984–87		
C. F. Rothe	1984–87		
F. L. Abel	1985–87		
H. S. Pitkow	1985–88		
W. S. Spielman	1985–88		

SGP, Society of General Physiologists; ASZ, Comparative Physiology Division of American Society of Zoologists.

strengthening the role of physiology departments and physiologists was in the realm of education. Acting in 1970 on the conviction that the activities of the Education Committee should be extended and strengthened (13), Council established an Office of Education within the Society's table of organization, and Orr Reynolds was appointed education officer. On the retirement of Daggs at the end of 1972, Reynolds, while remaining education officer, was also appointed executive secretary-treasurer. The existence of an education officer and Education Office permitted the Society's programs in education to maintain momentum and to grow in size and scope, with the support of grants and contracts obtained from foundations and governmental agencies.

During its most active period (1970–81) the Education Office maintained a staff of three people in addition to the education officer. Although the personnel of the office changed from year to year as the focus of activities changed, the staff usually included one person with physiological training, a graphic artist, and a clerical or administrative assistant. With the cessation of financial support for specific programs, however, the staff of the Education Office was disbanded in 1981.

Brochures on Careers in Physiology

A recurring mission of the Education Committee has been to prepare materials describing careers in physiology. Indeed the preparation of a very simple one-page handout for distribution to vocational counselors had been the newly created committee's initial activity in 1954. In 1960 a more extensive brochure was prepared, printed, and widely distributed on request to schools and individuals. It was criticized, however, as being too long, having no theme, and being not specific enough. Accordingly, as the supply was approaching exhaustion, Louise H. Marshall was appointed to the Education Committee with the responsibility of developing a series of illustrated brochures that would present the many-faceted career opportunities in physiology. The first, "Consider Physiology," appeared in 1963; copies were mailed to the 50,000 members of the American Association of Biology Teachers.

In addition to this general brochure, aimed mainly at high school students, a variety of folders was written by specialists actively working in each area. Another folder, largely inspired by Ray Daggs, "Choosing a Career," provided a timetable for making educational decisions all the way from high school to postdoctoral training. Yet another folder in the series, "Physiology in Colleges and Graduate Schools," detailed areas of training and requirements for the Ph.D. degree in physiology and provided a table listing institutions where specific training in subdisciplines of physiology could be obtained. (This table has been updated several times, most recently in 1983.)

In 1973 "Consider Physiology" was replaced with a completely new brochure. In "Careers in Physiology," F. N. Briggs emphasized the "intellectual pleasures" of a career in physiology and gave more information on the breadth of application of physiology. Finally, to update the material and approach (and to eliminate gender-oriented language), yet another new careers brochure was prepared in 1983 by the Education and the Career Opportunities in Physiology Committees under the guidance of David J. Ramsay. Ramsay also edited a collection of articles that appeared under the aegis of the Career Opportunities in Physiology Committee in *The Physiologist* between 1980 and 1983 (14).

Refresher Courses

The full-day refresher course at the autumn meeting has been, since 1954, the most enduring of all the Society's educational activities. The topic and the organizer for each refresher course are selected by the Education Committee (Table 2). The courses are designed to update a nonspecialist in the subject matter and techniques of a specific area of physiology and to provide information of particular use in teaching.

Teaching Sessions and the Learning Resources Center

Annual half-day teaching sessions at the spring meetings actually antedated the refresher courses; they were begun in 1951, before the existence of the Education Committee. Their object was to provide a forum for demonstration and discussion of teaching materials and techniques. Organized after 1953 by a chairman selected by the Education Committee, the annual session might deal with a general topic, a mode of instruction, or an education-related question (see Table 3).

Many of the presentations at the teaching session were made available to a wider audience through publication in *The Physiologist*. Moreover the 1973 teaching session, organized by Charles Tidball on computer-assisted education, was published by the Society as a separate booklet (19). In 1974, rather than a structured program with invited lectures, the teaching session consisted of voluntarily contributed papers on educational problems and innovations. (Interestingly, this marked a return to the

Table 2
REFRESHER COURSES, 1963–85

Title	Year	Organizer(s)
Comparative Physiology	1963	C. E. Lane
Physiological Psychology	1964	C. Pfaffmann
Use of Avian and Large Domestic Animals in Physiological Research	1965	J. E. Nellor
Mathematical and Computer Analyses of Circulatory and Body Fluid Control	1966	A. C. Guyton
Intercellular Communication	1967	K. Frank
Bioenergetics and Cell Organelles	1969	J. W. Green
Neuroendocrine Integration	1971	W. F. Ganong
Physiology of Human Reproduction	1972	E. Knobil
Physiologic Adaptations to the Environment	1973	B. Schmidt-Nielsen
The Structure and Function of the Visual System	1974	L. Wolbarsht
Principles of Organization in Sensorimotor Systems	1976	D. Humphrey
Mechanisms and Regulation of Gastrointestinal Secretion and Absorption	1977	L. Johnson
Smooth Muscle Physiology	1978	F. Fay
Some Aspects of Exercise Physiology	1979	E. Buskirk
Physiology of Calcium and Phosphate Regulation	1980	F. Knox
The Physiology of Aging	1981	P. Timiras
Microcirculation	1982	B. R. Duling
Physiology and Biochemistry of Receptors	1983	J. A. Spitzer
Anaerobic Threshold	1984	L. B. Gladden and D. R. Richardson
Exercise Physiology and Its Clinical Applications	1985	J. Naughton and D. Pendergast

format of the very first teaching session in 1951.) A similar unstructured program was scheduled for 1975, but because of a paucity of voluntary contributions submitted, the session was used instead for an "open-house," multiple-carousel presentation of the Society's recently completed first group of slide-tapes (see below). This was also done in 1977. (There was no teaching session in 1976.)

Then, in 1978, stimulated by the teaching exhibit at the IUPS Congress in Paris in 1977, the Education Committee decided on a new format. Exhibit hall space would be provided at both spring and fall meetings for the establishment of a learning resources center (LRC) in which members could present innovative teaching aids and methods. Abstracts of LRC presentations were grouped in the program and identified as "sponsored by the Education Committee of the APS." By 1980 this had proved successful enough that all FASEB societies were invited to make presentations at the LRC. Facilities made available at the LRC included poster boards, table space, telephonic connections for a computer terminal, a library of video tapes, and video cassette players.

Table 3
TEACHING SESSIONS, 1963–73

Title	Year	Organizer(s)
Analogue Computers in Physiology Teaching	1963	L. Peterson
Innovations in the Teaching of Physiology	1964	R. Forster
Graduate Teaching in Physiology	1965	C. Wilber
Animal Care Programs in Physiology Education	1966	B. Cohen
Impact of National Board Examinations on Teaching in Medical Schools	1967	W. Blake
Should Physiology Training Be Divorced from the Medical Curriculum?	1969	F. Hadley
[No title]	1970	A. Dawe
Scientific Journal Articles and the Computer	1971	S. Geiger
Demonstration of the Physiology of Drug Abuse	1972	H. Jones and O. Reynolds
Computer-Assisted Education	1973	C. Tidball

In the 1980 fall meeting, a new miniseminar format for facilitating educational information was also instituted by the Education Committee. Organized by John West, a program on the teaching of respiratory physiology included both a poster session and discussion in which different invited individuals presented their methods for the teaching of the subject to medical students. Similar sessions on other topics have been held in subsequent years as part of the LRC.

Tutorial Lectures

The refresher course and the teaching session were, for many years, the major educational activities sponsored by the Society at its meetings. Then at the request of Council, a series of tutorial lectures was instituted at the 1975 fall meeting, again in 1977, and annually thereafter. The purpose of these lectures is to present a critical review and analysis of the current state of knowledge about a field within physiology. Designed neither for the novice nor the expert in the field, the lectures are to develop the conceptual framework and background of the field through a critical review of essential experiments. The lecturers are selected by the Program Committee, with suggestions from the Education Committee. Since 1978 many of the lectures have been published in *The Physiology Teacher*.

Courses for Physicians

Every other year, from 1960 until 1970, the Education Committee sponsored intensive courses in physiology for physicians in collaboration with the American College of Physicians (ACP). Except for the first one, set up by Julius Comroe, the three-day courses were organized by Daniel Simmons and Charles Kleeman (and later by Simmons alone). They were held immediately preceding the annual ACP meetings in the same city and were taught by physiologists from all over the country, thus fulfilling one of the goals of the course—introducing well-known physiologists to practicing physicians. These courses were generally well attended and highly praised, and they were also a source of revenue for the Society; for example, registration for the 1962 course was approximately 250, and the APS netted $4,936.29. Indeed there was considerable interest on the part of ACP to make the course an annual event. Unfortunately by 1970 the attendance was much lower, and when the possibility of financial loss to the Society at future courses became a real concern, the Education Committee decided not to continue the program.

Laboratory Instruction in Physiology

During its early years the Education Committee supervised the compiling of two mimeographed collections of laboratory experiments for college courses in physiology—one in general physiology organized by Samuel R. Tipton and one in elementary human physiology organized by Charlotte Haywood. (Preparation of a third set of experiments for medical school physiology courses was considered but dropped.) By 1967 the $2,600 that had accumulated from the sale of these laboratory experiments were used for a revision of the laboratory experiments in general physiology. The revision, carried out by a committee headed by Tipton, although designed in part for the use of advanced undergraduate students, was planned to serve more as source material for college instructors in general physiology. It was not actually a laboratory manual, but rather a compilation of background material, exercises, and references, which would permit the instructor to compose his or her own set of experiments. Continuation of the program of providing teachers with laboratory exercises was one of the principal objectives of the Education Committee in inaugurating *The Physiology Teacher* in 1971 (see below).

Workshops for College Teachers of Physiology

In 1955 the Education Committee began sponsoring an annual two-week summer workshop for college teachers of physiology. Funded by the National Science Foundation (NSF), the workshops brought together a diverse group of teachers to share experiences and materials. Because of declining attendance, the program was canceled after the 1963 workshop. The idea of workshops was revived, however, in 1970 by a report of a Subcommittee on Physiology in Undergraduate Colleges chaired by David Bohr. This subcommittee had sent a questionnaire to 1,000 undergraduate departments of biology, in which they were asked to rank the order of their desires for various specified services by the Society. On the basis of the replies, one of the recommendations made by the Education Committee was to attempt to reinstitute some type of workshop program.

Council approved a pilot workshop, to be organized by Bohr and held at the University of Michigan over two weekends in November and December of that year. Thirty colleges within a radius of 200 miles of Ann Arbor were contacted, and thirty-four teachers attended. Although this pilot was deemed a success, no ongoing program was set up, largely because of problems concerning funding and lack of interest on the part of potential host institutions. Another variant on the earlier workshops was a short course on environmental physiology, a two-day meeting suggested and organized by John Anderson and held at the University of North Carolina (Chapel Hill) in 1973.

Visiting Scientist Program

The Visiting Scientist Program represented another attempt by the Education Committee to help college teachers of physiology. Funded by a National Institutes of Health (NIH) grant of $58,320 for eighteen months and $39,300 per year for four additional years, as requested by Ray Daggs for the committee, the Visiting Scientist Program was begun in January 1963. (This was, in certain respects, a continuation of the "visiting lecturers" activity initiated several years earlier by H. S. Mayerson and Arthur W. Martin, funded by NSF, but terminated in December 1962.) In the Visiting Scientist Program, experienced physiologists were sent to colleges that requested such visits. To begin the program, invitation questionnaires were mailed to scientists and potential host colleges; more than 500 individuals and 185 colleges responded. The visitors discussed with administrators, teachers, and students both general and specific scientific and teaching problems. In this manner, the gap between "high-powered researcher" and "isolated small-college teacher" was narrowed. So well received was this program that by 1 July 1963 fifty physiologists had visited sixty-nine colleges in thirty-two states.

By mid 1966, the Education Committee felt that the program had achieved its goals and that "a plateau had been reached" (i.e., most schools desiring a visit had been accommodated). Moreover NIH was requesting that if the program were to continue, the grant would have to be changed into a contract, which would have necessitated a cost-benefit bookkeeping difficult to apply to this program. For these reasons, the Education Committee decided to discontinue the program. During its four-year existence, the program had received 530 requests from 343 different schools, and 345 visits by 151 different scientists had been made to 209 schools in 42 states and the District of Columbia. The average visit was 1.7 days.

Since 1966 the Society has participated in three other types of visiting scientist programs. In 1968 Clifford Barger proposed that in lieu of the Porter Fellowship, the Porter Development Program be instituted with funds provided by the Harvard Apparatus Company. This program, described elsewhere in this volume, contained a provision for visiting lecturers at institutions of predominantly minority enrollment.

In 1977 the Federation of American Societies for Experimental Biology (FASEB, the Federation) accepted a continuing grant from the Burroughs Wellcome Fund to administer the Wellcome Visiting Professorships in the Basic Medical Sciences. This program provided for three such awards annually for each member society of FASEB. Each visiting professor spends two to five days at the host institution, engaged in teaching and discussion with students and faculty, and delivers a Wellcome Lecture on the subject of choice. Finally, in 1981 FASEB initiated the Visiting Scientists for Minority Institutions, funded by the Minority Access to Research Careers Program of the NIH. The number of such Visiting Scientists in Physiology is not specified, but approximately fifty such awards are made annually by the Federation, of which a sizable proportion is represented by physiologists.

Studies of Graduate and Medical Education

In the late 1960s a subcommittee of the Education Committee chaired by Walter C. Randall undertook to determine what recent changes in policies had occurred in graduate training programs, with the possibility of developing appropriate criteria or minimal standards for degree programs in physiology. A questionnaire was formulated and sent to forty-nine departments. The results provided valuable information concerning the philosophical objectives of such programs (16), the qualities sought in potential graduate students, levels of performance during graduate training, and efforts within departments to maintain suitable self-evaluation and upgrading procedures. It was decided not to attempt the creation of external criteria or standards.

This subcommittee also deliberated in 1966 on the question of the role of physiology in medical education (15). Spurred on by the concern that "medical physiology is in a state of flux and uncertainty," the subcommittee recommended that information be obtained about medical school teaching of physiology, that evaluation and accreditation of medical physiology teaching programs be considered, and that a symposium on the teaching of medical physiology be sponsored. Only the last recommendation was implemented, at the 1968 spring meeting. The recommendation for a questionnaire survey was dropped because it was believed that most of the information desired had become apparent during the development of this symposium. The question of evaluation of medical school departments of physiology was referred to the Association of Chairmen of Departments of Physiology. (Data derived from the annual survey undertaken by this group were published in *The Physiologist.*)

The Physiology Teacher

The recommendation of the Subcommittee on Physiology in Undergraduate Colleges for a workshop program was described earlier in this chapter. The other major recommendation of this subcommittee was that the Society publish a newsletter for undergraduate teachers of physiology. This was in response to a need expressed by college teachers of undergraduate courses in physiology for a source of information about current physiological topics, laboratory experiments and demonstrations, and a forum for the discussion of problems connected with the teaching of physiology. On the recommendation of the Education Committee, Council approved the expenditure of $1,500 for a pilot edition of this undergraduate newsletter, to be known as *The Physiology Teacher.*

The Physiology Teacher began as a quarterly in April 1971, initially under the editorship of Nancy Milburn (and subsequently Orr Reynolds). Each issue contained a detailed description of an experiment that could be conducted in most college settings, a question-and-answer section (short discussions of topics suitable for incorporation into physiology lectures), book reviews, and announcements of workshops and the Society's refresher courses. The initial effort was supported by a $3,000 grant from Harvard Apparatus Company, and by 1972 the project was self-supporting through its subscription and advertising fees.

Over the next few years, although comments from subscribers on its contents and usefulness were universally favorable, *The Physiology Teacher* was subject to several continuing problems. First was the lack of a sufficient number of manuscripts submitted voluntarily for publication. A second problem was whether certain articles should appear in *The Physiologist* or in *The Physiology Teacher*. (Some of the submitted articles that dealt with educational matters required their publication in *The Physiologist* because the members of the Society were entitled to receive them.) A third problem was the subscription level, which had stabilized at around 1,200 (cf. *The Physiologist* with a distribution of 6,000).

Since most subscribers to *The Physiology Teacher* were members of the Society and therefore also receiving *The Physiologist*, a logical solution to all these problems was to add *The Physiology Teacher* to *The Physiologist*; in reply to a questionnaire, ninety-two percent of the membership responding recommended this step. Accordingly, beginning in 1977, all recipients of *The Physiologist* also received *The Physiology Teacher*, inserted in all but the August (abstract) issue. Subscription to *The Physiology Teacher* separately was still available for nonmembers. By 1978, however, the shortage of material submitted to *The Physiology Teacher* had become severe, and a decision was made to increase the scope of the publication to make it more valuable to all teachers of physiology, not just to those functioning in a college setting.

One spinoff of *The Physiology Teacher* has been *Frontiers in the Teaching of Physiology: Computer Literacy and Simulation*, edited by C. S. Tidball and M. C. Shelesnyak (18). [This should not be confused with "Computer-Assisted Education," also edited by Tidball (19).] It consists of a series of articles that appeared in *The Physiology Teacher*, under the editorship of Tidball, with some additional new articles and a glossary of terms relevant to the field. This collection has also stimulated plans for other "reprint series" on various teaching and historical subjects.

Audiovisual Programs

In the early 1960s the Education Committee surveyed colleges and universities concerning their use of teaching films in physiology. This survey generated numerous requests for lists of films approved or recommended by the Society. A preliminary list was developed but never finalized; rather it was decided not to pursue the matter independently but to work with the American Association of Medical Colleges.

In 1969 a subcommittee of the Education Committee, chaired by L. M. N. Bach, again suggested that the Society obtain an annotated list of such materials. However, no action was taken on this suggestion for several years. At the 1969 fall meeting this subcommittee also showed eight teaching tapes and films for evaluation by attending physiologists and later published the results of the evaluation (8).

Then in 1972, under the leadership of Jack Kostyo and Orr Reynolds, the Education Committee launched what would become its most ambitious project to date. Existing audiovisual materials in the field of physiology were reviewed and evaluated and new audiovisual aids were produced. These projects were stimulated by the growing view of many physiologists, particularly those teaching in medical schools, that they could make more effective and efficient use of their teaching time through the use of good audiovisual materials. The overall task was to identify and evaluate currently available materials and then to produce new audiovisual materials for areas not adequately covered.

The project was initially (1972-73) funded by a contract, administered by Orr Reynolds, from the National Medical Audiovisual Center (NMAC). Over a thousand audiovisual materials collected by the staff of the Education Office were reviewed at FASEB headquarters by fourteen subspecialty review panels, each consisting of three to five experts. Of those examined, 263 were judged "adequate or better" for use in physiology courses at various specified academic levels. The review results were published as a supplement to The Physiologist and The Physiology Teacher (9); it was also available separately at nominal cost. From 1974 to 1975, with financial support from the National Fund for Medical Education (NFME), six new review panels were convened to review an additional 161 audiovisual materials, 94 of which were judged "adequate or better." These reviews were also published as a supplement to The Physiologist (3).

The review of existing audiovisual materials was only preliminary to the major purpose of the project, the production by the Society of audiovisual materials in areas not adequately covered by existing materials. A second contract (amounting ultimately to $150,000 over a 15-month period) from NMAC, in June 1973, provided funds for the production of sixteen slide-tape productions. The areas chosen were renal physiology, cardiac physiology, and temperature regulation. Through the Contract Steering Committee, individual members were selected for development of the slide-tapes, production guidelines were established, and an extensive review system utilizing faculty and students was set up at evaluation centers established at four departments of physiology. Duplication of the completed slide-tapes was to be performed at NMAC, and sales were to be handled by the General Services Administration of the federal government.

The preparation of the first sixteen slide-tapes proceeded well until mid 1974 when difficulties were encountered with NMAC. In particular, modifications of the completed slide-tapes by the staff of NMAC were thought by the Education Committee to be damaging the scientific and educational quality of the presentations and to be unacceptable to the authors and the Society. Moreover it became clear that there could be an interminable delay before these materials were marketed by the federal government. After vigorous protest and negotiations, APS received permission to complete production and to market the slide-tapes itself. (Two of the slide-tapes on temperature regulation were never marketed.) An arrangement for production and marketing was negotiated with the AV/MD Division of Steven K. Herlitz, Inc. (AV/MD), under which APS would recover thirty percent of the gross sales, and the first slide-tapes became available in the fall of 1975.

While these events were in progress, another group of four slide-tapes, in the area of acid-base physiology, was begun under another contract (for $83,071) from NMAC. In issuing this last contract, NMAC emphasized that they did not intend to fund any further slide-tapes, since it was their policy to provide financial support to professional groups only to launch audiovisual programs. It seemed unlikely that private foundation support could be found to subsidize further efforts, yet the Education Committee believed it would be unfortunate to terminate the activity, given that APS had acquired an effective mechanism and staff for producing these materials. Accordingly the committee recommended to Council that financial support be provided from Society sources for a period sufficient to determine whether the program was indeed of value and could be financially self-sufficient. Council approved this request in spring 1976.

As a consequence, elaborate plans were made by the Education Committee to provide an "in-house" mechanism for the preparation and quality control of slide-tape materials, and series on renal pathophysiology and on cardiac electrophysiology were then begun. A series on peripheral circulation was also later approved. The management of this program, including identification of authors in areas designated by the Education Committee and coordination of the scheduling of workshops, as well as script and art preparation, peer review, and student evaluation, was charged to the new Audiovisual Production Subcommittee, chaired initially by Robert Gunn.

Council, in March 1976, voted a maximal financial support of $150,000 until September 1978, with the expectation that the program would be self-supporting by then and that these funds would, in all probability, be recovered through sales of the materials. At the end of 1977, when the accumulated deficit of the program to the Society was $59,942, the Education Committee projected that this deficit would increase to $68,000 by the end of 1978 and would begin to decrease in 1979. Given the fact that the total deficit was less than half of that approved by Council in 1975, the committee requested that Council extend support beyond the original date of September 1978, and Council so acted in spring 1978.

Unfortunately these projections proved to be overly optimistic, and the deficit continued to increase. It was approximately $98,000 at the end of 1978 and was projected to reach approximately $143,000 by the end of 1980. The major reason for the discrepancy between the early estimate and the actual amount was the low production rate of slide-tapes, since sales of available tapes were reasonably close to original predictions. The system set up for developing the materials did not function as efficiently as originally predicted for many reasons, but several interrelated common denominators were apparent: 1) the absence of a single full-time individual to act as manager or editor of all aspects of the entire series, both intellectually and financially; and 2) too little early decision making on content before major production staff involvement, which led to costly delays and reworking of material.

At its August 1979 meeting, the Education Committee decided to recommend to Council a moratorium on production of new slide-tapes, beginning at the end of 1980. This date was chosen for several reasons. First, the program would have come close to the maximal financial commitment authorized by Council; even should it break even with continued production beyond 1980, the deficit would almost certainly remain very close to this maximal level for several years. Secondly, the period from August 1979 to 1980 would permit the orderly completion of those

Table 4
SLIDE-TAPES PRODUCED BY THE AUDIOVISUAL PRODUCTION SUBCOMMITTEE

	Author(s)
Renal physiology	
Body Fluids, Parts I–III	H. S. Frazier
The Kidney and Sodium Balance	F. G. Knox
Understanding Renal Hemodynamics, Parts I and II	L. G. Navar
Renal Clearance, Parts I and II	J. M. Ginsburg
Regulation of Renal Potassium Excretion	F. S. Wright
Cardiac physiology	
The Cardiac Pump: Structure and Mechanism	F. N. Briggs
Regulation of Myocardial Performance	F. N. Briggs
Heart Muscle Mechanics	G. N. Pollack
Mechanics of the Intact Heart	W. W. Parmley
Cardiac Contractility: The Ventricular Function Curve	W. W. Parmley
Indices of Performance: Clinical Application	W. W. Parmley
Acid-base physiology	
Buffer Chemistry: Physiological Applications	E. Schneider
The Role of Respiration in Hydrogen Ion Metabolism	A. B. Otis
The Role of Hemoglobin in Hydrogen Ion Metabolism	A. B. Otis
Renal Regulation of Hydrogen Ion Metabolism	J. Cohen
Disturbances of Hydrogen Ion Regulation	R. Tannen
Renal pathophysiology	
Metabolic Acidosis	H. L. Bleich
Hyponatremia	H. S. Frazier
Edema	C. H. Coggins
Disorders of Potassium Balance	F. S. Wright
Electrophysiology of the heart	
Basic Membrane Physiology, Parts I and II	H. A. Fozzard
Cardiac Cellular Electrophysiology	H. A. Fozzard
The Electrical Anatomy of the Heart, Parts I and II	R. W. Childers
Electrocardiography: Fundamental Theory, Parts I and II	M. F. Arnsdorf
Electrocardiography: Applied Theory, Parts I and II	M. F. Arnsdorf
Peripheral circulation	
Hemodynamics	S. Chien
Arterial and Venous Systems, Parts I and II	S. Chien
Exchange 1: The Capillaries and Solute Diffusion	B. R. Duling and P. C. Johnson
Exchange 2: Filtration and Absorption of Fluids	B. R. Duling and P. C. Johnson
Local Control. Part 1: Myogenic Mechanisms	B. R. Duling and P. C. Johnson
Neural Control	P. C. Johnson

materials then in production. Council approved this recommendation, and all production was completed by summer 1981. As of the end of 1985, although the deficit incurred by the program was $112,000, it was hoped that this would be fully recovered from sales over time. In Table 4, the forty-three slide-tapes produced by this program are listed.

In addition to these slide-tapes produced by the Education Committee, the Society also had "reviewed and recommended" a series of "Illustrated Lectures in Neurophysiology" by Beverly Bishop, and two series of "Illustrated Lectures in Respiratory Physiology" by John B. West. These were marketed by AV/MD to supplement APS-produced material.

Educational Objectives

As required by the original NMAC contract for the audiovisual project, a set of educational objectives (actually a list of topics) was developed by the Contract Steering Committee. The framework used by the committee was a set of objectives developed in 1969 by Noe Naeraa and his colleagues at the University of Aarhus, Denmark, and translated by the British Life Assurance Trust for Health Education. The list was to serve as a guide for the selection of topics for future audiovisual materials, but because it might be helpful to some college teachers lacking extensive experience in designing physiology courses, the Education Committee in 1983 authorized its publication and the printing of 3,000 copies (financed by contract funds), which were mailed as a supplement to *The Physiology Teacher* (2).

As noted above, however, these so-called objectives were really a list of topics and not stated in behavioral terms. In 1974 the Physiology Department of Aarhus extensively revised and extended the original set of objectives and presented them in behavioral terms. The Education Committee then supervised the translation of this revised listing. Although it was recognized that this reflected the views of only one department, it was felt that they were an excellent example of behaviorally stated objectives, which could be used as an information base by departments undertaking development of their own set of objectives. They were published in 1978 as a supplement to *The Physiologist* (4).

Educational Materials Review Board

In 1975 the Education Committee established an Educational Materials Review Board (EMRB) to provide peer review of educational materials produced both by the Society (e.g., several of the slide-tapes) and by others. The Society membership responded enthusiastically to the request for service on the EMRB, which came to have more than one hundred members. The major task of the EMRB was to submit, once a year, abstracts of valuable educational materials and review articles that would be of interest to teachers of physiology. These abstracts were published in *The Physiology Teacher* for five consecutive years (1976–80), but by the early 1980s the number of abstracts being submitted had dwindled and this activity ceased.

Continuing Medical Education Projects

As described earlier in this chapter, the Society in various ways has attempted to provide educational programs for clinicians. Examples are the courses for physicians and the articles in the series known originally as *Physiology for Physicians*. A major new venture in this realm was undertaken in 1981, when the Education Committee, through Arthur Vander and Orr Reynolds, applied for and received accreditation of the Society to grant continuing medical education (CME) credits. The eligible activities include not only the slide-tapes, but the Society's publications, scientific sessions, refresher courses, and tutorial lectures. The first fruit of this accreditation was the obtaining of CME credits by thirty physicians at the 1981 fall meeting.

Accreditation also stimulated a variety of new projects by the Education Committee. One task, supervised by Joel Michael, was publication in *The Physiologist* of a series of reviews of recent advances in physiology aimed at physicians, accompanied by a set of questions for submission for CME credits (5).

Computer-based Education

In a variety of ways previously mentioned in this chapter, the APS has sought to facilitate the use of computers in the teaching of physiology, and this goal has, at the time of this writing, become a central focus of the Education Committee. Moreover *The Physiology Teacher* has been increasingly devoted to this subject, and a survey of physiology departments, carried out by Carl Rothe, showed a high proportion of departments engaged in or planning extensive use of computer-based education (CBE) (17). "Computers in Physiology Teaching: How Can APS Help?" was presented during the 1983 FASEB Meeting to apprise members of the status of CBE and to poll them as to their present and future needs in this area and how they thought the APS could help (10). A large workshop on CBE was also held at the 1984 FASEB Meeting (11, 12). An effort is presently underway, in cooperation with the FASEB Office of Computer Services, to catalog existing CBE materials in the biomedical disciplines. A mechanism has been proposed whereby the APS, either through endorsement or actual participation, could help investigators obtain funds to support scientifically based research on the teaching of physiology. The goal of these efforts is to encourage all sorts of endeavors in the area of CBE.

High School Education

As we have seen, although the educational activities of the APS have been aimed at all academic levels, from colleges to practicing clinicians, no major projects have ever been implemented for educational levels below that of the colleges (excluding, of course, providing career information in the various APS brochures). This was not for lack of trying, however, and for the sake of completeness it is worth briefly mentioning the extensive efforts, from 1971 to 1976, of the Education Committee, chaired by Jack Kostyo, and of the education officer, Orr Reynolds, to achieve such a program.

For some time the Education Committee had been interested in contributing to education at the secondary school level. In 1971 an ad hoc committee, chaired by Ingrith Deyrup-Olsen, met to explore this question, as well as the special problems of teaching physiology to particular undergraduate and paramedical groups. One major proposal of this committee was to organize regional resource centers, where professional physiologists could help foster the improvement of physiology teaching (both in secondary schools and colleges) in their geographical areas. A second major proposal was the more extensive use of publications, especially *The Physiology Teacher*, to update and upgrade physiology teaching at these levels. A third proposal was for the Society to help develop modern secondary-level curricular materials emphasizing human physiology in the area of health and hygiene. In particular, the committee recommended a collaborative effort with the Biological Sciences Curriculum Study (BSCS), an organization with extensive experience in developing high school curricular materials. (The Education Committee had previously worked with BSCS in 1965–66 to provide reviews of texts then in development.)

In 1972 the Education Committee acted on this third proposal and decided to investigate providing an educational package in human physiology or biology that could be used as an alternative track in the high school biology curriculum. The idea of collaboration was explored with BSCS and was enthusiastically received. A joint APS/BSCS proposal was drafted (supervised by a subcommittee headed by Patricia Farnsworth) in which APS was to provide the scientific expertise and BSCS was to provide the educational specialists and technical machinery for the job. Unfortunately, despite the best efforts of Kostyo and Reynolds during the next few years, funding for this proposal could not be obtained from the various governmental agencies and foundations approached. Since then the desirability of producing educational materials for high schools has continued to be raised by the Education Committee, but no viable proposal has been forthcoming. It is to be hoped that this gap in the Society's educational activities will be filled in the future.

Section on the Teaching of Physiology

In April 1985, on the initiative of Harold Modell, a meeting was held to organize a Section on the Teaching of Physiology. Arguing the need for such a section, Modell wrote:

> The excitement generated by discussion in the Learning Resources area of the meetings suggests that the time has come to provide a broader forum of exchanging ideas relevant to physiology teaching. A group of APS members feel that a section within the Society is needed to guide movement in this direction (6).

A petition signed by over a hundred members was presented to Council, and the section was formally approved in October 1985. The new section is intended to complement the activities of the Education Committee. One of its goals is to bring teachers of undergraduate physiology into more active involvement in the Society and thereby to strengthen recruitment of physiologists at the undergraduate level

(7). Thus the Section on the Teaching of Physiology addresses in a new way a long-time concern of the Society.

Conclusion

No evaluation of results of the Society's efforts on education can be made.... Certain results can be seen in the changes of attitudes on the part of Society members, tending toward the feeling that teaching is an important activity. The [Education] Committee believes that teaching can take its place side by side with research in the ambition and thinking of physiologists in general.

These were among the conclusions of E. F. Adolph's review of the educational activities of the Society from 1937 to 1962 (1). They seem appropriate for the past quarter century as well.

Perhaps most striking have been the hundreds of persons who have participated in the Society's educational activities—as members of committees; as organizers of teaching sessions, refresher courses, workshops, and the like; and as tutorial lecturers, visiting lecturers, and participants in the slide-tape program. Also impressive is the number of educational programs, at various academic levels and of great diversity and scope. At least one of these programs (e.g., the refresher course) has been retained virtually unchanged over the entire period covered by this chapter, whereas others, such as the teaching sessions and LRC, have evolved while retaining their basic functions. Still others (e.g., the slide-tape program) were self-limited projects, which achieved a particular purpose and were then discontinued.

However, one cannot help but note the number of projects (e.g., the Educational Materials Review Board) that seem to fade away and the equally large number of excellent ideas that appear in the minutes of the Education Committee meetings but are never acted on or implemented. As a former chairman of the Education Committee and as the Society's only education officer to date (1985), we have recognized the difficulties both of *1*) initiating major new projects that often require application for grants or contracts, and *2*) maintaining continuity and momentum in old ones when the major vehicles for doing so are a committee that can meet only once or twice a year and an education officer who also has the many other duties of executive secretary-treasurer. Perhaps sometime in the future the Society will find it possible once again to employ a full-time education officer, with an office that can be funded largely by outside grants and contracts.

REFERENCES

The major sources used in preparing this chapter are the minutes of the Education Committee meetings, including many subcommittee reports appended to these minutes and the reports to Council by this committee. These materials are all on file at APS headquarters in Bethesda. Also, beginning in 1978, the reports to Council have been published in *The Physiologist.*

1. ADOLPH, E. F. Educational activities of the Society. In: *History of the American Physiological Society: The Third Quarter Century, 1937–1962*, edited by W. O. Fenn. Washington, DC: Am. Physiol. Soc., 1963, p. 146–154.

2. ANONYMOUS. Educational objectives in physiology. *Physiol. Teach. Suppl.* 2(3), 1973.

3. ANONYMOUS. Second issue: audiovisual aids useful in the teaching of physiology. *Physiologist Suppl.* 18, 1975.

4. ANONYMOUS. Objectives for a course in physiology at the Institute of Physiology, University of Aarhus, Denmark. *Physiol. Teach. Suppl.* and *Physiologist Suppl.* August 1978. (Translated by R. B. Gunn and J. Funder.)

5. ANONYMOUS. CME Category I Credits. *Physiologist* 28: 46, 1985.

6. ANONYMOUS. Establishment of a physiology teaching section within APS is proposed. *Physiologist* 28: 29, 1985.

7. ANONYMOUS. 134th APS Business Meeting. *Physiologist* 28: 465–467, 1985.

8. BACH, L. M. N. Physiologists view physiological teaching: an account of peer evaluations of videotapes and films for physiological teaching. *Physiologist* 13: 89–111, 1970.

9. DE HART, W. D., AND R. J. SIEGEL. Audiovisual aids useful in the teaching of physiology. *Physiologist Suppl.* 16, 1973.

10. MICHAEL, J. A. Computers in physiology teaching: how can APS help? *Physiologist* 26: 323–325, 1983.

11. MICHAEL, J. A. Workshop on computer-based education. *Physiologist* 27: 358–359, 1984.

12. MICHAEL, J. A. Computer-simulated physiology experiments: where are we coming from and where might we go? *Physiologist* 27: 434–436, 1984.

13. PROSSER, C. L. President's letter to membership with request for replies. *Physiologist* 12: 433–435, 1969.

14. RAMSAY, D. J. (Editor). *Careers in Physiology.* Bethesda, MD: Am. Physiol. Soc., n.d.

15. RANDALL, W. C. The role of physiology in medical education. *Physiologist* 10: 46–48, 1967.

16. RANDALL, W. C., AND M. P. KAYE. Minimal standards for graduate education in physiology. *Physiologist* 13: 411–416, 1970.

17. ROTHE, C. F. Trends in physiology teaching laboratories for medical students—1982. *Physiologist* 26: 148–149, 1983.

18. TIDBALL, C. S., AND M. C. SHELESNYAK (Editors). *Frontiers in the Teaching of Physiology: Computer Literacy and Simulation.* Bethesda, MD: Am. Physiol. Soc., 1981.

19. TIDBALL, C. S., ET AL. Computer-assisted education. *Physiologist* 16: 594–683, 1973.

CHAPTER 13

Finances

ARTHUR C. GUYTON

A long history of healthy financial affairs has enabled the American Physiological Society (APS, the Society) to undertake many projects that otherwise would not have been possible. The most important among these is the extensive program for publication of journals; others of note include *1*) publication of one of the most important of all bioscience handbook series; *2*) publication of monographs and special books; *3*) purchase of Beaumont House and the surrounding property, which was later turned over to the Federation of American Societies for Experimental Biology (FASEB, the Federation); *4*) providing much of the financing for the Lee Building on the Beaumont property, which is now the principal office building of APS and the other societies of FASEB; and *5*) development of an extensive educational slide-tape series. Furthermore the strong financial position of the Society has also enabled APS to take a lead in many of the operational functions of FASEB, as well as to carry on special beneficial programs for its own members, such as the recent establishment of an office to deal with the plethora of animal experimentation legislation.

Annual Expenditures

In Table 1 the annual expenditures of APS for Society operations and for publications are listed for 1963 and 1984. It is clear that the expenditures have increased approximately tenfold in these twenty-one years. (Not shown is an additional $272,000 in 1984, which is an assessment against APS for the operations of FASEB. Most of this amount, however, is returned to the Society in the form of profits from the annual meeting of FASEB.)

Most of the increase in Society expenses occurred between 1970 and 1982, a period of rampant inflation throughout the United States. At the same time, however, the Society was increasing services to its members; in addition its membership approximately doubled between 1963 and 1984. Among the most important of the

Table 1
ANNUAL EXPENDITURES

	1963	1984
Society operations	$ 49,860	$ 480,600
Publication operations	397,351	3,862,593
Total	447,211	4,343,193

additional services are the establishment and operation of sections in APS, initiation of a program for long-term financial development, and establishment of the office that deals with animal experimentation legislation. In a typical year the more important items among the Society's expenses include salaries (ca. 33% of the total), cost of publishing *The Physiologist* (ca. 14%), business office expenses (ca. 7%), hotel and travel (ca. 9%), rent (ca. 6%), professional services (ca. 7%), and addressing, mailing, telephone, printing, supplies, and equipment.

The expenses for the publications operation increased only about threefold between 1963 and 1975, partly because of inflation, but also because of a moderate increase in the number of pages published in the journals. Beginning in 1976, however, with the decision to sectionalize the journals, the publication expenditures increased another threefold by 1984. Here again the increase was caused partly by inflation, but to a greater extent by the approximately 130% increase in the total number of pages published in the journals after sectionalization. It should be noted that special efforts to provide better cost accounting and more efficiency in the publication process have made it possible to maintain a per page cost of publication that has changed very little in the last fifteen years.

Dues

The changes in the amounts of dues assessed the three categories of membership are shown in Table 2 for 1963 and 1984. The very large increase in associate members' dues occurred when they became recognized as members of FASEB; they are now assessed an extra fee for the benefits they receive from the Federation. The dues for the regular members have increased slightly more than the degree of inflation. However, as already noted, the Society is also providing important new services.

Operating Capital and Capital Reserves

The equity positions of the two major functional units of APS are shown in Table 3 for 1963 and 1984. For Society operations, approximately $90,000 were available at the beginning of 1963, but only $25,000 at the beginning of 1984. There has been no attempt to build up reserves for Society operations, because these are funded almost entirely from the annual dues.

On the other hand, the publications of the Society require a large amount of operating capital and capital reserves, which have grown approximately fourfold between 1963 and 1984. Beginning at about $800,000 in 1963, by 1984 they totaled almost $3,500,000. Of this amount, approximately $850,000–900,000 are in the form

Table 2
DUES

	1963	1984
Regular	$15	$80
Associate	5	55
Student	10*	15

* Beginning in 1977.

Table 3
OPERATING CAPITAL AND CAPITAL RESERVES

	1963	1984
Society operations	$ 89,992	$ 25,484
Publication operations	801,933	3,443,786
Total	891,925	3,469,270

Office equipment and furnishings not included.

of inventory, consisting mainly of volumes in the *Handbook of Physiology* series, but including also other special publications. The remaining $2,600,000 market value is mainly in the form of long-term investments. Through the years, the Finance Committee has attempted to maintain enough capital reserves to fund at least one year of publications operations in case of dire need. Unfortunately, with the tremendous expansion of our publications in the past few years, especially after sectionalization of the journals, it has not been possible to maintain the reserves at a level that might be desirable. The reserves have been accumulated through the years from two sources: profits on a few of the publications operations and income from the investments, some of which have been quite lucrative.

Special Funds

APS administers several special funds, some of which are memorial funds (see Table 4).

Profits from the Harvard Apparatus Company support the Porter Fund, which is presently used for the Porter Physiology Development Program. Awards are granted to stimulate and assist in the improvement of underdeveloped departments of physiology, especially in colleges and medical schools with predominantly minority enrollment.

The Perkins Fund, established in memory of John F. Perkins in 1967, assists in international cultural exchange by providing support for families of visiting physiologists.

The tum Suden Fund was established in 1976 as a bequest in the will of Caroline tum Suden, a longtime member of APS. The Bowditch Lectureship, the Ray G. Daggs Award, and the Caroline tum Sudem Professional Opportunity Awards and the Caroline tum Suden Travel Fellowships are all supported by this fund.

The International Union of Physiological Sciences (IUPS) Fund represents monies collected for use in connection with the meetings of IUPS held each three years. These funds are derived from multiple sources, and they are used to defray expenses

Table 4
SPECIAL FUNDS

	Year Established	Amount	
		Initial	1983
Porter	before 1963	$ 7,571*	$271,976
Perkins	1967	20,500	122,249
tum Suden	1976	93,371	154,637
IUPS	1970	82,192	93,598
Total		203,634	642,460

IUPS, International Union of Physiological Sciences. * 1963.

Table 5
APS FINANCE COMMITTEE, 1963–84

	Year			Year	
	Member	Chairman		Member	Chairman
H. Davis	1961–63	1962–63	C. F. Code	1970–76	
A. W. Martin	1961–64	1963–64	J. R. Brobeck	1973–75	
L. N. Katz	1962–65	1964–65		1980–81	
H. E. Essex	1963–64		D. C. Tosteson	1976–79	
H. S. Mayerson	1964–67	1965–67	A. C. Guyton	1976–82	1977–82
S. M. Tenney	1964–67		E. Knobil	1977–78	
E. B. Brown, Jr.	1965–67		R. E. Forster	1978–83	1982–83
	1971–77	1973–77	J. L. Kostyo	1979–80	
J. M. Brookhart	1967–73	1967–73		1981–82	
E. E. Selkurt	1967–71		P. C. Johnson	1982–85	1983–84
R. M. Berne	1967–70		E. H. Wood	1982–85	
	1975–76		F. J. Haddy	1983–86	

related to the international meeting, especially the cost of travel for young investigators.

Finance Committee

Members of the APS Finance Committee from 1963 to 1984 are listed in Table 5. In addition to the committee members, three other persons have been instrumental in guiding the Society's financial affairs: the two executive secretary-treasurers, Ray Daggs (1957–72) and Orr Reynolds (1973–84), and the Society's business manager, Walter Sonnenberg (1966–86).

CHAPTER 14

Women in Physiology

TOBY A. APPEL, MARIE M. CASSIDY, AND M. ELIZABETH TIDBALL

Since the early years of the American Physiological Society (APS, the Society) there have been women in physiology. The first woman member of APS, Ida Hyde, was elected in 1902. One of her papers, the result of work done in William T. Porter's laboratory at Harvard Medical School, appeared in the first volume of the *American Journal of Physiology* in 1898. Through the first seventy-five years of the Society's history, women were regularly elected members, participated on programs, and published in the Society's journals. However, not until the 1970s, when the role of women in the sciences became of general concern (5), did they begin to take an important role in the management of the Society.

Compared with other scientific fields, and especially with other branches of biology, physiology has always been a relatively difficult field for women. Few women found a place on the staff of medical schools, and when they did, they often remained for decades at the level of research associate or assistant professor. Departments of physiology in medical schools, with the single exception of the Woman's Medical College of Pennsylvania (now Medical College of Pennsylvania) had few women faculty members. Positions outside medical schools that allowed for the possibility of independent research were more limited than in such associated fields as biochemistry, nutrition, or anatomy. The women's colleges, however, did provide an important locus for physiology. Several had departments of physiology chaired by women, among them Mount Holyoke, Vassar, Goucher, and Wellesley (zoology and physiology). Physiology was also taught at Bryn Mawr, but for the most part by a succession of men, the earliest of whom was Frederic S. Lee, who was on the faculty there when he was elected a member of APS in 1888. Many of the early women members of the APS were first attracted to physiology as undergraduates at the women's colleges.

The number of women members of APS grew slowly but consistently through the decades after the admission of Ida Hyde in 1902 and of the second woman, Mabel Purefoy FitzGerald (then residing in the United States), in 1913. By 1930 twenty-six women had been elected to membership (Table 1).

Table 1
WOMEN ELECTED TO APS
MEMBERSHIP, 1902–30

	Year Elected
Ida Henrietta Hyde	1902
Mabel Purefoy FitzGerald	1913
Jessie Luella King	1914
Katherine Rotan Drinker	1915
Maud L. Menten	1915
Rosalind Wulzen	1916
Lillian Mary Moore	1919
Willey Glover Denis	1920
Mary Elizabeth Collett	1921
Helen Copeland Coombs	1921
Mary Swartz Rose	1921
Janet Howell Clark	1922
Matilda Moldenhauer Brooks	1923
Ann Stone Minot	1923
Florence Sabin	1923
Margaret Meta Kunde	1924
Jane Sands Robb	1924
Helen Bourquin	1925
Frieda Robscheit-Robbins	1925
Esther Greisheimer	1925
Marian Irwin Osterhaut	1927
Marie Hinrichs	1928
Erma Anita Smith	1928
Abby Howe Turner	1928
Anna M. Baetjer	1929
Mary Elizabeth Marsh	1929

Although the number of women elected in each decade showed a steady increase, the percentage of women elected did not. The total number of members elected each year grew ever larger, but the entrance of women into APS did not keep pace. It may come as a surprise that the five-year period in which the highest percentage of those elected were women was not 1975–79 or 1980–84, but rather 1920–24. After reaching a high point in the 1920s and 1930s, the percentage by five-year intervals dropped and did not rise substantially until the 1970s (Table 2). The percentage of women among current regular members of APS in 1985 is still a scant ten percent, though it is above eleven percent when all types of membership are considered.

APS, like many scientific societies, for a long time ignored its women members. Although women routinely presented ten-minute contributed papers on APS programs, frequently submitted articles to the *American Journal of Physiology*, and were occasionally chosen as Porter fellows, they otherwise participated little in Society affairs. In the minutes of the Society in the first seventy-five years, women are largely invisible. They were not elected officers, rarely were placed on editorial boards, and were not considered when committees were chosen. At business

Table 2
NUMBER OF WOMEN ELECTED TO REGULAR
MEMBERSHIP BY FIVE-YEAR INTERVALS

	Total Elected	No. of Women	%
1901–04	38	1	2.6
1905–09	53	0	0.0
1910–14	73	2	2.7
1915–19	80	4	5.0
1920–24	87	11	12.6
1925–29	98	8	8.2
1930–34	168	19	11.3
1935–39	166	11	6.6
1940–44	215	12	5.9
1945–49	349	21	6.0
1950–54	324	16	4.9
1955–59	550	39	7.1
1960–64	831	48	5.8
1965–69	1,063	59	5.6
1970–74	787	82	10.4
1975–79	1,110	124	11.2
1980–84	1,138	129	11.3

meetings, there is no record of their having made motions or otherwise spoken. When surveys of physiology were carried out by the Society in 1945 and in the 1950s, women were not considered as a category, and therefore the surveys contain almost no information on the status of women in the field. APS career brochures of the 1960s mentioned women but tended to typecast them as research assistants and informed them that "promotion may be slower than for a man with her background since she is more likely to resign for marriage." This last sentence was written by a woman and at the time (1963) represented reality.

Women on APS Committees

The limited participation of women in APS activities was in keeping with prevailing cultural attitudes about women, but it also reflected the Society's image of itself. Before the 1950s, the Society considered itself a strictly research organization, not actively concerned with the "profession" of physiology. Because the Society's activities were very circumscribed, aside from the Boards of Editors of the various APS journals and the Board of Publication Trustees, there were no APS standing committees before 1951. Short-term committees were chosen by Council for specified purposes, usually to examine an issue and to report back at the next meeting of Council. There was a general belief that officers and appointees to boards and committees of the Society should be reserved for the best men available—in practice, for leading researchers from prominent laboratories. Consequently, because members of these committees were almost always men with professorships at major medical centers, most of the men who were members of APS, as well as all of the women, were excluded.

The first APS committee to which women were appointed was appropriately the Education Committee established in 1953, for it was part of the mission of that committee to reach out to a broad spectrum of physiologists. In the 1950s, Charlotte Haywood, professor of physiology at Mount Holyoke, and Louise Palmer Wilson, professor of zoology at Wellesley, headed subcommittees to produce sets of undergraduate classroom experiments in human and general physiology. Louise H. Marshall, then at the National Institutes of Health (NIH), was the first woman to be elected a full member of an APS committee. She served on the Education Committee from 1961 to 1969 and recalls that she placed her research program at NIH in jeopardy because she spent so much of her time revising the APS career brochures. Only two other women can be identified on APS committees in the 1960s, Ingrith Deyrup-Olsen on the Education Committee and Neena B. Schwartz on the Program Advisory Committee (the equivalent of the present Program Executive Committee). It was not until the 1970s that women began actually to chair committees—Beverly Bishop for the Membership Advisory Committee and Helene Cecil for the Animal Care and Experimentation Committee—and that women were appointed as members in considerable numbers. That has been to an important extent the result of the establishment of the Committee on Committees, advocated by the Task Force on Women in Physiology (see below).

In APS publications also, only a few women have played significant roles to date. Perhaps the most important woman in the history of APS publications is a nonscientist, Sara Leslie, who, beginning as secretary of publications under Milton Lee in 1947, succeeded Lee as managing editor of APS journals in 1966, a post she held until her retirement in 1974. The first woman to be appointed to an Editorial Board (the equivalent of associate editor today) was Frances A. Hellebrandt, then professor of physical medicine at the Medical College of Virginia, who was a member of the initial Board of the *Journal of Applied Physiology* organized in 1948. In the 1950s, Williamina Himwich, noted for her research in biological psychiatry, prepared index volumes for APS publications and served for a number of years as the APS representative to the American Documentation Institute. When section editors were first established for the *American Journal of Physiology* in 1962, Jane A. Russell, associate professor of biochemistry at Emory University, was named the first editor for endocrinology. Since the 1970s several more women have served as associate editors. Only this year (1985), however, has the first woman, Jean McE. Marshall, been appointed to the APS Publications Committee.

Election of Bodil Schmidt-Nielsen

In the one hundred-year history of APS, only one woman has served as president or member of Council—Bodil Schmidt-Nielsen (president, 1975–76). Her election to Council in 1971 is a fascinating story, for initially it was the result of the deliberate efforts of a group of women physiologists in Washington, D.C.

The group originated in 1970 from informal discussions at the Federation of American Societies for Experimental Biology (FASEB, the Federation) Meeting in

Atlantic City. Esther Hardenbergh, Louise H. Marshall, Florence K. Millar, Willie White Smith, M. Elizabeth Tidball, and Marion Webster met on several occasions for lunch to discuss what might be done to further the position of women in APS. Because no woman had ever been elected to Council, they focused on the goal of bringing about the nomination of an appropriate woman in the next election. Although, according to the bylaws, the complete process of nominations and elections was to take place at the business meeting, in practice the nominations were informally mailed in advance. The women chose their candidate, Bodil Schmidt-Nielsen, with care and persuaded her to stand for office. They then solicited the support of a broad variety of physiologists, men and women, thirty-nine in all, to attach their names to a letter sent to all APS members in January 1971. The letter requested members to nominate Schmidt-Nielsen to Council and concluded:

> You may not be aware that the American Physiological Society, founded December 30, 1887, has never had a woman on Council or in an elective office. This letter represents an attempt to make available to the Society greater contributions from its women members. We believe that Dr. Bodil Schmidt-Nielsen would be an excellent member of Council and hope that you will support her nomination.

She was said to have received more nominations than anyone had ever received and was easily elected on the first ballot at the business meeting that April. Subsequently she was elected to the presidency.

The election of Bodil Schmidt-Nielsen to Council coincided with a general awakening of interest in the role of women in science, and her presence on Council no doubt contributed to the tremendous increase in the number of women joining the Society in the 1970s. Also indicative of the consciousness raising that year was the founding of the Association of Women in Science (AWIS) at the FASEB Meeting in Chicago in 1971. This organization now numbers close to 4,000 members and includes women in all the sciences. Among its early presidents were several APS members, including founding copresidents Judith Pool and Neena B. Schwartz, Estelle Ramey, Marion Webster, and Judith A. Ramaley.

Task Force on Women in Physiology, 1973–79

The Task Force on Women in Physiology was an outgrowth of the same informal group of Washington physiologists. It emerged as the result of a letter to President Robert Berne in 1973 from M. Elizabeth Tidball (professor of physiology at the George Washington University Medical Center and one of the members of the luncheon group) in which she requested the establishment of an official group within APS to benefit women. Council preferred at this time to set up a task force rather than a standing committee. Tidball, chairman and mainstay of the task force, tried to choose members representing a variety of career pathways: Elizabeth Carlsen Gerst, then taking time out to raise a family; Virginia M. Fiske, a representative of physiology in the women's colleges (Wellesley); Florence K. Millar, a researcher at NIH; June N. Barker, a full-time researcher outside government; and Ray G. Daggs, recently retired as executive secretary-treasurer of APS (8).

The following Statement of Purpose (6) was adopted at the task force's first meeting in March 1974:

1. To explore ways of encouraging, at all levels, the study of physiology by women.
2. To determine the demographic characteristics of women as physiologists.
3. To assist in bringing names of qualified women to the attention of the Council.
4. To serve as a resource for information regarding women physiologists and as a liaison with other APS committees and scientific groups.
5. To define special problems as appropriate.

Aware of the almost complete lack of information concerning the career patterns of women physiologists, the task force set out to identify women physiologists in and out of APS and to develop a comprehensive questionnaire. Although this ambitious project was not carried out, it did encourage APS to undertake a new computerized survey of all its members. Although that survey did not ask many of the sociological questions the task force had hoped to include, for the first time the number of women in the Society was regularly reported in the annual membership statistics.

The task force was more successful in giving visibility to women in physiology, encouraging more women to enter physiology and bringing about increased participation of women in APS committees. At the 1974 fall meeting in Albany, the task force sponsored a Round Table on Women in Science, and at the 1975 spring meeting, there were displays of literature and informal group discussions (7). It also worked behind the scenes for changes in the structure of the Society.

At its first meeting in March 1974, attended by APS President Daniel C. Tosteson, the question of affirmative action was posed and the idea of establishing a Committee on Committees was promoted "as a means of bringing a greater diversity of APS members, men and women, into the working structure of the Society." This committee, established in April 1974 to nominate members of standing committees to Council, has always had a woman among its members. It has succeeded not only in bringing more women onto committees, but in enlisting the participation of a greater variety of physiologists in general—by scientific specialty, by age and rank, by ethnic origin, by geographic location, and by place and type of employment. The task force, toward the end of its tenure, recommended the formation of a standing Committee on Career Opportunities in Physiology, established in April 1979. At least half the members of this seven-member committee are to be under the age of forty. One of its projects was to revise the APS career brochure, already changed since the 1960s, and, among other modifications, to eliminate the last vestiges of sexist language. It has also held a valuable series of symposia on career opportunities in physiology at APS meetings, since published in *The Physiologist* and as a separate volume.

The task force was also concerned with election procedures, and in particular with a system of nomination by mail that would be fair and open. At this time an ad hoc Committee on Election Procedures was considering methods of carrying out the entire election process by mail. The present system of election of officers by preferential mail ballot, adopted in 1974, was first suggested by Tidball (1). Since 1914 officers had been elected at business meetings by a series of ballots, with each ballot further reducing the number of candidates for office. Much of the meeting was taken up by elections, and only that small portion of the membership in attendance could register its choice. In the new system a blank nomination form is first sent to all members. Then, according to present procedure, the top eight candidates for Council and the top four candidates for president elect are placed on a mail ballot. To cast a valid vote, a member must rank order all candidates for each office. The system was thus intended to approximate the older system of multiple rounds of balloting at the business meetings. In practice, the system has proved cumbersome and has not lived up to the expectations of some of the women who initially supported it. This system, though highly democratic, makes election of anyone who represents any sort of minority exceedingly difficult. Though women have appeared on these ballots, none has been elected.

Committee on Women in Physiology

The task force members, feeling that they had reached at least some of their original goals and hoping that women would be viewed henceforth as full partners in all the Society's activities, decided to request their discharge as a task force in 1979. The Council, under President David Bohr, graciously moved to grant the request with many thanks. The following year, at the FASEB Meeting in Anaheim, at the initiative of Marie M. Cassidy, a group of concerned women members met to discuss issues relevant to the training, education, and employment of women physiologists. Despite the gains of the previous decade, they still perceived a relative lack of women in APS and on APS committees. They therefore agreed to form an ad hoc Caucus on Women in Physiology to address this and other problems. The members were Marie M. Cassidy, Paula T. Beall, Allahverdi Farmanfarmaian, Rita Guttman, Margaret C. Neville, and Janett Trubatch. A major function of the caucus was to increase awareness of the need for greater participation of women in APS activities. A logo was designed under which several reports appeared in *The Physiologist.* At the federation meetings beginning in 1981 a Women Scientists' Lounge was set up. That same year, the caucus petitioned Council to create a Section on Women in Physiology to provide support to women physiologists and also to serve as a focal point for sex-related research. Council, in general sympathetic to the goals of the caucus, felt that these two functions should be separated and voted that they would be willing to entertain a proposal to establish a standing Committee on Women in Physiology. This committee was established in 1982 with Cassidy as its first chairman. Since 1985 the chairman has been Helen J. Cooke (9–11).

Caroline tum Suden Awards

One of the committee's first concerns was the establishment of an appropriate memorial to Caroline tum Suden, an early woman member of APS who in her will had bequeathed $100,000 to APS, the largest single bequest in the history of the Society.

Caroline tum Suden's career, like that of many early women members of APS, was a relatively obscure one, marked by neither recognition nor advancement. She was born in San Francisco in 1900, graduated from Berkeley, and received an M.A. degree at Columbia and a Ph.D. degree in physiology from Boston University in 1933. She became a member of APS in 1936. She remained associated with the Boston University School of Medicine through 1947 but never rose above the level of instructor. Her research, much of it carried out as junior author with her former thesis advisor, Leland C. Wyman, was concerned with adrenal function. Their studies, most of which were published in the *American Journal of Physiology*, dealt with temperature regulation, blood volume, blood sugar and vasomotor responses, susceptibility to histamine in totally or partially adrenalectomized rats, and the homotransplantation of adrenal cortical tissue. After three years (1947–50) as an assistant professor in the Department of Physiology at Mount Holyoke College, she was invited by Bruce Dill in 1950 to join the staff of the Army Chemical Center at Edgewood Arsenal, Maryland, where she worked mainly in the field of neurobiology. She died in 1976, two years after her retirement. She is remembered by colleagues as an able and a meticulous researcher. Her entire career was one of devotion to physiological research (2, 3).

The income from the Caroline tum Suden bequest now funds the Daggs Award, the Bowditch Lecture, and the Caroline tum Suden Travel Fellowships, which provide funds for outstanding investigators to take part in APS-sponsored symposia. The Women's Caucus felt that some of the bequest ought to be used "in a way that will both honor her and benefit the present and future female membership of the Society." A proposal was submitted in 1982 to fund annually six Caroline tum Suden Professional Opportunity Awards. At first Council considered establishing the awards for women alone, but Cassidy and others, concerned with preventing "ghettoization" of women in science, argued that this would be a mistake. Council wisely reversed its decision and opened the awards to both young men and women; resulting experience has proved that the award is considered prestigious, if junior, in merit.

Recipients of the Caroline tum Suden Awards are to be graduate students or postdoctoral fellows who wish to present a paper at the APS/FASEB spring meeting and utilize the FASEB Placement Service. They are selected on the basis of their abstracts by members of the Committee on Women in Physiology and members of the Career Opportunities in Physiology Committee. The first awards were made in 1983. About half the awardees have been women (Table 3).

Another major activity of the Committee on Women in Physiology has been the sponsorship of a series of lectures at the annual FASEB meetings on topics of interest to women scientists. At the first lecture in 1983, Sandra Tangri of American University and the Urban Institute addressed the problem of "Dual Career Marriages." In 1984 Mary Ann Frey, director of the Biomedical Research and Environmental Laboratories at the Kennedy Space Center, spoke on "Physiological Adaptability of Women to

Table 3
CAROLINE TUM SUDEN
PROFESSIONAL
OPPORTUNITY AWARDS,
1983–86

	Year
James Blank	1983
Reed Hoyt	1983
Valerie Kalter	1983
Robert Knabb	1983
Jeri Taylor	1983
Virginia Zinmeister	1983
Steven I. Bellin	1984
Andrew S. Greene	1984
Linda E. Kupfer	1984
Mu-En Lee	1984
Robabeh Moussavi	1984
Bruce A. Webster	1984
Lynn M. Baxendale	1985
Frederic R. Boockfor	1985
Hannah V. Carey	1985
Michael A. Gropper	1985
Robert T. Mallet	1985
Karen M. Wilson	1985
Mark S. Alsberge	1986
Cathy A. Bruner	1986
Katherine J Lucchesi	1986
Rick G Schnellman	1986
Kathleen A. Thompson	1986
Margaret R. Warner	1986

Gravitational Space Flight." In 1985 Cathy Olsen of the National Science Foundation gave a presentation on "Sex, Hormones and Behavior." Other activities of the committee have included sponsoring the Women's Lounge at Federation meetings, providing advice and counsel to the American Physical Society and the Society for Neurosciences as they were setting up women's committees, and, with the aid of the APS staff, responding to a large volume of mail from undergraduate, graduate, and postdoctoral women seeking career advice. In 1986 the committee cosponsored with the Centennial Committee an exhibit on early women members of APS prepared by Toby Appel.

In the future, increasing cooperation among committees on women in the various FASEB societies is anticipated. A survey of other FASEB societies revealed that all but the American Institute of Nutrition and the American Association of Pathologists now (1986) have women's committees. (An explanation may be that these two societies have substantial numbers of women members, and therefore there is little perception of minority status.) Beginning in 1986 the responsibility for the annual lecture series is being rotated among the societies. The 1986 lecture was sponsored by the American Society for Pharmacology and Experimental Therapeutics. Whereas the chairmen of the Committees on Women in FASEB societies have been in fairly constant communication during the past few years, it appears desirable to develop a FASEB-wide advisory body on issues of general importance to women scientists.

Expansion of the role of women in APS from 1970 to the present coincided with an era of general awakening of interest in women in the professions and owed much to initiatives taken by women members of APS themselves. In turn, the more active participation of women in the Society was part of a more general transformation in the 1970s, which brought into the Society and into its governance a much greater diversity of physiologists.

Nonetheless, even with the gains of the past fifteen years, in 1986 it continues to be difficult for women to have an impact as scientists generally and on the profession of physiology in particular. It is striking, for example, that in 1986 there is only one woman chairing a department of physiology in a medical school. Paul Ricoeur (4) has noted:

> the social game going on behind the scene of the scientific world helps us to recognize a social institution in science itself, not only an objective way of research but a social institution which is neither innocent nor powerless.

It is that social institution that most women and many men hope will change by broadening its acknowledgment of what constitutes valid and valued professional activity, thereby encouraging greater participation on the part of all motivated and talented persons, among them women. It is further hoped that members of APS, in tune with both time present and time future, will be among those who lead the way into an equitable and humane twenty-first century. Such an approach will neither diminish nor preclude the emergence, cultivation, and recognition of true scientific excellence, as is sometimes tacitly assumed. Rather it will endow the second century of American physiology with renewed vigor and vision as the dominant basic biomedical discipline.

This chapter is based in part on materials from the APS Archives, including APS Council minutes, Ray G. Daggs' papers, and papers of the Task Force on Women in Physiology.

REFERENCES

1. BERNE, R. M. Proposed new bylaw for election of officers. *Physiologist* 17: 2–5, 1974.
2. DILL, D. B. Caroline tum-Suden. *Physiologist* 20(5):14, 1977.
3. GUTTMAN, R. Biographical notice—Caroline tum Suden. *Physiologist* 23(6): 25–26, 1980.
4. RICOEUR, P. Response to Tidball "Views from the sciences." *Natl. Inst. Campus Ministries J.* 6: 46, 1981.
5. RUSSO, N. F., AND M. M. CASSIDY. Women in science and technology. In: *Sage Yearbooks in Women's Policy Studies. Women in Washington: Advocates for Public Policy*, edited by I. Tinker. Beverly Hills, CA: Sage, 1983, vol. 7, p. 250–262.
6. TIDBALL, M. E. Report of the Task Force on Women in Physiology. *Physiologist* 17: 135–137, 1974.
7. TIDBALL, M. E. Women in the biosciences—a brief primer. *Physiologist* 18: 1–36, 1975.
8. TIDBALL, M. E. A brief history of the Task Force on Women in Physiology (APS). *Physiologist* 23(5): 12–13, 1980.
9. Women in physiology. *Physiologist* 23(6): 25–26, 1980.
10. Women in physiology. *Physiologist* 24(3): 45–46, 1981.
11. Women in physiology. *Physiologist* 26: 135–136, 1983.

CHAPTER 15

Use and Care of Laboratory Animals

HELENE C. CECIL AND WILLIAM M. SAMUELS

The history of the American Physiological Society (APS, the Society) and its members is closely associated with defense of the use of animals in research and the moral principles involved in the use of animals for experimental purposes. In 1913 A. J. Carlson proposed the following resolution:

1. We, the members of The Federation of American Societies for Experimental Biology—comprising the American Physiological Society, The American Society of Biological Chemists, The American Society for Pharmacology and Experimental Therapeutics, and The American Society for Experimental Pathology—in convention assembled, hereby express our accord with the declaration of the recent International Medical Congress and other authoritative medical organizations, in favor of the scientific method designated properly animal experimentation but sometimes vivisection.

2. We point to the remarkable and innumerable achievements by means of animal experimentation in the past in advancing the knowledge of biological laws and devising methods of procedure for the cure of disease and for the prevention of suffering in human beings and in lower animals. We emphasize the necessity of animal experimentation in continuing similar beneficent work in the future.

3. We are firmly opposed to cruelty to animals. We heartily support all humane efforts to prevent the wanton infliction of pain. The vast majority of experiments of animals need not be and, in fact, are not accompanied by any pain whatsoever. Under the regulations already in force, which reduce discomfort to the least possible amount and which require the decision of doubtful cases by the respon-

Editors' note: The first segment of this chapter provides a historical overview of the Society's role in issues regarding the care, use, and treatment of laboratory animals. It was prepared by Helene Cecil, chairwoman of the Society's Animal Care and Experimentation Committee, during much of the time these issues were growing to their current levels of intensity. In the second segment William Samuels explores the emergence of the animal rights movement in the United States and its effect on the Society and traces the federal government's increasing involvement with laboratory animal issues and the Society's role in the design of congressional legislation.

sible laboratory director, the performance of those rare experiments which involve pain is, we believe, justifiable.

4. We regret the widespread lack of information regarding the aims, the achievements and procedures of animal experimentation. We deplore the persistent misrepresentation of these aims, achievements, and procedures by those who are opposed to this scientific method. We protest against the frequent denunciations of self-sacrificing, high-minded men of science who are devoting their lives to the welfare of mankind in efforts to solve the complicated problems of living beings and their diseases.

In 1909 Walter B. Cannon formulated a code of laboratory procedures regarding animals that was adopted and enforced in the laboratories of medical schools (3). These rules formulated by Cannon were the basis of the "Guiding Principles in the Care and Use of Animals" (Guiding Principles) proposed by the APS Committee on the Use and Care of Animals and adopted by the Council in 1953. Since then the Guiding Principles have been revised three times, most recently in 1980, but retain the essential elements of the original statement. The Society has proclaimed that *1*) adherence to the Guiding Principles is an obligation of membership in the Society and *2*) articles based on experiments violating the Guiding Principles shall not be acceptable for publication in the journals of the Society (1). Since 1981 the Program Advisory Committee has been reviewing abstracts of papers to be presented at meetings to ensure that no violations relate to the treatment of animals. In addition, abstracts must include a statement certifying adherence to APS's Guiding Principles. Other scientific societies also use the Guiding Principles as their criteria for animal experimentation. These Guiding Principles, along with A. J. Carlson's resolution of 1913, were reaffirmed by the Federation of American Societies for Experimental Biology (FASEB, the Federation) in 1980.

In 1963 the "Guide for Laboratory Animal Facilities and Care," prepared by a National Institutes of Health (NIH) Animal Care Panel, was adopted by the Institute of Laboratory Animal Resources of the National Research Council (NRC) and was subsequently revised (1965, 1968, 1972, 1978, and 1985). The title was changed in 1972 to "Guide for the Care and Use of Laboratory Animals," and it became commonly known as the NIH Guide. The APS Council endorsed the NIH Guide in 1963 and strongly recommended that members of the Society bring this excellent brochure to the attention of their universities and institutions (2). The 1980 revision of the Society's Guiding Principles incorporated a statement on the use of the NIH Guide, which has evolved into the most significant document used in the oversight of research animal facilities. Now any institution with NIH funding must abide by the specifications in the NIH Guide. In addition the NIH Guide is also the reference manual used during the American Association for the Accreditation of Laboratory Animal Care (AAALAC) inspections for accreditation of facilities. Therefore, because of the importance of the NIH Guide, in 1983 APS responded to a call for statements on preparation of its sixth edition. The APS Committee on Animal Care and Experimentation reviewed the fifth edition and prepared for the Society a statement that endorsed the NIH Guide and detailed suggested changes to be considered during the revision process. Helene C. Cecil presented APS's testimony at a public hearing held by the Institute of Laboratory Animal Resources (NRC-National Academy of Sciences) in Washington, D.C., in May 1983. Additional testimony was given at public hearings in San Francisco by David J. Ramsay and in Chicago by Walter C. Randall.

Although APS's "Guiding Principles in the Care and Use of Animals" has served the Society continuously since its adoption in 1953, the committee that proposed it has not. The Committee on the Use and Care of Animals (Table 1) was established in 1952, and its activities during the 1950s are documented by W. O. Fenn (4a). Bennett J. Cohen chaired the committee from 1961 to 1967 and, in addition, was active in organizations devoted to the establishment of standards for care of laboratory animals [Animal Care Panel, AAALAC, American Association for Laboratory Animal Science (AALAS), and National Society for Medical Research (NSMR)]. Through his participation in these organizations he kept the APS Council apprised of the high level of activity in deliberations on animal care that occurred during the 1960s. Perhaps because the APS Council felt that Cohen's participation with these organizations could fulfill the functions of the APS Committee on the Use and Care of Animals, at its meeting in February 1967 Council discussed disbanding the committee and establishing a strong Public Affairs Committee instead.

The prevailing issue concerning laboratory use of animals at that Council meeting appeared to be the Society's editorial policy. The APS journals carried the statement, "Papers about which question is raised as to the proper use and care of experimental animals will be referred to the Committee on the Use and Care of Animals." The Council decided that rather than refer questionable manuscripts to that committee, the Publications Committee should make the final decision about publication. At the following Council meeting in April 1967, the Committee on the Use and Care of Animals was abolished. Cohen, however, would remain as APS representative to

Table 1
COMMITTEE ON THE USE AND CARE
OF ANIMALS, 1952–67

	Year	
	Member	Chairman
H. E. Essex	1952–56	1952–56
H. W. Davenport	1952–59	1956–57
J. Haldi	1952–55	
W. F. Hamilton	1952–58	
R. F. Pitts	1957–59	1957–59
A. B. Otis	1958–61	1959–61
A. Hemingway	1959–62	
R. Galambos	1959–63	
B. J. Cohen	1961–67	1961–67
E. Knobil	1962–68*	
P. D. MacLean	1963–66	
R. W. Doty	1966–69*	

Committee not listed in 1967–68 and thereafter.

NSMR. Apparently Council felt that the defense of the use of animals in research was adequately covered by the publication of the NIH Guide in 1963, the passage of the Animal Welfare Act in 1966 (discussed later in this chapter), the adoption of APS's Guiding Principles, and the active participation of Cohen and other Society members in AAALAC, AALAS, and NSMR.

With the abolishment of the Committee on the Use and Care of Animals in 1967, the Public Affairs Committee was established. One of the charges to the Public Affairs Committee was "to deal with matters pertaining to animal legislation and advise Council of actions to be taken." During its formative years, however, the Public Affairs Committee did not report any activity in the area of animal legislation. Therefore in April 1971 the Council decided that with the passage of the Animal Welfare Act of 1970, the Society needed a committee to integrate legislation on animal care with the guidelines for publications and also to advise the Society on the ethics of animal experimentation and care. The Council reestablished the committee as the Committee on Animal Care and Experimentation (Table 2), and Harold R. Parker, the Society's representative to NSMR, was selected to chair the

Table 2
COMMITTEE ON ANIMAL CARE AND EXPERIMENTATION, 1971–86

	Year	
	Member	Chairman
H. R. Parker	1971–74	1971–74
E. T. Angelakos	1971–74	
L. S. Lilienfield	1971–75	
W. J. Tietz, Jr.	1971–77	1974–77
G. H. Stabenfeldt	1974–77	
J. W. Crowell	1974–76	
K. A. Gaar, Jr.	1975–78	1977–78
M. H. Hast	1976–79	
	1980–82	
C. E. Cornelius	1977–80	
D. E. Donald	1977–80	
H. C. Cecil	1978–84	1978–84
R. L. Hazelwood	1979–82	
D. Robertshaw	1980–83	
O. A. Smith	1980–84	
L. J. Ramazzotto	1981–86	
H. H. Erickson	1982–85	
J. A. Krasney	1982–85	
I. J. Fox	1983–86	1984–86
J. F. Amend	1984–87	
J. H. U. Brown	1984–87	
T. N. Pullman	1984–87	
N. Granger	1985–88	
I. Zucker	1985–	
W. M. Samuels*	1982–	

* Ex officio.

committee. The committee resumed its role of evaluating manuscripts when referees questioned whether the methods used conformed to APS's Guiding Principles. From 1959, therefore, with the adoption of the Guiding Principles, physiologists have been expected to conform to guidelines established by APS. Now with the passage of the Animal Welfare Act of 1970 the physiologist must conform to legal, as well as moral, guidelines. The report from the committee in 1974 again urged all investigators to become familiar with the standards of the Animal Welfare Act published in December 1971.

The role of the committee today is very similar to that of 1971: it is responsible for *1*) revising the Guiding Principles, *2*) evaluating manuscripts referred to it by the Publications Committee, and *3*) keeping the Council and Society apprised of legislative and public concerns for laboratory animal welfare. The chairman of the committee also served as the APS representative to the NSMR until 1985 when the NSMR joined with the Association for Biomedical Research to become the National Association for Biomedical Research (NABR). APS became a member of NABR in 1986. In 1981 the committee recommended that APS apply for membership on the AAALAC Board of Trustees, and since that time the Society has had a designated representative to AAALAC's board (O. A. Smith, 1981–83; L. J. Ramazzotto, 1983–86). In addition Ramazzotto was elected in 1984 to a four-year term on AAALAC's council.

Although the Society has maintained intrasociety activity through review of manuscripts and abstracts for adherence to the Guiding Principles, external activities also played an important role during the 1980s. These activities included *1*) attending and participating in conferences (or studies) on issues concerning laboratory animals, *2*) monitoring and responding to activism and militancy of animal rights groups, and *3*) responding to proposed legislation, both state and federal (see below). Physiologists (as individuals or sponsored by their institution), as well as the Society, demonstrated through these activities that they could not be complacent about animal experimentation but must interact with the public on the animal issue.

Conferences and Studies

Only a brief description of conferences and studies will be given, because it is difficult to outline all the involvement of physiologists as participants, attendees, and reviewers.

The symposium Trends in Bioassay Methodology: in Vivo, in Vitro, and Mathematical Approaches (15), organized by the NIH in collaboration with the other agencies participating in the National Toxicology Program, was held 18–20 February 1981. It was the first in a series of government-sponsored symposia and studies resulting from pressures to require researchers to replace animal experimentation with alternative methods. The goal was to describe state-of-the-art bioassay from a variety of perspectives. The technical presentations focused on the advantages, limitations, and prospects for commonly used research and testing methods. For those assay tasks that require the use of live animals, the participants gave particular attention to the feasibility of using in vitro or mathematical methods.

The First Conference on Scientific Perspectives in Animal Welfare (4), sponsored by the Scientist's Center for Animal Welfare, was convened 11–13 November 1981. Invited speakers discussed the responsibilities of the individual investigator, the institution, the funding agency, and the editorial reviewers in animal experimentation. These same topics were also discussed in workshops during the conference. Many APS members participated.

The National Symposium on Imperatives in Research Animal Use: Scientific Needs and Animal Welfare (14) was held 11–12 April 1984. Benefits of animal research to clinical medicine, moral stewardship and responsibility of investigators, legislative and institutional standards for care and use of animals, model systems for development of new information in biomedicine, and understanding and controlling pain were discussed.

In a study originated in 1983, NIH asked NRC for advice regarding the potential of using microorganisms, tissue cultures, invertebrate animals, and mathematical formulations as models in biomedical research (12). The NRC committee focused on six areas of basic research, which were examined in a series of workshops in 1984: cellular immunology, learning, development, regulation, diseases and aging, and mathematical modeling.

At the request of Senator Orrin Hatch, chairman of the Senate Committee on Labor and Human Resources, the Office of Technology Assessment analyzed the scientific, regulatory, economic, legal, and ethical considerations involved in alternative technologies in biomedical and behavioral research, toxicity testing, and education. The 441-page report (13) was published in 1985.

A committee of the NRC's Commission on Life Sciences (CLS) will examine issues such as public concerns about animal use and treatment, benefits derived by humans and animals from research with animals, scientific and technical developments that may affect the use of animals for experimental purposes, and effectiveness of current regulatory and self-regulatory or voluntary guidelines for animal care and use. Concurrently a national survey of the extent of laboratory animal use, facilities, and resources will be conducted by a committee of NRC's Institute of Laboratory Animal Resources. This survey is expected to be an important source of information for the CLS committee. These studies will result in two reports: one will present the findings of the national survey, and the other will assess the issues surrounding the use of animals in research. Support for the studies is provided by federal agencies, pharmaceutical companies, and private organizations. The studies are scheduled to be completed in 1987.

The above studies and conferences emphasize the challenge animal rights groups have made to legislators, the federal government, and the biomedical community. The response to the challenge has been to give balanced comprehensive facts about the use of laboratory animals. Meanwhile, as they continue to defend the use of animals in research, physiologists also strive for scientific evidence on evaluation and alleviation of pain and stress in animals. APS has contributed to symposia and publications in these areas as follows.

The symposium Animal Pain, Perception and Alleviation, presented at the Sixty-sixth Annual Meeting of FASEB, 20–21 April 1982, was published by APS in 1983 (10). It specifically addressed pain in animals to establish the factual background on acute pain.

A number of APS members participated in the International Conference on Animal Stress held 6–9 July 1983 in Sacramento, California. The symposium, which addressed new directions in defining and evaluating the effect of stress in animals, was published by APS in 1985 (11).

A series of contributed papers reviewing the health benefits from experimental research with the cat (16), dog (6), mouse (8), rabbit (5), rat (7), and primate (9) was first published in *The Physiologist* and subsequently as a single volume by the Foundation for Biomedical Research. The papers are intended to convey a sense of the value of biomedical research, as well as the necessity for the use of animals in research, and to serve as background documents for the scientist or writer who seeks to improve the public's understanding of medical research.

APS and the Animal Rights Movement

Probably history will mark 11 September 1981 as the beginning of animal rights as a viable movement in the United States. On that day a volunteer laboratory animal assistant publicly charged a Silver Spring, Maryland, scientist with abuse of research monkeys. The incident had legal, political, and social ramifications regarding the rights of animals and swept APS into the mainstream of this international phenomenon.

In the United States the animal rights movement, by and large, is led by two groups: one overt, one covert. The overt group is People for the Ethical Treatment of Animals (PETA); the covert group is the Animal Liberation Front (ALF). Despite the denials by PETA, the suspicion remains that the leadership of the two groups is one and the same.

The founder of PETA, Alex Pacheco, as a twenty-three-year-old college student, volunteered his services as a laboratory animal assistant at the Institute for Behavioral Research in Silver Spring. Four months later Pacheco made public accusations of animal abuse that led to the confiscation of seventeen monkeys by Montgomery County Police and the criminal prosecution of Edward Taub for 119 counts of violation of the animal cruelty statute of the State of Maryland. Moreover NIH suspended the $117,000 that remained unspent from a $200,000 grant supporting Taub's research.

A five-day trial of Taub in the county's district court, heard by a judge without a jury, brought a parade of witnesses from both the animal protection organizations and the scientific community. Three weeks later the judge acquitted Taub on all but six counts; he ruled that Taub was guilty of failing to provide adequate veterinary medical care for six deafferented monkeys. Taub was fined a total of $3,000.

On an appeal a jury acquitted Taub on five of these six counts, and the Maryland Court of Appeals dismissed the sixth count on the grounds that the state courts had no legal jurisdiction, inasmuch as animal research is not included in the statute concerned with animal cruelty. (A two-year lobbying effort by The Humane Society of the United States and its state and local affiliates resulted in the Maryland General Assembly's amending the law in 1985 to include laboratory animals under the state's animal cruelty statute.)

In August 1983, Taub, a member of APS since 1971, requested that APS support him in obtaining restoration of his status as a research investigator. Despite his having been relieved of criminal charges, his grant had been terminated by NIH. The APS Council, to investigate the circumstances surrounding this affair, formed a special committee with APS former President Earl H. Wood as its chairman.

The committee's investigation extended over a period of three months and included a review of hundreds of pages of documentation, such as court records and animal facility inspection reports by the U.S. Department of Agriculture (USDA) and NIH. The committee also interviewed seventeen persons directly involved in the affair and conducted site visits at NIH and its animal facility at Poolesville, Maryland, and at the Institute for Behavioral Research.

After reviewing the special committee's report the Council issued the following four-point statement:

1. Although the controversy now seems to involve only one investigator, we [Council] are persuaded that the disposition of this matter will seriously affect biomedical research throughout the nation, if not the world, for years to come. It is therefore an appropriate matter for concern and participation by the scientific community and particularly biomedical societies such as APS.
2. APS and others in the scientific community should express serious concern to NIH, USDA, and other appropriate agencies if conditions warranting termination of a grant were allowed to exist for years while numerous inspection and evaluation reports did not note these conditions so corrective actions could be taken. However, if the deficiencies were of recent origin, the suspension and ultimate termination of the grant without adequate notice or opportunity to remedy is unwarranted. Scientists have the right to be able to rely upon inspection and site visit reports to assure an effective job of self-policing.
3. Whenever research dollars are authorized for animal research, the appropriation should include funds to provide and maintain adequate animal facilities to meet the standards of the NIH Guide for the Care and Treatment of Laboratory Animals and the Animal Welfare Act. To this end, NIH, APS, other scientific societies, and humane societies should work together to secure the availability of funds for research animal facilities.
4. Because suspension and termination of the NIH grant appears to have been motivated and unduly influenced by unusual political and societal pressures, APS, in concert with other societies, should encourage NIH to enable Dr. Taub to complete the research previously approved by peer review, with appropriate conditions imposed to ensure that the handling of the animals is beyond reproach.

PETA continued to pursue its primary objective—custody of the confiscated monkeys—with the filing of a civil suit in federal court. Joining PETA in the suit were the International Primate Protection League, the Association for Animal Law Enforcement, and seven animal rights activists. By using the arguments employed for gaining custody of battered children, the suit claimed that the animal rights organizations had a legal claim because the organizations and individuals filing the suit had established a bond with the monkeys. Since the time of the confiscation the animal rights supporters had been visiting the monkeys at an NIH facility several times each week and bringing fresh fruit, toys, and a television set to improve the monkeys' environment and well-being.

A federal magistrate reviewing the suit for the Federal District Court recommended that the suit be dismissed on the grounds that neither animal rights groups nor individual supporters of the animal rights movement had been granted standing by the courts and therefore were not eligible to file a civil suit. (In general terms, standing is the recognition granted by a court to private citizens and organizations as plaintiffs with legally protectable and tangible interests at stake in litigation.)

In an effort to overturn the lower-court decision and win recognition of standing with the courts, the ruling was appealed to the Federal Court of Appeals in Richmond. The appeal, however, caused an unexpected effect, a unification of the scientific community for the first time. Before the appeal was filed, the scientific community, by and large, had shrugged off raids, demonstrations, and theft of laboratory animals as little more than acts of vandalism. By contrast, the question of granting standing to animal rights groups and individuals became of significant concern. It would open the courts to more suits, inasmuch as there would be a legal recognition that animal rights advocates have a guardian relationship with laboratory animals. In support of the defendants—the Institute for Behavioral Research, which owns the monkeys, and NIH, which has been maintaining the monkeys—an amicus curiae brief was filed with the appeals court on the behalf of sixty-nine scientific organizations, including APS, which was one of the six members of the coordinating committee that initiated the friend-of-the-court brief. In September 1986 the Federal Court of Appeals upheld the Federal District Court's judgment and a month later rejected the plantiff's petition for a rehearing.

APS and Legislation Concerning Laboratory Animals

During the fourth of six consecutive Friday afternoon meetings in 1982 in the U.S. House of Representatives' Cannon Office Building, Francis Haddy made public the Society's first legislative proposal concerning laboratory animals. The purpose of the meetings was to try to obtain from opposing factions some agreement on animal welfare legislation proposed by Pennsylvania Democratic Congressman Doug Walgren of Pittsburgh. Organizations invited to the meetings, in addition to APS, were the American Humane Association, the American Psychological Association, the Animal Welfare Institute, the Association of American Medical Colleges, the Humane Society of the United States, PETA, and NSMR. Representing APS were Haddy, the Society's president; Orr Reynolds, executive secretary-treasurer; and

William Samuels, the Society's consultant on matters dealing with laboratory animals and government relations.

Walgren's legislation for laboratory animal "reform" stemmed from the incident several months earlier in Silver Spring, Maryland, as described above. The essence of Walgren's legislation was to charge a second federal agency (NIH) with the responsibility of regulating the care, use, and treatment of laboratory animals. These regulations would cover all animals used in research supported by U.S. Public Health Service grant funds and would be in addition to the standards for animal care established by the USDA as mandated by the Animal Welfare Act. Both the research and education communities strongly opposed the legislation, despite Walgren's willingness to delete from the bill several controversial provisions, including mandatory accreditation of all animal laboratories by AAALAC and use of forty to sixty percent of all NIH animal research grant monies to develop alternative nonanimal methods.

In the first three Friday afternoon sessions, there was little agreement between the researchers and educators and the animal welfare advocates. The latter believed that the proposed legislation already had been compromised by the deletion of mandatory funds for development of nonanimal methods and of the requirement that all laboratory animal facilities be accredited. The fourth session showed no promise of change in the attitudes displayed in the previous sessions until Haddy spoke:

> We already have one federal agency [USDA] charged by the Animal Welfare Act to monitor the care, use, and treatment of laboratory animals and since no one has yet to prove that the Animal Welfare Act is broken, it makes absolutely no sense to reinvent a duplicate program. What APS proposes is that we work to improve the shortcomings of the present system by getting the Congress to amend the Animal Welfare Act rather than legislating a duplicate system.

Christine Stevens, president of the Animal Welfare Institute, immediately agreed with the APS proposal, and others voiced qualified support for the concept. Only PETA opposed totally the proposition. Thus, by Haddy's proposal, a concept was initiated that culminated forty-six months later when the Congress finally amended the Animal Welfare Act in December 1985.

Stevens' support for the APS proposal came as no surprise to anyone knowing the historical aspects of the Animal Welfare Act, which was enacted in 1966 as the Laboratory Animal Welfare Act and was renamed the Animal Welfare Act in 1970 when the Congress amended the law to include zoos, circuses, carnivals, exhibitions, and wholesale pet dealers. Then it was primarily Stevens' efforts that persuaded the Congress that there was a need for federal monitoring of the care, use, and treatment of animals. Prior to this, each state set its own standards for animal welfare.

In the early 1960s Stevens successfully promoted the enactment of the Humane Slaughter Act, and since that time she has promoted the need for federal standards for laboratory animals. She gained much of her political clout from her husband, Roger Stevens, a former chairman of the National Finance Committee of the National Democratic Party, and developed much of her concern for animals from visiting the laboratory of her father, Robert Gesell, when he was chairman of the Department of Physiology at the University of Michigan. (Gesell, a member of APS, criticized his colleagues at the 1952 business meeting of the Society for improper and unjustified use of animals in biological and medical research.)

Thus the concept fostered by the Society forged an unusual alliance, inasmuch as APS was joined in its effort by the one person who was primarily responsible for the federal laws governing the care of laboratory animals and whose father had quarreled openly with his APS colleagues for alleged abuses of laboratory animals (see chapt. 5).

Throughout the remainder of the year APS and its allies in the research and education communities continued their coordinated efforts to modify the Walgren bill, which was gaining support in the Congress because of an extensive letter-writing campaign by animal rights advocates. Moreover the members of Congress did not fail to notice that the volume of mail they were receiving on the issues of animal rights and animal research exceeded what they received about the war in Vietnam. This barrage of mail put the Walgren bill on a fast track within the House Committee on Science and Technology, which approved the bill in August. But the mail also aroused the interest of the House Subcommittee on Health and the Environment, which did not hold its hearings on the bill until December. As a result the bill died in the subcommittee when the Ninety-seventh Congress adjourned two weeks later.

Meanwhile, both APS and the Animal Welfare Institute independently sought congressional support to amend the Animal Welfare Act, but they found little interest while the Walgren bill was moving ahead.

When the Walgren bill died with the adjournment of the Ninety-seventh Congress in December, the concept was immediately taken up by Republican Senator Robert Dole of Kansas. Dole had been the sponsor of the Senate version of the Walgren bill during the Ninety-seventh Congress. He was now willing to provide the Ninety-eighth Congress with an alternative to laboratory animal legislation. Despite APS's strong interest in amending the then-current law rather than in creating new legislative authorities to govern the care and use of laboratory animals, the draft amendments proposed by Dole and communicated to APS fell short of what the Society believed were viable modifications to the Animal Welfare Act. Walter Randall, who was then APS president, Haddy, Reynolds, and Samuels met several times in the early weeks of 1983 to determine how the legislation must be changed before the Society could support the bill. Consequently, when Dole introduced the Improved Standards for Laboratory Animals Act in March, most of the Society's substantive recommendations were incorporated in the bill, and APS Council gave the proposal its qualified support pending some technical changes in the bill's language. In a move to broaden the support base for Dole's proposal, APS sought FASEB's endorsement of the Council's action but was rebuffed by the FASEB Board, because the American Society for Pharmacology and Experimental Therapeutics opposed any amendments to the Animal Welfare Act.[1]

In July, Dole conducted hearings on his proposed amendments to the Animal Welfare Act, and Randall testified that although the Society continued to support the concept, it was withholding its complete support until language was changed in four sections of the bill. Of most concern to the Society was identification of the Institutional Animal Care Committee's "nonaffiliated member," who, according to legislation, was to represent "concern for the welfare of the animal subjects." Not only did this portray a public image that researchers and educators do not have concern for their laboratory animals, but it also implied that the nonaffiliated person must be an animal welfare protectionist. The Society had been an early proponent of institutional animal care committees with a nonaffiliated member, but not in those terms.

At the hearing only APS and the American Institute of Biological Sciences testified in support of the concept, whereas the other research and educational organizations spoke against the need for any kind of legislation involving laboratory animals. The USDA also spoke against the bill; it claimed that its Animal and Plant Health Inspection Service already had too many responsibilities under the Animal Welfare Act. The animal welfare organizations gave the proposal mixed reviews.

After the hearing the issue of laboratory animal reform slipped into a lull that lasted into the second session of the Ninety-eighth Congress. Then California Democratic Congressman George E. Brown, Jr., of Riverside introduced a companion to Dole's bill, and Walgren successfully moved to have this proposal added as a rider to the NIH authorization renewal bill. In the Senate, Orrin Hatch, Utah Republican, and Edward Kennedy, Massachusetts Democrat, initiated bipartisan legislation calling for a national study on the use and need of laboratory animals for research, education, and testing, while two Republican Congressmen, Edward Madigan of Illinois and James Broyhill of North Carolina, promoted a similar study in the House.

Also heightening congressional interest in laboratory animal reforms in the summer of 1984 were two highly publicized raids at the University of Pennsylvania, staged by the ALF. In the first raid the vandals stole sixty hours of video tapes on head injury research involving the use of nonhuman primates. The tapes were distilled into a twenty-eight-minute display of alleged animal cruelty that was shown widely on both local and network television. The second raid six weeks later resulted in the theft of thirteen animals. To counter this stepped-up vandalism by the ALF, the Society, in its second legislative proposal concerning laboratory animals, proposed that the Congress enact legislation to make ransacking or stealing from federally funded research laboratories a federal crime. The proposal was approved by the APS Council and endorsed by the Society's membership at its fall meeting in August 1984.

Because of this renewed interest in laboratory animals in the Congress and with a December adjournment less than four months away, the sponsors of animal welfare legislation were looking for fast tracks. Brown was the first to move; he held hearings on his bill in September. This hearing, however, differed from the Dole hearing (described above) in that the research and education organizations joined with APS in support of amending the Animal Welfare Act as a viable alternative to enacting new legislative authorities, although they gave only qualified support for Brown's bill until changes in the bill's language were agreed on, including the definition of the nonaffiliated member of the Institutional Animal Care Committee.

As at the Dole hearing, Randall presented the Society's position before the Brown committee. This time, however, Randall made public the APS proposal that raids on federally funded research institutions be made a federal crime. Both the Association of American Medical Colleges and the NSMR endorsed the proposal, whereas the animal welfare organizations berated it.

Meanwhile in the Senate a standoff developed when its members could not decide which bill to consider—the Dole or the Hatch-Kennedy bill. In the House, Madigan and Broyhill, senior members of the Committee on Energy and Commerce, presented their animal study bill before their own committee, which was in the process of reconsidering the NIH authorization renewal. The study bill was offered as a substitute amendment for the Walgren rider, which had been added several weeks before. When the committee failed to accept the study as a substitute for the Walgren bill, the study was also added as a rider and was included, along with the Walgren provisions, in the House-passed version of the NIH renewal authorizations. In the late fall the Senate adopted the House bill, and in December the Congress passed the NIH renewal bill, but the President vetoed it after the Congress had adjourned. Thus the study bills, the Walgren bill, and the Dole-Brown Animal Welfare Act amendments all died with the Ninety-eighth Congress.

When the Ninety-ninth Congress convened in January 1985, Dole was quick to reintroduce his legislative proposal. The language of the bill had been revised to accommodate the concerns of APS and its colleagial societies, including the definition of the nonaffiliated member of the Institutional Animal Care Committee.

Brown, however, did not rush to resubmit legislation, that is, not until April, when ALF raided a major research facility in his home district (the University of California at Riverside) and caused more than $683,000 in damage and stole 467 laboratory animals. Soon after the raid Brown not only introduced a companion bill to the revised Dole bill, but he also introduced the APS proposal that would make such raids, thefts, and vandalism a federal crime. The purpose of this latter bill was to permit authorities to follow vandals across state lines. Since 1982, animal rights raiders had staged four raids a year on academic research institutions and had taken both stolen items and laboratory animals to other states.

Although the proposal that such raids be made a federal offense did not receive support from within the Congress, it may have given the ALF and others second thoughts about conducting more raids. (The Riverside raid was the only invasion of a laboratory in 1985, possibly because the raiders were concerned that to continue their raiding of animal laboratories would generate the congressional support the bill had heretofore failed to receive. In fact ALF did not conduct another raid until a few days after the Ninety-ninth Congress adjourned in October 1985. In that incident, at the laboratories at the University of Oregon, 156 laboratory animals were stolen.)

Also in the spring of 1985 the House again considered the renewal authorization for NIH. This time, however, the bill's only laboratory animal reference was the Walgren proposal. Interest in a study of laboratory animal use and need subsided in both the House and Senate, because the Congressional Office of Technology Assessment had begun its own study of these issues.

By mid autumn both the NIH authorization renewal bill and the amendments to the Animal Welfare Act again were on fast tracks; the latter was added as a rider to the 1985 farm bill. In November the Congress passed the NIH reauthorization bill, and the President again vetoed the bill. This time, however, the Congress was still in session and overrode the veto; thus the Walgren provisions became federal law.[2]

The major terms of the Walgren provisions direct the U.S. Department of Health and Human Services, through the NIH director, to establish guidelines for laboratory animal care and treatment. Also a recipient of U.S. Public Health Service funds must have an Institutional Animal Care Committee, and this committee must provide NIH with proper assurances that the institution is in compliance with the guidelines. The NIH director also is charged with creating a plan for conducting both research and training in valid and reliable alternatives to animal models and in methods that will reduce the numbers of animals used and also lessen any pain and distress laboratory animals may experience.

A few weeks after the NIH renewal was passed, the farm bill was also passed and signed by the President. The farm bill included the amendments to the Animal Welfare Act originally proposed forty-six months earlier by APS. Of the nine sections of the act that were amended, two sections involve directly the care, use, and treatment of laboratory animals. The first requires that all institutions provide adequate veterinary care to assure appropriate use of anesthetics, analgesics, tranquilizing drugs, and euthanasia and to ensure that animal pain and distress are minimized. Moreover the attending veterinarian is to determine proper exercise for all laboratory dogs and to assure that nonhuman primates are maintained in an adequate psychological environment.

Secondly, all institutions using animals must have institutional animal care committees to ensure that the standards and guidelines for laboratory animal use and treatment are being met. The language in this section was changed so that all members of the committee are to be responsible for "society's concerns for the welfare of the animal subjects."

Other amendments to the act include mandatory annual inspection of animal facilities by the USDA Animal and Plant Health Inspection Service; annual institutional training programs on animal care for all persons involved in the use of laboratory animals; creation of an information clearinghouse at the National Agricultural Library on methods of research and alternatives to animal models; increased penalties for institutions and individuals found to be in violation of the act; and consultation among the secretaries of the Departments of Agriculture and Health and Human Services before any new standards, guidelines, or regulations involving the care, use, or treatment of animals are promulgated.

With the passage of laboratory animal legislation by the enactment of both the NIH and the farm bill, the Congress has relieved itself of the pressures it has felt since the incident at Silver Spring in September 1981. The belief that the pressure is off the Congress is supported by the current shift in the momentum for laboratory animal legislative reform to state houses and municipal seats of government, where animal rights activists are conducting extensive campaigns for laws that would prohibit the release of any animal from a pound or shelter for any purpose other than pet adoption. The Humane Society of the United States, which is sponsoring a coalition of national, state, and local animal welfare groups, has as its goal the enactment of such laws in all fifty states before the year 2000 and predicts that forty percent of the goal will be reached before 1990. Twelve states already have adopted such laws. Moreover this coalition has further stated that its eventual goal is the abolition of all animal use for the purposes of research, education, or testing. Thus, as APS begins its second century, most of the issues surrounding the use of laboratory animals remain unchanged. Only the location of the debate has been shifted.

NOTES

[1] Several weeks later Walgren reintroduced his bill, in which he reinstated some of the provisions he had deleted in the previous Congress. This time, however, there would be no companion bill in the Senate.

[2] Normally the promulgation of regulations follows the enactment of legislation. This time, however, the U.S. Public Health Service adopted regulations tailored to the proposed Walgren amendments one year before the legislation was enacted. Both the law and the regulations became effective on 31 December 1985.

REFERENCES

1. ANONYMOUS. The care and use of animals. *Physiologist* 2(4): 47, 1959

2. ANONYMOUS. Guide for laboratory animal facilities and care. *Physiologist* 6: 334, 1963.

3. BENISON, J. In defense of medical research. *Harvard Med. Alum. Bull.* 44: 16–23, 1970.

4. DODDS, W. J., AND F. B. ORLANS (Editors). *Scientific Perspectives on Animal Welfare.* New York: Academic, 1983.

4a. FENN, W. O. *History of the American Physiological Society: The Third Quarter Century, 1937–1962.* Washington, DC: Am. Physiol. Soc., 1963.

5. FOX, R. R. Health benefits in animal research. The rabbit as a research subject. *Physiologist* 27: 393–402, 1984.

6. GAY, W. I. Health benefits in animal research. The dog as a research subject. *Physiologist* 27: 133–141, 1984.

7. GILL, T. J., III. Health benefits in animal research. The rat in biomedical research. *Physiologist* 28: 9–17, 1985.

8. JONAS, A. M. Health benefits in animal research. The mouse in biomedical research. *Physiologist* 27: 330–346, 1984.

9. KING, F. A., AND C. J. YARBROUGH. Health benefits in animal research. Medical and behavioral benefits from primate research. *Physiologist* 28: 75–87, 1985.

10. KITCHELL, R. I., AND H. H. ERICKSON (Editors). *Animal Pain, Perception and Alleviation.* Bethesda, MD: Am. Physiol. Soc., 1983.

11. MOBERG, G. P. (Editor). *Animal Stress.* Bethesda, MD: Am. Physiol. Soc., 1985.

12. NATIONAL RESEARCH COUNCIL. *Models for Biomedical Research: A New Perspective.* Washington, DC: Natl. Acad. Press, 1985.

13. U.S. CONGRESS, OFFICE OF TECHNOLOGY ASSESSMENT. *Alternatives to Animal Use in Research Testing and Education.* Washington, DC: US Govt. Printing Office, 1986. [Publ. no. OTA-BA-273.]

14. U.S. DEPARTMENT OF HEALTH AND HUMAN SERVICES, PUBLIC HEALTH SERVICE, NATIONAL INSTITUTES OF HEALTH. *National Symposium on Imperatives in Research Animal Use: Scientific Needs and Animal Welfare.* Washington, DC: US Govt. Printing Office, 1985. [NIH publ. no. 85-2746.]

15. U.S. DEPARTMENT OF HEALTH AND HUMAN SERVICES, PUBLIC HEALTH SERVICE, NATIONAL INSTITUTES OF HEALTH. *Trends in Bioassay Methodology: in Vivo, in Vitro, and Mathematical Approaches.* Washington, DC: US Govt. Printing Office, 1981. [NIH publ. no. 82-2382.]

16. WARFIELD, M. S., AND W. I. GAY. Health benefits in animals. The cat as a research subject. *Physiologist* 27: 177–189, 1984.

CHAPTER 16

FASEB and the APS

JOHN R. BROBECK

A history of any organization should give an account of the plans and accomplishments of its leaders and the major problems they encountered during their tenure of office. For the American Physiological Society (the Society, APS) and the Federation of American Societies for Experimental Biology (the Federation, FASEB), their leaders for the past forty years have been first of all their executive officers. Milton O. Lee was appointed to this position in both organizations in 1947 and remained in charge of FASEB operations until 1965. Near the middle of his term, in 1956, he relinquished responsibility as executive secretary of APS on the appointment of Ray G. Daggs as executive secretary-treasurer. Lee continued, however, as managing editor of the APS journals until his retirement in 1965. Ray Daggs served the Society until 1972, when he was succeeded by Orr E. Reynolds, a member of APS since 1948.

Milton Lee was elected to membership in the APS in 1927. His research interests were in endocrinology, metabolism, and nutrition. He was the only physiologist who has been director of FASEB. When he retired from the position, a nominating committee was appointed under the chairmanship of Philip Handler, with John Brookhart serving as APS representative. Joseph F. A. McManus, then a professor at Indiana University at Bloomington, and for 1962–63 the president of the American Society for Experimental Pathology, was selected for the FASEB directorship. He held the office until 1970, when he resigned to become dean of the College of Medicine of the Medical University of South Carolina. Four years later he relinquished the deanship and returned to the discipline of pathology as a professor in that department.

To fill the position when McManus left, E. H. Lennett, chairman of the Federation Board, appointed a nominating committee with Lauren A. Woods as chairman and John Brobeck as APS representative. Their final reporting was delayed longer than had been anticipated. Consequently, on departure of McManus from the Beaumont campus, Stanford Moore, then chairman of the Federation Board, served also as acting executive director.

In 1971 the Federation appointed as executive director Eugene L. Hess, by training a chemist and a member of the American Society of Biological Chemists from 1958. He had been a senior member of the scientific staff at the Worcester Foundation for Experimental Biology for eight years and then had devoted six years to administrative positions at the National Science Foundation. He held the office at FASEB for eight years. His successor was chosen by a nominating committee with G. Barry Pierce as chairman and William F. Ganong as APS representative. They proposed the name of Robert W. Krauss, dean of the College of Science at Oregon State University in Corvallis, who took office at the Federation in 1979. Krauss previously had been professor of plant physiology and chairman of the Department of Botany at the University of Maryland.

In the nearly forty years since Lee became its first executive officer, the Federation has changed in many ways, and not only by growing in size and complexity. If one of Lee's principal objectives was scientific communication and publication, particularly in the journals of FASEB and APS, that of McManus was to strengthen the influence of the Federation in national scientific affairs. Members of the constituent societies, especially the APS, sometimes did not fully agree with the tactics he employed toward this goal. On his resignation at least a few members of the Federation Board set out to bring all programs of FASEB more directly under the influence of the six societies. This was done by appointment of a six-member advisory committee for those offices of FASEB that did not already have one. Each year as he was preparing a new budget, Eugene Hess often said that his life would be much easier and finances would be brighter if he did not have to attend and pay the expenses of so many committee meetings. The committees remain in place, however, as evidence of the interest of the several societies and of their intention to be at least informed about all FASEB activities.

Hess brought to the office of executive director a significant change when he revised the form in which finances were projected and recorded. For the first time it became possible to know just how much income each of the Federation's programs was producing, what its costs were, and whether it was being managed efficiently. Hess has written that the most important reward of being in office was interchange with the society executive officers in residence at the Beaumont campus, as well as with officers elected by the societies, during their terms on the Federation Board. In reference to the APS, he mentioned Orr Reynolds for "his ability to communicate, his intrinsic fair mindedness, and his innate receptiveness to new ideas." He also wrote of the favorable relations between the Society and the Federation during 1975–76, when Arthur Guyton, past president of APS, was chairman of the Federation Board. It was during the tenure of Hess that poster sessions were introduced to help minimize disadvantages of the growing size of annual meetings, the public information and public affairs activities of FASEB were expanded, the first congressional fellow was appointed, and the mission of the Life Sciences Research Office was clarified. In 1974 the Gregory Pincus Memorial Award was established, in 1975 the 3M Life Sciences Award was begun, and in 1977 the Wellcome Visiting Professorships in the Basic Medical Sciences were instituted. It was also while Hess was in office that the question of equity of the societies in Federation resources was debated and finally settled (see below).

Hess left office with relationships between FASEB and the societies less tangled than they had been for some years. He steered the Federation through the hazards of permitting member societies to meet separately from the annual meeting (see later), and he introduced effective cost accounting. He was not able to persuade all critics that *Federation Proceedings* is a worthwhile publication that every member must subscribe to, but this is a chronic complaint that antedates the appointment of McManus and continues to appear on the desk of Krauss. Finally, when Hess retired there was much concern about the finances of FASEB. Attendance at annual meetings was either stable or declining, costs were rising (especially labor costs at the meetings), the intentions of exhibitors and advertisers were uncertain, and there was reason to fear that soon the income from meetings would no longer cover those operating costs of FASEB for which the societies were ultimately responsible.

As noted below, Krauss has been able to handle the income/cost problem more favorably than anyone believed to be possible. He revised the manner in which "overhead" is assigned to the several programs of FASEB and the societies and thus made the Office of Scientific Meetings more competitive with outside offices providing similar services. Moreover by careful management he restrained rising costs while income was being improved. Simultaneously all FASEB departments expanded their services to member societies, as well as to new scientific societies that came to reside on the Beaumont campus.

A major new program Krauss originated is the FASEB Summer Research Conferences. Held in an academic setting in Vermont, somewhat in the pattern of the Gordon Research Conferences, the FASEB program began in 1982. During their first year the conferences attracted some 900 participants, from nearly every state and many foreign countries. By 1984 there were 1,301 in attendance. Topics selected are more definitely biological than those of the Gordon Conferences and are chosen to represent interests of members of the FASEB constituent societies. Following the pattern reemphasized during Hess's tenure, these conferences are guided by a Scientific Advisory Committee; Jack L. Kostyo was the APS representative for 1982–83 and 1983–84, and Michael J. Jackson was the APS member for 1984–85.

In any consideration of relations between FASEB and APS, the Beaumont campus must be mentioned. Wallace Fenn referred briefly to the purchase of the property in 1954, with most of the interesting details of the negotiations included in his chapter on publications (1). Older members of the APS still feel that other societies of FASEB should feel grateful because the APS, via the Board of Publication Trustees, was willing to underwrite much of the cost of purchasing the property, its mansion, and outbuildings, until an occupancy permit could be obtained and title could be transferred to the Federation a few months later. In less than ten years, however, what had seemed to be commodious facilities had proved to be inadequate for growing needs of the societies and FASEB, and the first wing of what became the Milton O. Lee Building was dedicated in 1963. Additional wings were occupied in 1965 and 1967. Ten years later a need for additional space was perceived, and architect's plans and a model in a contemporary style were authorized. Financing and construction, however, were not approved by the Federation Board at that time. Finally, in 1978 another proposal for additional space was drafted, in the form of two additional wings to the Lee Building in corresponding architectural style. A Development Fund Committee was appointed to oversee the securing of financing. Earl H. Wood and John Brobeck represented the APS on this committee, the principal function of which was to solicit gifts from foundations and corporations.

The two new wings are expected to increase total usable space from about 48,000 to 78,000 square feet. Of this amount the APS, the tenant with the largest amount of space, will increase its occupancy from about 8,000 to 10,000 square feet. Other societies will increase their utilization modestly; FASEB will go from about 20,000 to 27,000 square feet, and some additional space would be available for new tenant societies that might wish to take advantage of Beaumont's location and facilities. On 21 June 1984, the Federation Board voted unanimously to proceed immediately with the construction of both wings at a cost not to exceed $3,600,000.

By 1964 the total registration of members and scientific nonmembers at the FASEB annual meeting had reached nearly 12,500 persons. It continued to rise until 1968, when it attained something more than 15,500. In 1969, however, representatives of the American Society of Biological Chemists (ASBC) informed the Federation Board that in 1971 they proposed to hold a separate annual meeting in San Francisco. In spite of a long-standing complaint that the FASEB annual meeting had become "too big," the possibility that its size would be reduced by the absence of one of the societies unsettled the officers of the other five. The ASBC meeting in San Francisco turned out to be highly successful—scientifically, socially, and financially. It brought two problems to FASEB, however, and thus to the other societies. Income from the annual meeting ordinarily supported the ongoing activities of FASEB. But at first none of the ASBC meeting income reverted to FASEB, and so there was apprehension about how nonmeeting costs of the Federation would be paid when ASBC met separately, as they expected to do from time to time. Eventually, through intermediation first by Hess and then by Krauss as executive director, and in fulfillment of a commitment made by Stanford Moore in 1969 that "a formula would be worked out whereby the income to the Federation would not be decreased because of this innovation in the Meeting," an equitable plan for assigning costs was agreed on. In brief the six societies are now assessed on a per member basis for nonmeeting expenses for the Federation. Then net income from each joint meeting is returned on a proportional basis to those societies that participated in the meeting. By 1983 what at first seemed to be an insoluble problem had been worked out so that each participating society received an income allocation equal to its assessment for that year.

The second problem created by the success of the first ASBC independent annual meeting was the possibility that member societies, specifically the ASBC but also perhaps the APS, might be well advised to withdraw from the Federation to manage more economically and efficiently their own meetings and business affairs. In the usual sense of the word the Federation cannot be said to be "wealthy," and there has never been any serious question about its not-for-profit status. But the development of commercial and residential properties in the Bethesda area, the construction of the three wings of the Milton O. Lee Building, and generalized inflation of property values have greatly increased the market value of Federation holdings. These, together with investments belonging to the Operating Reserve Fund and the Funded Depreciation Reserve, give the impression that FASEB has become at least potentially well endowed. In discussions of what would happen if a society should withdraw, it was assumed by nearly everyone that it would be entitled to some share of this wealth, inasmuch as the six societies together were judged to have created it. The fact that this opinion was widely held, however, solved no problem. Rather, it intensified the dilemma. It appeared that if a major society should decide to leave, the assets of the Federation would have to be liquidated to provide a fair dowry.

A further complication would arise if any other society should wish to join the Federation. If the six member societies each shared in some proportional way in FASEB's assets, the addition of any new society would have the effect of diluting the equity of the original six. How could a new society join unless it was willing and able to contribute its share to the capital resources?

After innumerable hours, throughout many months, spent in discussions and even argument, the issue was abruptly settled, in a manner both surprising and conclusive. Asked to collaborate on a document that stated just what was the equity of the several societies and what each might count on if it should withdraw, the lawyers representing FASEB, those employed by APS, and those of ASBC simultaneously advised their clients that no society had any equity whatsoever. The resources of FASEB belong to the Federation in its own right as a legally constituted scientific and educational society.

So clear and convincing were these opinions that almost at once everyone involved in the negotiations felt a sense of relief that there was, after all the fuss, no problem. This brought a new spirit of harmony and cooperation to meetings of the Federation Board and its Executive Committee. It seems questionable, of course, whether the some 20,000 members of the six constituent societies were very much caught up in these disputes. Yet it was the consequent atmosphere of collegiality among officers on the Federation Board and Executive Committee that permitted other, less threatening issues to be handled expeditiously and that eventually yielded the situation where by 1983 the Federation could once again operate with no net cost to the several societies.

REFERENCE

1. FENN, W. O. *History of the American Physiological Society: The Third Quarter Century, 1937–1962.* Washington, DC: Am. Physiol. Soc., 1963.

CHAPTER 17

International Physiology

KNUT SCHMIDT-NIELSEN

Almost a century ago, in 1889, the first International Physiological Congress was held in Basel, Switzerland, two years after the American Physiological Society (APS, the Society) was founded in 1887. During these nearly one hundred years, American physiology has played an increasingly important role in the development of modern physiology, not only in North America but internationally as well.

Although international congresses have been held for almost a century, the formal organization of international physiology dates back only thirty-three years, to 1953, when the International Union of Physiological Sciences (IUPS) was formed. American physiologists, notably members of the APS Council, played a leading role in the conferences that led to the formal establishment of IUPS. Yet the informality of the earlier years of international meetings and conferences is not well known today and deserves a brief account.

How did the international physiological congresses begin? The first congress was the result of a letter that the Physiological Society (United Kingdom), in 1887, addressed to 109 leading physiologists in different countries. Membership in the congress was to be open to professors and less senior teachers of medical and natural sciences in universities and other scientific institutions and to all who had done original research in human or animal physiology or who were concerned with biological problems of general interest. The official list of participants in this congress numbered 126.

The eight persons who had signed the invitation to the first congress formed a committee that later was supplemented with representatives from a larger geographical area, and this group grew into a self-perpetuating International Committee, later called the Permanent Committee. By the sixth congress, in Brussels in 1904, the Permanent Committee had grown through additions without any simultaneous subtraction and now had twenty-four members, of whom four were designated as general secretaries. The number attending this congress was 256, barely more than ten times the size of the Permanent Committee.

From their beginning the international congresses have for nearly a century continued at three-year intervals, interrupted only by the two major world wars. The first congress after World War I was held in Paris in 1920, and the first after World War II was held in Oxford in 1947, the venue that had been selected by the Permanent Committee at the Zurich congress in 1938.

APS and the Formation of an International Union

After World War II there was a great spirit of international cooperation, and there was increasing interest in forming an international society or union of physiology. The congresses had grown in size, with substantially more than a thousand participants, and at the Oxford congress (1947) the need for a more formal organization to arrange the congresses was discussed.

The program at Oxford listed an invited speaker, J. Needham, who was to give an evening lecture on "International Scientific Cooperation and the Value of an International Union of Physiology." However, he did not appear, and the lecture was not given.

One of the proponents of a union, Henry C. Bazett, expressed his disappointment in a letter to Wallace Fenn (10 September 1947):

> In regard to the International Congress, nothing was done about the future possible organization of an international union of physiology since a talk by Needham on the situation which was scheduled did not come off.

Bazett urged the formation of an international union before the next congress to be held in Copenhagen in 1950. "In my opinion," he wrote, "it is essential to maintain some permanent international organization." (Bazett was later elected president of APS and assumed office in 1950.)

Within APS there was growing interest in forming a union. In 1948 the minutes of the APS Council meeting (14 March 1948) refer to a report by Bazett:

> Since there exists no formal organization and the representation is so large the problem of taking care of international groups has become acute. At the present time the Local Committee, set up at different meetings, has complete charge of the organization of the international congresses and there is no way the American Physiological Society can have a vote in organizing the congresses.

A motion was passed to establish a committee "to consider the problem of an international union or other international body and to recommend the steps the American Physiological Society should take in setting up such an organization." The president of APS, Wallace Fenn, appointed Bazett (chairman), A. C. Ivy, and H. S. Gasser to this committee.

Bazett's committee presented a resolution favoring the organization of an International Union of Physiological Sciences to the APS Fall Meeting in Minneapolis in September 1948, and this resolution from the committee was approved by the APS membership.

Bazett had visited E. D. Adrian, who was the foreign secretary of the Physiological Society and also a member of the Permanent Committee. After this visit, Adrian wrote (in 1948) to Göran Liljestrand (Stockholm), secretary of the Permanent Committee, that

the American Physiological Society (and the Canadian) have been agitating for an International Union of Physiology. Lundsgaard is anxious to have one in time for the Congress in Denmark, and although I do not think it will help us much, I do not feel that we ought to hold out against the formation of the Union. I have got the Physiological Society here to agree to join if one is formed.

Nevertheless it was Adrian who drafted the first proposed statutes for IUPS. Although he had used statutes of other unions in the International Council of Scientific Unions (ICSU) as samples, the proposed statutes made the union, like the Permanent Committee, self-perpetuating. ICSU therefore refused to accept IUPS under this "somewhat undemocratic and nonrepresentative proposal."

At the APS Spring Meeting in 1950, a subcommittee was appointed to join Bazett, Gasser, and Ivy at the International Congress in Copenhagen to plan further toward an international union. Members of the APS subcommittee under Bazett's chairmanship were A. J. Carlson, W. O. Fenn, M. B. Visscher, Eugene Landis, David Bruce Dill, Ralph Gerard, and Fred Hitchcock.

The tragedy was that Bazett, who had assumed the presidency of APS at its April 1950 meeting, did not himself reach Copenhagen. He died of a heart attack on board the ship during the Atlantic crossing.

At the business meeting of the 1950 congress in Copenhagen, Adrian spoke and circulated printed copies of the proposed statutes. Changes were made in the membership of the Permanent Committee "that would enable it to act most effectively as an interim body. Since the U.S.A. group had been especially active in regard to the Union, Visscher was made Secretary."

The report continued:

Dr. Adrian finally called on Dr. Gerard to speak briefly on the Union and the interest of the APS in it. The general notion of a Union was received enthusiastically by the Congress, but no formal vote was taken since further action will be taken by the International (Permanent) Committee through the National Societies.

The wording "Physiological Sciences" was used in the union's name in the hope of leaving the door open for collateral groups, especially pharmacology and biochemistry. Later, in 1967, Fenn wrote, "this device proved quite inadequate for the purpose and now . . . there are separate unions of biochemistry, biophysics, pharmacology, and nutrition."

Visscher reported to the 1951 APS Spring Meeting and circulated to the membership the proposed statutes for organization of the union. At the 1951 fall meeting his report was followed by discussion of the statutes and of the best means of adhering to the proposed union.

At its spring meeting in 1952, APS approved a resolution that "the American Physiological Society become a founding member of the International Union of Physiological Sciences and participate in the final drafting and ratification of its statutes." The Society voted to send five delegates to the 1953 Physiological Congress in Montreal, one of them its president and the others to be appointed by the Society on the recommendation of the president with advice and consent of the Council. The Society also passed a resolution that

the Delegates of the APS at the ICUPS [sic] together with the Council of this Society be requested to study the relationship, if one is to be established, between this Society and the U.S. National Research Council in connection with adherence to the IUPS

and that a recommendation for action be presented to the next meeting.

In late August 1953 five delegates from the United States (E. F. Adolph, ex officio; C. H. Best; C. F. Schmidt; F. O. Schmitt; and M. B. Visscher) met with the organizing assembly for IUPS in Montreal. They represented physiology in the United States and not APS as a society. IUPS was formally established in Montreal on 2 September 1953. A council was elected for three years with Best as president and Visscher as secretary.

Why did APS not represent American physiology in the newly formed union? The IUPS statutes allowed each country only one adhering organization, and the method of adherence was therefore an important consideration. If APS itself was the adhering body, U.S. members would be limited to members of the Society. If, on the other hand, the APS affiliated through the National Research Council (NRC) of the National Academy of Sciences (NAS), the door would remain open to groups of pharmacologists, biologists, and other physiological scientists.

The APS committee therefore recommended adherence through a National Committee for Physiology, set up under the aegis of NRC. The recommendation explained that the affiliating societies would retain direct connection with the union through their nomination of members of the National Committee. At its fall meeting in 1954, the Society accepted the recommendation of the APS Committee on Adherence to the IUPS.

APS and Its Contributions to International Physiology

APS and its members have in a number of ways made immense contributions to international physiology. The greatest contribution has probably been through active participation in the scientific programs of the international congresses. Participation has been facilitated by financial support and travel stipends to individual scientists for congress attendance and participation. A major effort went into the organization of the XXIV IUPS Congress in Washington, D.C., in 1968. This effort and the active contributions of many APS members made the congress a great success.

Another area of contribution has been through the support of a variety of publications. This includes publication of books; financial assistance for an IUPS newsletter that later was discontinued; a brief period when the IUPS newsletter was distributed as an insert in *The Physiologist*; and in 1984 an agreement between APS and IUPS for the publication of a review journal, *News in Physiological Sciences*. Finally APS members have had a considerable direct influence on the activities of IUPS through the many members who have served on IUPS Council and a variety of IUPS commissions.

Congress Participation and Travel Support

APS is by far the largest organization of physiologists in any of the approximately fifty countries that adhere to IUPS. If for no other reason, the mere number of U.S. physiologists is sufficient to be an influential factor in each of the congresses. Most U.S. physiologists belong to APS, but there is no simple way to determine the precise contributions they have made to the international congresses. There are, however, several areas in which the influence and good judgment of APS and its members can be traced back to what is known as the U.S. National Committee for IUPS. Before discussing these contributions, the existence of this committee needs a few words of explanation.

As noted above the official U.S. adherence to IUPS is through the NRC-NAS, which appoints a National Committee to advise on official relationships with IUPS. APS has had the privilege of more members simultaneously serving on this committee (three or four) than any other supporting society. Furthermore the representatives of other societies, such as the Society of General Physiologists or the American Society of Zoologists, have often been members also of APS. Thus APS has had a substantial influence on the transactions of this committee, and many chairmen of the committee have been current or past presidents of APS. The membership of the U.S. National Committee for IUPS from 1955 to 1986 is listed in Table 1.

The influence of the U.S. National Committee in international affairs has been in three major areas: *1*) the representation and votes of the U.S. delegates to the IUPS General Assembly, *2*) advice and proposals submitted for congress programs, and *3*) financial support of U.S. participation in congresses through its travel program.

Over the last thirty years the travel program has awarded a total of close to $900,000 to 1,210 U.S. physiologists for participation in international physiological congresses (Table 2). The travel stipends have been awarded on the recommendation of a subgroup of the U.S. National Committee, which has carefully evaluated each application. Over the years more than fifty percent of the applicants have received travel awards.

The applications, as well as the awards, have included physiologists from Canada, and occasionally from Mexico. In other words, the awards have covered the same geographical area, North America, that pertained to the ordinary membership of APS (until the recent change in APS statutes).

Why were no awards given for travel to the 1968 congress in Washington or to the 1986 congress in Vancouver? In each case the decision can easily be explained.

In 1968 it was assumed that many U.S. scientists would have a fairly short distance to travel, and furthermore many were supported by federal grants (National Institutes of Health and/or National Science Foundation) that permitted the use of grant funds for domestic travel. Financing the Washington congress was a major task, and all efforts were concentrated on financing the congress itself, rather than a separate travel program.

For three consecutive years before the Washington congress, APS had asked its

Table 1
U.S. NATIONAL COMMITTEE FOR IUPS, 1955–86

	Chairman	Members		Chairman	Members
1955* and 1956	M. B. Visscher	K. K. Chen			J. R. Brobeck
		W. O. Fenn			J. E. Dowling
		R. W. Gerard			A. C. Guyton
		C. L. Prosser			J. F. Hoffman
		C. F. Schmidt			H. Rahn
		M. H. Seevers			K. Schmidt-Nielsen
		J. A. Shannon	1977	W. F. Ganong	J. R. Brobeck
		H. B. Steinbach			J. M. Brookhart
1959	W. O. Fenn	K. K. Chen			J. S. Cook
		R. W. Gerard			F. J. Haddy
		O. Krayer			A. Kleinzeller
		W. D. McElroy			J. R. Pappenheimer
		C. F. Schmidt			K. Schmidt-Nielsen
		J. A. Shannon			D. C. Tosteson
		M. B. Visscher	1980	O. E. Reynolds	J. B. Bassingthwaighte
		V. J. Wulff			D. H. Cohen
1962	R. W. Gerard	K. K. Chen			J. S. Cook
		W. O. Fenn			R. E. Forster II
		O. Krayer			W. F. Ganong
		H. S. Mayerson			A. Kleinzeller
		W. D. McElroy			E. Knobil
		C. F. Schmidt			C. P. Mangum
		J. A. Shannon			J. R. Pappenheimer†
		V. J. Wulff			K. Schmidt-Nielsen†
1965	M. B. Visscher	J. M. Brookhart			E. H. Wood
		J. H. Comroe, Jr.	1983	J. B. Bassingthwaighte	M. V. L. Bennett
		W. O. Fenn			D. H. Cohen
		L. S. Goodman			A. P. Fishman
		O. Krayer			F. J. Haddy
		H. S. Mayerson			E. Knobil
		W. D. McElroy			C. P. Mangum
		C. F. Schmidt			L. J. Mullins
		V. J. Wulff			O. E. Reynolds
1968	M. B. Visscher	J. M. Brookhart			K. Schmidt-Nielsen†
		L. D. Carlson			C. F. Stevens
		J. H. Comroe, Jr.			B. M. Twarog
		W. O. Fenn			H. Valtin†
		F. C. MacIntosh	1986	A. P. Fishman	M. V. L. Bennett
		W. D. McElroy			J. E. Dowling†
		H. Rahn			M. Frank†
		K. Schmidt-Nielsen			W. E. Gordon†
		V. J. Wulff			P. D. Harris
1971	J. M. Brookhart	A. C. Barger			T. R. Harris
		L. D. Carlson			F. G. Knox
		W. O. Fenn			M. C. Moore-Ede
		J. F. Hoffman			H. E. Morgan
		D. J. Ingle			S. K. Pierce
		K. Schmidt-Nielsen			H. V. Sparks, Jr.†
		H. Rahn			D. G. Stuart
		J. A. F. Stevenson			B. M. Twarog
		M. B. Visscher			H. Valtin†
		V. J. Wulff			J. B. West
1974	J. M. Brookhart	A. C. Barger			W. D. Willis, Jr.

* Organizational meeting.   † Ex officio.

Table 2
TRAVEL SUPPORT AWARDED TO NORTH AMERICAN PHYSIOLOGISTS
FOR ATTENDANCE AT IUPS CONGRESSES, 1956–86

Year	Congress	Location	No. of Awards	No. of Applications	Approx. Avg Award	Total of Awards
1956	XX	Brussells, Belgium	37		$ 500	$ 18,500
1959	XXI	Buenos Aires, Argentina	102	173	750	75,350
1962	XXII	Leiden, The Netherlands	154	216	600	87,755
1965	XXIII	Tokyo, Japan	150		750	113,640
1968	XXIV	Washington, DC, USA				
1971	XXV	Munich, FRG	132	318	400	47,305
1974	XXVI	New Delhi, India	78	271	1,350	106,245
1977	XXVII	Paris, France	167	345	650	107,000
1980	XXVIII	Budapest, Hungary	220	386	750	170,000
1983	XXIX	Sydney, Australia	170	411	1,000	172,900
1986	XXX	Vancouver, Canada				
Total			1,210	>2,307		$898,695

members for a voluntary contribution of ten dollars per year, in addition to the regular membership dues of fifteen dollars per year. The total came to $72,773. When the accounts for the congress were closed, it became evident that the congress had been a success also financially; the surplus amounted to about $87,000. APS Council in February 1969 decided to recommend to the congress Executive Committee that the surplus be invested and held in trust by APS, with the earnings to be used to provide travel by American physiologists to foreign congresses. Because the income from the capital investment would accumulate over a three-year period between each congress, the funds would be an important core for the initiation of later travel programs. One important aspect that might not be immediately obvious is that the use of these funds would be unrestricted, in the sense that they could be used to support scientists who, for some reason (e.g., federal employment), were ineligible for support from funds solicited from federal agencies, such as the National Institutes of Health or the National Science Foundation. This surplus from the Washington congress has remained as a trust fund administered by APS.

The decision not to undertake a travel support program for the Vancouver congress in 1986 may seem surprising. However, the decision was made after careful consideration. One major reason was that federal agencies that award research grants in the United States, notably the National Institutes of Health, do not consider Canada to be a foreign country, and travel to the Vancouver congress could then presumably be charged to research grants held by individual U.S. investigators. The U.S. National Committee therefore in December 1984 recommended that the interest accumulated from the trust fund by 1986 be transferred to the Organizing Committee for the Vancouver congress for unrestricted use but suggested that the funds should help reduce the cost of attending the congress for students and persons from abroad.

APS members can take pride in the long-lasting effects of their contributions in the years preceding the Washington congress; their contributions were mainly responsible for the surplus that has been so fruitfully used for travel to congresses for nearly two decades.

APS and the Washington Congress in 1968

During the first century of international physiological congresses, only four have been held in North America, two in the United States and two in Canada. The thirteenth congress was in Boston in 1929; the nineteenth in Montreal in 1953; the twenty-fourth in Washington, D.C., in 1968; and the most recent, the thirtieth, in Vancouver in 1986.

The 1968 congress in Washington involved a major contribution from APS and its members. The president of the congress was Wallace Fenn (president of APS, 1946–48), who later served as secretary and then as president of IUPS.

The initiative to organize a congress in Washington originated shortly after IUPS had been established in 1953. In 1956 APS Council discussed the desirability of inviting IUPS to hold a congress in the United States. For a return to North America, however, it seemed that 1959 would be too soon after the congress held in Montreal in 1953, but 1962 was considered a more suitable date. Nevertheless it was not until 1962 (during the congress in Leiden) that the IUPS General Assembly accepted the invitation from the United States, and because the lead time for a congress is six years, this acceptance was for a congress in 1968.

In 1965, three years before the congress, at the business meeting of APS it was voted to assess the members ten dollars per year for the next three years, in addition to the regular dues, which at that time were fifteen dollars per year. The funds were to be used to support the congress, which would be under the jurisdiction of the U.S. National Committee for IUPS. However, the organization of a congress is a major task, which was assumed by an Executive Committee of eight members: Fenn, president and chairman; Visscher, executive vice-president; R. E. Forster, finance chairman; J. R. Pappenheimer, program chairman; H. Rahn, secretary; and vice-presidents P. Bard, representing APS; D. W. Bishop, representing the Society for General Physiologists; and D. W. Bronk, representing NAS.

Fenn, Visscher, Forster, Pappenheimer, and Rahn, all members of APS, each headed one of the subcommittees for the congress. (Each of these men has served as APS president, as has Philip Bard, the representative of APS on the Executive Committee.) With this many distinguished APS members on the Congress Committee, the congress became an undertaking in which APS played a leading role.

Registration for the Washington congress included 3,673 active members and 379 affiliate members. Most of the affiliate members were spouses of the active members; that is, about one active member out of ten was accompanied by a spouse. Of the active members, 2,463 were from the United States, and the remaining 1,210 were from a total of fifty-six foreign countries.

The location of the congress within the United States had not been decided when the invitation was first presented to IUPS. As late as April 1964 New York City was considered the first choice and Washington, D.C., the second choice. However, after a small subcommittee had examined the facilities available in the two cities, the choice fell on Washington, D.C., where a number of large hotels located within easy walking distance of one another became the congress venue.

The congress was an immense success, both scientifically and socially. Even the weather cooperated admirably: the first day was stiflingly hot and humid, which showed the foreign visitors how insufferable Washington can be on a summer day; by the next day the weather had changed to cool and crisp air with a clear blue sky, so that walking about outdoors was a pleasure. After the demonstration of Washington weather at its worst, the participants enjoyed ideal weather conditions for the remainder of the congress.

Collaboration with IUPS in Publication Activities

APS has provided substantial support, both with finances and personnel, for a number of publication projects that relate to IUPS and other aspects of international physiology. Four major undertakings are worth mentioning: publication of books; support for an IUPS newsletter; production of a *World Directory of Physiologists*; and finally the recently established journal, *News in Physiological Sciences.*

Books

During the nineteenth congress in Montreal in 1953, Adrian chaired a Conference on the Future and Limitations of Physiology. Several of the participants prepared essays for this conference, and, supplemented with a few additional contributions, these essays were assembled into a book. *Perspectives in Physiology,* edited by Ilza Veith, then assistant professor of the history of medicine at the University of Chicago, was published the following year (1954) by APS. It contained five general essays by some of the world's most distinguished physiologists and, in addition, twelve very interesting historical descriptions of how physiology developed in the home countries of these twelve authors, ranging (alphabetically) from Argentina to Yugoslavia.

A second book, *Some Founders of Physiology,* was compiled and edited by Chauncey D. Leake and jointly sponsored by IUPS and APS. It contains a very brief note (and for many also a portrait) about each of more than 300 distinguished physiologists, ranging from Hippocrates and Aristotle through centuries of distinguished classical contributions to the beginning of the twentieth century. This book was prepared to be distributed to the participants in the XX IUPS Congress in Brussels in 1956. It is interesting to leaf through the pages of this book and come across classical names that we usually associate with major discoveries in physics and chemistry, but whose contributions nevertheless were essential to the progress of physiological science.

A most important historical document, *History of the International Congresses of Physiological Sciences, 1889–1968,* was produced with Fenn as editor and main contributor. It was sponsored by IUPS and bears the IUPS seal on its cover, but it was published in 1968 by APS and distributed to all participants at the Washington congress. The book contains a great deal of material of historical interest, including the original initiative for the physiological congresses and the letter of invitation from the Physiological Society to the 1889 congress.

APS also presented as a gift to all participants in the 1968 Washington congress a reprint copy of Walter B. Cannon's book, *The Way of an Investigator.* The cost of printing 5,000 copies of this book was estimated to be about $10,000, which was to be taken from the fund accumulated from the voluntary contributions of APS members.

Newsletter

An IUPS newsletter was first published in 1960, initially as a few mimeographed sheets. Publication of an IUPS newsletter twice a year was proposed at the Leiden congress in 1962; it was to be distributed free of charge to all member countries in any number requested. The intent was to keep all affiliated societies informed of IUPS affairs, meetings, and congresses and to publish special articles with news and other items contributed by corresponding editors.

By the late 1970s it was realized that the IUPS newsletter did not meet the needs of individual physiologists and was too expensive for the union to continue. APS considered that individual physiological societies could more effectively communicate with their members through their own society newsletters.

At this time APS generously offered to publish the IUPS newsletter as an insert in *The Physiologist.* This first twelve-page IUPS newsletter (vol. 7, no. 2) to be published as an insert in *The Physiologist* (vol. 23, no. 2) appeared in April 1980. The newsletter was placed in the center and printed on a different paper, so it could easily be pulled out and kept separate from *The Physiologist,* if so desired.

The union accepted the offer to continue this arrangement, which would entail no expense for IUPS. However, the IUPS council member who had agreed to supply information for the newsletter failed to do so, and thus the IUPS newsletter, which had never enjoyed much popularity, vanished without being greatly missed by the rank and file members of APS.

World Directory

An undertaking of much greater significance than the newsletter was, I believe, the publication of the *World Directory of Physiologists.* As early as 1956, Fenn proposed to the U.S. National Committee for IUPS the publication every three years of a comprehensive list of the departments, institutes, societies, and important committees in the physiological societies throughout the world, together with their officers, staff members, and, as far as possible, members. The plans progressed slowly, but when APS Council in 1975 agreed to preorder 5,000 copies of a contemplated world directory, the IUPS Secretariat undertook to collect the information for the first edition.

The names collected by the IUPS Secretariat were submitted to the APS office for assembly and computer printing. This first directory contained about 14,000 names and was available at the 1977 congress in Paris, but it was rather incomplete because the IUPS Secretariat had not obtained the necessary information from all member societies. The second edition, available at the 1980 congress in Budapest, was far more complete and contained approximately 26,000 names. Financially this directory has not made any profit, nor has it been a drain on the APS treasury; income from sales has roughly balanced the printing cost. However, without the help of APS and its administrative officers, notably Orr Reynolds, the directory would probably never have been possible.

In 1986 the IUPS Secretariat assembled information for an updated edition of the world directory, more complete than any of the preceding issues. The major expense of the printing was assumed by the Federation of American Societies for Experimental Biology (FASEB), with the initial expense expected to be recovered from sales. However, to protect FASEB against a possible loss in this venture, an anonymous physiologist has extended a personal guarantee of $10,000 against such losses.

News in Physiological Sciences

The most recent publication venture undertaken by APS in collaboration with IUPS is the journal *News in Physiological Sciences* (*NIPS*), which was first published in 1986. This journal has a long and rather complex history. In 1980 the IUPS Council had discussed a resumption of its defunct newsletter, to be published with a more scientific content of simple articles and reviews. IUPS entered into an agreement with a commercial publisher, but the contract later fell through.

In 1983, at the IUPS Congress in Sydney, the president of APS, A. P. Fishman, proposed that APS underwrite the major cost of starting a scientific review journal of the type contemplated by IUPS. This generous proposal was more favorable for IUPS than that simultaneously being negotiated with a major university press. Obviously IUPS could not by itself provide the funds to back a new journal of the projected scope, and IUPS Council readily accepted the offer. The result is the journal that is now distributed to all APS members at no charge.

APS and Service on IUPS Council and Commissions

APS members have influenced international physiology through service on IUPS Council (Table 3). The first president of IUPS, C. H. Best (1953), the codiscoverer of insulin, was a Canadian and an APS member.

The IUPS Constitution limits the number of persons from a country who can serve on IUPS Council at any given time to a maximum of two (of a total of fifteen Council members). Nevertheless, in some periods, as many as three APS members have served simultaneously, because they have represented two different countries, the United States and Canada.

Of the twelve APS members who over the years have helped direct the activities

Table 3
APS MEMBERS WHO HAVE SERVED ON IUPS COUNCIL, 1953–86

Year	President	Vice-president	Treasurer	Secretary	Councillor
1953–56	C. H. Best*			M. Visscher	
1956–59				M. Visscher	C. H. Best*
1959–62		C. F. Schmidt		W. O. Fenn	F. C. MacIntosh*
1962–65		C. F. Schmidt	F. C. MacIntosh*	W. O. Fenn	
1965–68			F. C. MacIntosh*		H. Rahn
1968–71	W. O. Fenn				H. Rahn
1971–74		H. Rahn			
1974–77			J. Brookhart		J. Pappenheimer
1977–80			J. Brookhart		J. Pappenheimer
1980–83	K. Schmidt-Nielsen				J. Pappenheimer
1983–86	K. Schmidt-Nielsen		H. Valtin		K. Krnjević*
1986–89			H. Valtin		R. Berne
					K. Krnjević*

* Canadian.

of international physiology through their service on IUPS Council, nine have served as IUPS officers. The influence of APS has thus been substantial, because one or more of its members has always been on the Executive Committee of IUPS.

Many of the international activities of IUPS are carried out by its various commissions. Twenty-seven such commissions, which cover various areas of physiology, ranging from auditory physiology to undersea physiology, work in a variety of ways, from arranging their own meetings and activities and suggesting programs and symposia for international congresses to producing a variety of publications, including symposium volumes that have become important reference works.

Each commission has, on the average, about eight members, ranging from as few as six to more than a dozen. At last count the twenty-seven commissions had a total of 219 members, of which fifty-five were from the United States. However, this is not a constant number, because the membership of the commissions changes and new commissions are formed as others are discontinued. Nevertheless the importance of the U.S. contribution to the commission activities is obvious from the fact that about twenty-five percent of their members are from the United States.

Other International Activities

APS and its members have always been interested in international activities, and many members have taught and done research in various parts of the world. A very important, but not fully appreciated, result of international contacts is direct collaboration between APS members and colleagues in other countries. Many excellent and important research projects have started at international congresses, stimulated by the personal contacts.

Although the ties between APS and the Physiological Society have been close since the founding of APS, it was not until 1985 that the two societies held their first joint meeting in Cambridge on 12–14 September. The program was organized by Michael Jackson and Anthony Angel.

An expression of the current interest in international contacts within APS is a program that Fishman, when he was APS president in 1983, initiated through the formation of an APS Committee on International Physiology.

An important change is to enlarge the membership of APS to include as regular members physiologists who reside in Central and South America. We have plans for a closer collaboration with the Physiological Society, including a bidirectional exchange of young scientists. There is also an attempt to implement a closer exchange program with colleagues in the Soviet Union. Finally a permanent link to international physiology will continue through the activities of the U.S. National Committee for IUPS.

It is a great satisfaction to see the many excellent results of the leadership of APS and its individual members in international collaboration, understanding, and friendship.

I am grateful to the following persons for much of the information in this chapter: Dr. Orr Reynolds, formerly APS executive secretary-treasurer; Dr. Toby Appel, APS historian/archivist; and June Ewing, director of the U.S. National Committee for IUPS.

CHAPTER 18

Neuroscience

ROBERT W. DOTY

From its inception the American Physiological Society (APS, the Society) has provided an attractive forum for neurophysiology and, until recently at least, has counted among its members most of the prominent neuroscientists of North America. The formation of the Society for Neuroscience in 1969 was thus a development of major significance in this last quarter century of the Society's history; neurophysiologists were drawn to meetings of the Society for Neuroscience and away from those of the APS. I will endeavor to trace some of the background to this development, as well as the Society's present endeavors to attract the neurophysiological community back to its meetings.

The Society, of course, has often had to contend with the "departure" of major interest groups (e.g., biochemists, pharmacologists, biophysicists, and general physiologists), but never in such numbers as the neurophysiologists in the 1970s. Actually the first departure was that of the psychologists (11). G. Stanley Hall and Joseph Jastrow, both founding members of the APS, became presidents of the American Psychological Association. Hall was its first president, as well as founder of the *American Journal of Psychology.* There is, however, a certain strength accruing to the Society in these departures of those with widely differing interests—the Society thus retains a clarity of purpose otherwise diluted in organizations that, like the American Psychological Association, seek to accommodate almost any interest and thereby become largely "umbrella" organizations. Had the Society retained such groups as the biochemists within its membership, it would itself have become the federation of societies for experimental biology, or physiology, if you will. Calling it all physiology, however, would not lessen the numbers or confine the diversity of expertise.

There is also a historical irony to the fact that the Society, whose centennial we now celebrate, was itself formed as something of a splinter group from a preexisting scientific society. H. H. Donaldson, a founding member of APS and in addition the organizer of the Society's first scientific meeting (1888) in Washington, D.C., described how dissatisfaction had arisen with the Natural History Section of the American Association for the Advancement of Science (AAAS) (5). There were too many amateurs, and meetings were held in the summer rather than over the Christmas holidays. Thus the American Society of Naturalists was founded.

> Soon fission began. The Geologists withdrew—next the Physiologists—then other groups, and so the parent Society of Naturalists kept on budding until it was reduced to something like a nucleus without cytoplasm (5).

Despite its propensity for "budding," the APS has not met that fate and has always remained a complete and thriving entity.

The historical roots of the Society for Neuroscience clearly lie with the "axonologists," a prodigiously creative group of neurophysiologists, who from 1930 to 1942 met for dinner and discussion at the annual meeting of the APS. Louise Marshall has perceptively limned many of the personalities and events associated with this group (16). Scientifically it was invested with the excitement emanating from rapid developments in electrophysiology, abetted by the new tools of vacuum tube amplifiers, the cathode ray tube, and such discoveries as single-unit activity and the human electroencephalogram. Among the group also were men such as Wallace O. Fenn, Ralph W. Gerard, and Francis O. Schmitt, who sustained active research programs in the biochemistry of nerve.

The interests of this group were undoubtedly reflected in the founding of the *Journal of Neurophysiology* in 1938 by John F. Fulton, Ralph W. Gerard, and Johannes G. Dusser de Barrene.[1] It was the first such journal devoted to basic neuroscience since Clarence Luther Herrick started the *Journal of Comparative Neurology* in 1891. The pace of initiating specialty journals in neuroscience quickened enormously after World War II. First was *Electroencephalography and Clinical Neurophysiology* in 1949 under the editorship of Herbert H. Jasper and W. Grey Walter, a journal sponsored by the International Federation of EEG Societies. The broad focus of this journal and its wide international purview, as noted below, ultimately played a cardinal role in setting the stage for the Society for Neuroscience. Likewise the *Journal of Neurochemistry*, founded in 1956 by a team of ten editors, enjoyed a wide international distribution. In 1959 William F. Windle founded *Experimental Neurology*, and in 1960 this and the four other journals just mentioned published a total of 2,655 pages. By 1983 more than twenty-five journals predominantly devoted to basic neuroscience were publishing well over 53,000 pages annually.

Public support of neuroscience also increased rapidly in this era, as outlined in the commemorative volume of the twenty-fifth anniversary of the National Institute of Neurological and Communicative Disorders and Stroke (9, 23). The APS was fully alert to these developments and indeed vigorously promoted them. The Board of Publication Trustees, consisting of Philip Bard, William F. Hamilton, and Maurice B. Visscher, chairman, made a most productive decision in sponsoring the *Handbook of Physiology* (17). The first three volumes on neurophysiology under the editorship of John Field, Horace W. Magoun, and Victor E. Hall practically defined the field of basic neuroscience (8).

Beginning in 1952 the Society also provided the unofficial home for the "neuro-physiology group," who, like their more distinguished predecessors, the axonolo-gists, gathered for talk and imbibition in association with the spring meeting of the Society. In his history of the preceding quarter century, Wallace Fenn gave a synopsis of the origination of this group by T. C. Ruch, L. M. N. Bach, and John Lilly, together with something of its early history (7). A highly successful undertaking, it attracted a large audience each year until meetings of the Society for Neuroscience began to erode its appeal. The effort was abandoned after the Federation of American Societies for Experimental Biology (FASEB) Spring Meeting in Atlantic City in 1973, and the "committee" then in charge, Harry D. Patton, M. G. F. Fuortes, and Karl Frank, turned over their small operating fund to the Society for Neuroscience (15).

The gathering momentum of neuroscience as a discipline was shaped and focused in the late 1950s and early 1960s with great success by Herbert H. Jasper, Horace W. ("Tid") Magoun, and Francis O. Schmitt. Magoun, later joined by his colleagues Donald B. Lindsley and John D. French, initiated a spectacularly successful neuro-science program leading to establishment of the Brain Research Institute at the Universitiy of California at Los Angeles, which became the mecca for postdoctoral fellows in neuroscience from all over the world (10). In addition, Magoun chaired the Josiah Macy Foundation Conferences on the central nervous system and behavior. The reports of these conferences, with their richly illustrated historical background, reopening of contacts with Eastern European colleagues, and emphasis on neural correlates of behavior, created a keen sense of excitement and continuity, despite the sometimes unfortunately chaotic nature of the verbatim text (2-4).

Frank Schmitt, a charter member of the axonologists and one of their more versatile members in that he pursued chemical and physicochemical as well as electrophysiological fundamentals of neural activity, organized the Neuroscience Research Program at the Massachusetts Institute of Technology. This became some-thing of a "think tank" for the discipline; the arrangement and consequences of neural activity from molecule to mind were critically examined, and these vigorous discussions were disseminated in a continuing series of publications. An extremely useful outgrowth of this activity has been the Neuroscience Study Programs, the first of which was held in 1966 (18); they now number four in all (19-21). In their ingeniously broad spectrum of topics, these volumes have directed interest into many channels of neuroscience; they stand as lucid summaries of present knowl-edge. To this scientific treasure has been added also Worden's irreplaceable collec-tion of historical vignettes (24).

Herbert Jasper's role extends far beyond his editorship of the *EEG Journal*. He became the executive secretary, in Paris, of a unique new body of scientists, the International Brain Research Organization (IBRO). IBRO ultimately provided the model, and thereby much of the stimulus, for the formation of the Society for Neuroscience. The first step toward IBRO occurred at a meeting of the International Federation of EEG Societies in 1955 in Marseilles, when Henri Gastaut, the local host, strongly abetted by Alfred Fessard, proposed to his Russian colleagues that the next meeting be held in Moscow under the sponsorship of the Soviet Academy of Sciences (13). This duly transpired in October 1958, and the outcome was the famous "Moscow Colloquium" (14). Even more significant, however, was the resolution, formulated by I. S. Beritashvili, H. H. Jasper, Henri Gastaut, and V. S. Rusinov, and unanimously passed at that colloquium, to the effect that a formal structure be devised to advance brain research throughout the world. The diligence and skill of Jasper, Fessard, and Heinrich Waelsch were largely responsible for the realization of this conception. While serving as executive secretary of IBRO, Jasper dealt with the sponsorship of UNESCO and, together with Wilder Penfield and F. C. MacIntosh, had IBRO incorporated formally by an act of the Canadian Parliament at Ottawa in October 1961. The bill read in part that the goals of the newly chartered organization were

> to foster throughout the world fundamental scientific research contributing to an understanding of the brain, normal and abnormal . . . to develop, support, coordinate, promote and undertake scientific research and education in all fields concerning the brain, and to study the impact of brain research on education, behaviour, and the welfare of man (12).

The membership structure of IBRO was distinctive in that it was divided into various areas of expertise: neuroanatomy, neurochemistry, neuroendocrinology, neuropharmacology, neurophysiology, neuropathology, behavioral sciences, and neurocommunications and biophysics. It thus effected an amalgamation of the many disciplines involved in study of the brain and behavior and in so doing provided the broadly based model later emulated by the Society for Neuroscience.

The group in the United States from which the Society for Neuroscience ultimately arose did not come into being until 1965, and at its inception there certainly was no expectation that it should function to produce a regional imitation of IBRO. Nonetheless, this is what transpired. A Committee on Brain Science was formed at the National Academy of Sciences (NAS) through the National Research Council (NRC) to prepare the United States' contribution to a worldwide IBRO survey of brain research and to review and advise the National Library of Medicine on indexing current literature. The committee sought input from numerous societies with interests relevant to its mission; in so doing it came to recognize the diffuseness of neuroscience, a part of many disciplines but lacking a focus of its own. Thus, by its fifth meeting, in June 1967, the idea had begun to crystallize that a single society along the multidisciplinary lines of IBRO itself might substantially strengthen the many disparate studies of the nervous system. Under the chairmanship of Neal E. Miller, and with the creative talent of Louise H. Marshall, its executive secretary, the committee's deliberations gradually evolved into definitive steps to form such a society. The major, and difficult, question was, What kind of a society should this be?

To this end colleagues throughout the country were contacted by Ralph Gerard, who assigned each a geographical region in which to sample opinion as to the desirability and/or possible structure of a society for the "brain sciences," as it was being called at that time. There was significant sentiment against forming a new society, one suspects largely from the more established scientists already satisfied with their professional ties. In any event, the idea being evolved initially was that there should be an extensive network of local organizations, much in the style of Sigma Xi, and that the "national group would build upon rather than supersede local groups" (R. Gerard, letter to Doty, 1968).

On Sunday, 25 August 1968 in Washington, D.C., the first day of the XXIV IUPS Congress, Gerard, together with several administrators from relevant government agencies, met with twenty of the regional representatives to discuss what course of action, if any, should be taken. There being obvious concurrence that some national organization for neuroscience was desirable, even though its ultimate format was still only vaguely perceived, Gerard appointed Edward R. Perl to act as chairman of a group to devise, in continuing consultation with as many colleagues as feasible, a specific plan of action. The members of this organizing group were selected by Gerard to be representative not only of geographic locality but of the scientific disciplines involved. Perl was at that time in the Department of Physiology at the University of Utah College of Medicine. Other members of the group were John M. Brookhart, Department of Physiology, University of Oregon Medical School; Robert W. Doty, Center for Brain Research, University of Rochester; Alfred Pope, Biochemistry, McLean Hospital, Belmont, Massachusetts; Vernon Rowland, Department of Psychiatry, Case Western Reserve University School of Medicine; James M. Sprague, Department of Anatomy, University of Pennsylvania School of Medicine; Robert L. Thompson, Department of Psychology, Hunter College, New York; and John E. Wilson, Department of Biochemistry, University of North Carolina School of Medicine.

This group met again on 7 December (Pearl Harbor Day) 1968, again in Washington, D.C., now with purposes much better resolved. It was clear that a national society would meet a number of envisioned needs, among them that of international liaison. Foremost, however, it should promote teaching and research and public understanding. With commendable skill Perl drafted a constitution for the new society that was then circulated to 250 "charter" members of the incipient society for comment and revision. The organizing group next met with the NAS-NRC Committee on Brain Science on 16 June 1969. The constitution was accepted by all, and Neal Miller, as chairman of the Empowering Committee, then declared the Society for Neuroscience formed, with the organizing group plus himself and Ralph Gerard as councillors of the new society. The immediate order of business was the election of Perl as president and Louise Marshall as secretary-treasurer, to serve until their successors might be chosen and qualified. The legal formalities were completed a few days later, 11 July 1969, when the Articles of Incorporation were recorded with the District of Columbia. Temporary office space was given the new society by the NAS, and Marjorie Wilson was recruited as executive secretary, a brilliant choice, as her talent, patience, unwavering dedication, and mature common sense were essential ingredients in the society's success in coping with its explosive growth during the first decade of its existence. By October letters soliciting membership were going out on society stationery.

Much of the initial membership recruitment came via the formation of local chapters, and this has remained a major feature of the Society for Neuroscience in many locales, even though it is largely ignored in others. The appointed Council met on 22 January 1970 to select a slate of candidates for the next officers of the society from the nominations that had been received. Vernon B. Mountcastle agreed to stand for the office of president and was duly elected, together with new members of Council. Ralph Gerard was designated honorary president, and Louise Marshall was named as a special consultant to the Council.

The first meeting of the newly elected officers was held in Atlantic City at the time of the FASEB meetings, 15 April 1970. John Brookhart, a former president of the APS (1965–66) was designated to keep President C. Ladd Prosser and members of Council of the APS informed of the plans and activities of the Society for Neuroscience. The likelihood that the new society would seriously diminish the vigor of neurophysiology within the APS was recognized by all, and with regret on both sides; yet planning went forward without rancor, indeed with cordiality and cooperation. The APS has consistently worked closely with the Society for Neuro-science, and many maintain membership in both societies. The spirit of good will is admirably manifested in the recent reworking of the neurophysiological sections of the *Handbook of Physiology* and in the continuing excellence and openness of the *Journal of Neurophysiology.*[2] Further important support was provided not long after Mountcastle took office as president of the Society for Neuroscience; it found a much needed home at Beaumont campus, which it has only recently had to leave when it outgrew the space available.

The Alfred P. Sloan Foundation provided a generous grant that enabled the new society to proceed with some confidence. There was, however, an interesting little episode concerning tax law. The grant stipulated that the recipient must have tax-exempt status, which had not yet been obtained. Thus, until this detail was tidied up, the officers of the society had to offer their personal guarantee to replace expended funds in the event the Internal Revenue Service (IRS) took some excep-tion to the society's application. There was a bit of silent relief when the IRS approval came through.

The expenditures did begin to mount, not only for the society office, but also as the first annual meeting was being organized. This meeting, 27–30 October 1971 in Washington, D.C., masterfully arranged by Henry G. Wagner, with Arthur A. Ward, Jr., as program chairman, was a resounding success, with 1,395 registrants. The future of the society was thus assured. It is gradually becoming apparent, however, that this very success may soon become a major problem! Membership has reached 9,500, and there were 8,000 registrants at the 1984 meeting in Anaheim, California, a throng that is leading some to start muttering about meetings being too large and hectic. It would not surprise me to see a decade hence that the neurophysiologists meet in one city and the neurochemists in another—shades of FASEB!

From the beginning the Society for Neuroscience has included many members from Canada and Mexico and is thus in actuality, like the APS, a North American society rather than a purely national group. The European response to IBRO at first was the formation of the European Brain and Behaviour Society, which was organized concurrently with the Society for Neuroscience. When the phenomenal success of the latter became apparent, however, a much more inclusive group came into being in 1977, the European Neuroscience Association, which is comparable to the Society for Neuroscience in its coverage. The corresponding group in Eastern Europe is Intermozg (meaning, approximately, "interbrain"), and there are now many other more nationally based neuroscience groups in Japan, Australia, and elsewhere.

As anticipated, the meetings of the Society for Neuroscience did reduce participation by neurophysiologists in the affairs of the APS. Thus, shortly after he assumed the office, President Daniel C. Tosteson appointed a committee, chaired by the late Edward V. Evarts, to examine what the Society's course should be in this regard. Their report in May 1974 recommended, among other things, the deliberate strengthening of neurophysiological contributions to the programs of APS meetings (6). This advice was followed. Under the imaginative initiative of George G. Somjen, with enthusiastic and able support from David O. Carpenter and Janett Trubatch, a continuing series of excellent neurophysiological symposia with wide appeal has been mounted at each of the APS meetings, beginning with the 1974 fall meeting. These activities, and the interest they generated, have in turn led to formation of a section on the nervous system within the Society (22), which was officially recognized by Council in April 1978 (1).

The formation of a special section for neurophysiology in effect reflects the continuing growth of the Society and has been followed by the establishment of many other sections of special interest and expertise within the APS. Such a course may be an inevitable consequence of ever-expanding knowledge, for none can pretend to mastery or even detailed comprehension of the almost mythically ingenious procedures currently employed to decipher and manipulate the processes of life. Through it all, physiology as a science seems secure; for whereas chemistry and anatomy discern the foundation and behavior measures the outcome, it is physiology that offers the insight into the daedal coupling, how chemical anatomy yields a behavioral synthesis.

I am grateful to Dr. Louise H. Marshall for her many cogent comments and corrections on an earlier version of this account and to Dr. Toby A. Appel, historian and archivist for APS, for her help in recounting recent developments.

NOTES

[1] In 1961 this journal was purchased by the Society from Charles C Thomas, publisher, and Yale University, which had acquired its interest as a bequest from the late John F. Fulton (ref. 7, p. 76).

[2] Purchase of this journal by the APS in 1961 had been motivated in part by a desire to provide for the needs and interests of its neurophysiologist members.

REFERENCES

1. ANONYMOUS. Charter for the section on the nervous system. *Physiologist* 21(3): 25, 1978.
2. BRAZIER, M. A. B. (Editor). *The Central Nervous System and Behavior: Transactions of the First Conference.* New York: Josiah Macy, Jr., Foundation, 1959.

3. BRAZIER, M. A. B. (Editor). *The Central Nervous System and Behavior: Transactions of the Second Conference.* New York: Josiah Macy, Jr., Foundation, 1959.

4. BRAZIER, M. A. B. (Editor). *The Central Nervous System and Behavior: Transactions of the Third Conference.* New York: Josiah Macy, Jr., Foundation, 1960.

5. DONALDSON, H. H. The early days of the American Physiological Society. *Science* 75: 599–601, 1932.

6. EVARTS, E. Report of task force on neurophysiology. *Physiologist* 17: 139–146, 1974.

7. FENN, W. O. *History of the American Physiological Society: The Third Quarter Century, 1937–1962.* Washington, DC: Am. Physiol. Soc., 1963.

8. FIELD, J., H. W. MAGOUN, AND V. E. HALL (Editors). *Handbook of Physiology. Neurophysiology.* Washington, DC: Am. Physiol. Soc., 1959 and 1960, vols. I–III.

9. FRANK, R. J., L. H. MARSHALL, AND H. W. MAGOUN. The neurosciences. In: *Advances in American Medicine: Essays at the Bicentennial*, edited by J. Z. Bowers and E. F. Purcell. New York: Josiah Macy, Jr., Foundation, 1976, vol. 2, p. 552–613.

10. FRENCH, J. D., D. B. LINDSLEY, AND H. W. MAGOUN. *An American Contribution to Neuroscience: The Brain Research Institute, UCLA, 1959–1984.* Los Angeles, CA: UCLA Publ. Serv., 1984.

11. HOWELL, W. H. The American Physiological Society during its first twenty-five years. In: *History of the American Physiological Society Semicentennial, 1887–1937.* Baltimore, MD: Am. Physiol. Soc., 1938, p. 1–89.

12. JASPER, H. H. The IBRO record. *IBRO News* 1: N2–N8, 1973.

13. JASPER, H. H. Reflections on the founding of IBRO. *IBRO News* 12: 6–8, 1984.

14. JASPER, H. H., AND G. D. SMIRNOV (Editors). The Moscow Colloquium on Electroencephalography of Higher Nervous Activity. *EEG Clin. Neurophysiol. Suppl.* 13, 1960.

15. MARSHALL, L. H. Announcement. *Neurosci. Newslett.* 4(4): 1, 1973.

16. MARSHALL, L. H. The fecundity of aggregates: the axonologists at Washington University, 1922–1942. *Perspect. Biol. Med.* 26: 613–636, 1983.

17. O'MALLEY, F. K., AND H. W. MAGOUN. The first American-based *Handbook of Physiology. Physiologist* 28: 35–39, 1985.

18. QUARTON, G. C., T. MELNECHUK, AND F. O. SCHMITT (Editors). *The Neurosciences.* New York: Rockefeller Univ. Press, 1967.

19. SCHMITT, F. O. (Editor). *The Neurosciences, Second Study Program.* New York: Rockefeller Univ. Press, 1970.

20. SCHMITT, F. O., AND F. G. WORDEN (Editors). *The Neurosciences, Third Study Program.* Cambridge, MA: MIT Press, 1974.

21. SCHMITT, F. O., AND F. G WORDEN (Editors). *The Neurosciences, Fourth Study Program.* Cambridge, MA: MIT Press, 1979.

22. SOMJEN, G. Announcement by the section on neurophysiology. *Physiologist* 19: 13, 1976.

23. TOWER, D. B. (Editor). *The Nervous System. The Basic Neurosciences.* New York: Raven, 1975, vol. 1.

24. WORDEN, F. G., J. P. SWAZEY, AND G. ADELMAN (Editors). *The Neurosciences: Paths of Discovery.* Cambridge, MA: MIT Press, 1975.

C H A P T E R 1 9

Sectionalization

JOHN S. COOK

Many physiologists would give as a second definition of their field the one applied by G. N. Lewis to physical chemistry: "That branch of science which includes everything that is interesting and excludes everything that is uninteresting." Such a definition implies a wide array of intellectually appealing subject matter, which in physiology is certainly self-evident to the practitioners. As the major organization in the United States representing the field, the American Physiological Society (APS, the Society) embraces many of the diverse areas understood by the word "physiology." From an organismic point of view, however, it is largely concerned with the physiology of vertebrates; invertebrates tend to be viewed mostly as model systems, although not invariably so (see below). Plants and microbes are almost totally neglected. Nevertheless vertebrate, and especially mammalian, physiology covers a very broad spectrum of science from the molecular biology of cells to behavior of whole animals, and all this is accommodated by the Society.

It is a curious fact that an organization as large and as diverse as the Society waited ninety years before it formally recognized its subsections. The reasons for this delay are not entirely apparent from the archives. The principal sentiment against sectionalization appears to have been the feeling in Council that the character of the Society should reflect the integrative quality of physiology rather than its specialties.

The idea of sectionalization is certainly not new. For many years other large scientific societies have been sectionalized, while administrative cohesiveness of the whole has been maintained. Within the Society there was reference to sectionalization as early as 1906. As will be discussed below, on many occasions, particularly in the last forty years, there repeatedly was pressure within the Society to recognize specialty subsections that might even include nonmembers of the parent Society. Not surprisingly these discussions frequently centered around the Society's efforts to prevent specialty groups from splitting off to form new national societies. For example, between 1946 and 1976, a period when the health-related sciences were

435

burgeoning, many new physiologically related societies were established in the United States. The response within APS was a continuing discussion of formal recognition for its subspecialties, but this did not happen until 1976. When sectionalization finally came about, it was as much because of necessities of programming meetings as any other factor. The actual sectionalization turned out to be a relatively simple matter, because for a long time the members, acting on their own and according to their own professional needs and interests, had been holding specialty meetings and dinners in conjunction with the Federation of American Societies for Experimental Biology (FASEB, the Federation) meetings. Identification of the then-current leaders of the individual groups was straightforward, and the organization of most sections required little more than drawing up a simple charter and giving the Society members as a whole the opportunity to identify themselves with the established sections. But there were some problems.

In this chapter some of the pressures for sectionalization, with special reference to the splitting off of a few representative societies, are described. This is followed by a brief description of the sections that are now constituted and their background and goals, with a final word on the developing role of the sections in the activities and governance of the Society.

Pressures for Sectionalization

When the process of sectionalization of the Society formally began in 1976, the idea had been discussed in Council and by the members on many occasions for seventy years. Two main pressures eventually brought it about: *1)* the recurrent problem of the appearance of new societies with substantial physiological components that threatened to draw members from APS and *2)* the problem of programming the Society meetings, which were increasingly unwieldy. The Program Committee clearly needed the help and expertise of specialists with credentials certified by their peers.

In reference to new societies, the list published by Orr Reynolds in *The Physiologist* in 1976 may be accepted as the definitive record. In it he cited ten associations for which more than half the founding committee members were also members of APS. But several of these groups and one or two other sister societies have had histories worthy of special comment in the context of sectionalization.

The first to peel off were the biochemists. The prime mover in this event was John J. Abel (Johns Hopkins). In his correspondence (1906) suggesting the move to some of his colleagues, Russell H. Chittenden and Lafayette B. Mendel (both at Yale) expressed misgivings and suggested, apparently for the first time in APS records, a subsection of the Society for physiological chemistry. Chittenden of course had been one of the founders of APS and was its president from 1896 to 1904. Abel's answer guardedly suggested that there was not much sympathy for chemistry in APS at the time and that in any event he was not proposing "a secession from the Physiology Society," but rather an independent addition, whereby the biochemists could meet together, and not in competition, with the physiologists. In his later (1945) history of the American Society of Biological Chemists (ASBC), Chittenden reiterated that among the physiologists early in this century there had been a stronger representation of those interested in the "physical" aspects of physiology than of those interested in the "chemical" aspects. ASBC was founded in December 1906. Despite his earlier proposal for a chemistry subsection of APS, Chittenden was among the founders of ASBC and became its first president. FASEB was created in 1913 with both the physiologists and biochemists among the charter member societies.

As Abel had suggested, the two societies among others met together amicably until 1971. For some time, however, many biochemists had felt that the meetings were too large and that in any case they had closer association with the biophysicists, microbiologists, and cell biologists than with some of the other Federation societies. The political unrest of 1968 finally triggered separate meetings. After the turbulent Democratic National Convention with its attendant street riots and police actions, many biochemists balked at convening with FASEB in Mayor Daley's Chicago in 1971. FASEB nevertheless kept its appointment with the city, but the biochemists met independently later in the year in San Francisco. Having enjoyed the somewhat smaller meeting, they now meet independently of APS more often than not. Physiologists currently seem to have a good deal more interest in chemistry than they were said to have had in 1906, and the separation from the biochemists is generally regarded as a serious loss. Because of concern within the Society over this schism, in 1977 during the presidency of Fran Ganong a Task Force on Physiological Chemistry was appointed with Robert E. Fellows as chairman. Sectionalization of the Society had just begun at the time. The charge to the task force was to explore the profitability of developing a section on physiological chemistry. After due consideration of the question the task force decided that the newly forming Sections on Cell and General Physiology and on Endocrinology and Metabolism would very much overlap the functions of such a biochemical section, and the idea was shelved.

Another sister society to APS, the American Association of Immunologists (AAI), was formed in 1913. Historically immunology has been associated with microbiology, and the founders of AAI were not predominantly members of APS. The subsequent history of the field has placed immunology squarely in the camp of physiology, and happily AAI has met regularly with APS, except in 1978 and 1984 when AAI met with the biochemists. Although no subsection for immunology appears to have been contemplated by APS Council, the physiologists have had the good fortune not to have lost major contact with the immunologists at the annual meetings.

With the expansion of science after World War II, all the FASEB societies, including APS, continued to grow, and the meetings became progressively larger, to the dismay of many members. Inevitably a number of new societies, founded by scientists who wished to preserve their identity and meet in more intimate surroundings, appeared. An interesting example of the conflict of loyalties felt by even the staunchest members of APS when a new society was formed appears in a 1945 correspondence between L. V. Heilbrunn and Wallace O. Fenn. Heilbrunn wrote that he and a number of other physiologists at the Marine Biological Laboratory in Woods Hole (W. Amberson, S. C. Brooks, E. N. Harvey, M. H. Jacobs, R. S. Lillie, A. K. Parpart, and Eric Ponder) were interested in forming the group that in 1946 became the Society of General Physiologists (SGP), and he asked if Fenn were interested in becoming a charter member. Although the possibility of the new society's meeting with the Federation was brought up in the letter, APS was not specifically mentioned. At the time of the correspondence Fenn was secretary of APS, and his response was a plea to Heilbrunn not to fractionate APS, together with a number of arguments against such a move. Nevertheless, Fenn ended by saying that if the new society were formed despite his plea, "I should probably feel unhappy to be left out of it." Shades of Russell Chittenden.

Fenn's arguments to Heilbrunn, primarily that splinter societies weaken the parent society and are likely to be weak in themselves, recur repeatedly in the subsequent minutes of APS Council. He wrote, "The chemists in this country have a single strong and influential society and profit accordingly." Fenn stated that there is room for, and the APS constitution permits, the formation of a subgroup that could have its own regional meetings if it chose, and as an enticement, he indicated that such a subgroup could have its abstracts printed and distributed by the Federation. He went on to say presciently that a "reorganization of this sort might pave the way for the incorporation of other groups such as the endocrinologists and the neurophysiologists as subsections, each with a separate chairman." Fenn's suggestions were by and large endorsed by Council, although with misgivings about sectionalization, and Council even considered taking over the *Journal of General Physiology* as part of the package, but these ideas were not to prevail. The Woods Hole group preferred meetings that were smaller and more informal than those of the Federation, although they did meet with APS in the fall of 1955. They also liked meetings that covered more closely defined topics and that included plant and microbial physiology, which were hardly represented in APS. In comparison with some of the later splinter societies, SGP has to this day retained the small and informal format. Many of its members hold dual membership with APS and see no conflict of interest. The two organizations have and serve different functions. Even so, as late as 1961 the hope was expressed in APS Council that the general physiologists might be enticed back to APS, with appropriate sectionalization of the parent Society again a salient consideration.

Several years after the formation of SGP a similar problem for APS arose with the biophysicists, and APS's response says a good deal about the climate for sectionalization within the Society at that time. In the mid 1950s several groups met independently, at first unaware of each other, to consider organizing American biophysicists into a single association. Two of these groups were of particular interest from APS's point of view. One arose from a new National Institutes of Health (NIH) Study Section for Biophysics and Biophysical Chemistry, chaired by F. O. Schmitt. Schmitt, an APS member, established biophysics at the Massachusetts Institute of Technology.

The major organizer of the second group was Wilbur Selle, who convened a number of interested persons during the 1955 APS Fall Meeting, held at Tufts. APS Council had been apprised of this meeting and sent as representatives President William Hamilton and Executive Secretary Milton O. Lee. (President-elect Alan C. Burton, himself recognized as a distinguished biophysicist, was delegated to attend a similar meeting at the same time in Ann Arbor.) At the Tufts meeting the biophysicists pointed out that they shared many common interests, and they looked for closer ties with many persons who were not in any sense physiologists. They also expressed some dissatisfaction with the reception of their papers in the Society's journals and discussed the possibility of founding their own journal. Hamilton and Lee urged them not to break away and offered a package of concessions, including the possibility of a biophysics section of the *American Journal of Physiology.*

Council at this fall APS meeting and at the following spring meeting took extraordinary steps to hold the biophysicists. One was a change in the bylaws for membership eligibility, with the addition to the word "Physiology" the words "and/or Biophysics," capitalized. Milton Lee drew up a plan for "divisions" within the Society. Such divisions were to have their own chairmen, who would be ex officio vice-presidents of the Society and members of Council. The divisions were to draw up their own bylaws, provided that they were not inconsistent with those of the Society. They were authorized to establish their own journals and to nominate the editors and members of the editorial boards. The plan was both very general and complete and in many ways presaged the organization of the sections that was to come about twenty years later. Looking beyond the immediate problem, Council appointed committees to negotiate not only with the biophysicists but also with the general physiologists and neurophysiologists. In the fall, William Hamilton wrote a "President's News Letter" (1955) describing all the disadvantages of separatism and polled a number of biophysicists to determine their dissatisfactions in an effort to address them directly. Alan Burton later described all these moves as a "panic reaction."

Nevertheless the biophysicists moved ahead cautiously. They held a sort of trial meeting, the National Biophysical Conference, in Columbus, Ohio, in the spring of 1957. The meeting was organized by E. C. Pollard, Samuel Talbot, Otto H. Schmitt, and K. S. Cole; Talbot, Schmitt, and Cole were also members of APS at the time. At the organizational meeting a temporary council was elected; there were twenty members, nine of whom were also APS members. The Biophysical Society thus came into being, and two years later so did the *Biophysical Journal.* The biophysicists saw their interests extending into fields of engineering and physics, and indeed part of the organizers' early intention was to incorporate the new society within the American Institute of Physics. In any case, by no means did all biophysicists consider themselves physiologists as well.

In the fall of 1956, before this National Biophysical Conference, Alan Burton wrote a "President's News Letter" to the APS membership in which he outlined a number of reasons why, on second thought, it was not a bad or even a threatening idea for the biophysicists to organize separately. One of his reasons was, "After all, most of us in the Society have more than one affiliation, and feel that our contribution to the Society is the greater because of it." He concluded that

> the future of our Society will be best served if we continue to emphasize integration rather than specialization. . . . Specialization will look after itself. The nurture of generalization, of cross-fertilization of ideas, of integration, of holistic philosophy of physiology becomes all the more important.

Milton Lee's plan for divisions in 1955 came to naught.

Twice in 1960 another organizing group met to form what became the American Society for Cell Biology (ASCB). Of the twenty-one participants, only three were members of APS: Teru Hayashi (muscle), Arthur K. Solomon (transport physiology and biophysics), and Paul Weiss (developmental biology). The formation of this society followed by several years the establishment in 1955 of a Rockefeller journal, the *Journal of Biophysical and Biochemical Cytology,* and two of its editors, George Palade and Keith Porter, were on the ASCB Organizing Committee. The journal, in turn, had been created when electron microscopy was becoming a major technique in a number of laboratories, and there was a need for better quality in the publication, not to say interpretation, of electron micrographs. This journal became the *Journal of Cell Biology* in 1962. From APS's point of view, the new ASCB was a sleeper. At the time of ASCB's founding, APS had other major internal concerns, and because there were few APS members on the Founding Committee the APS governance may have regarded the new group as an association of cytologists. Whatever their thoughts, the Council minutes show no concern. What happened is history. Form follows function and function follows form. As subcellular structures became more clearly defined at increasingly higher resolution, the cell biologists became more deeply involved in subcellular physiology. In 1970 Ladd Prosser, then president of APS, voiced his dismay that organelle physiology had become almost unrepresented in APS, a situation that is little improved today. There is no doubt that many of the major advances in modern molecular and cellular physiology are introduced at ASCB rather than at APS meetings. Yet even in hindsight there is no reason to believe that the formation of a section of subcellular physiology within APS would have prevented the self-identification of cell biology as a field in itself; the consequences, however, have not been trivial. Cell biology is indeed universally recognized as an entity, with ASCB its American spokesorganization. To a large extent APS has consequently lost an opportunity to represent an attractive, exciting, and extraordinarily important segment of modern physiology. For the past several years the editors of the *American Journal of Physiology: Cell Physiology (AJP: Cell Physiology)* have been making a conscious effort to have the field well covered in their segment of the Society's journals, but so have the editors of the *Journal of Biological Chemistry* and the *Journal of Cellular Physiology,* as well as the *Journal of Cell Biology.* Consequently APS's influence in the field remains regrettably minor.

Also in 1960 the neurophysiologists in APS wanted a somewhat greater voice in their own affairs, especially regarding publication. In March 1961 Society President

Julius Comroe, Executive Secretary-Treasurer Ray Daggs, and others met with Karl Frank at his home in Bethesda to discuss the issue. Frank polled the identifiable neurophysiologists in the Society, about 250 scientists, and reported back to Council that of the 90 who responded, they were 2.5 to 1 in favor of forming a section within the Society and 4 to 1 against forming a new society. (The most positive response to Frank's poll was a seven-to-one endorsement that the Society acquire the *Journal of Neurophysiology*, an idea that was being much discussed between the Board of Publication Trustees and the Council at the time. The acquisition of the journal in 1962 is described in the chapter by A. P. Fishman and S. R. Geiger in this volume.) Meanwhile Ray Daggs was scrutinizing the organization of two large sectionalized societies, the American Psychological Association and the American Chemical Society, and preparing a report for Council on how sections might function. More or less simultaneously with all the above, Ted Bullock, himself a neurophysiologist, was polling another constituency of the Society with a view toward forming a section on comparative physiology. This is described below in the context of that section's origins. The outcome in 1961 was the same in both cases: the Society represented physiologists first and specialists second, and the integrative quality of physiology was again thought to be best served without sectionalization.

Eight years later, in 1969, the Society for Neuroscience (SN) was incorporated. SN has become the largest of the offshoot groups. With 9,000 members, it is now substantially larger than APS itself. Like the Biophysical Society, its components reach beyond physiology per se and include biophysics and behavior, anatomy and chemistry, and pathology and pharmacology. The origins of this society and its cordial relations with APS, although it drew off many members and depleted APS meetings of neurophysiologists, are detailed in the chapter in this volume by Robert Doty. At the time of SN's founding, Ladd Prosser, who also had a sizable professional interest in neurophysiology himself, was president of APS. In the 1950s Prosser had served on a committee of the American Society of Zoologists (ASZ) that had overseen its sectionalization, including a division of comparative physiology. In his view the divisions were responsive to members' needs and the move had strengthened the ASZ. He again urged a similar move for APS and drew up a new bylaw for the purpose, which with Council's approval was taken to the Society's business meeting for discussion. The pros and cons in the discussion were not recorded, but at its conclusion the bylaw was voted down.

The survey above does not include all the societies with major physiological components. Among others are the Endocrine Society, the Microcirculatory Society, and the Biomedical Engineering Society. The latter two meet either just before or in conjunction with the APS spring meetings, as guest societies, and associations with the parent Society remain close. This survey intends to show, by a few examples, how APS did or did not perceive new societies as threats to its ability to represent the broad scope of physiology in the United States. The major, possibly the only, avenue available to APS to prevent splintering was offering the various groups more autonomy in the form of subsections, including their own journals. Until the entire

Society was sectionalized, in each case the offer appears to have been tendered to the offshoot groups alone. Although APS has had little success in stemming the organization of these societies, in most cases good rapport has been maintained, and APS has never stopped growing.

A different sort of pressure calling for recognizable sections in the Society arose in connection with programming the large meetings. In 1975–76, when Bodil Schmidt-Nielsen was president, Maurice Goodman was chairman of the Program Committee. His committee wrestled with the then-current problem that too few people representing too narrow a range of expertise had to set up a program, including symposia and workshops, for the entire Society. The committee felt that the process involved too much ad hoc and informal consultation and that the process could be improved with more formal contacts with the subdisciplines of the Society. As reported in *The Physiologist* in May 1976, "The Program Committee recognizes that greater sensitivity to the programming needs of specialty groups within the Society may combat the growing trend toward fragmentation within Physiology." A Task Force on Scientific Programs was formed to consider how to handle the problem. Its members were Alfred Fishman (chairman), Moe Goodman, Jere Mead, and George Somjen, with Bodil Schmidt-Nielsen and Orr Reynolds ex officio. The committee met in July 1976 at the Mount Desert Island Biological Laboratory, where Schmidt-Nielsen was in residence. They formulated a recommendation that is now practice: a small Program Executive Committee to oversee the process and report to Council and a larger Program Advisory Committee composed of representatives of specialty groups. Sectionalization of the *American Journal of Physiology* into separate component journals was just getting under way in 1976, and the first sections of the Society were also organizing. There were thus sources for the selection and appointment of appropriate individuals to the new advisory committee. At first ten such groups were recognized; currently on the Program Advisory Committee there are fifteen, appointed by their respective sections or groups.

One innovation, the institution of poster sessions in the mid 1970s, has improved programming and appealed to members, who were becoming increasingly dissatisfied with unwieldy meetings. Scrambling to hear widely separated ten-minute papers has been substantially reduced, and the number-independent concentration of presentations in a single area has greatly enhanced one-on-one interactions with interested colleagues.

In addition to the breaking away of new societies and the need for improved programming, another pressure was exerted by some members. APS had always accommodated the informal groups that held dinners or other get-togethers at APS meetings, but participation in several of these sessions was by invitation and/or election only, and the number of participants was restricted. Some Society members were understandably miffed by their exclusion from these discussions, minisymposia, and workshops in their field, and they were resentful that APS at least appeared to support these elitist activities by listing them in the program and arranging meeting space for them. Letters of complaint were addressed to Council or the officers. To deal effectively with this problem, the Society formally recognized the major groups as sections, with the understanding that the meetings of each section were open to all members of the Society who expressed interest.

It Finally Happens

A little like Topsy, sectionalization just "grow'd" over a period of several years before it was formally recognized by Council. In 1974 incoming President Arthur Guyton, in a document to Council entitled "APS as Umbrella Organization," suggested that any specialty group of fifty or more scientists could become a section with an appropriate specialty name and that the sections could elect officers, provide input to the programs, set their own standards for membership, collect dues, nominate members to the Society's standing committees, recommend editors or editorial board members to the Publications Committee, and very importantly "contract with the parent APS for services that APS can reasonably provide." Once again Council decided not to present the document to the membership as a bylaw but instructed Orr Reynolds, who had succeeded Ray Daggs as executive secretary-treasurer in 1973, to revise the proposal into an "information statement" for publication in *The Physiologist*. This was done, and a list of services available from the Society to the specialty groups was provided; it included arranging for meetings, maintaining membership lists, centralizing mailings and dues collections, and establishing liaison with appropriate Society committees.

The following year, during Bodil Schmidt-Nielsen's presidency, several groups appealed to Council for more formal status. These included the clinical physiologists, neurophysiologists, gastrointestinal physiologists, comparative physiologists, and the "epithelial transport workers." All were encouraged to develop their own organizational plan and to interact with the Society through the executive secretary-treasurer's office, and Council again elected to wait and follow developments before changing the constitution or bylaws.

The gastrointestinal group was the first section to achieve formal recognition. Since its charter was to become the model that later sections would follow, Kenneth Hubel, chairman of the Gastrointestinal Steering Committee in 1975, worked closely with Council and the executive secretary-treasurer in drawing up the document. The Society's attorney was also consulted to make sure that no incompatible children were being fostered within the parent Society. It was established that the charters should not be called "constitutions" but "statements of organization and procedures" and that each should contain a disclaimer disavowing conflict with the Society's own constitution, bylaws, and operational guide. Beyond these points there was a strong sentiment in Council that each section should set up its own rules without interference. Another minor issue, resolved in 1980, was that with sections having become Sections, this name was to be reserved for these specialty organizations. The designation of other groups, like the Section on Physiology in Clinical Sciences, which had different functions, was changed, in the case cited to the Committee on Clinical Sciences. There was no revisionist way to alter the names of the many sections that had come and gone on an informal basis in the earlier decades.

From the outset the sections functioned well, particularly with respect to pro-gramming meetings, although there were a few problems. By 1979, during the presidency of David Bohr, Council was ready to consider the experiment a success, and under the urging of President-elect Ernst Knobil the bylaws were amended to read:

> Article X. Society Sections and Affiliations. *Section 1. Society Sections.* Upon accept-ance of a Statement of Organization and Procedures by Council, any group of members of the Society may form a section which encompasses an area of physiology. Such sections shall:
> a. Advise the Society on matters of interest to the specialty group represented by the section.
> b. Assist the Society in organization of scientific meetings.
> c. Nominate individuals for membership on Society committees.
> d. Be open to all members of the Society expressing interest in section membership.
>   The Executive Secretary-Treasurer shall provide assistance to sections in the carrying out of section business.
>   Nothing in a section's Statement of Organization and Procedures can be construed as contradictory to the Constitution and Bylaws or Operational Guide of the Society.

By Council action in October 1979 the number of members required to form a section was set at 100, and each section is to be recognized for no more than five years. This second point was adopted because many specialty groups in the past had outlived their usefulness and faded away. The five-year limitation enables Council to ascertain that the sections are alive and well. The new bylaw was presented to the membership at the business meeting in the spring of 1980, and sections became, at last, official. In 1983 the role of the sections was strengthened by including one member from each on a Section Advisory Committee (see below).

Virtually all the charters include the purposes of advising the Society on "matters of interest" to the section and assisting the Society in organizing scientific sessions, symposia, and other programs of interest to the section (Table 1). They all explicitly state that membership is open, and they describe the officers and their means of election. Some have dues, some definitely state that "dues will not be assessed," and some equivocate on this point. They all end with a minor variant on the obligatory disclaimer of conflict of interest with the Society's own rules.

Gastrointestinal Section

At the conclusion of the scientific sessions on gastrointestinal physiology at the 1950 FASEB Meeting, Charles F. Code proposed to those in attendance that a gastroenterological section be formed within APS to hold an annual dinner at the spring meeting. The suggestion was enthusiastically adopted, and a Steering Com-mittee was formed, composed of F. R. Steggerda and M. H. F. Friedman with Charles Code as chairman. The first such dinner meeting was held in May 1951 in Cleveland.

Table 1
SECTIONS AND GROUPS,* 1986

	Year†
Gastrointestinal	1976
Renal	1977
Environmental, Thermal, and Exercise Physiology	1977
Comparative Physiology	1977
Nervous System	1978
Cardiovascular	1980
Endocrinology and Metabolism	1980
Respiration	1980
Cell and General Physiology	1981
Epithelial Transport Group	
Muscle (MYOBIO) Group	
Neural Control and Autonomic Regulation	1981
Water and Electrolyte Homeostasis	1982
History of Physiology	1984
Teaching of Physiology	1985

* Groups listed are those represented on the Program Advisory Committee.   † Year in which charter, or Statement of Organization and Procedures, was approved.

The group established the rule that, apart from the speaker's presentation, the business of the section must be accomplished in less than ten minutes at the annual meeting.

At the twenty-fifth anniversary of the annual dinner in 1975, Charles Code made a challenging speech suggesting that the "rapid, bang-bang sequence of ten-minute presentations" did not serve the best interests of the gastrointestinal physiologists and that the ever-expanding Federation was overdiluting the interesting and even hard-to-find papers. He pointed out that the annual meeting of the American Gastroenterological Association had become the "bright spot" for those in the field, but he also held that APS and the Federation could be a focal point of gastrointestinal physiology if appropriate steps were taken. Essentially he proposed that the gastrointestinal physiologists become a corporate body with clout that could have greater impact on the parent organizations.

That same year the gastrointestinal physiologists applied to Council for more input into the selection of session chairmen and sponsorship of symposia at APS meetings. Council encouraged the section to organize, with these activities a prime purpose of the organization. As described above, Kenneth Hubel and his Gastrointestinal Steering Committee developed with Council and the lawyers the first Statement of Organization and Procedures. Because everyone favored this development from the start, all potential problems were worked out easily and rapidly, to everyone's mild astonishment. The statement lists among its purposes sponsorship of the annual dinners and organization of an annual symposium. It was approved by Council in April 1976.

In this first section's charter it is explicitly stated that "Intersociety Membership in to those with an interest in gastrointestinal physiology but who are not members of the American Physiological Society. Intersociety Members share all the rights and responsibilities of Regular Members," the latter presumably within the section's purview. The intersociety provision was included in a number of the charters drawn up later by other sections and has presented a problem when the sections' activities extend to Society functions beyond holding a scientific session or dinner (see below).

Recently the section has instituted an awards program in which the speaker at the annual dinner is recognized for his/her contributions to the field. The program was expanded in 1986 to add two awards for younger investigators, based on the best abstracts in gastrointestinal physiology that have a graduate student or postdoctoral fellow as first author. In addition, the section has taken an active role in organizing FASEB Summer Research Conferences in Gastrointestinal Differentiation (1985) and Adaptation (scheduled for 1987).

Renal Section

The second section to have its charter accepted by Council, the Renal Section, like several other sections, developed from its annual dinner at the spring meetings. The original Steering Committee composed of Frank Knox (chairman), Sidney Solomon (secretary), and Reinier Beeuwkes III (treasurer) drew up the Statement of Organization and Procedures, which was accepted by Council in March 1977.

Like other major organs and organ systems, the kidney is the central focus of a separate large society, the American Society of Nephrology (ASN), which has a much broader scope than just physiology and includes pathology, pharmacology, and, as a major component, clinical studies as well. Despite the fact that the ASN meetings are an attractive drawing card for renal physiologists, the Renal Section of APS has been a vigorous group, sponsoring large and well-attended symposia at the spring meetings. Quite properly the emphasis tends to be on kidney function per se more than the kidney as part of the whole organism. This emphasis led to the formation of yet another renal-related section in 1982 (see below).

The charter for the Renal Section includes the continuance of the annual dinner, at which since 1979 the section has given young investigators awards for the best papers brought to APS meetings.

Environmental, Thermal, and Exercise Physiology Section

Of the two precursors to this section, the Temperature Regulation Group and the Physiology of Exercise Group, the former is the more venerable. It was founded by James D. Hardy in 1948 with the expectation, based on the experience of the axonologists, that it might last five to ten years. In 1986 the Temperature Regulation Dinner held its thirty-ninth consecutive banquet cum speaker. Hardy was interested not only in temperature regulation but in all aspects of environmental physiology, including life at extreme temperatures and humidities, thermal tolerance, and gravitational effects. These interests found expression both in the talks and in selection of later chairmen of the Temperature Regulation Dinner.

The Physiology of Exercise Group appears to have originated in the mid 1960s, although the exact date is not clear. Many of its members also belonged to the Temperature Regulation Group. When sectionalization of the Society began in 1976 several members of the two groups decided to form a section for their joint interests. This informal committee, including Ellsworth Buskirk, Ethan Nadel, Steven Horvath, and Mel Fregly, presented their Statement of Organization and Procedures to Council, which approved it in the fall of 1977.

In addition to the usual input into programming, one of the activities of the section is the annual Harwood Belding Award to a graduate student who presents the best paper at either the fall or spring meeting. The award, begun in 1975, is presented at the Temperature Regulation Dinner. Despite their amalgamation into a single section, the Temperature Regulation Dinner and the Physiology of Exercise Mixer remain separate but still overlap in the cast of participants. Care is taken in electing the Steering Committee to ensure the adequate representation of both, and joint business meetings have been organized.

Many members of the section belong to other sections as well, especially cardiovascular and respiration. For a function-oriented section, this is listed by a surprisingly large number of Society members as "primary interest"; it is larger than the Renal, Gastrointestinal, or Cell and General Physiology Sections, among others.

Comparative Physiology Section

The original Comparative Physiology Group was convened by Knut Schmidt-Nielsen in a mailing in late 1950, and the group had its first dinner meeting at the FASEB Meeting in Cleveland in 1951. Donald R. Griffin and Per F. Scholander were the speakers. To avoid any formal organization, it was decided that the speakers at one meeting would arrange for the speakers and other details for the following meeting; thus a self-perpetuating officer-free nonorganization was successfully created. At first this was an "invitation only" dinner to keep it small and informal, but at the urging of E. F. Adolph it was opened to everyone in 1958.

By 1961 the meetings were becoming large; Schmidt-Nielsen's mailing list had reached 120, and the intimacy of the early meetings was imperiled. At this time Ted Bullock, who was certainly a supporter of the group, was also becoming concerned over the possible development of a futile rivalry with ASZ. Bullock was chairman that year of the Division of Comparative Physiology of ASZ and saw clearly that the two groups had very similar goals and were likely to compete for speakers and for the members' attention. But the overlap in membership of the two groups was by no means total. It was also true that a session on "comparative physiology" at either meeting drew a mixed bag of contributions and that such sessions were much more profitable if they focused on a common theme. All this suggested to Bullock that an organization was needed to coordinate the interests of the comparative physiologists, and he proposed that a section be formed in APS that would merge and be congruent with the equivalent section in ASZ. Membership in either society would be sufficient qualification for membership in the intersociety section. A poll of those on Schmidt-

Nielsen's mailing list showed a very favorable response to the idea. The proposal was brought to Council simultaneously with Karl Frank's proposal for a neurophysiology section, and there was an extensive correspondence between APS President-elect Horace Davenport, ASZ President Ladd Prosser, Bullock, Ray Daggs, and others. Daggs surveyed sectionalization of other societies at this time, and the issue was put on Council's agenda for the fall of 1961. At that meeting, however, other matters occupied Council's time, and the issue was not resolved either then or at the subsequent Council meetings. It seems simply to have gone away.

In 1975 the issue was revived in a letter from John Lawrence of ASZ to President Bodil Schmidt-Nielsen, who was enthusiastic about intersociety ties. With Council's approval she appointed a Task Force on Comparative Physiology, cochaired by Frank Conte (APS) and Mike Greenberg (ASZ). In August 1976 both societies met simultaneously in Philadelphia, and the task force held its first meeting. In addition to the cochairmen, the Steering Committee included Ronald Alvarado, Howard Bern, John Brobeck, Jim Cameron, Ann Kammer, John Roberts, and Knut Schmidt-Nielsen, the latter a member of the International Commission on Comparative Physiology.

The task force had several major concerns. First, it was felt that comparative physiology as a discipline needed greater visibility in the predominantly medically oriented APS, and the task force drew up the Statement of Organization and Procedures for becoming an APS section. This was approved in the fall of 1977. Secondly, it was felt that the field needed more outlet for publications of high standards. With the establishment of *AJP: Regulatory, Integrative and Comparative Physiology*, edited by Eugene Yates, Frank Conte and Bodil Schmidt-Nielsen were appointed associate editors. The task force also was concerned with meetings and programming.

After its formation as a section, the group decided to alternate its focus between the fall meetings in even years and spring meetings in odd years. They organized joint fall meetings with APS and ASZ (Division of Comparative Physiology and Biochemistry) in St. Louis in 1978, in Toronto in 1980, and in San Diego in 1982. This arrangement, however, has not been without its difficulties, since ASZ traditionally holds its annual meeting during Christmas week, and the two meetings so close together are somewhat in competition. The issue is still being worked out. The members feel that the joint meetings are very strong.

A Comparative Physiology Dinner was held in Philadelphia in 1976 but has since faded away. In New Orleans at the 1986 fall meeting the dinner was revived. At that time the section presented the first annual young investigator prize, the Per F. Scholander Award, named in honor of an outstanding investigator who was also speaker at the first dinner in 1951.

Section on the Nervous System

Neurophysiology in all its aspects was a major component of APS for the first three-quarters of its first hundred years, and many distinguished neurophysiologists have served as Society presidents. In 1952 a Neurophysiology Group was formed; L.

M. N. Bach, H. W. Magoun, and Harold Himwich were elected as the first Steering Committee. After one less than totally successful dinner meeting ("postprandial ebb of intellectual function [was] quite evident") in Chicago in 1953, the group adopted the tradition of meeting the day before the beginning of the Federation meeting. An afternoon symposium was followed by a cocktail party, and the symposium topics were chosen to cover the full range of neurophysiology every few years. There was already some dissatisfaction in the Society about closed meetings of special-interest groups, and the Neurophysiology Group opted to keep their meetings open, in part to attract younger participants. In a further effort to avoid elitism, there were no officers. A Steering Committee of two was elected to select the symposium moderator, handle the menial organizational tasks, and otherwise stay out of sight.

In 1961, Karl Frank's questionnaire (described above) identified 250 neurophysiologists in APS. According to a survey made by Ray Daggs in the same year, neurophysiologists comprised 13.3% of the regular membership, a close second in size to the cardiovascular component.

As already noted, the formation of the Biophysical Society in 1956 drew off many of the axonologists and synaptologists, followed in 1969 by a major siphoning off of neurophysiology from APS at the time of the formation of SN. This had been long foreseen by Council, and the acquisition of the *Journal of Neurophysiology* in 1962 had been in part an effort by the Society to retain a major strength in the area.

In 1974 while Dan Tosteson was president, a Task Force on Neurophysiology was established (Edward Evarts, chairman) to explore how to revitalize the Society's role in the field. Tosteson emphasized in his charge to the task force that its role was not "to combat the formation of new societies," which would take place of its own accord, but to ensure that neurobiology would retain its proper place in the context of physiology and the APS. In its first report the task force specifically stated that the formation of a section on neurophysiology would not be advisable. After consulting with many people, the task force focused on improving the programs. It seemed unrealistic to compete with the fall meeting of SN, and the task force's aim was to enhance neurophysiology at the Society's spring meeting. Reporting for the task force a year later, George Somjen stated that the task force further felt that the traditional symposium areas were antiquated and that a particularly fruitful area for exploration could be interface symposia in which the relation of neurophysiology to other fields of physiology could be developed. Some suggestions included synaptogenesis, involving tissue culturists, and the neural control of renal function. By 1977 Somjen was able to report that both the interest and attendance at the newer, smaller symposia were noticeably improved, and, in an apparent change of heart within the task force, he asked Council for advice on the formation of a section on neurophysiology. An Organizing Committee was formed under David Carpenter, and the Statement of Organization and Procedures was presented to Council in the spring of 1978. By common consent the name of the section was changed to Section on the Nervous System, and the charter was then approved.

At the 1985 spring meetings in Anaheim the section scheduled five of the forty-two APS symposia, and together with the Section on Neural Control and Autonomic Regulation accounted for about eight percent of the contributed presentations. The current chairman of the section, Janett Trubatch, reports a dramatic increase over the past few years in the participation of the section in APS affairs. It is to be hoped that with the new emphasis on the nervous system as an interactive component in other areas of physiology, this very important field will flourish as the integral part of the Society, and of physiology, that it is.

Cardiovascular Section

Originally established in 1933, the group that became the Cardiovascular Section has the longest continuous identity of any in the APS. (This group had been preceded by the axonologists, who were organized by Ralph Gerard in 1930 but who disbanded by common consent, having become too large, in 1942.)

About twenty members of APS and their guests with an interest in the circulatory system attended an organizational dinner in Cincinnati in 1933 and elected an organizing committee: Carl Wiggers, W. J. Meek, and Louis N. Katz. Those attending proposed some very zany names for the group, including the "Cordisanguinarians" or the "Hales and Harveys." Yandell Henderson in absentia won but could not collect the prize (free dinner) for "Hearties," but in the end Louis Katz's somewhat more prosaic "Circulation Section" became the real winner by common usage. The organizing committee later met and established the Circulation Section and the annual Circulation Dinner, with a discussion of "electrical currents created by the heart" the subject of the first meeting in Chicago in 1934. Circulation Dinners have continued since that time. The meetings were open to members of the section and guests and other members of the Society "on application to the limit of available accommodations." Active membership was limited to one hundred and was terminated for any member who failed to appear for two consecutive years. Excused from this latter restriction were a number of senior members who were admitted to the category of distinguished members. In about 1962 the Circulation Section began organization of an open symposium as part of the regular Society program.

In 1962 W. F. Hamilton described this group as "lots of fun and quite indestructible," and this has continued to be the case. In 1980 a committee composed of James Davis, Eric Feigl, and Kiichi Sagawa published the Statement of Organization and Procedures for a new Cardiovascular Section, which formally replaced the informal Circulation Section, and they issued an open invitation to all members of the Society to join. Unlike some of the other sections, their charter makes no allowance for admitting other than APS members. Also unique is the recognition of fellows of the section, and with the adoption of the charter in Anaheim in 1980 all the previously elected regular and distinguished members of the expiring Circulation Section automatically became fellows. The number of fellows is limited to five percent of the regular membership of APS.

In addition to organizing the symposia the section confers two annual awards: the Carl J. Wiggers Award (established in 1965) to a senior cardiovascular scientist "whose investigations have made a lasting impact on the field" and the Harold T. Lamport Award (established in 1978) to a "promising young investigator for excellence in recent research." The section charter specifically identifies its membership as a "candidate pool for the Wiggers Award."

At this writing the Cardiovascular Section is easily the largest in the Society. In a 1986 poll requesting the Society's members to identify their section of primary interest, about half of the members responded, and one-fifth of them indicated the Cardiovascular Section. Among other things the large size reflects the diversity of cardiovascular physiology. Only a year after the adoption of the initial charter it was amended at the Atlanta meeting in 1981 to recognize and formalize the creation of subsections. The impetus behind the amendment was the fact that the Cardiac Mechanics Group, chaired by Karl T. Weber, and the Splanchnic Circulation Group, chaired by C. C. Chou, had for several years been conducting successful meetings of their own, and it was generally felt that their inclusion in the larger section would strengthen all the groups. The charter change is written in very general terms so that new subsections can be added without further amendment.

Again because of its size and diversity the section appealed to Council for greater representation on the Program Advisory Committee. The section was comprised of more than twenty percent of the Society's members but had less than ten percent of the representation on the Program Advisory Committee, and no one person could adequately represent the scope of the section in programming. The Council felt in general that section issues could be worked out at the section level and brought to all-Society committee meetings by a single person, but Council approved two members, one ex officio, from the Cardiovascular Section on the Program Advisory Committee.

Section on Endocrinology and Metabolism

There were several progenitors to this section, but none with the continuity of the Circulation Section transmogrified into the Cardiovascular Section. The Metabolic Discussion Group, formed in 1934 with Samuel Soskin as the principal organizer and longtime head, was originally an APS group. For many years it held dinner meetings, generally on the Wednesday night of Federation Week, with a speaker on some topic of metabolism or endocrinology. Many of the group's members were biochemists, and in 1973 the leadership passed from APS member Ed Masoro to ASBC member Dick Hanson. When the ASBC began to meet independently of APS, the Metabolism Dinners became associated with the ASBC rather than the APS meetings. Since the formation of the section in 1980, at least once when the two societies met together the section dinner and the Metabolism Dinner became a joint affair.

For a while in the 1960s and 1970s the Society listed as one of its ancillary groups the more specifically oriented Adrenal Physiology Group, but this association was no longer to be found in the Society's 1976 compilation.

Another progenitor to the Endocrinology and Metabolism Section was the Neuroendocrine Discussion Group, formed in 1956, again with an annual dinner at the spring APS meeting. For a time the group seriously considered founding its own society, but this idea seems to have faded after the group was able to arrange with the APS Program Committee for neuroendocrine sessions to be scheduled on the day of their dinner. Nevertheless, after the rise of SN at about the same time that the biochemists formed their own society, endocrinology and metabolism, at least in some aspects, were much weakened in APS.

As president of the Society in 1980, Ernie Knobil, himself an endocrinologist, continued to urge the formation of identifiable sections to strengthen the Society as a whole. With his encouragement, S. M. McCann chaired an organizing committee, with members Bob Fellows, Henry Friesen, Howard Morgan, Ed Masoro, and Jimmy D. Neill, which drew up the Statement of Organization and Procedures for the Endocrinology and Metabolism Section that was adopted in 1980. The charter included the admissibility of intersociety members. At its first meeting in Atlanta in 1981 the section, chaired by Ed Masoro, decided explicitly to develop plans to recruit members from ASBC, the American Institute of Nutrition, the American Diabetes Association, and the American Society for Clinical Investigation. Exactly how these plans are developed may have to be modified if, as described below, APS membership becomes a prerequisite for section membership.

The section remains vigorous in the promotion of symposia and its annual dinner. In 1986 the Section on Endocrinology and Metabolism was virtually tied with the Respiration Section as the second largest in the Society.

Respiration Section

By 1950 a profusion of relevant equations in the literature put studies on respiration and pulmonary mechanics on a stronger quantitative basis than ever before. Lack of standardization in the nomenclature contributed more to the confusion than the profusion. Respiratory physiologists complained that they had great difficulty in deciphering their colleagues' papers. To deal with this problem, John Pappenheimer convened a group of "breathers" at a dinner in Atlantic City during the 1950 FASEB Meeting. Present were Comroe, Cournand, Ferguson, Filley, Fowler, Gray, Helmholtz, Otis, Pappenheimer, Rahn, and Riley. From this dinner meeting emerged a set of definitions and symbols (*Federation Proc.* 7: 602, 1950) that quickly became standard for the field.

The meeting was such a success that it was followed by a series of "nonannual" dinner meetings, with afterdinner speakers, that continue to this day. Among the factors that have contributed to the nonannual frequency of the dinners are similar dinners held in conjunction with the triennial international IUPS meetings, starting at Leiden in 1962. The U.S. nonannual dinners have always been open and well attended, with a good deal of banter and informal camaraderie. Formalization of the proceedings was resisted. There were no officers; for the next dinner only a chairperson, who was responsible for all the organization, was selected.

Given this "unstructured structure," some of the regular participants (they could hardly be called members because there was nothing to be a member of) were reluctant to tie themselves down to the formality of a section. Nevertheless a section organizational meeting was called in April 1978, again at the FASEB Meetings in Atlantic City. The ad hoc committee for organization was chaired by Sami Said and included also Lynne Reid (chairperson for the Respiration Dinner that year) and Norman Staub. The Statement of Organization and Procedures was presented to Council by Robert Hyatt and approved in April 1980. It explicitly includes as one of its purposes the perpetuation of the Respiration Dinner "to coordinate activities with physiologists working primarily in other systems that are related to the respiratory system," and the dinners continue unabated.

Section on Cell and General Physiology

Because of interest expressed by the Program Committee in biophysics in 1975, Council attempted to establish a Task Force on Cell Physiology. Several persons were contacted to organize such a task force, but none felt in a position to undertake the commitment at the time; more or less by default Robert Forster became the de facto head. Despite a large mailing the subject did not generate much enthusiasm among the membership. Several years later President Ernst Knobil urged John Cook to set up a section for "cellular physiology and membrane transport." I found thirteen or so physiologists in a variety of subfields, including epithelial transport (see below), who were willing to have their names listed as members of an Organizing Committee. This committee never met and from the outset never planned to meet, as its members were well aware. The point was to have a broad spectrum for name (and subspecialty) recognition to attract as many potential section members as possible to the organizational meeting in April 1981 in Atlanta. A Statement of Organization and Procedures had been drawn up by Marie Cassidy and me and was presented to the seventy-five or so who attended. Thom McManus pointed out that "cellular physiology," which was part of the original name, meant "physiology that was made up of cells," and with the name of the section therefore duly changed to Cell and General Physiology, the charter was adopted and accepted by Council on the following day. A curious misprint in the published statement says that "a quorum shall consist of fewer than 3 members of the Steering Committee," but this "rule" seems not to have created any special problems.

In addition to programming functions, the section holds an annual dinner with a distinguished speaker at the FASEB meetings and, with the help of a few pharmaceutical companies, gives as many as three awards to young investigators for their presentations at the meetings. Long-range goals for this section include the establishment of better communications with other societies, notably ASCB, the Biophysical Society, and SN.

Epithelial Transport Group

The Epithelial Transport Group was established in 1975 largely through the efforts of Sandy Helman who had contacted 300 to 400 people. Over 300 potential members attended the informal spring meeting in Atlantic City that year, and many more asked to be placed on the mailing list. At the 1975 fall meeting of Council, Helman was invited to describe this group and was encouraged by Arthur Guyton to form a section, tentatively referred to as the "Epithelial Transport Workers," a name that is sometimes convoluted to the "Transport Workers Union" and is still commonly used by its members. The major outcome of this interchange was that the Society took charge of maintaining the mailing list and helped in setting up the group's meetings. In 1979–80 Ernie Wright was chairman of the "section," and a set of bylaws was drawn up and submitted to Council. But at that time Council was already committed to the establishment of a Section on Cell and General Physiology, which the epithelial group was encouraged to join. This was not popularly received by the Transport Workers, and the group has maintained its own identity, with its own Steering Committee and annual evening meeting during FASEB Week. However, many of the group's members have indeed joined the Cell and General Physiology Section as well. Although the Epithelial Transport Group has had a major concern with the Society's programming, since they have a representative on the Program Advisory Committee, there has not been a strong resurgent move within the group to become a more formal section. Interestingly, Joe Saunders, in reporting on the 1985 spring meeting, noted that in the entire Society, epithelial transport was the single topic category that attracted the largest number of abstracts.

Muscle Group (MYOBIO)

In 1984 a group of muscle physiologists in APS recognized that despite the existence of the Cell and General Physiology Section, muscle physiology was not a strongly identifiable presence at the meetings and, if anything, was in fact dwindling. Much of the reason for this was of course that muscle physiology was strongly represented in the Biophysical Society. To enhance the field's representation in APS, especially as an important physiological subject compared with the ever-finer analysis of the biophysical approach, Marion Siegman took the lead in establishing the Muscle Group. Her inspiration came from a very successful MYOBIO research group drawn from the various universities in and around Philadelphia, from which the Pennsylvania Muscle Institute had arisen. The original Steering Committee consisted of three persons from each of the three principal areas of skeletal, smooth, and cardiac muscle. The group's goals include the planning of symposia and workshops dealing with themes that will attract investigators from other disciplines, like biophysics, biochemistry, and pharmacology, and a "breakaway from the very orthodox system orientation" as an impetus for new avenues of investigation. The first social meeting of the group was with the Cell and General Physiology Section in Anaheim in 1985, and they have sponsored or cosponsored a number of symposia at subsequent meetings.

Although initially formed independently of the Cell and General Physiology Section, at the 1986 Spring APS Meeting the group became a subgroup of that section, and its chief officer meets with the council of the section to plan programs and other activities. In addition, the group has representatives on both the Program and Section Advisory Committees.

Neural Control and Autonomic Regulation Section

Even before sectionalization began, an informal group within the Society was concerned with problems of how the brain controls various visceral functions, particularly in the cardiovascular system. The group met at an annual cocktail party and held informal discussions, without formal presentations, of recent work. An important characteristic of the loose structure has been that it enables graduate students and postdoctoral fellows to meet socially and exchange ideas with more established investigators. John Manning and Larry Schramm played significant roles in keeping this group going.

After sectionalization was instituted many of the members of this group found it difficult to identify solely with some of the more narrowly defined specialties, like the Cardiovascular Section or Section on the Nervous System. Under Manning's leadership the Statement of Organization and Procedures for the Neural Control and Autonomic Regulation Section was drawn up and approved by Council in the fall of 1981. The section especially encourages interactions with anatomists, pharmacologists, and others in the contributed sessions and of course has Program and Section Advisory Committee representatives as well.

One of the smaller "primary interest" sections, it is functionally oriented and cuts across more than one organ system; many of its members identify their primary interest as cardiovascular or nervous system and thus belong to more than one section. Its total membership is several times its "primary interest" listing.

Section on Water and Electrolyte Homeostasis

During the early stages of sectionalization some of the Society's councillors were concerned that a proliferation of sections corresponding to very specific subsubspecialties might develop. The Parietal Cell Club and the Red Cell Club, for example, were very active groups, but it was not clear that a section for every cell type was the way to go. It was suggested in Council that perhaps the sections should be limited to five or six, corresponding to the sections of the journals, but this suggestion was not adopted. Nevertheless by 1982 there was a section of the Society corresponding to each section of the *American Journal of Physiology* with the exception of half of *AJP: Regulatory, Integrative and Comparative Physiology*, where there was the Section on Comparative Physiology, but none for regulatory or integrative physiology. This of course had been a factor in the formation of the Neural Control and Autonomic Regulation Section (see above).

A Biological Regulation and Control Group existed from 1968 to 1972. E. F. Adolph and David A. Robinson spoke at their 1968 dinner. The interest of this group was in a functional systems analysis approach to physiology rather than the more classical organ systems approach. In 1972, after a number of successful meetings, Lloyd Partridge, in a letter to the participants, disbanded the group on the grounds that what they advocated had largely been accomplished: "Today elementary concepts of systems analysis have become well enough known that they are introduced in most major textbooks of physiology." Therefore for 1972 he had "found no speakers and chosen no wine," and he ended his letter with a plea for the teaching of physiology entirely from the systems analysis point of view.

A number of Society members naturally remain very interested in regulation and integration; one group has been interested in salt and water homeostasis in particular. Some already were identified with the Cardiovascular or Renal Section or the Epithelial Transport Group but did not find in the symposia or meetings of these groups the "big picture" that they were concerned with. Others, with particular interests in subjects like water intake and thirst, salt appetite, and the endocrine control of fluid distributions, found no existing sections that exactly met their needs. Finally they organized a task force made up of Steve Bealer, John Claybaugh, Allen Cowley, Joan Crofton, Kim Johnson, David Ramsay, Wilbur Sawyer, and David Yang, with Leonard Share as chairman. Share, who is the 1986 head of the Section on Water and Electrolyte Homeostasis, is credited with having done most or all of the initial legwork that included drawing up the Statement of Organization and Procedures.

The section considers its only significant role the planning and organization of symposia, and the section's Program Advisory Committee representative, Bob Shade, has been extolled for doing this job exceptionally well and with vigor and enthusiasm. The business meeting of the section takes place after one of the scientific sessions; there is no dinner or other get-together.

Nobody has yet solved the problem of simultaneous horizontal and vertical organization. The Section on Water and Electrolyte Homeostasis, like several of the other sections described above, deals with multiorgan functions, whereas a number of the other major sections deal with the complexity of functions of a single organ. When in January 1986 there was a call from the executive secretary-treasurer for Society members to identify the section of their primary interest, many fewer persons designated water and electrolyte homeostasis than had voted in the previous year's election of section officers. Apparently many of the members, and there were several hundred who responded to the initial call for section organization, continue to identify themselves first with the heart, the kidney, or the endocrines. The section may thus fall below the critical mass of one hundred and have to be renamed as a group. As long as they continue to have good symposia, the members by and large appear to be equanimitous about such an eventuality.

Section on the History of Physiology

Everybody is interested in history—well, almost everybody. Historical articles had long been a feature in *The Physiologist.* As the Society's centennial approached, the interest in history picked up accordingly. An important additional factor was the appointment of Toby Appel as the Society's archivist.

In 1983 John West thought it would be appropriate to bring together the diverse interests in the history of physiology. With the support of Clifford Barger, Ralph Kellogg, Daniel Gilbert, and Louise Marshall, he proposed the establishment of a section that was duly organized at the 1984 spring meeting in St. Louis. Toby Appel has been a more than expert secretary to the section. An interesting point in the goals, as given in the Statement of Organization and Procedures, is to "work toward the establishment of a center for the history of physiology." Of course a major goal is to have input into the programming, and beginning in 1985 history has been listed as a topic category in the calls for papers at APS meetings.

In their early meetings the section heard presentations by Clifford Barger on Walter B. Cannon; by Robert G. Frank (Medical History Division, University of California, Los Angeles) on "Innocents Abroad," describing the influence of their European training on early members of the Society; and by Bodil Schmidt-Nielsen on her parents, August and Marie Krogh. The section also cosponsored the symposium on William Beaumont at the 1985 fall meeting.

An interest in history is being encouraged by the Society in the form of the Orr Reynolds Award, announced in 1985, for the best historical article on some aspect of the field. Pending peer review, the article is to be published in the Society journal most relevant to the topic. The selection committee for this award consists of three APS members, one of whom is to be a professional historian. The committee is to be appointed by Council in consultation with the chairman of the Section on History of Physiology.

Teaching of Physiology Section

Teaching is an important responsibility for most members of the Society and a prime responsibility for some. Under the leadership of Society President Ladd Prosser in 1969, Council endorsed the formation of an Office of Education, and in his fall newsletter, published in *The Physiologist*, Prosser called for nominations. Orr Reynolds was recruited for the position, and even after he later became the Society's executive secretary-treasurer on Ray Daggs' retirement in 1973 he continued to edit *The Physiology Teacher*, published in *The Physiologist*. In later years much of the responsibility for editing *The Physiology Teacher* was taken over by Executive Editor M. C. Shelesnyak. Not widely appreciated is the fact that articles in *The Physiology Teacher* were peer reviewed.

Organized in 1953 by Edward Adolph, the Education Committee included Ladd Prosser and William Amberson. This early committee set up a program of workshops for college teachers and obtained NIH support, but after a few years the program lost NIH support for bureaucratic reasons, and the program ended. In more recent years the Education Committee's major responsibilities have been the continuing education of Society members, developing career opportunities literature for recruiting purposes, and developing materials, such as audiovisual aids. Harold I. Modell, a member of this committee in 1983–86, had long been interested in new teaching methods for graduate and undergraduate students of physiology. Modell, together with Joel Michael and others, has believed that there is a real need and role for a section on teaching, complementing and not competing with the Education Committee in its functions, and open to all members of the Society. This view was reinforced by the vigorous discussions at the Learning Resource Center areas at the spring FASEB meetings, involving physiologists with very diverse interests but common problems in teaching. In October 1984 Modell polled 200 members of APS who had, on a Society questionnaire, listed teaching as their primary responsibility. Receiving more than the requisite one hundred positive responses, he called an organizational meeting for a section at the 1985 spring meeting. A charter for the section was drawn up and approved by Council at the 1985 fall meeting. The section held its first business meeting in St. Louis in April 1986. Like the other sections the Teaching of Physiology Section has representatives on both the Section and Program Advisory Committees.

The section cites as its goals that

the group can provide input to Council with respect to matters of interest to physiologists engaged in classroom teaching and to the Program Advisory Committee with respect to organizing and presenting open communications sessions, symposia, and other programs of interest to physiologists engaged in classroom teaching. The intent of the Section is to complement the standing APS Education Committee and provide a primary affiliation for physiologists whose major effort is directed toward teaching.

The section has another goal. In January 1986 *The Physiologist* was reorganized into a newsletter, edited by the executive secretary-treasurer, and general articles on leading-edge research in physiology are being solicited and published in a new journal, *News in Physiological Sciences* (*NIPS*). Articles on history that had been part of *The Physiologist* are to be submitted to the appropriate section of the *American Journal of Physiology*. In all this *The Physiology Teacher* has disappeared. The Teaching of Physiology Section is exploring means for the publication of new and effective teaching techniques and a forum for discussion of teaching issues at all levels.

In addition to the above, one of the important motivations behind the section is the recognition that physiologists of the future need to be attracted to the field at the undergraduate level. The section wishes to assist and encourage undergraduate teachers with the most effective means possible in developing their curricula and in stimulating their students. In responses to a recent letter from Walter Randall, head of the Long-Range Planning Committee, several sections have emphasized the importance of this goal as a Society function. This laudable aim deserves the support of all members of the Society.

Future Role of the Sections

Having been well established for good and sufficient reasons, the sections have by and large been functioning well for their originally designated purposes, or at least for their own purposes. From the Society's point of view the matter has not been quite so simple. The original idea was that sections within the Society would allow total or partial autonomy to the many self-identifiable internal groups and thus preserve the cohesiveness of the Society as the principal physiological spokesorganization in the United States. But once the sections were formed, established their own programs, and gained direct input into the Society's programs, there remained the important question of how their individual aspirations could contribute to, and be integrated with, the Society as a whole. Council minutes record one Society officer as noting that "we have the sections and now we must figure out what to do with them." The matter has not been finally resolved, but several mechanisms have been initiated or are at least under consideration.

Representation of each section (and some of the groups) on the Program Advisory Committee has been generally useful and helpful in programming symposia at Society meetings. But the Society has interests and responsibilities in addition to holding meetings; involving the sections in these other activities is an issue that was faced by the Long-Range Planning Task Force, appointed in 1980 with Robert M. Berne as chairman. The task force reported to Council at a special meeting in November 1983 during Al Fishman's presidency. Among its recommendations were the appointment of a standing Long-Range Planning Committee and the establishment of the Section Advisory Committee, the latter composed of the section chairmen "or the equivalent." Council approved both, and charged the Section Advisory Committee with interacting with the Long-Range Planning Committee and also with meeting with Council at the spring meeting. Past-president Walter Randall was selected as chairman of the Long-Range Planning Committee and Marion Siegman as chairman and Janett Trubatch as recorder of the Section Advisory Committee. The committee's charge is to uncover at the section level and bring to Council issues of general importance to the sections, a sort of democratization of the Society from the grass-roots level. Perceived dangers in this way of doing business are infrequency in the Section Advisory Committee meetings and problems of continuity with an ever-changing set of representatives. It may thus be difficult for the group to deal with long-standing issues, and there is the possibility of its tending to focus on more short-term problems. It is to reduce the potentially ad hoc nature of issues that may arise and to broaden its concerns that the Section Advisory Committee is to meet on a regular basis with the Long-Range Planning Committee, the latter being more an arm of Council. Two members of the Section Advisory Committee also meet with Council. Finally and crucially, the section representatives on the committee are expected to report back to, and bring instructions from, their respective sections and section Steering Committees, which in turn are expected to meet on a regular basis, presumably but not necessarily at the annual spring meeting. The Long-Range Planning Committee has recently recommended to Council that the sections individually report to Council on a regular basis to ensure that they are still active and have a democratically elected set of officers. (The former contingency has already been allowed for in at least one of the section charters: if the Cell and General Physiology Section does not meet for three consecutive years it is automatically dissolved and its assets revert to the Society. In addition, Council's five-year recognition rule covers the same point.)

At the level of the individual members, concerns of a particularly sectional nature

("Why aren't there more people like us on Council?" and "Is the Society too medically oriented?") can be raised and discussed in the sectional meetings. Since many Society members belong to more than one section, there remains the possibility, probably not too dangerous, that a small number of persons could bring an issue to many sections and, if it comes to a vote, could be overrepresented in the Society as a whole. In part to avoid this contingency, the executive secretary-treasurer, Martin Frank, is attempting to have each member of the Society identify his/her section of primary interest where the vote would count and to enable members who do not belong to any section to vote at large. The possibility of voting through the sections on Society issues has raised yet another potential problem in that some of the section charters explicitly do not require membership in the Society for membership in the section. The Long-Range Planning Committee has therefore recommended to Council that Society membership be made a requirement for section membership; some section charters may have to be amended. At this writing the Council has not acted on the recommendations.

Thus a moderately elaborate machinery is in place for enhancing, through the sectional organization, the members' input into the workings and the governance of the Society. To this writer the mechanism has at least one significant meritorious aspect. It has always been possible of course for any individual member to raise an issue at the business meetings, and frequently these issues have been favorably acted on if the climate was right. But sometimes the climate of a business meeting has not been right, and time is always short, and less than ten percent of the membership of the Society is present in the best cases. Raising issues through the sections allows for preliminary discussions, refining of ideas, and drumming up of support. More mature proposals can thus be brought to the Society as a whole. The remaining major problem would appear to be that of logistics. Section meetings are infrequent and usually parallel in time with, rather than prior to, Council meetings and the business meetings. How to raise issues and have them acted on in a timely manner through the new mechanism is not yet entirely clear but is being addressed by the Long-Range Planning Committee and the Section Advisory Committee. The machinery is in place, and responsible persons are seeking the means to oil it.

There is a small irony here. In approving sectionalization in the mid 1970s, Council repeatedly avoided handing down rules for how it should be done. The sections were encouraged to organize in the way each saw best for its own purposes. Now that the question of a governance role has come up, several section representatives have expressed concern over the fact that section organization and operation are indeed not standardized and feel that they should be for an optimal input to the Society's business.

In its early meetings the Section Advisory Committee has recognized the need for interacting with Society committees other than Council. They propose to meet as well with the Program Advisory Committee in the planning of meetings. They also anticipate interaction with the Publications Committee to have more input into the editorial policies and editorial staffing of the sectionalized journals. The latter goal is written into most if not all of the charters of the individual sections, but the means for implementation have not been well worked out. The Section Advisory Committee can perform a real service if it establishes the way for all the sections.

The history of sectionalization is certainly not complete. There remain a number of subspecialty groups that continue to meet with the Federation and that are not full-blown sections and may never be. For example, independently of the Cell and General Physiology Section, the Red Cell Club remains autonomous and reasonably unstructured; it meets twice a year, the spring session being frequently but not always on the Sunday before the beginning of the FASEB meeting. Similarly the Liver Dinner remains independent of the Gastrointestinal Section, although both are represented by a single component of the Society's journals. It is doubtful that either of these groups, and there are others of a similar cast, will become sections, since their members can find representation through the somewhat larger and broader existing sections. If not, the system as established is flexible enough for them to find recognition, particularly in programming, as did the Epithelial Transport and Muscle Groups. It is clear that the mood of Council has done a turnabout in the past decade and is now willing to admit that, for all of the breadth and integrative capacity of physiology and for all of the wide-ranging interests of physiologists, in the nitty-gritty analysis the individual scientists are specialists and will remain so.

Large societies, like graveyards and curricula, are not easily moved. Whether sectionalization of APS will turn out to be a revolution in the Society's governance and function, and if so whether it will be the best of all possible revolutions, are not yet apparent. The early signs are positive.

The story of the various informal groups in the Society between 1937 and 1962 is to be found in the chapter on ancillary groups in *The History of the American Physiological Society: The Third Quarter Century* by Wallace O. Fenn, a delightful and informative book published by the Society in 1963. It is filled with interesting anecdotes that have not been repeated here (e.g., for some reason, Joseph Erlanger, a Nobel Prize winner and former president of the Society, was given the soubriquet "The Nude of Ranvier"). Most of the hard facts about the sections have been published in various issues of *The Physiologist*, and the Council's discussions and agonizings over the emergence of new societies and the benefits or otherwise of sectionalization are to be found in the Council's minutes.

In compiling this account I have been in contact with dozens of colleagues and friends, old and new, and many too many to be listed here. I am grateful to all of them. In addition to what is in these pages, which is as accurate as I have been able to ascertain, my contacts occasionally gave me interesting but manifestly unprintable items that made the fact gathering more entertaining than might appear from the present account. One particularly tempting suggestion was that this chapter be written as a critical sociological essay on the variety of people and personalities who like to involve themselves in the running of scientific societies. I am not sure that this would have been what the editors had in mind.

People at the Society's office who have been exceptionally helpful are Orr Reynolds, Lorraine Tucker, and the Society's historian/archivist, Toby Appel. I thank all three of you.

CHAPTER 20

Porter Physiology Development Program

A. CLIFFORD BARGER

As we approach the one hundredth anniversary of the founding of the American Physiological Society (APS, the Society) we should recall an unusual and significant event that occurred at the Fiftieth Anniversary Celebration in Baltimore, Maryland, on 1 April 1938. President Walter E. Garrey, acting on a vote of Council, introduced William Townsend Porter as honorary president and toastmaster with the following warm words of praise (1):

There remains to me the very great privilege of introducing to you the toastmaster of the evening. The name William Townsend Porter is well known to all of you. Nevertheless I wish to take this opportunity for an expression of the eminent part he has played in the development of American physiology. Before the organization of our Society he had established the first laboratory of physiology west of the Atlantic Seaboard at the old St. Louis Medical School, now Washington University. He early established an enviable reputation as a meticulous and penetrating investigator and as a stimulating teacher at the Harvard Medical School. As the first managing editor of the *American Journal of Physiology* he placed it on the highest scientific and literary plane. For those of us beginning to contribute at that time, I might almost say he taught us how to write and made us like it. I for one will ever be grateful for his pertinent advice and his constructive criticism. Interested always in the development of laboratory teaching, Professor Porter made the designing of apparatus a hobby and through his foundation of the Harvard Apparatus Company he made it possible for American physiological laboratories to obtain apparatus adequate to their needs and at prices within their means. Through his generosity these two foundations are now operated not only for the intellectual but also for the material advantage of the American Physiological Society. They have made possible the establishment of the annual Porter Fellowship. Few indeed are the men who have had American physiology so near the heart. It is with satisfying pleasure that I present to you the toastmaster of the evening, our honorary president, Professor William Townsend Porter.

The ovation that ensued deeply touched Porter, who responded:

> Your generous applause moves almost to tears this aged man and poor. It is the more welcome because for many years I have not enjoyed the pleasure and the great benefit of these meetings. My relation to the Society has been that of a flying buttress— I have helped from the outside in my small way.
>
> I am pleased, too, because you escape the error of all my earlier life. I used to put the head above the heart. You are more wise. You applaud, not the intellect, but the wish to be of service.

The Porter Fellowship mentioned by President Garrey and funded by proceeds from the Harvard Apparatus Company was first awarded by Council in 1921. From 1952 to 1962 a selection was made annually by a regular Porter Fellowship Committee appointed by Council. (For members of the committee and Porter Fellows, see ref. 1.) For the period 1962–66 the members of the Porter Fellowship Committee are listed in Table 1; the Porter Fellows are listed in Table 2. For reasons outlined below, the Porter Fellowship Committee was dissolved in 1966.

After World War II, as the fellowship programs of the National Institutes of Health, the National Science Foundation, and the National Research Council expanded, the number of applications for the Porter Fellowship dwindled. Moreover no Black had been a Porter Fellow in the nearly half century of awards. In 1965, after discussions with Edward P. Radford and Edward W. Hawthorne, I approached John M. Brookhart, then president of the Society, with a proposal that the funds from the Harvard Apparatus Company be used to attract more Blacks to physiology. With the enthusiastic support of Brookhart and of Council, Hawthorne and I presented the proposal to the directors of the Harvard Apparatus Company, who enthusiastically approved

Table 1
PORTER FELLOWSHIP COMMITTEE, 1962–66

	Year	
	Member	Chairman
R. S. Alexander	1960–63	1962–63
J. K. Hampton	1961–67	1963–64
J. F. Perkins, Jr.	1962–65	1963–64
D. W. Rennie	1963–66	
C. C. Hunt	1965–66	
W. F. H. M. Mommaerts	1966	

Table 2
PORTER FELLOWS, 1962–68

	Year
Joel E. Brown	1962–63
Marshall Elzinga	1963–64
Fred E. Wilson	1964–65
Michael F. C. Crick	1965–66
Ann Martin Graybiel	1966–68

it. The company agreed to increase its yearly grant to the Society to "encourage and assist young men and women of promise in the study of physiology." It was understood that for the immediate term the money would be used to increase the number of minority physiologists. The agreement was much broader than the previous program; not only were funds provided for predoctoral fellowships, but some money was also made available for postdoctoral fellowships for physiologists teaching in schools with predominantly black students, or planning to teach there, who might wish to obtain additional experience in research and teaching. Since a department of physiology in a school with predominantly black enrollment with a strong modern laboratory program would be more likely to attract students to the field of physiology, the Harvard Apparatus Company also agreed to make grants of their laboratory equipment to selected schools. Finally, funds were provided for Visiting Porter Lecturers to teach for periods of days to months at schools with predominantly black students.

In 1934, the year the Harvard Apparatus Company became a nonprofit organization, the Society elected its first black member, Joseph L. Johnson. Johnson did his doctoral work under Anton J. Carlson at the University of Chicago and then became head of the Department of Physiology at Howard and dean of the Medical School. The second black member, elected in 1948, was also from Howard, Walter M. Booker. The third black member, elected in 1950, was Daniel Rolfe, chairman of the Department of Physiology and dean at Meharry Medical College. Thus, in sixteen years three Blacks had been elected to membership in the APS. At the time Council began deliberations concerning the Porter Physiology Development Program (1965), the Society had eight regular and associate black members.

In 1966 President Robert F. Forster appointed an ad hoc committee under the aegis of the Education Committee to establish the Porter Physiology Development Program: Philip Bard, Howard E. Morgan, David F. Opdyke, Arthur B. Otis, and Edward P. Radford, with Edward W. Hawthorne and A. Clifford Barger as cochairmen. In 1968 the status of the Porter Physiology Development Program was changed to that of a standing committee with the same membership. The members of the Society who have served on the committee are listed in Table 3.

Table 4 lists recipients of Porter Physiology Development Committee funds. Four additional predoctoral fellows are currently enrolled in Ph.D. programs.

The Porter Physiology Development Committee has also encouraged local collaborative efforts between schools with predominantly black and those with predominantly white enrollments and has supplied financial assistance for honoraria and laboratory equipment. The faculty of Emory University School of Medicine has been extremely helpful in establishing an honors physiology course at Spelman College for black students in the Atlanta educational complex—Spelman, Morehouse, Morris Brown, and Clark Colleges and Atlanta University. Alexander M. McPhedran and Elbert P. Tuttle of Emory worked closely with William B. LeFlore of Spelman to establish the Atlanta Consortium. More recently the successive chairmen of the Department of Physiology at Emory (Jack L. Kostyo, John W. Manning, and Robert

Table 3
PORTER DEVELOPMENT COMMITTEE, 1968–84

	Year		Year
A. C. Barger*	1968–86	M. Lieberman	1974–77
E. W. Hawthorne*	1968–86	J. Kostyo	1976–79
E. P. Radford, Jr.	1968–79	B. Rennick	1977–80
D. F. Opdyke	1968–71	E. L. Ison-Franklin	1978–87
A. B. Otis	1968–70	S. Solomon	1978–84
H. E. Morgan	1968–70	C. E. McCormick	1979–82
P. Bard	1968–69	W. N. Stainsby	1979–82
C. Russ	1969–81	J. Santos-Martinez	1980–83
H. V. Sparks, Jr.	1970–78	J. W. Manning	1981–87
	1982–85	J. M. Horowitz	1982–85
W. Tong	1970–73	J. M. Stinson	1983–86
W. C. Foster	1971–74	E. L. Pautler	1984–87
R. D. Berlin	1973–76		

* Cochairman, 1968–86.

Table 4
PORTER FELLOWS, 1968–86

	Year		Year
Postdoctoral fellows		Predoctoral fellows	
Kenneth Olden	1970	Joseph Hinds	1967–71
James Townsel	1971–72	John C. S. Fray	1968–75
Mary O. Pinkett	1972	Russell J. Tearney	1969–72
	1974	Nathaniel G. Pitts	1972–74
Renty Franklin	1974	Jean Flagg	1973–76
Hardin Jones, Jr.	1978	Pamela Gunter-Smith	1973–77
Jose Garcia-Arraras	1981–83		
Nelson Escobales	1982–84		
Jorge R. Mancillas	1983–85		

B. Gunn) and their staffs have collaborated with LeFlore and the faculty at Spelman. For the past several years two former Porter Predoctoral Fellows, Pamela Gunter-Smith of the Armed Forces Radiobiology Research Institute and John C. S. Fray of the University of Massachusetts Medical School, have served as Visiting Porter Lecturers at Spelman, the first such products of the Porter Physiology Development Program. A similar consortium has been established in New Orleans with Nicholas R. DiLuzio and the faculty at Tulane Medical School, John J. Spitzer and his staff at Louisiana State University Medical School, and Joyce M. Verrett at the Division of Natural Sciences at Dillard University.

In a further attempt to stimulate minority students to select research careers in physiology, a summer undergraduate research internship program for minority students has been started at Michigan State University under the guidance of Robert P. Pittman, Harvey V. Sparks, Jr., and Merle K. Heidemann of the Department of Physiology. A similar program for Native American undergraduates has been established in the Department of Physiology of the College of Veterinary Medicine at

Colorado State University at Fort Collins under the aegis of David Robertshaw and Eugene L. Pautler.

At the centennial of the APS, it is appropriate to express the appreciation of the Society for the generous financial contributions of William Townsend Porter and the Harvard Apparatus Foundation for gifts that now total over $750,000.

REFERENCES

1. FENN, W. O. *History of the American Physiological Society: The Third Quarter Century, 1937–1962.* Washington, DC: Am. Physiol. Soc., 1963, p. 133.
2. HOWELL, W. H., AND C. W. GREENE. *History of the American Physiological Society Semicentennial, 1887–1937.* Baltimore, MD: Am. Physiol. Soc., 1938, p. 193.

CHAPTER 21

Perkins Memorial Fund

JOHN PAPPENHEIMER

John Forbes Perkins, Jr., was professor of physiology at the University of Chicago and a devoted member of the American Physiological Society (APS, the Society) for twenty years before his untimely death at the age of fifty-seven in 1966. His principal research interests were in respiratory physiology, but he also published papers on smooth muscle and on temperature regulation. The APS *Handbook of Physiology* section on respiration is graced with his scholarly article on the history of respiratory physiology that includes a remarkable collection of portraits of famous physiologists of the seventeenth and eighteenth centuries (1). John Perkins and his wife, Frances, had a wide circle of friends within APS, but in addition they were key members of the international family of respiratory physiologists, and their home in Chicago was a haven for many foreign guests. They were keenly aware of the problems faced by young foreign physiologists who could not afford to bring their families with them during a year's fellowship in the United States. It was natural therefore for Mrs. Perkins to think of a memorial to her husband that would aid the *families* of visiting physiologists. She discussed her ideas with friends in APS, including several past presidents and members of Council. As a result of these discussions, APS agreed to administer a memorial fund established with the broad mandate "to encourage enterprises in Physiology which have cultural as well as scientific merit as judged by Council." It was recognized that the nature of such enterprises may change from time to time with the progress of science and with the changes in the relation of science to society. As an initial step, however, APS agreed to initiate a program of supplemental support for the families of visiting scientists as originally proposed by Mrs. Perkins.

The Perkins Memorial Fellowships were announced in *The Physiologist* (2) and in *Respiration Physiology* (3). The announcements read, in part:

This fund is designed to provide supplementary support for the families of foreign physiologists who have arranged for fellowships or for sabbatical leave to carry out scientific work in the United States. It is the interest of this Fund to develop the full potentialities for cultural benefit associated with this type of scientific exchange.

Preference will be given to foreign physiologists in the fields of respiratory or neurophysiology who have arranged to work in the United States; in exceptional cases, however, the fund may be used in support of United States physiologists who have arranged to work in a foreign laboratory and who can demonstrate the need for additional support to include their families for the purposes of cultural exchange.

Each application for an award will be made jointly by the visiting scientist and his host. Ordinarily the joint applicants will have made financial arrangements for the visiting scientist himself before applying to the Perkins Fund for family support. The application will contain an account of these arrangements, together with a description of the proposed scientific work and a brief account of how the visitor and his family intend to make use of the cultural benefits.

The Perkins Award Committee, appointed by Council, consists of two past presidents of the Society and two members of Council, with the executive secretary-treasurer of APS and a representative of the Perkins family acting ex officio. An initial gift from the Perkins family was sufficient to start the program in 1967, partly from capital and partly from income on investments. Subsequent contributions were made to the fund by friends and members of the Perkins family until, in 1977, the interest on investments was sufficient to provide supplemental support for two to four families annually at the rate of $2,000–8,000, depending on need. From 1967 to the present (1984) there have been awards to twenty-eight families of physiologists, including sixty-six children from sixteen different countries (Table 1). The awards are unique in that they are designed primarily for the families of scientists who have already received some financial support on the basis of scientific merit. Thus review of each application by the Perkins Award Committee is primarily concerned with interests of the spouse, number and ages of children and plans for their schooling, the *lack* of previous experience in the host country, and the competence and willingness of the host to assist the visitor in participating in cultural exchanges.

There is no reason to doubt the success of this program in the period 1967–84. Almost all recipients of Perkins Awards have attended meetings of APS and have published papers in APS journals based on research conducted during their visits. The long-term cultural benefits can perhaps be judged by letters of appreciation received from the recipients or their hosts. Recipients are invited to write a brief informal report of both scientific and nonscientific aspects of their visit. In general these reports are written six months to two years after the recipient has returned home, when the family has had some time to review the experience with some perspective. Space does not permit inclusion of these letters in this report, but a sample of their flavor may be obtained from the following excerpts:

> Not the least of the benefits arising from your support was the contact which my wife and family made with American people and institutions. My wife is a magistrate at home and she was treated with the greatest kindness and courtesy in Tucson by the local judges who took a lot of trouble to show her your legal system in operation. The children too benefited a lot in being able to mix with American children at school.

Table 1
JOHN F. PERKINS, JR., FELLOWSHIP AWARDS

Recipients	No. of Children	Country	Host	Host Institution	Tenure
Dr. and Mrs. A. Chinet	1	Switzerland	L. E. Farhi	State University of New York at Buffalo	1968–69
Dr. and Mrs. R. Holland	4	Australia	R. E. Forster	University of Pennsylvania	1969–70
Dr. and Mrs. K. Norum	3	Norway	J. A. Glomset	University of Washington	1969–70
Dr. and Mrs. S. Nosaka	2	Japan	S. C. Wang	Columbia University College of Physicians & Surgeons	1969–70
Dr. and Mrs. D. Denison	3	England	J. B. West	University of California, San Diego	1970
Dr. and Mrs. Y. Ohta	3	Japan	L. E. Farhi	State University of New York at Buffalo	1970
Dr. and Mrs. J. Wieth	2	Denmark	D. C. Tosteson	Duke University Medical Center	1970–71
Dr. and Mrs. J. Kay	2	England	R. F. Grover	University of Colorado Medical Center	1971–72
Dr. and Mrs. D. Robertshaw	0	Kenya	C. R. Taylor	Harvard University	1972
Dr. and Mrs. K. Pleschka	3	West Germany	S. C. Wang	Columbia University College of Physicians & Surgeons	1972–73
Dr. and Mrs. M. Zade-Oppen	2	Sweden	D. C. Tosteson	Duke University Medical Center	1972–73
Dr. and Mrs. A. Malan	2	France	H. Rahn	State University of New York at Buffalo	1972–73
Dr. and Mrs. R. Cross	3	Australia	A. C. Barger	Harvard University	1973
Dr. and Mrs. A. Taylor	5	England	D. G. Stuart	University of Arizona	1974–75
Dr. and Mrs. A. Kunz	5	United States	P. Scheid	Max Planck Institüt, Göttingen, West Germany	1974–75
Dr. and Mrs. J. Sinclair	4	New Zealand	S. M. Tenney	Dartmouth College Medical School	1974–75
Dr. and Mrs. J. Miczoch	1	Australia	R. F. Grover	University of Colorado Medical Center	1975–76
Dr. and Mrs. L. Hermansen	4	Norway	L. B. Rowell	University of Washington	1975–76
Dr. and Mrs. G. Mooneh	2	Belgium	P. G. Nelson	National Institute of Child Health and Human Development	1976–77
Dr. and Mrs. B. J. B. Grant	3	England	J. B. West	University of California, San Diego	1977–78
Dr. and Mrs. S. Koyama	2	Japan	J. W. Manning	Emory University	1978–79
Dr. and Mrs. E. Petersen	1	England	K. Wasserman	University of California, Los Angeles	1978–79
Dr. and Mrs. W. St. John	1	United States	A. L. Bianchi	Marseilles, France	1980–81
Dr. and Mrs. Y. Gottlieb	2	Israel	G. Fischbach	Harvard Medical School	1980–81
Dr. and Mrs. R. Arieli	4	Israel	H. D. Van Liew	State University of New York at Buffalo	1980–81
Dr. and Mrs. Shao-Yung Huang	0	China	J. T. Reeves	University of Colorado Medical Center	1981–82
Dr. and Mrs. M. Kneussl	1	Austria	H. Kazemi	Harvard Medical School	1982–83
Dr. and Mrs. K. Koike	1	Japan	N. Staub	University of California, San Francisco	1982–83

To my great pleasure I also made the acquaintance of a young Ph.D. from Loyola University, Chicago, who has similar interests to my own in neural control of respiration. . . . After my return I arranged a Fellowship Stipend for him from the Max Planck Society for a period of two years. Recalling that the purpose of the Perkins Award is to promote cultural and scientific exchange I am very happy to act in accordance with this plan.

In concluding this report, we should like to express our sincere respect on the unique idea on which the Perkins Fund is founded and our cordial acknowledgment for the personal courtesy which Mrs. Perkins and her family showed to us. I would like to add that it was the experience in the States supported by the Perkins Fellowship that motivated me to change to a pure physiologist. . . . I had been an Associate of Pathology, Faculty of Medicine; now I am Associate in the Division of Neurophysiology, Institute for Brain Research.

My visit to the United States was extremely successful. . . . As outcome four papers have been accepted or published in your journals. . . . I trust you have received a separate letter from my wife. . . . We were very proud of our trip to the East Coast as guests of ——. We were pleased to be the visitors at the White House and the Capitol Hill. Now I have returned home (Shanghai) for more than one year. No doubt all my experiences in the United States is of great importance to my future research here. I have summarized all the experiences in the News Communications of the Chinese Physiological Society and a talk to the annual meeting of Respiratory Physiologists, on which occasion my Chinese colleagues shared my joyfulness.

[and from the host] I know for a fact that they have been wonderful ambassadors for the People's Republic of China to this country. . . . I have gotten reports from all sides and believe that not only did they enjoy themselves but that the U. S. has been more than pleased to host them. In these days of international strife it is wonderful to see Foundations dedicated to the principle of international reconciliation . . . the importance goes far beyond the mere scientific implications.

A complete set of letters from recipients is on file in the office of the executive secretary-treasurer of APS, and excerpts are available in a pamphlet describing the Perkins Memorial Fund published by APS in 1977. The letters are indeed a tribute to the program, and many of them provide interesting accounts of contemporary physiology in the United States as viewed by young scientists visiting from abroad. In several instances the recipients of Perkins Awards have arranged for continuing collaboration with their American hosts in their own country, thus extending the cultural benefits of the program.

Success of the Perkins Award Program has been catalyzed by the continued personal interest of the Perkins family and the active participation of friends of John Perkins. This personal aspect must inevitably come to an end, but we can hope that some of the traditions established in the first twenty years of operation will contribute to continued success of the program in the future.

REFERENCES

1. PERKINS, J. F. Historical development of respiratory physiology. In: *Handbook of Physiology. Respiration.* Washington, DC: Am. Physiol. Soc., 1964, sect. 3, vol. I, chapt. 1, p. 1–62.
2. *Physiologist* 11: 89–90, 1967.
3. *Respir. Physiol.* 4: 281, 1968.

CHAPTER 22

Awards and Honors

TOBY A. APPEL

Bowditch Lecture

The Bowditch Lecture (2), named in honor of the first president of the American Physiological Society (APS, the Society) has been given annually at the APS fall meeting since 1956. The initial funds for the lecture were supplied by an anonymous contribution resulting from the initiative of L. N. Katz. APS Council voted in the fall of 1955 that the funds be used to furnish an honorarium for a lectureship at the fall meeting, to be known as the Henry Pickering Bowditch Lecture. Council's approval of the plan was to be contingent on the contributor's agreement. By tradition the Bowditch Lecturer, who is to be under forty years of age, has been named by the president of the Society from among APS members who have done outstanding work. The first Bowditch Lecture, "Role of the red blood corpuscles in the regulation of renal blood flow and glomular filtration rate," was given by John R. Pappenheimer, Jr., at the Eighth Fall Meeting in Rochester, New York (1956). (See Table 1 for a list of Bowditch Lectures, 1956–86.) The Bowditch Lecture is now supported by the unrestricted bequest of Caroline tum Suden.

Ray G. Daggs Award

The Ray G. Daggs Award (1), named in honor of the Society's first full-time executive secretary-treasurer (1956–72), was announced by APS President Robert Berne in 1973. It is presented annually at the spring meeting "to a physiologist who is judged to have provided distinguished service to the science of physiology and to the American Physiological Society." The awardee is determined by a three-member committee appointed by Council (see Table 2). The Daggs Award is now funded by the unrestricted bequest of Caroline tum Suden.

Caroline tum Suden Professional Opportunity Awards

The Caroline tum Suden Professional Opportunity Awards, established by APS in 1982, are funded by the bequest of Caroline tum Suden, a member of the Society

Table 1
BOWDITCH LECTURES, 1956–86

Year	Lecturer	Title	Publication
1956	J. R. Pappenheimer	Role of the red blood corpuscles in the regulation of renal blood flow and glomerular filtration rate	*Physiologist* 1(3): 8–24, 1958
1957	B. Schmidt-Nielsen	The resourcefulness of nature in physiological adaptation to the environment	*Physiologist* 1(2): 4–20, 1958
1958	A. B. DuBois	New concepts in cardiopulmonary physiology developed by use of the body plethysmograph	*Physiologist* 2(2): 8–23, 1959
1959	L. M. Beidler	Physiology of taste	*Physiologist* 3(1): 5–12, 1960
1960	C. W. Gottschalk	Micropuncture studies of tubular function in the mammalian kidney	*Physiologist* 4(1): 35–55, 1961
1961	J. A. Clements	Surface phenomena in relation to pulmonary function	*Physiologist* 5: 11–28, 1962
1962	T. H. Wilson	Intestinal absorption of vitamin B_{12}	*Physiologist* 6: 11–26, 1963
1963	E. M. Renkin	Transport of large molecules across capillary walls	*Physiologist* 7: 12–28, 76, 1964
1964	D. L. Gilbert	Atmosphere and oxygen	*Physiologist* 8: 9–34, 1965
1965	E. Knobil	The pituitary growth hormone: an adventure in physiology	*Physiologist* 9: 25–44, 1966
1966	D. H. Hubel	Effects of distortion of sensory input on the visual system of kittens	*Physiologist* 10: 17–45, 1967
1967	P. F. Curran	Coupling between transport processes in intestine	*Physiologist* 11: 3–23, 1968
1968	E. Braunwald	The determinants of myocardial oxygen consumption	*Physiologist* 12: 65–93, 455, 1969
1969	J. Urquhart III	Blood-borne signals, the measuring and modelling of humoral communication and control	*Physiologist* 13: 7–41, 1970
1970	D. Kennedy	Crayfish interneurons	*Physiologist* 14: 5–30, 1971
1971	J. A. Herd	The physiology of strong emotions: Cannon's scientific legacy re-examined	*Physiologist* 15: 5–16, 1972
1972	F. Strumwasser	Neural and humoral factors in the temporal organization of behavior	*Physiologist* 16: 9–42, 1973
1973	R. R. Llinas	Motor aspects of cerebellar control	*Physiologist* 17: 19–46, 1974
1974	C. M. Armstrong	Ionic pores and gates in nerve membranes	*Physiologist* 18: 93–98, 1975
1975	T. G. Coleman	From Aristotle to modern computers: the role of theories in biological research	*Physiologist* 18: 509–518, 1975
1976	J. M. Diamond	The epithelial junction: bridge, gate and fence	*Physiologist* 20(1): 10–18, 1977
1977	F. G. Knox	The intrarenal metabolism of phosphate	*Physiologist* 20(6): 25–31, 1977
1978	J. B. Martin	Brain mechanisms for integration of growth	*Physiologist* 22(1): 23–29, 1979
1979	J. H. Szurszewski	Electrophysical events which regulate cellular and molecular events of gastrointestinal smooth muscle	

Table 1—*Continued*

Year	Lecturer	Title	Publication
1980	F. J. Karsch	Seasonal reproduction: a saga of reversible fertility	*Physiologist* 23(6): 29–38, 1980
1981	B. K. Gilbert	New computer technologies and their potential for expanded vistas in biomedicine	
1982	K. Hermsmeyer	Electrogenic ion pumps and other determinants of membrane potential in vascular muscle	*Physiologist* 25: 454–465, 1982
1983	D. R. Kostreva	Functional mapping of cardiovascular reflexes and the heart using the 2-[^{14}C] deoxyglucose	*Physiologist* 26: 333–350, 1983
1984	W. W. Chin	Glycoprotein hormone genes: hormonal regulation of expression	
1985	M. C. Moore-Ede	Physiology of the circadian timing system	*Am. J. Physiol.* 250 (*Regulatory Integrative Comp. Physiol* 19): R735–R752, 1986
1986	Y. E. Goldman	Probing the mechanochemistry of muscle contraction	

Table 2
RECIPIENTS OF THE RAY G.
DAGGS AWARD, 1974–86

Year	Recipient
1974	J. H. Brookhart
1975	M. B. Visscher
1976	J. D. Hardy
1977	J. H. Comroe
1978	H. Rahn
1979	J. R. Pappenheimer
1980	J. R. Brobeck
1981	A. C. Guyton
1982	R. W. Berliner
1983	C. L. Prosser
1984	E. F. Adolph
1985	A. C. Barger
1986	D. B. Dill

from 1936 to 1976. The awards provide funds for junior physiologists, graduate students, or postdoctoral fellows to present a paper at the APS/FASEB meeting and to utilize the FASEB Placement Service. Recipients are chosen by the Committee on Women in Physiology on the basis of abstracts submitted in advance of the meeting. (The history of this award and a list of recipients are found in the chapter on women in physiology.)

Physiology in Perspective:
Walter B. Cannon Memorial Lecture

The Walter B. Cannon Memorial Lecture, established in 1985, is sponsored by the Grass Foundation. It honors Walter B. Cannon, president of APS from 1913 to

Table 3
CANNON LECTURES, 1985–86

Year	Lecturer	Title
1985	R. Guillemin	The language of polypeptides and the wisdom of the body
1986	R. J. Havel	The lipid transport function of lipoproteins in mammalian blood plasma

Table 4
PHYSIOLOGY IN PERSPECTIVE LECTURES, 1983–84

Year	Lecturer	Title
1983	F. E. Bloom	Functional significance of neurotransmitter diversity
1984	J. R. Vane	Prostacyclin—from discovery to the clinic

1916 and one of this century's most distinguished physiologists. This plenary lecture is given annually by an outstanding physiologist at the APS spring meeting on a subject that addresses some aspect of the concept of homeostasis as enunciated in Cannon's classic work, *The Wisdom of the Body* (1932). The lecturer is selected by the current APS president (see Table 3). The Cannon Lecture is a continuation of the Physiology in Perspective Lecture established by Council in 1983 (see Table 4).

Orr Reynolds Award

The Orr Reynolds Award (3, 4) for the best article submitted by a member of the Society on some aspect of the history of physiology will be given annually at the spring business meeting. It was established at the 1985 spring meeting of Council to honor Orr E. Reynolds, who was retiring later that year after twelve years of service to the Society as executive secretary-treasurer (1973–85). The award is an appropriate recognition of Reynolds' long-time interest in and encouragement of the history of physiology. Selection will be made by a committee appointed by Council in consultation with the chairman of the Section on the History of Physiology. It is hoped that after peer review, the article will be published in an APS journal.

G. Edgar Folk, Jr., Senior Physiologist Fund

The G. Edgar Folk, Jr., Senior Physiologist Fund (5) was established through the generosity of the family and students of G. Edgar Folk, Jr., a member of APS since 1953. It provides modest but helpful assistance to senior physiologists seventy years or older who no longer have grant funds available to them. Awards might be used for presenting a paper at an APS meeting, engaging in a modest series of experiments, or completing a manuscript. The award was first announced at the 1986 spring business meeting in St. Louis. Recipients will be selected throughout the year with the assistance of the Senior Physiologists Committee. Names of awardees will not be made public.

REFERENCES

1. *Physiologist* 16: 111, 1973.
2. *Physiologist* 21(6): 4, 1978.
3. *Physiologist* 28: 133, 1985.
4. *Physiologist* 29: 15, 1986.
5. *Physiologist* 29: 42, 1986.

CHAPTER 23

Centennial Celebration

ARTHUR OTIS AND MARTIN FRANK

A history is usually an after-the-fact account. Because this chapter is the story of an event that has not yet taken place, it is more a progress report and lacks the perspective of a historical essay. However, it may demonstrate some of the problems that can arise in planning and arranging a special birthday party for a society of over 6,500 members.

The principal sources used in writing this chapter are minutes of Council meetings, minutes of the Centennial Celebration Committee (CCC) meetings, items in *The Physiologist*, and files related to departmental histories. These, of course, form only a partial record of all the thought, discussion, planning, and effort that went into the Centennial Celebration. Everyone who has participated in committee meetings knows that some of the most interesting, most revealing, and certainly most colorful deliberations are never entered in the formal records.

History of the Centennial Celebration

The Centennial Celebration seems to have had its inception in April 1977 at the American Physiological Society (APS, the Society) Spring Meeting in Chicago. During that meeting Orr Reynolds, executive secretary-treasurer, reported to Council (Council minutes, spring 1977) that in a forthcoming issue of *The Physiologist*, the Society would publish a historical article by D. B. Dill about L. J. Henderson (2) and on the facing page, an announcement that the centennial of the Society would occur in 1987.

At the same meeting, a rather bemused member, Arthur Otis, whose torpid sense of history had been somewhat aroused by events of the recent national bicentennial year, appeared before Council to make a brief statement and present the following in writing:

Ten years hence our Society will celebrate its 100th birthday.

In anticipation of this occasion, I would urge Council now to begin plans for the preparation of a suitable publication dealing with the history of Physiology in America, and especially the history of the first hundred years of the American Physiological Society. Such a publication could be a comprehensive account by a single writer or a collection of papers dealing with various aspects of our history or both, depending on the availability and capability of suitable authors.

At any rate, I believe if such a project is started early enough a fine and valuable publication will result.

Council responded by proposing that an ad hoc committee consisting of Ralph Kellogg, Arthur Otis, and S. Marsh Tenney be appointed and that the Committee on Committees nominate additional members and recommend a chairman. Earl Wood, chairman of the Committee on Committees, in a letter dated 21 September 1977, conveyed this instruction to the committee membership with the realistic admonishment that

some of the members and possibly the chairman of this committee should be in an age group for which there is a high probability they will be active in society affairs ten years from now.

Subsequently, in May 1978, President William F. Ganong appointed an APS Centennial Celebration Committee consisting of seven members. The first meeting of this committee was held during the APS Fall Meeting in St. Louis in October 1978. In the meantime, M. C. Shelesnyak, working at Beaumont House, had written two documents: a "Draft Proposal for Organization Duties and Responsibilities of the Committee for the Commemorative Program for the Centennial Year of the American Physiological Society, 1986–1987" and a proposal for a "Commemorative Program for the Centennial Year of the American Physiological Society."

These documents, consisting of some seventeen typewritten pages, were comprehensive, imaginative, and ambitious. They served as a more than sufficient agenda for the committee meeting. According to Shelesnyak's proposal, the primary objective of the CCC was

to utilize the Centennial Year activities to make the scientific community and the lay public aware of the history of the American Physiological Society (and essentially the history of American physiology); aware of the nature of physiology and its role in understanding biological processes; how physiology and physiologists promote human welfare directly, and through stimulating advances in medicine, agriculture, public health, environmental protection and other physiologically based sciences.

His documents also proposed the establishment of a task force to initiate, develop, organize, and implement the detailed operations of the program and to organize and supervise the execution of specific projects. The CCC was to guide the task force, and the task force was to "aid the CC [CCC] in the execution of the CC's duties and responsibilities."

The committee was somewhat overwhelmed by the scope of the program proposal and was concerned that they might become overextended. The committee, however, did support the concept that the commemorative program go beyond the Society history book and a centennial meeting, to use the occasion to develop a dynamic projection of physiology and its contributions to Science and Society.

At the same time, committee members urged that the publication of histories of departments of physiology in the several schools and universities be given a high priority.

The committee strongly endorsed the task force concept and proposed Shelesnyak as task force director. Council voted unanimously that he be appointed for a one-year term. At the 1979 spring meeting, Council, acting on a recommendation by the committee, unanimously voted to extend this appointment for the duration of the project. Shelesnyak served in this capacity until 1985, and his contributions to the development of the Centennial Celebration were invaluable. He not only advanced many imaginative ideas but, with Orr Reynolds, who became task force director in 1985, provided a continuity of effort in the central office that the committee would have been unable to sustain by itself.

Thanks in large part to Shelesnyak's efforts, by the beginning of 1980 the general concept could be stated in the following document:

The American Physiological Society

Centennial—1987

The American Physiological Society was founded in 1887. The Society's plan to have a Centennial Celebration was announced at the general meeting in Atlantic City in 1978. Since that time, the Centennial Commemorative Committee was activated and a Task Force was formed.

We believe it is appropriate at this time to spell out the general concept of the Centennial Commemorative Celebration.

The objective of the Program is: "To utilize the Centennial Year activities to make the scientific community and the lay public aware of the history of the American Physiological Society (and essentially the history of American physiology); aware of the nature of physiology and its role in understanding biological processes; in doing so, how physiology and physiologists promote human welfare, directly and through stimulating advances in medicine, agriculture, public health, environmental protection and other physiologically based sciences" (from "Organization Duties and Responsibilities of the Committee for the Commemorative Program for the Centennial Year of the American Physiological Society, 1986–1987").

The period for commemoration will begin with the 30th International Congress of Physiological Sciences in 1986 and end with a Centennial Meeting (The APS Fall Meeting) in 1987.

Projects on: publications (histories, biographical directory, vignettes, etc.), Centennial Lectures, exhibits, promotional activities, fundings have been planned. The Committee and its Task Force are working to implement these various projects.

An essential part of the program for developing and carrying out the Centennial Commemorative is to have the entire membership of the Society become involved in the various activities. It is also the hope of the Centennial Celebration Committee to enlist active participation and involvement of interested historians, biomedical scientists, social scientists, scholars in arts and humanities, and professionals in government and industry.

M. C. Shelesnyak, Task Force Director

Committee on the Commemorative Program

for the Centennial Year

Centennial Celebration Committee

As with all activities necessitating the formulation of a committee, a name had to be attached to that committee for reference in correspondence and on meeting agendas. Through a period of fits and starts, the committee was called by various names, the earliest being the Centennial History Committee, in a letter from Earl Wood to the members of the Committee on Committees dated 21 September 1977. The name that finally persisted, however, was that used by APS President William F. Ganong in 1978, the Centennial Celebration Committee or CCC. Its duties, established by Council in April 1978 and amended in April 1980, were described as follows:

> The membership consists of the Chairman, and members as selected by Council, plus, in ex-officio status: the President and immediate Past-President, the Executive Secretary-Treasurer, the Director of the Task Force, and chairmen of Society committees that have direct interest in the program. These committees include the Publications, Education, Program, and Finance Committees, and the Committee on Senior Physiologists.
>
> The primary duty of the Committee is to utilize the Centennial Year activities to make the scientific community and the lay public aware of: *1*) the history of the American Physiological Society (and essentially the history of American physiology); *2*) the nature of physiology and its role in understanding biological processes; and *3*) the ways in which physiology and physiologists promote human welfare directly, and through stimulating advances in medicine, agriculture, public health, environmental protection, and other physiologically based sciences.
>
> The Committee will meet at the Spring and Fall Meetings of the Society and at such times as essential.
>
> The Committee will cease functioning within 24 months after the Centennial Meeting of the APS in 1987.

The members of the CCC, chaired successively by Earl Wood (1978–80), Peter Chevalier (1980–84), and Alfred P. Fishman (1984–87), are listed in Table 1.

Table 1
CENTENNIAL CELEBRATION COMMITTEE

	Year				Year	
	Member	Chairman			Member	Chairman
E. H. Wood	1978–80	1978–80		A. P. Fishman†	1979–83	1984–87
H. W. Davenport	1978–80				1983–87	
R. H. Kellogg	1978–87			D. L. Gilbert	1983–87	
G. K. Moe	1978–80			H. V. Sparks, Jr.	1985–87	
A. B. Otis	1978–87			N. C. Staub	1985–87	
M. C. Shelesnyak*	1978–87			J. B. West	1985–87	
P. A. Chevalier	1979–87	1980–84		O. E. Reynolds†‡	1978–87	
J. D. Neill	1979–81			M. Kafka†	1979–84	
L. L. Langley	1980–87			H. E. Morgan†	1982–85	
S. Ochs	1980–87			P. C. Johnson†	1985–87	
R. J. T. Joy	1981–87			M. Frank†	1985–87	
J. S. Cowan	1982–87					

* Task force director (1978–85).   † Ex officio.   ‡ Task force director (1985–87).

Site of the Centennial Meeting

Although the most important aspect of the centennial is perhaps the fostering of an awareness, attitude, or frame of mind among members of the Society, the crowning event is a centennial meeting or birthday party. Suggestions were received from various sources as to suitable locations for this gala occasion.

John Pappenheimer, in a letter to President Ganong, suggested Boston because of Bowditch's role in the founding of the Society. The CCC, at its Dallas meeting, discussed Washington, D.C., New York, Philadelphia, Boston, and St. Louis as possible sites.

In February 1980, at the request of Francis Haddy, John Bullard, assistant dean of the Uniformed Services University of the Health Sciences at Bethesda, Maryland, invited the Society to use the facilities on the campus for their meeting. The CCC recommended the Washington, D.C.-Bethesda location, but Council deferred action until 1981 when it approved the site unanimously and specified that the meeting be held in late August 1987. At the same time, Council indicated that the centennial year should begin with the XXX International Union of Physiological Sciences (IUPS) Meeting, scheduled for Vancouver, British Columbia, in 1986.

A potential problem arose when it was discovered that the Federation of American Societies for Experimental Biology (FASEB, the Federation) was planning to hold its 1987 spring meeting in Washington, D.C., and expected APS to hold its centennial meeting in conjunction with the FASEB meeting. Furthermore the other constituent societies of FASEB desired and expected to participate in our celebration. This caused considerable consternation, because it was feared that our birthday party might become lost, or at best overwhelmed and diluted, in the melee that might result from meeting with the larger group.

Calmness eventually prevailed, however, when it was realized that perhaps the younger societies of FASEB were acting not as ill-mannered intruders but rather as ebullient children eager to participate in their mother's birthday party. Some reassurance was found in a letter to Orr Reynolds from Robert W. Krauss, executive director of FASEB, dated 20 August 1981, in which he stated, "We will do everything we can to cooperate with APS in arranging a cordial program that will accomplish your wishes." When Council met in the fall of 1981, the minutes show that

> it was moved, seconded, and passed unanimously that if FASEB meets in Washington, D.C., in 1987, there be a firm agreement that APS shall have control of two days, and that the theme of the meeting shall be devoted to the Centennial of Biological Sciences; if this is not agreeable, the APS shall request FASEB to change the site of its 1987 meeting.

As a result, it was decided that the APS centennial year would be celebrated by events occurring at three separate meetings: the Vancouver meeting, 12–20 July 1986; the major Centennial Celebration Meeting, scheduled for the FASEB Meeting, 29 March to 3 April 1987 in Washington, D.C.; and the APS fall meeting, scheduled for 11–16 October 1987 in San Diego, California.

Centennial Theme

In the spring of 1982, Council, acting on the recommendation of the CCC, passed unanimously a motion that the theme for the 1987 centennial be "A Century of Progress in Physiology." This seems to have had its roots in a response to the invitation from FASEB that APS hold its 1987 Centennial Meeting within the framework of the FASEB meeting. In response to the invitation, APS President Francis Haddy had written the following to the chairman of the Federation Board, Earl Wood, on 26 October 1981:

> In response to this invitation, APS proposes that development of the theme program for the 1987 meeting be the responsibility of the APS. The APS Program and Centennial Committees would work with the program committees of other participating societies to develop a coherent thematic program emphasizing current advances in experimental biology and the historical roots of these contemporary programs in physiological developments of former years. The title of the theme might be something like "A Century of Progress in Experimental Biology."

Not until the fall of 1984 were plans started to bring to fruition these theme symposia for "A Century of Progress." At that time, Michael Jackson, the theme organizer, convened a meeting of representatives of the FASEB societies during the April 1985 meeting in Anaheim, California. The original plans included six theme symposia, one organized by each of the societies. Each symposium would deal with subjects of contemporary interest and begin with an oral presentation on what was known about the subject in the 1880s. The initial intentions, however, were gradually modified so that the current centennial theme was developed.

Michael Jackson, chairman of the APS Program Executive Committee, served also as chairman of the FASEB Organizing Committee. The representatives of the six societies who served on the Centennial Theme Committee viewed their task as identifying topics that could be portrayed against a historical background, with full regard for their future prospects. The proposed program was designed to have plenary lecturers in the morning and afternoon as components of sessions drawn from contributed abstracts. The task of identifying the abstracts that would be included within the respective programs was the responsibility of the several contributing societies. Each of the seven plenary lecturers was a nationally and internationally recognized contributor to experimental biology, in a field that drew on the past and looked toward the future (Table 2).

The first four plenary lectures were scheduled from 11:30 to 12:30 in the morning, at the end of contributed papers sessions. The remaining lectures were to be presented from 12:30 to 1:30 in the afternoon to initiate the program derived from abstracts selected for that session.

Finances

The formulation and especially the implementation of a balanced budget is a major problem for many organizations, and one with which governing bodies are often reluctant to come to grips. A budget for the centennial proved to be no

Table 2

PLENARY LECTURES PRESENTED AT APS CENTENNIAL MEETING

Title of Lecture	Lecturer	Affiliation
Developmental Regulation of Mammalian Genes	David Baltimore	Whitehead Institute of Biomedical Research, Cambridge, MA
The Vitamin D Story: A Success of Basic Science in the Treatment of Disease	Hector F. DeLuca	University of Wisconsin, Madison
Misplacing Oncogenes. Studies Using Transgenic Mice	Philip Leder	Harvard Medical School, Boston, MA
Inflammation and the Mechanism of Anti-inflammatory Drugs	John Vane	St. Bartholomew's Hospital Medical College, London
Neurotransmitters: Past, Present, and Future Directions	Floyd E. Bloom	Scripps Research Institute, La Jolla, CA
Control of Pituitary Functions: One Hundred Years of Progress	Roger Guilleman	Salk Institute, San Diego, CA
Neuroreceptors	Solomon H. Snyder	Johns Hopkins School of Medicine, Baltimore, MD

exception to this rule. The first inkling of financial considerations for the centennial by Council appears in the minutes of its 1977 fall meeting: "A means of raising funds would be to strike a medal for sale at $50 to members."

In the spring of 1979 the CCC chairman, Earl Wood, presented to Council a budget of $27,400 for operation of the Centennial Task Force in 1980. This was tentatively approved by Council and forwarded to the Finance Committee. As a method of generating income, Council agreed that members, when paying their annual dues, should have the option of making a voluntary donation to the Centennial Celebration Fund, and in the fall of 1980 Council authorized giving centennial jewelry to members who made a donation.

A year later a budget for the centennial still had not been established. Wood recommended a dues increase specifically designated to be applied to costs of the centennial, but no action was taken by Council.

At the 1982 spring meeting, CCC Chairman Peter Chevalier reminded Council that the fund for the centennial was still very small. Orr Reynolds presented a budget proposal for 1982–87, with a total amount of $314,000. In addition $385,000 was estimated for costs of books planned as part of the centennial, but it was anticipated that this amount would be largely recovered from sales. As a direct tax to raise funds to meet centennial costs, Council voted to assess a twenty-dollar abstract fee at each fall meeting beginning in 1983, the proceeds to be earmarked for the Centennial Celebration.

In the fall of 1982 Council approved a recommendation from the CCC that the five-dollar assessment for travel funds to attend the 1986 IUPS Congress be diverted to the centennial budget and voted to request that FASEB contribute five dollars from each registration fee of the 1987 spring meeting to the APS centennial. Council also agreed to include, in its 1983 budget, funds to support a full-time archivist.

No important budgetary advances seem to have been made during 1983. Council minutes of the 1984 spring meeting, however, reported as follows:

> The long-term Centennial budget projects a cumulative deficit for activities of approximately $35,000 through 1986. It is anticipated assessing registrants $5 for the Centennial at the 1987 meeting, which will no doubt be used for special lecturers, and does not provide for any commemorative material for the registrants. Centennial publications are expected to cost approximately $120,000 with sales of about $20,000. Although it is hoped to obtain grant funds to offset some of the Centennial expenses, it cannot be counted on at this time. Therefore, Council must consider other means of funding the Centennial beyond that already budgeted from dues, voluntary assessments, abstract handling fee, and income from the Centennial meeting.

As a follow-up to the concern about the impending centennial budget, it was stated in the Council minutes that

> it is important to have a clear and concise budget for Council to consider at the August meeting. At that time it must be decided specifically what it wants to do, how much it will cost, and what other sources of income and funding.

At the August 1984 meeting, the members of Council expressed concern about the slow progress of the planning for the Society's Centennial Celebration. It was felt that the funds allocated at the spring meeting were inadequate if the Society truly wished to have a first-class birthday party. Alfred Fishman expressed strongly his belief that the APS centennial should be an extravaganza with events such as an opening reception at the Corcoran in the evening, attendance at a symphony performance, and a large exhibit at the Smithsonian. To commemorate this occasion, a medallion, briefcase, APS history, and other memorabilia should be provided in the registration packet. During the ensuing discussion at the Council meeting, estimated costs for the various activities were set forth for discussion by the various Council members, with an ultimate budget of $400,000–500,000 being required to undertake all the activities desired. Instead of a suggested $5 surcharge, it was proposed that FASEB be asked to add $20 to the registration fee, which would realize $200,000 for 10,000 attendees. A motion was unanimously passed by Council, which approved the transfer over the next two years (1984–85) of up to $200,000 from the earnings of the Publications Contingency Reserve Fund to be set aside for the Centennial Celebration. Overall the August 1984 meeting of Council proved to be the benchmark at which financing for the centennial meeting was finally mobilized. At that meeting Council passed a motion that commitments should be consistent with the planned budget of $650,000, $350,000 of which was already authorized by Council, including $150,000 in the current budget and $200,000 from the income from the Publications Contingency Reserve Fund. Finally the centennial was about to roll. The members of Council expressed their pleasure with a number of exciting program plans and the budgetary planning that would enable the Society to make its centennial a notable occasion.

Although the August 1984 Council committed funds of up to $650,000 for the Centennial Celebration, they identified sources of only the first $350,000. It was therefore incumbent on Council and the CCC to identify other mechanisms to solicit funds for the forthcoming meeting. It was also necessary for Fishman to convince a committee consisting of Max Milner, Henry Metzger, and Harold Hardeman, the FASEB Centennial Committee, that indeed the registration fee for the spring meeting should include a fifteen-dollar surcharge. The fact that FASEB would also be celebrating its seventy-fifth birthday during the APS centennial contributed to the agreement by the FASEB Centennial Committee that a fifteen-dollar surcharge be added to the registration fee and that the talents and personnel of FASEB and convention planning be made available for the occasion to help make the joint APS/FASEB meeting a success.

Norman Staub, member of the CCC, proposed the development of an APS Second Century Founders Club consisting of five tiers corresponding to five levels of support, each named after one of the five founders. It was hoped that something special would be planned for each group, perhaps a party in 1987 for guests and large donors. Staub, with the assistance of William Samuels, APS public affairs consultant, accepted responsibility for organizing the drive and asked for suggestions for memorabilia to be offered to the various levels of donors. The Founders Club was promoted with a brochure distributed to the total APS membership that allowed each individual to decide privately on the level of contribution that should be made to the Society in support of the Centennial Celebration.

Fund-raising, however, did not stop here. A second effort was started in March 1986 to seek approximately $65,000 from corporations for support of the twenty-five foreign guests invited to participate in the centennial. The proposals sought support for one or more of these invited guests. An additional means of fund-raising included the striking of a centennial medallion as well as the manufacture of other memorabilia, as described below.

Society Archivist

As the various historical projects developed, it became apparent, especially to the central office, that the Society could make good use of the services of a professional historian and archivist. Council realized, however, that the employment of a historian/archivist would require additional financial resources from the Society and discussed the matter at length at its 1981 fall meeting. At that meeting, it was moved and seconded that Council approve a one-year appointment of an archivist, which would be financed by an additional five-dollar increase in dues. However, the motion was tabled for consideration at a later time. In the interim, Council voted to employ Grace Gurtowski, FASEB librarian, on a part-time basis to help organize the Society's archives. Specifically, Mrs. Gurtowski sorted the old boxes of membership cards and application forms, set up folders, and entered members' names in the computer to provide the Society with a computer file and printout of all past members of APS. In addition, she helped sort and file photographs, which provided APS with an extensive and usable collection. Her assistance in organizing the Society's archives was indispensable when Toby Appel was finally hired to become the Society's archivist/historian. In the fall of 1982 Council determined that the Society probably could afford a full-time archivist, and in September 1983 Appel joined the APS staff. Her responsibilities have included organizing the archives and photograph collection, compiling the roster of past members of APS, assisting authors of chapters of the fourth quarter century history, reworking the history of the first seventy-five years, preparing centennial exhibits for meetings, and assisting members with other centennial-related projects.

Centennial Logo

The centennial logo was designed through the initiative of Peter Chevalier, chairman of the CCC from 1980 to 1984. Basically it is the APS seal with the word "Centennial" substituted for "Founded in 1887." It first appeared in 1982 on two supplements to *The Physiologist*, as well as on the cover of the reprinted volume, *Circulation of the Blood: Men and Ideas*, which appeared in 1982. Since that time the logo has appeared on letterheads, in the APS journals, on centennial flags and medallions, and in numerous other forms of announcements and memorabilia commemorating the forthcoming celebration.

Departmental Histories

The departmental histories project was initiated by Orr Reynolds, on the suggestion of Arthur Otis, through a memo dated 14 July 1977 to chairmen of physiology departments. It was enthusiastically endorsed by the CCC at its first meeting, and Horace Davenport was requested to develop guidelines for compiling such histories (1). At the spring 1979 meeting of the CCC, Arthur Otis was nominated as an associate editor of *The Physiologist* to handle departmental histories, and at the spring 1980 meeting he was asked to write a letter to department chairmen urging the submission of manuscripts. Such a letter was mailed in September 1980 with a follow-up in August 1984. To date, numerous departmental histories have been submitted, and most have been published in *The Physiologist* (Table 3). The latest group of manuscripts was received early in 1986 and published as a supplement to the October 1986 issue of *The Physiologist*. The final supplement contained nine histories, which brought the total number of departmental histories published to thirty-seven.

History of the American Physiological Society

Because the first fifty years of the Society's history had already been written by William H. Howell and Charles W. Greene and that of the subsequent twenty-five years by Wallace O. Fenn, it was almost taken for granted that a history of the fourth quarter century would be prepared for the centennial.

In the summer of 1980 John Brobeck accepted responsibility for editing a history of the 1963–87 period and invited Orr Reynolds to serve as coeditor. Because the

Table 3
DEPARTMENTAL HISTORIES

Institution	Years	Author	Reference
Univ. of Wisconsin, Madison	1906–1952	W. B. Youmans	*Physiologist* 18: 99–103, 1975
Southern Illinois Univ., Carbondale	1947–1977	H. M. Kaplan	*Physiologist* 20(5): 26–27, 1977
Univ. of Saskatchewan, Saskatoon	1926–1978	L. B. Jaques	*Physiologist* 21(5): 12–14, 1978
Marshall Univ. School of Medicine, Huntington, WV	1976–1980	E. Aserinsky	*Physiologist* 25: 39, 1982
Universidad Central de Caribe, Cayey, PR	1976–1980	J. Santos-Martinez	*Physiologist* 25: 39, 1982
Oral Roberts Univ., Tulsa, OK	1977–1981	E. B. Brown, Jr.	*Physiologist* 25: 40–41, 1982
Univ. of Michigan, Ann Arbor	1850–1923	H. Davenport	*Physiologist Suppl.* 25: 1–96, 1982
Uniformed Services Univ. of Health Sciences, Bethesda, MD	1976–1982	F. J. Haddy	*Physiologist* 25: 414, 1982
Medical College of Ohio, Toledo	1964–1982	A. V. McGrady	*Physiologist* 25: 414–425, 1982
Temple Univ., Philadelphia, PA	1863–1982	M. F. Tansy	*Physiologist* 25: 416–418, 1982
Cornell Univ., College of Veterinary Medicine, Ithaca, NY*	1868–1960	W. Hansel et al.	*Physiologist* 25: 469–474, 1982
Medical Laboratories of the Army Chemical Corps, Edgewood, MD	1946–1961	D. B. Dill	*Physiologist* 25: 474–478, 1982
New York Univ. School of Medicine, New York	1841–1960	H. Chasis	*Physiologist* 26: 64–70, 1983
Meharry Medical College, Nashville, TN	1876–1981	J. N. Stinson	*Physiologist* 26: 119–120, 1983
Northwestern Univ., Chicago, IL	1894–1982	J. H. Annegers	*Physiologist* 26: 120–121, 1983
Mount Desert Island Biological Laboratory, Salsbury Cove, ME	1926–1983	B. M. Schmidt-Nielsen	*Physiologist* 26: 260–266, 1983
Oregon Health Sciences Univ. School of Dentistry, Portland	1899–1980	L. H. Elwell	*Physiologist* 26: 269, 1983
Univ. of Cincinnati, Cincinnati, OH	1819–1983	D. L. Kline	*Physiologist* 26: 366–368, 1983
Boston Univ., Boston, MA	1873–1948	E. R. Loew	*Physiologist* 27: 4–12, 1984
Jefferson Medical College, Philadelphia, PA	1842–1982	L. M. Rosenfeld	*Physiologist* 27: 113–127, 1984
Indiana Univ., Bloomington	1854–1983	R. S. Elizondo et al.	*Physiologist* 27: 319–324, 1984
Indiana Univ. School of Medicine, Indianapolis	1958–1983	E. E. Selkurt	*Physiologist* 27: 325–329, 1984
Medical Univ. of South Carolina, Charleston	1824–1983	A. D. Horres	*Physiologist* 27: 385–389, 1984
Creighton Univ. School of Medicine, Omaha, NE	1892–1985	D. F. Magee	*Physiologist* 28: 139–140, 1985
Univ. of Virginia, Charlottesville	1825–1985	G. C. Pitts	*Physiologist* 28: 402–406, 1985
Univ. of Kentucky, Lexington	1890–1985	L. L. Boyarsky	*Physiologist* 28: 482–484, 1985

Table 3—Continued

Institution	Years	Author	Reference
Univ. of Oklahoma Health Sciences Center, Oklahoma City	1898–1985	A. K. Weiss	*Physiologist* 28: 485–490, 1985
Ohio State Univ., Columbus	1879–1985	M. A. Lessler and F. A. Hitchcock	*Physiologist* 28: 491–501, 1985
State Univ. of New York, Buffalo	1846–1986	H. Rahn	*Physiologist Suppl.* 29: 1–6, 1986
Univ. of California, Berkeley		N. Pace	*Physiologist Suppl.* 29: 7–20, 1986
Univ. of British Columbia, Vancouver		D. H. Copp	*Physiologist Suppl.* 29: 21–26, 1986
Univ. of Florida College of Medicine, Gainesville	1956–1981	A. B. Otis	*Physiologist Suppl.* 29: 27–33, 1986
Univ. of South Florida College of Medicine, Tampa		C. H. Baker	*Physiologist Suppl.* 29: 34, 1986
Philadelphia College of Osteopathic Medicine, Philadelphia, PA		M. H. F. Friedman	*Physiologist Suppl.* 29: 35–45, 1986
Univ. of California, Davis		A. H. Smith et al.	*Physiologist Suppl.* 29: 46–57, 1986
U.S. Army Research Institute, Natick, MA		R. Francesconi et al.	*Physiologist Suppl.* 29: 58–62, 1986
Medical College of Wisconsin, Madison	1921–1985	J. J. Smith	*Physiologist Suppl.* 29: 63–73, 1986
Dartmouth Medical School, Hanover, NH	1797–1983	S. M. Tenney	Archives only
Univ. of Mississippi, Oxford and Jackson	1903–1984	A. C. Guyton	Archives only
San Juan Bautista School of Medicine, San Juan, PR	1978–1986	C. Font	Archives only

* See also ref. 3.

two volumes describing the first seventy-five years were out of print, the possibility of reissuing them was also considered. However, current plans are for Toby Appel to summarize and rewrite the history of the first seventy-five years and to include it in the volume encompassing the last twenty-five years. Also to be considered for inclusion in the history is a centennial roster listing all past and present members of APS. The various chapters for the centennial volume were contributed by members of the Society who have been actively involved in the development of activities in the various areas of the Society.

History of Physiology in America

Although at first there was some confusion between the notion of a history of physiology in America and that of a history of APS, it soon became apparent that a clear distinction could be made between the two. It was also recognized that the writing of a comprehensive history of physiology in America would be a formidable task that would probably require the time and talents of a professional historian.

In the autumn of 1980 Gerald Geison, professor of history of science, Princeton

University, communicated with the Society that he would be interested in writing a history up to 1940, and in April 1981 Council accepted a $2,000 grant from Interdisciplinary Communications Associates in support of the preliminary data collection. In the collection of the data, Geison quickly discovered that the task of writing the history of American physiology to 1940 was immense and not likely to be accomplished by one person to any degree of satisfaction in the time allotted. After some discussion, Geison came up with an alternate plan making use of the talents of many historians for producing a volume in time for the centennial. To help finance the program, Geison and Appel applied to the National Library of Medicine Extramural Programs for a small grant to hold a workshop on the history of American physiology to produce an edited volume of historical articles by 1987. The project, to cover as much of the history of American physiology as possible, would be accomplished by assigning appropriate topics to historians of science and physiologists who were already working on the history of American physiology. Each participant would write a paper to be brought to the workshop at the National Library of Medicine on 16 and 17 January 1986. Unfortunately the National Library of Medicine was unable to provide resources for the meeting, and it became apparent that if the symposium and production of the volume were to take place, it would have to occur under the auspices (and financial support) of APS. To ensure completion of the project, APS allocated $14,000 in support of the symposium on the history of physiology in America. As a result, the participants produced the chapters for *Physiology in the American Context, 1850–1940*, which will be available in 1987.

Centennial Book Series. People and Ideas

The Publications Committee, together with the CCC, announced in 1982 a projected series of books dealing with people and ideas in physiology to be published with an imprint of the centennial logo. This series of books was designed to include the period since 1940, because Geison's original proposal extended only up to that date. The volume on which this series is modeled is *Circulation of the Blood: Men and Ideas*, as noted above. Volumes are expected to be completed on endocrinology, membranes and transport, and renal physiology during the centennial year.

Lectures

Shelesnyak's original extensive program proposal included lectures as an important part of the celebration. Subsequently, at the fall 1980 CCC meeting, Sidney Ochs proposed a series of lectures on historical subjects to be given at Society meetings and eventually published as a book. He presented a list of general and specific subjects that would be appropriate for such a series and later submitted a list of possible lecturers. His proposal was promptly endorsed by the committee; for various reasons, however, including the reluctance of some invitees and the fact that scheduling of lectures must be referred to the Program Committee, implementation did not occur immediately. The series finally started at the 1982 fall meeting.

This and other lectures that may be considered a part of the Centennial Lecture Series are listed below. A lecture given by Ralph Kellogg at the 1980 fall meeting was viewed as the prototype for the series entitled "Centennial Lectures." Other lectures in the series are listed in Table 4.

In addition to this Centennial Lecture Series it was proposed that lectures related to the historical development of physiology in various countries as well as in various disciplines be presented at the IUPS meeting in Vancouver. For the 1987 finale, the opening plenary session will feature brief greetings from distinguished foreign guests from England, France, and Germany, the three nations in which the APS founders received their training. The presentations will address the historical and present collegiality that has typified relations between physiologists in the United States and these countries. Symposia, exhibits, and plenary sessions are scheduled to deal with the frontiers of the physiological sciences. In each instance, a Janus-like approach will be adopted, with each timely topic portrayed against a historical background with full regard for its future prospect. A special issue of *Federation Proceedings* tracing the origins of the FASEB societies to APS will be provided to the attendees. In addition, many of the symposium organizers included one speaker to provide a historical perspective to the session.

Centennial Memorabilia

Through the initiative of Orr E. Reynolds on the suggestion from several members of the CCC, items of centennial jewelry were designed and produced for sale or made available as rewards for donations to the Centennial Celebration Fund. A retired jeweler in Bethesda, Maryland, Roland St. Laurent, arranged for manufacture

Table 4
CENTENNIAL LECTURE SERIES

Title	Lecturer	Meeting
Review of the History of Respiratory Gases	R. Kellogg	Fall 1980
The Growth of Ideas About Pulmonary Circulation and Edema	A. Fishman	Spring 1982
A Perspective of Respiratory Mechanics	A. B. Otis	Fall 1982
Stephen Hales—Neglected Respiratory Physiologist	J. West	Spring 1984
The Life and Times of Walter Bradford Cannon: A Progress Report	A. C. Barger	Fall 1984
Development of Our Knowledge of the Pancreas as an Endocrine Gland	R. Levine	Spring 1985
"Innocents Abroad"? American Physiologists in European Laboratories	R. G. Frank, Jr.	Spring 1985
William Beaumont's World*		Fall 1985
August Krogh and Capillary Physiology	B. Schmidt-Nielsen	Spring 1986
The Founding Fathers of the American Physiological Society†		Spring 1987

* Symposium organized by R. J. T. Joy in honor of the bicentennial of the birth of Beaumont.
† Symposium organized by D. L. Gilbert.

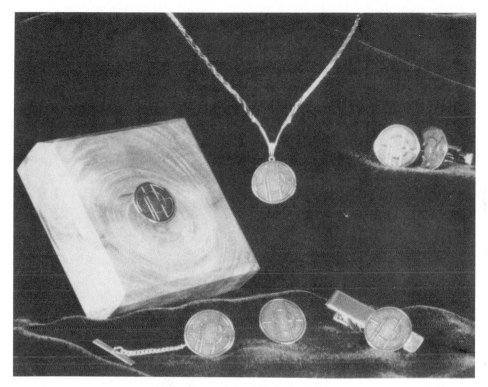

Fig. 1. Centennial jewelry

of the jewelry without charging the Society for his efforts. The attractive collection, including a paperweight, pendant/charm, tie clasp, tie tack, cuff links, and lapel pin (Fig. 1), appeared in 1981. The central portion of the centennial logo or APS seal was the dominant feature of each item.

Although a medal was mentioned as a fund-raising tool for the centennial as early as 1977, this concept was not pursued further until August 1983, when Reynolds presented a proposal from Ralph Sonnenschein, an early sponsor of the idea. Initial investment was estimated at approximately $8,000, an amount about which the CCC was not overly enthusiastic. Although it was anticipated that funds would be recovered from sales, it was difficult to determine the number of medals that would be purchased by the membership. Before Council was willing to allocate funds for such a project, it was necessary for Reynolds to solicit information from the membership, through an announcement in *The Physiologist*, to determine the degree of interest and the approximate price they might be willing to pay for a centennial medallion.

At the April 1985 meeting of the CCC, the question of a medallion occasioned considerable debate. Sonnenschein circulated sketches of the proposed medal, designed at no cost to APS with the hope that it would be purchased. The design was provided by Medallic Art Company and was based on rough sketches submitted by Sonnenschein. Although Reynolds was in favor of having 500 medals made, the consensus of the group was that the project would be too expensive and that the medals would not be ordered by APS members. Council did decide, however, to seek additional estimates for a centennial medallion. Two such estimates proved to be less expensive than that of Medallic Art, and it was decided that the project should go forward with the purchase of 1,000 medallions for sale and distribution. The medallion chosen was designed by Susie Mann, APS art editor, and produced by Medalcraft through Windsor Associates. It features the faces of the five founders of the Society and the centennial logo on the reverse (Fig. 2).

The design selected for this medallion proved to be versatile. It has been used in various publications and announcements of the centennial meeting, as well as in preparation of other memorabilia for distribution and sale to the membership (Fig. 3). A cup, plate, and mug, as well as a tile, all include portions of the front and back of the medallion. In addition, a new T-shirt that features the faces of the five founders has been designed. It complements two other T-shirts that were sold to members to raise funds for the centennial meeting: a blue golf shirt with the centennial logo over the left breast and a T-shirt bearing the anatomy of the thorax on the front and the abdomen on the back and imprinted with "Physiologists know the inside story."

Other Projects

The Centennial Celebration eventually will come to focus on the activities of Founders Week. The Society has extended its hand of friendship to selected members of the international physiological community by inviting them to attend the forthcoming celebration. The centennial guests (Table 5), representing a number of countries, will be asked to participate in the plenary lectures, symposia, and activities of the week. An opening session will include recognition of the invited guests and brief remarks from three of them representing England, France, and

Fig. 2. Centennial medallion

Fig. 3. Centennial Founders set

Germany, countries in which the founders for APS received their training in the science. After the opening session, a reception for all persons attending the Federation meeting in Washington, D.C., will include music by the U.S. Marine Band, as well as food and drink. The Association of Chairmen of Departments of Physiology and APS will cosponsor a panel discussion on "The New Golden Age of Physiology: Perspectives on the Future of Physiology." Similarly the National Institutes of Health and APS will commemorate their joint Centennial Celebration by organizing a panel discussion on biomedical research during the second century. This will be followed by a joint reception to honor the National Institutes of Health and APS. The Presidents' Dinner, scheduled for Tuesday evening, 31 March, will honor the Society's past presidents and their roles in the growth of APS and the discipline of physiology. On the last evening of Founders Week a closing lecture and reception will be sponsored by the U.S. National Committee of IUPS at the National Academy of Sciences for APS members attending the meetings. In addition, tickets have been reserved for an evening at the Kennedy Center for the Performing Arts.

While physiologists, in general, make efforts to recognize the founding of APS and its one hundred-year history, attempts are also under way to acquire recognition for the event from the White House and from Congress. Overtures by William Samuels and Martin Frank to meet and exchange correspondence with a special

Table 5
FOREIGN GUESTS INVITED TO THE APS CENTENNIAL MEETING

	Affiliation	Country
Emilio Agostoni	Univ. of Milan	Italy
Anthony Angel	The University, Sheffield	United Kingdom
Knut Aukland	Univ. of Bergen	Norway
Christian Crone	Univ. of Copenhagen	Denmark
Pierre Dejours	Centre National de la Recherche Scientifique, Strasbourg	France
T. P. Feng	Shanghai Institute of Physiology	People's Republic of China
Bjorn U. G. Folkow	Univ. of Göteborg	Sweden
Oleg G. Gazenko	Institute of Biomedical Problems, Moscow	USSR
Andrew F. Huxley	Trinity College, Cambridge	United Kingdom
Susan D. Iverson	Merck, Sharpe & Dohme, Essex	United Kingdom
Dora Jassik-Gerschenfeld	Univ. Pierre et Marie Curie, Paris	France
Kjell Johansen	Univ. of Aarhus	Denmark
Bernard Katz	Univ. College of London	United Kingdom
Paul I. Korner	Baker Medical Research Institute, Prahan	Australia
Rita Levi-Montalcini	Laboratorio di Biologia Cellulare, Rome	Italy
Anthony D. C. MacKnight	Univ. of Otago, Dunedin	New Zealand
Carlos Monge	Univ. Peruana Cayetano Heredia, Lima	Peru
Johannes Piiper	Max-Planck Institüt für experimentelle Medizin, Göttingen	West Germany
Berry Pinchow	Ben-Gurion Univ. of the Negev, Sede-Boqer	Israel
Irene I. Schulz	Max-Planck Institüt für Biophysik, Frankfurt	West Germany
Andrzej M. Trzebski	Institute of Physiological Sciences, Warsaw	Poland
Karl J. Ullrich	Max-Planck Institüt für Biophysik, Frankfurt	West Germany
John Vane	St. Bartholomew's Hospital Medical College, London	United Kingdom
Ewald R. Weibel	Univ. Bern	Switzerland
John G. Widdicombe	St. George's Hospital, London	United Kingdom

assistant to the president of the United States have resulted in a positive feedback. The most likely options seem to be the appearance of the vice-president of the United States at a plenary session of the centennial meeting. The White House also expressed interest in adding the APS centennial plate to its collection, as well as in accepting a centennial medallion. It is hoped that these will be presented to the White House during American Physiologists Week. Also a medallion will be carried into space on a NASA space shuttle flight by APS member Robert Phillips, who will be a payload specialist on one of the flights in 1988. In addition, during the Ninety-ninth Congress, Congressman Doug Walgren introduced H. R. Resolution 677 and Senator Robert Dole introduced a similar resolution in the Senate designating the week beginning 29 March 1987 as American Physiologists Week. The resolution was unanimously approved by the Senate before adjournment, but it fell short of the necessary cosponsors in the House. Both Senator Dole and Congressman Walgren have agreed to introduce the resolution again during the One Hundredth Congress. It is hoped that the APS membership will convince their senators and representatives to cosponsor the resolutions so they can be approved by Congress. Should the resolution be approved, the President will probably sign the proclamation in the spring of 1987.

Conclusion

A birthday party is a very special occasion. To arrange a successful one requires much effort. Not only must the sequence of events be carefully planned, but somehow a way must be found to pay for the balloons and cake and ice cream. If all goes well, it is worth the price. A birthday party evokes a gamut of emotions. The general atmosphere is one of gaiety, but there are more serious moments. It is a time for looking back and giving thanks for the pleasures and accomplishments of the past. It is also a time for looking forward hopefully to achievements of the future. It is with this outlook that we celebrate "A Century of Progress in Physiology."

REFERENCES

1. DAVENPORT, H. W. Some notes on preparing a history of a department of physiology. *Physiologist* 22(1): 30–31, 1979.
2. DILL, D. B. L. J. Henderson, his transition from physical chemist to physiologist; his qualities as a man. *Physiologist* 20(2): 1–15, 1977.
3. MARSHALL, L. H. More on Burt Green Wilder at Cornell. *Physiologist* 26: 361–363, 1983.

APPENDIX 1

Members of Council, 1887–1987

1888[1]

H. P. Bowditch, President
H. N. Martin, Secretary-Treasurer
J. G. Curtis
H. Sewall
H. C Wood

1889

S. W. Mitchell, President
H. N. Martin, Secretary-Treasurer
H. P. Bowditch
J. G. Curtis
H. C Wood

1890

S. W. Mitchell, President
H. N. Martin, Secretary-Treasurer
H. P. Bowditch
J. G. Curtis
H. H. Donaldson

1891

H. P. Bowditch, President
H. N. Martin, Secretary-Treasurer
R. H. Chittenden
J. G. Curtis
H. H. Donaldson

1892

H. P. Bowditch, President
H. N. Martin, Secretary-Treasurer
R. H. Chittenden
J. G. Curtis
W. H. Howell

1893

H. P. Bowditch, President
W. P. Lombard, Secretary-Treasurer
R. H. Chittenden

J. G. Curtis
W. H. Howell

1894

H. P. Bowditch, President
W. P. Lombard, Secretary-Treasurer
R. H. Chittenden
W. H. Howell
J. W. Warren

1895

H. P. Bowditch, President
F. S. Lee, Secretary-Treasurer
R. H. Chittenden
W. H. Howell
W. P. Lombard

1896

R. H. Chittenden, President
F. S. Lee, Secretary-Treasurer
H. P. Bowditch
W. H. Howell
J. W. Warren

1897–98

R. H. Chittenden, President
F. S. Lee, Secretary-Treasurer
H. P. Bowditch
W. H. Howell
W. P. Lombard

1899–1903

R. H. Chittenden, President
F. S. Lee, Secretary-Treasurer
W. H. Howell
W. P. Lombard
W. T. Porter

1904

R. H. Chittenden, President
W. T. Porter, Secretary-Treasurer
W. H. Howell
F. S. Lee
W. P. Lombard

1905

W. H. Howell, President
L. B. Mendel, Secretary
W. B. Cannon, Treasurer
R. H. Chittenden
S. J. Meltzer

1906

W. H. Howell, President
L. B. Mendel, Secretary
W. B. Cannon, Treasurer
A. B. Macallum
S. J. Meltzer

1907

W. H. Howell, President
L. B. Mendel, Secretary
W. B. Cannon, Treasurer
J. J. Abel
G. Lusk

1908

W. H. Howell, President
R. Hunt, Secretary
W. B. Cannon, Treasurer
J. J. Abel
G. Lusk

1909

W. H. Howell, President
R. Hunt, Secretary
W. B. Cannon, Treasurer
A. J. Carlson
W. P. Lombard

1910

W. H. Howell, President
A. J. Carlson, Secretary
W. B. Cannon, Treasurer

J. Erlanger
F. S. Lee

1911–12

S. J. Meltzer, President
A. J. Carlson, Secretary
W. B. Cannon, Treasurer
J. Erlanger
F. S. Lee

1913

S. J. Meltzer, President
A. J. Carlson, Secretary
J. Erlanger, Treasurer
W. B. Cannon
F. S. Lee

1914

W. B. Cannon, President
A. J. Carlson, Secretary
J. Erlanger, Treasurer
F. S. Lee
S. J. Meltzer

1915

W. B. Cannon, President
C. W. Greene, Secretary
J. Erlanger, Treasurer
W. E. Garrey
W. H. Howell
J. J. R. Macleod
W. J. Meek

1916

W. B. Cannon, President
C. W. Greene, Secretary
J. Erlanger, Treasurer
W. H. Howell
W. E. Garrey
J. J. R. Macleod
W. J. Meek

1917–1918[2]

F. S. Lee, President
C. W. Greene, Secretary
J. Erlanger, Treasurer
W. B. Cannon

W. H. Howell
J. J. R. Macleod
W. J. Meek

1919

W. P. Lombard, President
C. W. Greene, Secretary
J. Erlanger, Treasurer
W. B. Cannon
Y. Henderson
J. J. R. Macleod
W. J. Meek

1920

W. P. Lombard, President
C. W. Greene, Secretary
J. Erlanger, Treasurer
W. B. Cannon
Y. Henderson
J. J. R. Macleod
C. J. Wiggers

1921–22

J. J. R. Macleod, President
C. W. Greene, Secretary
J. Erlanger, Treasurer
A. J. Carlson
J. A. E. Eyster
Y. Henderson
C. J. Wiggers

1923

A. J. Carlson, President
C. W. Greene, Secretary
J. Erlanger, Treasurer
J. A. E. Eyster
A. B. Luckhardt
J. R. Murlin
C. J. Wiggers

1924

A. J. Carlson, President
W. J. Meek, Secretary
C. K. Drinker, Treasurer
J. A. E. Eyster
W. E. Garrey

A. B. Luckhardt
J. R. Murlin

1925

A. J. Carlson, President
W. J. Meek, Secretary
A. Forbes, Treasurer
J. Erlanger
J. A. E. Eyster
W. E. Garrey
J. R. Murlin

1926–27[3]

J. Erlanger, President
W. J. Meek, Secretary
A. Forbes, Treasurer
W. E. Garrey
A. B. Luckhardt
J. R. Murlin
C. J. Wiggers

1927–28

J. Erlanger, President
W. J. Meek, Secretary
A. Forbes, Treasurer
W. E. Garrey
R. Gesell
A. B. Luckhardt
C. J. Wiggers

1928–29[3]

J. Erlanger, President
W. J. Meek, Secretary
A. Forbes, Treasurer
A. J. Carlson
R. Gesell
A. B. Luckhardt
C. J. Wiggers

1929–30

W. J. Meek, President
A. C. Redfield, Secretary
A. Forbes, Treasurer
A. J. Carlson
R. Gesell
J. R. Murlin
C. J. Wiggers

1930–31

W. J. Meek, President
A. B. Luckhardt, Secretary
A. Forbes, Treasurer
A. J. Carlson
R. Gesell
E. G. Martin
J. R. Murlin

1931–32

W. J. Meek, President
A. B. Luckhardt, Secretary
A. Forbes, Treasurer
A. J. Carlson
E. G. Martin
J. R. Murlin
J. Tait

1932–33

A. B. Luckhardt, President
F. C. Mann, Secretary
A. Forbes, Treasurer
E. G. Martin
W. J. Meek
J. R. Murlin
J. Tait

1933–34

A. B. Luckhardt, President
F. C. Mann, Secretary
A. Forbes, Treasurer
H. S. Gasser
E. G. Martin
W. J. Meek
J. Tait

1934–35

C. W. Greene, President
F. C. Mann, Secretary
A. Forbes, Treasurer
H. S. Gasser
A. B. Luckhardt
W. J. Meek
J. Tait

1935–36

F. C. Mann, President

A. C. Ivy, Secretary
A. Forbes, Treasurer
C. H. Best
H. S. Gasser
A. B. Luckhardt
W. J. Meek

1936–37

F. C. Mann, President
A. C. Ivy, Secretary
W. O. Fenn, Treasurer
P. Bard
C. H. Best
H. S. Gasser
A. B. Luckhardt

1937–38

W. T. Porter, Honorary President
W. E. Garrey, President
A. C. Ivy, Secretary
W. O. Fenn, Treasurer
P. Bard
C. H. Best
H. S. Gasser
A. B. Luckhardt

1938–39

W. E. Garrey, President
A. C. Ivy, Secretary
W. O. Fenn, Treasurer
P. Bard
C. H. Best
H. S. Gasser
A. B. Luckhardt

1939–40

A. C. Ivy, President
P. Bard, Secretary
W. O. Fenn, Treasurer
C. H. Best
H. S. Gasser
A. B. Luckhardt
M. B. Visscher

1940–41

A. C. Ivy, President
P. Bard, Secretary

C. J. Wiggers, Treasurer
C. H. Best
H. S. Gasser
A. B. Luckhardt
M. B. Visscher

1941–42

P. Bard, President
C. J. Wiggers, Secretary
H. Davis, Treasurer
C. H. Best
H. E. Essex
A. B. Luckhardt
M. B. Visscher

1942–46[4]

P. Bard, President
W. O. Fenn, Secretary
H. Davis, Treasurer
C. H. Best
H. E. Essex
W. F. Hamilton
M. B. Visscher

1946–48

W. O. Fenn, President
M. B. Visscher, Secretary
D. B. Dill, Treasurer
II. C. Bazett
H. E. Essex
W. F. Hamilton
E. M. Landis

1948–49

M. B. Visscher, President
C. J. Wiggers, President Elect
W. O. Fenn, Past President
H. C. Bazett
D. B. Dill
W. F. Hamilton
E. M. Landis

1949–50

C. J. Wiggers, President
H. C. Bazett, President Elect
M. B. Visscher, Past President
E. F. Adolph

D. B. Dill
R. W. Gerard
E. M. Landis

1950–51

H. C. Bazett, President (April–July 1950)
D. B. Dill, President (July 1950–April 1951)
C. J. Wiggers, Past President
E. F. Adolph
R. W. Gerard
E. M. Landis
F. A. Hitchcock

1951–52[5]

R. W. Gerard, President
E. M. Landis, President Elect
D. B. Dill, Past President
E. F. Adolph
H. W. Davenport
H. E. Essex
F. A. Hitchcock

1952–53

E. M. Landis, President
E. F. Adolph, President Elect
R. W. Gerard, Past President
H. W. Davenport
H. E. Essex
F. A. Hitchcock
L. N. Katz

1953–54

E. F. Adolph, President
H. E. Essex, President Elect
E. M. Landis, Past President
A. C. Burton
H. W. Davenport
F. A. Hitchcock
L. N. Katz

1954–55

H. E. Essex, President
W. F. Hamilton, President Elect
E. F. Adolph, Past President
A. C. Burton
H. W. Davenport

F. A. Hitchcock
L. N. Katz

1955–56

W. F. Hamilton, President
A. C. Burton, President Elect
H. E. Essex, Past President
H. W. Davenport
F. A. Hitchcock
L. N. Katz
R. F. Pitts

1956–57

A. C. Burton, President
L. N. Katz, President Elect
W. F. Hamilton, Past President
J. H. Comroe, Jr.
H. W. Davenport
H. Davis
R. F. Pitts

1957–58

L. N. Katz, President
H. Davis, President Elect
A. C. Burton, Past President
J. H. Comroe, Jr.
J. D. Hardy
H. S. Mayerson
R. F. Pitts

1958–59

H. Davis, President
R. F. Pitts, President Elect
L. N. Katz, Past President
J. H. Comroe, Jr.
J. D. Hardy
H. S. Mayerson
T. C. Ruch

1959–60

R. F. Pitts, President
J. H. Comroe, Jr., President Elect
H. Davis, Past President
H. W. Davenport
J. D. Hardy
H. S. Mayerson
T. C. Ruch

1960–61

J. H. Comroe, Jr., President
H. W. Davenport, President Elect
R. F. Pitts, Past President
J. M. Brookhart
H. S. Mayerson
H. Rahn
T. C. Ruch

1961–62

H. W. Davenport, President
H. S. Mayerson, President Elect
J. H. Comroe, Jr., Past President
J. M. Brookhart
J. R. Pappenheimer
H. Rahn
T. C. Ruch

1962–63

H. S. Mayerson, President
H. Rahn, President Elect
H. W. Davenport, Past President
J. M. Brookhart
J. D. Hardy
J. R. Pappenheimer
T. C. Ruch

1963–64

H. Rahn, President
J. R. Pappenheimer, President Elect
H. S. Mayerson, Past President
J. M. Brookhart
K. S. Cole
R. E. Forster
J. D. Hardy

1964–65

J. R. Pappenheimer, President
J. M. Brookhart, President Elect
H. Rahn, Past President
L. D. Carlson
K. S. Cole
R. E. Forster
J. D. Hardy

1965–66

J. M. Brookhart, President

R. E. Forster, President Elect
J. R. Pappenheimer, Past President
R. W. Berliner
L. D. Carlson
J. D. Hardy
H. D. Patton

1966–67

R. E. Forster, President
R. W. Berliner, President Elect
J. M. Brookhart, Past President
L. D. Carlson
J. D. Hardy
C. L. Prosser
H. D. Patton

1967–68

R. W. Berliner, President
L. D. Carlson, President Elect
R. E. Forster, Past President
J. R. Brobeck
J. D. Hardy
H. D. Patton
C. L. Prosser

1968–69

L. D. Carlson, President
C. L. Prosser, President Elect
R. W. Berliner, Past President
A. C. Barger
J. R. Brobeck
H. D. Patton
D. C. Tosteson

1969–70

C. L. Prosser, President
A. C. Barger, President Elect
L. D. Carlson, Past President
J. R. Brobeck
E. Knobil
H. D. Patton
D. C. Tosteson

1970–71

A. C. Barger, President
J. R. Brobeck, President Elect
C. L. Prosser, Past President

R. M. Berne
E. Knobil
H. D. Patton
D. C. Tosteson

1971–72

J. R. Brobeck, President
R. M. Berne, President Elect
A. C. Barger, Past President
E. Knobil
B. M. Schmidt-Nielsen
E. E. Selkurt
D. C. Tosteson

1972–73

R. M. Berne, President
D. C. Tosteson, President Elect
J. R. Brobeck, Past President
A. C. Guyton
J. Mead
B. M. Schmidt-Nielsen
E. E. Selkurt

1973–74

D. C. Tosteson, President
A. C. Guyton, President Elect
R. M. Berne, Past President
D. F. Bohr
J. Mead
B. M. Schmidt-Nielsen
E. E. Selkurt

1974–75

A. C. Guyton, President
B. M. Schmidt-Nielsen, President Elect
D. C. Tosteson, Past President
D. F. Bohr
P. F. Curran
J. O. Davis
J. Mead

1975–76

B. M. Schmidt-Nielsen, President
E. E. Selkurt, President Elect
A. C. Guyton, Past President
D. F. Bohr
J. O. Davis

W. F. Ganong
J. Mead

1976–77

E. E. Selkurt, President
W. F. Ganong, President Elect
B. M. Schmidt-Nielsen, Past President
J. O. Davis
F. J. Haddy
J. Mead
W. C. Randall

1977–78

W. F. Ganong, President
D. F. Bohr, President Elect
E. E. Selkurt, Past President
J. O. Davis
F. J. Haddy
W. C. Randall
E. H. Wood

1978–79

D. F. Bohr, President
E. Knobil, President Elect
W. F. Ganong, Past President
F. J. Haddy
P. C. Johnson
W. C. Randall
E. H. Wood

1979–80

E. Knobil, President
E. H. Wood, President Elect
D. F. Bohr, Past President
L. Farhi
P. C. Johnson
S. McCann
W. C. Randall

1980–81

E. H. Wood, President
F. J. Haddy, President Elect
E. Knobil, Past President
L. Farhi
P. C. Johnson
J. L. Kostyo
S. McCann

1981–82

F. J. Haddy, President
W. C. Randall, President Elect
E. H. Wood, Past President
L. Farhi
P. C. Johnson
J. L. Kostyo
J. B. West

1982–83

W. C. Randall, President
A. P. Fishman, President Elect
F. J. Haddy, Past President
L. Farhi
F. G. Knox
J. L. Kostyo
J. B. West

1983–84

A. P. Fishman, President
J. B. West, President Elect
W. C. Randall, Past President
F. G. Knox
J. L. Kostyo
H. E. Morgan
N. C. Staub

1984–85

J. B. West, President
H. E. Morgan, President Elect
A. P. Fishman, Past President
F. G. Knox
H. V. Sparks, Jr.
N. C. Staub
A. E. Taylor

1985–86

H. E. Morgan, President
F. G. Knox, President Elect
J. B. West, Past President
S. Chien
H. V. Sparks, Jr.
N. C. Staub
A. E. Taylor

1986–87

F. G. Knox, President

H. V. Sparks, Jr., President Elect
H. E. Morgan, Past President
S. Chien
J. A. Nadel
N. C. Staub
A. E. Taylor

Executive Secretary-Treasurer
  (ex officio member of Council)
M. O. Lee, 1947–56
R. G. Daggs, 1956–72
O. E. Reynolds, 1973–85
M. Frank, 1985–

NOTES

[1] Year denotes year of service. Officers were elected at the December annual meeting to serve the following year. The officers for 1888 held an initial meeting on 30 December 1887.

[2] There were no meetings held in 1918 and two in 1919. Lee served as president through the first meeting in 1919.

[3] No meetings were held in 1926. After 1927 meetings were held in the spring. Officers were elected at the annual meeting, took office at the end of the meeting, and served through the following meeting.

[4] No meetings or elections were held in 1943, 1944, or 1945.

[5] From 1951 terms of newly elected officers began on 1 July.

APPENDIX 2

Meetings, 1887–1987

Organizational Meeting, College of Physicians and Surgeons, New York, NY, 30 December 1887

1st Special Meeting, Washington, DC, 18–19 September 1888

1st Annual Meeting, Jefferson Medical College and University of Pennsylvania, Philadelphia, PA, 29–30 December 1888

2nd Annual Meeting, College of Physicians and Surgeons, New York, NY, 27–28 December 1889

3rd Annual Meeting, Harvard Medical School, Boston, MA, 30 December 1890

2nd Special Meeting, Washington, DC, 22 September 1891

4th Annual Meeting, Academy of Natural Sciences, Philadelphia, PA, 29 December 1891

5th Annual Meeting, Princeton University, Princeton, NJ, 27–28 December 1892

6th Annual Meeting, Yale University, New Haven, CT, 28–29 December 1893

3rd Special Meeting, Washington, DC, 29 May to 1 June 1894

7th Annual Meeting, Johns Hopkins University, Baltimore, MD, 27–28 December 1894

8th Annual Meeting, University of Pennsylvania and Jefferson Medical College, Philadelphia, PA, 27–28 December 1895

9th Annual Meeting, Harvard University, Boston and Cambridge, MA, 29–30 December 1896

4th Special Meeting, Washington, DC, 4–6 May 1897

10th Annual Meeting, Cornell University and New York State Veterinary College, Ithaca, NY, 28–29 December 1897

11th Annual Meeting, College of Physicians and Surgeons, Columbia University, and the University and Bellevue Hospital Medical College, New York, NY, 28–30 December 1898

12th Annual Meeting, Yale University, New Haven, CT, 27–29 December 1899

5th Special Meeting, Washington, DC (Columbian Medical College), 1–2 May 1900

13th Annual Meeting, Johns Hopkins University, Baltimore, MD, 27–28 December 1900

14th Annual Meeting, University of Chicago, Chicago, IL, 30–31 December 1901

15th Annual Meeting, Columbian Medical College, Washington, DC, 30–31 December 1902

6th Special Meeting, Washington, DC (Columbian Medical College), 12 May 1903

16th Annual Meeting, University of Pennsylvania and Jefferson Medical College, Philadelphia, PA, 29–30 December 1903

17th Annual Meeting, University of Pennsylvania and Jefferson Medical College, Philadelphia, PA, 27–28 December 1904

18th Annual Meeting, University of Michigan, Ann Arbor, MI, 28–29 December 1905

19th Annual Meeting, University and Bellevue Hospital Medical College, College of Physicians and Surgeons, and Rockefeller Institute for Medical Research, New York, NY, 27–29 December 1906

7th Special Meeting, Washington, DC (George Washington University), 7 and 9 May 1907

20th Annual Meeting, University of Chicago, Chicago, IL, 31 December 1907 to 2 January 1908

21st Annual Meeting, Johns Hopkins University, Baltimore, MD, 29–31 December 1908

22nd Annual Meeting, Harvard University, Boston, MA, 28–30 December 1909

23rd Annual Meeting, Yale University, New Haven, CT, 27–30 December 1910

24th Annual Meeting, Johns Hopkins University, Baltimore, MD, and George Washington University, Washington, DC, 26–29 December 1911

25th Annual Meeting, Western Reserve Medical School, Cleveland, OH, 30 December 1912 to 1 January 1913

26th Annual Meeting, Jefferson Medical College and the University of Pennsylvania, Philadelphia, PA, 28–31 December 1913

27th Annual Meeting, Washington University, St. Louis, MO, 28–31 December 1914

28th Annual Meeting, Harvard University, Boston, MA, 26–29 December 1915

29th Annual Meeting, Cornell University Medical College, New York, NY, 28–30 December 1916

30th Annual Meeting, University of Minnesota, Minneapolis, and the Mayo Foundation for Medical Research, Rochester, MN, 27–29 December 1917

31st Annual Meeting, Johns Hopkins University, Baltimore, MD, 24–26 April 1919

32nd Annual Meeting, University of Cincinnati, Cincinnati, OH, 29–31 December 1919

33rd Annual Meeting, University of Chicago, Chicago, IL, 28–30 December 1920

34th Annual Meeting, Yale University, New Haven, CT, 27–30 December 1921

35th Annual Meeting, University of Toronto, Toronto, Ontario, Canada, 27–29 December 1922

36th Annual Meeting, Washington University and St. Louis University, St. Louis, MO, 27–29 December 1923

37th Annual Meeting, Washington, DC, 29–31 December 1924

38th Annual Meeting, Western Reserve University, Cleveland, OH, 28–30 December 1925

39th Annual Meeting, University of Rochester, Rochester, NY, 14–16 April 1927

40th Annual Meeting, University of Michigan, Ann Arbor, MI, 12–14 April 1928

41st Annual Meeting combined with XIII International Physiological Congress, Harvard University, Boston and Cambridge, MA, 19–23 August 1929

42nd Annual Meeting, University of Chicago, Chicago, IL, 26–29 March 1930

43rd Annual Meeting, McGill University, Montreal, Quebec, Canada, 8–11 April 1931

44th Annual Meeting, University of Pennsylvania, Philadelphia, PA, 27–30 April 1932

45th Annual Meeting, Cincinnati, OH, 10–12 April 1933

46th Annual Meeting, New York, NY, 28–31 March 1934

47th Annual Meeting, Detroit, MI, 10–13 April 1935

48th Annual Meeting, Washington, DC, 25–28 March 1936

49th Annual Meeting, Memphis, TN, 21–24 April 1937

50th Annual Meeting, Baltimore, MD, 30 March to 2 April 1938

51st Annual Meeting, Toronto, Ontario, Canada, 26–29 April 1939

52nd Annual Meeting, New Orleans, LA, 13–16 March 1940

53rd Annual Meeting, Chicago, IL, 15–19 April 1941

54th Annual Meeting, Boston, MA, 30 March to 4 April 1942[1]

55th Annual Meeting, Atlantic City, NJ, 11–15 March 1946

56th Annual Meeting, Chicago, IL, 18–22 May 1947

57th Annual Meeting, Atlantic City, NJ, 15–19 March 1948

1st Fall Meeting, University of Minnesota, Minneapolis, MN, 15–18 September 1948

58th Spring Meeting, Detroit, MI, 18–22 April 1949

2nd Fall Meeting, University of Georgia, Augusta, GA, 14–17 September 1949

59th Spring Meeting, Atlantic City, NJ, 17–21 April 1950

3rd Fall Meeting, Ohio State University, Columbus, OH, 14–16 September 1950

60th Spring Meeting, Cleveland, OH, 29 April to 3 May 1951

4th Fall Meeting, University of Utah, Salt Lake City, UT, 6–8 September 1951

61st Spring Meeting, New York, NY, 13–18 April 1952

5th Fall Meeting, Tulane University, New Orleans, LA, 4–6 September 1952

62nd Spring Meeting, Chicago, IL, 5–10 April 1953[2]

63rd Spring Meeting, Atlantic City, NJ, 11–15 April 1954

6th Fall Meeting, University of Wisconsin, Madison, WI, 8–10 September 1954

64th Spring Meeting, San Francisco, CA, 10–15 April 1955

7th Fall Meeting, Tufts University, Medford, and Marine Biological Laboratory, Woods Hole, MA, 6–9 September 1955

65th Spring Meeting, Atlantic City, NJ, 16 20 April 1956

8th Fall Meeting, University of Rochester, Rochester, NY, 4–7 September 1956

66th Spring Meeting, Chicago, IL, 15–19 April 1957

9th Fall Meeting, State University of Iowa, Iowa City, IA, 3–6 September 1957

67th Spring Meeting, Philadelphia, PA, 14–18 April 1958

10th Fall Meeting, University of Western Ontario, London, Ontario, Canada, 2–5 September 1958

68th Spring Meeting, Atlantic City, NJ, 13–17 April 1959

11th Fall Meeting, University of Illinois, Urbana, IL, 8–11 September 1959

69th Spring Meeting, Chicago, IL, 11–15 April 1960

12th Fall Meeting, Stanford University and University of California, San Francisco, CA, 23–26 August 1960

70th Spring Meeting, Atlantic City, NJ, 10–14 April 1961

13th Fall Meeting, Indiana University, Bloomington, IN, 5–8 September 1961

71st Spring Meeting, Atlantic City, NJ, 14–19 April 1962

14th Fall Meeting, University of Buffalo, Buffalo, NY, 28–31 August 1962

72nd Spring Meeting, Atlantic City, NJ, 16–20 April 1963

15th Fall Meeting, University of Miami, Coral Gables, FL, 27–30 August 1963

73rd Spring Meeting, Chicago, IL, 12–17 April 1964

16th Fall Meeting, Brown University, Providence, RI, 7–11 September 1964

74th Spring Meeting, Atlantic City, NJ, 9–14 April 1965

17th Fall Meeting, University of California at Los Angeles, Los Angeles, CA, 23–27 August 1965

75th Spring Meeting, Atlantic City, NJ, 11–16 April 1966

18th Fall Meeting, Baylor University College of Medicine and the University of Texas Dental Branch, Houston, TX, 29 August to 2 September 1966

76th Spring Meeting, Chicago, IL, 16–21 April 1967

19th Fall Meeting, Howard University, Washington, DC, 21–25 August 1967

77th Spring Meeting, Atlantic City, NJ, 15–20 April 1968[3]

78th Spring Meeting, Atlantic City, NJ, 14–18 April 1969

20th Fall Meeting, University of California at Davis, Davis, CA, 25–29 August 1969

79th Spring Meeting, Atlantic City, NJ, 12–17 April 1970

21st Fall Meeting, Indiana University, Bloomington, IN, 27 August to 3 September 1970

80th Spring Meeting, Chicago, IL, 12–17 April 1971

22nd Fall Meeting, University of Kansas, Lawrence and Kansas City, KS, 16–19 August 1971

81st Spring Meeting, Atlantic City, NJ, 9–14 April 1972

23rd Fall Meeting, Pennsylvania State University, University Park, PA, 27–31 August 1972

82nd Spring Meeting, Atlantic City, NJ, 15–20 April 1973

24th Fall Meeting, University of Rochester, Rochester, NY, 20–24 August 1973

83rd Spring Meeting, Atlantic City, NJ, 8–12 April 1974

25th Fall Meeting, State University of New York at Albany and Albany Medical College of Union University, Albany, NY, 12–16 August 1974

84th Spring Meeting, Atlantic City, NJ, 13–18 April 1975

26th Fall Meeting, San Francisco, CA, 6–10 October 1975

85th Spring Meeting, Anaheim, CA, 11–16 April 1976

27th Fall Meeting, University of Pennsylvania, Philadelphia, PA, 15–20 August 1976

86th Spring Meeting, Chicago, IL, 1–8 April 1977

28th Fall Meeting, Hollywood Beach, FL, 9–14 October 1977

87th Spring Meeting, Atlantic City, NJ, 9–14 April 1978

29th Fall Meeting, St. Louis, MO, 22–27 October 1978

88th Spring Meeting, Dallas, TX, 6–10 April 1979

Specialty Meeting, "Relation Between Brain Neurotransmitters and Endocrine Function," Michigan State University, East Lansing, MI, 22–24 August 1979

30th Fall Meeting, New Orleans, LA, 15–19 October 1979

89th Spring Meeting, Anaheim, CA, 13–18 April 1980

31st Fall Meeting, Toronto, Ontario, Canada, 12–17 October 1980

90th Spring Meeting, Atlanta, GA, 12–17 April 1981

32nd Fall Meeting, Cincinnati, OH, 11–16 October 1981

91st Spring Meeting, New Orleans, LA, 20–23 April 1982

33rd Fall Meeting, San Diego, CA, 10–15 October 1982

92nd Spring Meeting, Chicago, IL, 10–15 April 1983

34th Fall Meeting, Honolulu, HI, 20–24 August 1983

93rd Spring Meeting, St. Louis, MO, 1–6 April 1984

35th Fall Meeting, Lexington, KY, 26–30, August 1984

94th Spring Meeting, Anaheim, CA, 21–26 April 1985

Joint Meeting with the Physiological Society, Cambridge University, Cambridge, United Kingdom, 12–14 September 1985

36th Fall Meeting, Niagara Falls, NY, 13–18 October 1985

95th Spring Meeting, St. Louis, MO, 13–18 April 1986

37th Fall Meeting, New Orleans, LA, 5–10 October 1986

96th Spring Meeting, Washington, DC, 29 March to 3 April 1987

NOTES

[1] There were no regular meetings in 1943, 1944, or 1945. Council met in Rochester, NY, 18–19 May 1945.

[2] There was no regular fall meeting in 1953 because of the XIX International Congress of Physiological Sciences, Montreal, 31 August to 5 September 1954.

[3] There was no regular fall meeting in 1968 because of the XXIV International Congress of Physiological Sciences, Washington, DC, 25–31 August 1968.

APPENDIX 3

Sources

This appendix summarizes, for the benefit of present and future historians, how and where to find basic information about the membership, officers, scientific programs, and activities of the American Physiological Society (APS, the Society).

Published Histories

Prior to the present volume, there have been two published histories of the Society:

HOWELL, W. H., AND C. W. GREENE. *History of the American Physiological Society Semicentennial, 1887–1937.* Baltimore, MD: Am. Physiol. Soc., 1938.

FENN, W. O. *History of the American Physiological Society: The Third Quarter Century, 1937–1962.* Washington, DC: Am. Physiol. Soc., 1963.

Many articles on the current activities and history of APS have appeared in the Society's house organ, *The Physiologist,* published since 1957. Indexes have appeared in volumes 24(3): i–xvii, 1981 (for vols. 1–22, 1957–79); 26: 56–61, 1983 (for vols. 23–25, 1980–82); 26: 397–400, 1983; 27: 441–443, 1984; and 28: 531–532, 1985.

APS Archives

The Society has minutes of Council and business meetings going back to the organizational meeting of APS in 1887, as well as considerable additional archival material, particularly for the period since 1947. Manuscript records of the Society are described in:

APPEL, T. A. The archives of the American Physiological Society. *Physiologist* 27: 131–132, 1984.

Membership of APS

From 1888 through 1913 the Society published annual (or nearly annual) directories containing the constitution and lists of officers and members. An incomplete set of these brochures is found in the APS Archives. From 1914, the Federation of American Societies for Experimental Biology (FASEB, the Federation) published the constitution and directory of APS members annually in the *Federation Yearbook.* A collection of the annual yearbooks of the Federation from 1914 through 1941 is available in the Federation Library. From 1942 through 1963 the constitutions, lists of officers, past officers, and committee members, and a directory of members (including year of election) of the Federation societies were published in *Federation Proceedings.* Since 1964 this information has been published annually in *Federation of American Societies for Experimental Biology. Directory of Members.* A complete set since 1964 is available in the APS Archives or in the Federation Library.

Members elected each year for the first fifty years (1887–1937) of the Society's history are found in Howell and Greene (above). Members elected each year from 1938 through 1962 are found in Fenn (above). Since 1958 newly elected members have been listed twice yearly in *The Physiologist.* Deceased members, listed by Howell and Greene and by Fenn, are also published twice yearly in *The Physiologist.*

Scientific Meetings

Programs of the earliest annual meetings of APS appear not to have been printed. The papers presented are noted in the minutes of the Society. A few printed programs for the period before 1914 are in the APS Archives. From 1914 through 1941 programs were published separately by the Federation. A nearly complete set is available in the APS Archives. From 1942 through 1961 programs of the Federation meetings (APS spring meetings) were included in *Federation Proceedings.* Since 1962 programs have been published separately by the Federation. Complete sets of programs since 1964 are in the APS Archives and in the Federation Library. Programs of fall meetings have been published separately by APS. A full set of programs, from 1948 to the present, is available in the APS Archives. Unlike other early biological societies, APS did not publish its proceedings (report of the business meeting) until the establishment of *The Physiologist* in 1957.

Abstracts of papers presented at the annual meeting were published beginning in 1896 (for the 1895 meeting) in *Science.* From 1898 (beginning with the 1897 meeting) through 1941 abstracts were published in the *American Journal of Physiology.* From 1942 on, abstracts of the spring meeting have been published in *Federation Proceedings.* Abstracts of the APS fall meeting were published in the *American Journal of Physiology* from 1948 through 1956. Since 1957 fall meeting abstracts have been published in *The Physiologist.*

Publications

There is little material in the APS Archives on publications before 1914. For the period after 1914, annual reports of the managing editor, minutes of the Board of Publication Trustees (1956–61), and minutes of the Publications Committee, as well as selected correspondence, can be found in the archives. Names of editors and members of editorial boards are listed in the annual Federation directories. "Publications of the American Physiological Society," an alphabetical cumulative list of editors and members of editorial boards (1933–74), detailed chronologies of each of the APS publications, and a list of all books and special publications of APS, was compiled by Sara Leslie in 1974 and is available in the APS Archives.

Considerable material on the policies and history of journals and books published by APS can be found in *The Physiologist* and also in editorials and announcements in the various APS journals since 1977. The decision to reorganize the APS journals in 1977 is discussed in:

FISHMAN, A. P., AND S. R. GEIGER. Multiple choice, a publications option for a diverse science. *Scholarly Publ.* 9: 243–251, 1978.

Index